Intelligent Data Analysis

Springer

Berlin
Heidelberg
New York
Hong Kong
London
Milan
Paris
Tokyo

Editors
Michael Berthold • David J. Hand

Intelligent Data Analysis

An Introduction

2nd revised and extended Edition

Editors

Michael Berthold
Tripos, Inc.
Data Analysis Research Lab.
601 Gateway Blvd., Suite 720
South San Francisco, CA 94080
USA

berthold@tripos.com

David J. Hand
Department of Mathematics
Imperial College
Huxley Building
180 Queen's Gate
London, SW7 2BZ
UK

d.j.hand@ic.ac.uk

With 140 Figures, 22 in color
and 50 Tables

Library of Congress Cataloging-in-Publication Data
Intelligent data analyis: an introduction/editors, Michael Berthold, David Hand. –
2nd rev. and extended ed.
p. cm.
Includes bibliographical references and index.
ISBN 3-540-43060-1 (acid-free paper)
1. Mathematical statistics. 2. Mathematical statistics–Data processing. 3. Artificial
intelligence. I. Berthold, M. (Michael) II. Hand, D. J.

QA276.I565 2003
519.5–dc21

2003041211

ACM Subject Classification (1998): I.2, H.3, G.3, I.5.1, I.4, J.2, J.1, J.3, F.4.1, F.1

ISBN 3-540-43060-1 Springer-Verlag Berlin Heidelberg New York

This work is subject to copyright. All rights are reserved, whether the whole or part of the material is concerned, specifically the rights of translation, reprinting, reuse of illustrations, recitation, broadcasting, reproduction on microfilm or in any other way, and storage in data banks. Duplication of this publication or parts thereof is permitted only under the provisions of the German Copyright Law of September 9, 1965, in its current version, and permission for use must always be obtained from Springer-Verlag. Violations are liable for prosecution under the German Copyright Law.

Springer-Verlag Berlin Heidelberg New York
is a member of BertelsmannSpringer Science+Business Media GmbH
http://www.springer.de

© Springer-Verlag Berlin Heidelberg 1999, 2003
Printed in Germany

The use of general descriptive names, trademarks, etc. in this publication does not imply, even in the absence of a specific statement, that such names are exempt from the relevant protective laws and regulations and therefore free for general use.

Typesetting: Camera-ready by the editors
Cover Design: KünkelLopka, Heidelberg
Printed on acid-free paper 45/3142SR – 5 4 3 2 1 0

Preface to the Second Edition

We were pleasantly surprised by the success of the first edition of this book. Many of our colleagues have started to use it for teaching purposes, and feedback from industrial researchers has also shown that it is useful for practitioners. So, when Springer-Verlag approached us and asked us to revise the material for a second edition, we gladly took the opportunity to rearrange some of the existing material and to invite new authors to write two new chapters. These additional chapters cover material that has attracted considerable attention since the first edition of the book appeared. They deal with kernel methods and support vector machines on the one hand, and visualization on the other. Kernel methods represent a relatively new technology, but one which is showing great promise. Visualization methods have been around in some form or other ever since data analysis began, but are currently experiencing a renaissance in response to the increase, in numbers and size, of large data sets. In addition the chapter on rule induction has been replaced with a new version, covering this topic in much more detail.

As research continues, and new tools and methods for data analysis continue to be developed, so it becomes ever more difficult to cover all of the important techniques. Indeed, we are probably further from this goal than we were with the original edition – too many new fields have emerged over the past three years. However, we believe that this revision still provides a solid basis for anyone interested in the analysis of real data.

We are very grateful to the authors of the new chapters for working with us to an extremely tight schedule. We also would like to thank the authors of the existing chapters for spending so much time carefully revising and updating their chapters. And, again, all this would not have been possible without the help of many people, including Olfa Nasraoui, Ashley Morris, and Jim Farrand.

Once again, we owe especial thanks to Alfred Hofmann and Ingeborg Mayer of Springer-Verlag, for their continued support for this book and their patience with various delays during the preparation of this second edition.

November 2002

South San Francisco, CA, USA Michael R. Berthold
London, UK David J. Hand

Preface to the First Edition

The obvious question, when confronted with a book with the title of this one, is why "intelligent" data analysis? The answer is that modern data analysis uses tools developed by a wide variety of intellectual communities and that "intelligent data analysis", or IDA, has been adopted as an overall term. It should be taken to imply the intelligent application of data analytic tools, and also the application of "intelligent" data analytic tools, computer programs which probe more deeply into structure than first generation methods. These aspects reflect the distinct influences of statistics and machine learning on the subject matter.

The importance of intelligent data analysis arises from the fact that the modern world is a data-driven world. We are surrounded by data, numerical and otherwise, which must be analysed and processed to convert it into *information* which informs, instructs, answers, or otherwise aids understanding and decision making. The quantity of such data is huge and growing, the number of sources is effectively unlimited, and the range of areas covered is vast: industrial, commercial, financial, and scientific activities are all generating such data.

The origin of this book was a wish to have a single introductory source to which we could direct students, rather than having to direct them to multiple sources. However, it soon became apparent that wider interest existed, and that potential readers other than our students would appreciate a compilation of some of the most important tools of intelligent data analysis. Such readers include people from a wide variety of backgrounds and positions who find themselves confronted by the need to make sense of data.

Given the wide range of topics we hoped to cover, we rapidly abandoned the idea of writing the entire volume ourselves, and instead decided to invite appropriate experts to contribute separate chapters. We did, however, make considerable efforts to ensure that these chapters complemented and built on each other, so that a rounded picture resulted. We are especially grateful to the authors for their patience in putting up with repeated requests for revision so as to make the chapters meld better.

In a volume such as this there are many people whose names do not explicitly appear as contributors, but without whom the work would be of substantially reduced quality. These people include Jay Diamond, Matt Easley, Sibylle Frank, Steven Greenberg, Thomas Hofmann, Joy Hollenback, Joe Iwanski, Carlo Marchesi, Roger Mitton, Vanessa Robins, Nancy Shaw, and Camille Sinanan for their painstaking proofreading and other help, as well as Stefan Wrobel, Chris Roadknight and Dominic Palmer-Brown for stimulating discussions and contributions which, though not appearing in print, have led to critical reassessment of how we thought some of the material should be presented.

Finally, we owe especial thanks to Alfred Hofmann from Springer-Verlag, for his enthusiasm and support for this book right from the start.

February 1999

Berkeley, California Michael Berthold
London, United Kingdom David J. Hand

Table of Contents

Chapter 1
Introduction

David J. Hand
Imperial College, United Kingdom

1.1. Why "Intelligent Data Analysis"?

It must be obvious to everyone - to everyone who is reading this book, at least - that progress in computer technology is radically altering human life. Some of the changes are subtle and concealed. The microprocessors that control traffic lights or dishwashers, are examples. But others are overt and striking. The very word processor on which I am creating this chapter could not have been imagined 50 years ago; speech recognition devices, such as are now available for attachment to PCs, could have been imagined, but no-one would have had any idea of how to build such a thing.

This book is about one of those overt and striking changes: the way in which computer technology is enabling us to answer questions which would have defied an answer, perhaps even have defied a formulation, only a few decades ago. In particular, this book is about a technology which rides on top of the progress in electronic and computer hardware: the technology of data analysis.

It is fair to say that modern data analysis is a very different kind of animal from anything which existed prior to about 1950. Indeed, it is no exaggeration to say that modern data is a very different kind of animal from anything which existed before. We will discuss in some detail exactly what is meant by data in the modern world in Section 1.3 but, to get the ball rolling, it seems more convenient to begin, in this section, by briefly examining the notion of "intelligent data analysis". Why analyse data? Why is this book concerned with "intelligent" data analysis? What is the alternative to "intelligent" data analysis? And so on. In between these two sections, in Section 1.2, we will look at the cause of all this change: the computer and its impact.

To get started, we will assume in this opening section that "data" simply comprise a collection of numerical values recording the magnitudes of various attributes of the objects under study. Then "data analysis" describes the processing of those data. Of course, one does not set out simply to analyse data. One always has some objective in mind: one wants to answer certain questions. These questions might be high level general questions, perhaps exploratory: for example, are there any interesting structures in the data? Are any records anomalous? Can we summarise the data in a convenient way? Or the questions might be more specifically confirmatory: Is this group different from that one? Does this attribute change over time? Can we predict the value of this attribute from the measured values of these? And so on.

Orthogonal to the exploratory/confirmatory distinction, we can also distinguish between descriptive and inferential analyses. A descriptive (or summarising) analysis is aimed at making a statement about the data set to hand. This might consist of observations on the entirety of a population (all employees of a corporation, all species of beetle which live in some locality), with the aim being to answer questions about that population: what is the proportion of females? How many of the beetle species have never been observed elsewhere? In contrast, an inferential analysis is aimed at trying to draw conclusions which have more general validity. What can we say about the likely proportion of females next year? Is the number of beetle species in this locality declining? Often inferential studies are based on samples from some population, and the aim is to try to make some general statement about the broader population, most (or some) of which has not been observed. Often it is not possible to observe all of the population (indeed, this may not always be well-defined - the population of London changes minute by minute).

The sorts of tools required for exploratory and confirmatory analyses differ, just as they do for descriptive and inferential analyses. Of course, there is considerable overlap - we are, at base, analysing data. Often, moreover, a tool which appears common is used in different ways. Take something as basic as the mean of a sample as an illustration. As a description of the sample, this is fixed and accurate and is the value - assuming no errors in the computation, of course. On the other hand, as a value derived in an inferential process, it is an estimate of the parameter of some distribution. The fact that it is based on a sample - that it is an estimate - means that it is not really what we are interested in. In some sense we expect it to be incorrect, to be subject to change (if we had taken a different sample, for example, we would expect it to be different), and to have distributional properties in its own right. The single number which has emerged from the computational process of calculating the mean will be used in different ways according to whether one is interested in description or inference. The fact that the mean of sample A is larger than the mean of sample B is an observed fact - and if someone asks which sample has the larger mean we reply "A". This may be different from what we would reply to the question "Which population has the larger mean, that from which A was drawn or that from which B was drawn?" This is an inferential question, and the variability in the data (as measured by, for example, the standard deviations of the samples) may mean we have no confidence at all that the mean of one population is larger than that of the other.

Given the above, a possible definition of data analysis is the process of computing various summaries and derived values from the given collection of data. The word "process" is important here. There is, in some quarters, an apparent belief that data analysis simply consists of picking and applying a tool to match the presenting problem. This is a misconception, based on an artificial idealisation of the world. Indeed, the misconception has even been dignified with a name: it is called the cookbook fallacy, based on the mistaken idea that one simply picks an appropriate recipe from one's collection. (And not the idea that one cooks one's data!) There are several reasons why this is incorrect. One is that data analysis is not simply a collection of isolated tools, each completely different from the other, simply lying around waiting to be matched to the problem. Rather the tools of data analysis have complex interrelationships: analysis of variance is a linear model, as is regression analysis; linear models are a special case of generalised linear models (which generalise from straightforward linearity), and also of the general linear model (a multivariate extension); logistic regression is a generalised linear model and is also a simple form of neural network; generalised additive models generalise in a different way; nonparametric methods relax some of the assumptions of classical parametric tests, but in doing so alter the hypotheses being tested in subtle ways; and so one can go on.

A second reason that the cookbook fallacy is incorrect lies in its notion of matching a problem to a technique. Only very rarely is a research question stated sufficiently precisely that a single and simple application of one method will suffice. In fact, what happens in practice is that data analysis is an iterative process. One studies the data, examines it using some analytic technique, decides to look at it another way, perhaps modifying it in the process by transformation or partitioning, and then goes back to the beginning and applies another data analytic tool. This can go round and round many times. Each technique is being used to probe a slightly different aspect of the data - to ask a slightly different question of the data. Several authors have attempted to formalise this process (for example [236, 408, 427]). Often the process throws up aspects of the data that have not been considered before, so that other analytic chains are started. What is essentially being described here is a voyage of discovery - and it is this sense of discovery and investigation which makes modern data analysis so exciting. It was this which led a geologist colleague of mine to comment that he envied us data analysts. He and other similar experts had to spend the time and tedium collecting the data, but the data analysts were necessarily in at the kill, when the exciting structures were laid bare. Note the contrast between this notion, that modern data analysis is the most exciting of disciplines, and the lay view of statistics - that it is a dry and tedious subject suited only to those who couldn't stand the excitement of accountancy as a profession. The explanation for the mismatch lies in the fact that the lay view is a historical view. A view based on the perception that data analysts spend their time scratching away at columns of figures (with a quill pen, no doubt!). This fails to take into account the changes we referred to at the start of this chapter: the impact of the computer in removing the drudgery and tedium. The quill pen has been replaced by a computer. The days of mindless calculation replaced by a command to the machine - which then effortlessly, accurately, and probably effectively instantaneously carries out the calculation. We shall return to this in Section 1.2.

One reason that the word "intelligent" appears in the title of this book is also implicit in the previous paragraph: the repeated application of methods, as one attempts to tease out the structure, to understand what is going on, and to refine the questions that the researchers are seeking to answer, requires painstaking care and, above all, intelligence. "Intelligent" data analysis is not a haphazard application of statistical and machine learning tools, not a random walk through the space of analytic techniques, but a carefully planned and considered process of deciding what will be most useful and revealing.

1.2. How the Computer Is Changing Things/the Merger of Disciplines

Intelligent data analysis has its origins in various disciplines. If I were to single out two as the most important, I would choose statistics and machine learning. Of these, of course, statistics is the older - machines which can have a hope of learning have not been around for that long. But the mere fact of the youth of machine learning does not mean that it does not have its own culture, its own interests, emphases, aims, and objectives which are not always in line with those of statistics. This fact, that these two disciplines at the heart of intelligent data analysis have differences, has led to a creative tension, which has benefited the development of data analytic tools.

Statistics has its roots in mathematics. Indeed many statisticians still regard it as fundamentally a branch of mathematics. In my view this has been detrimental to the development of the discipline (see, for example [243], and other papers in that issue). Of course, it is true that statistics is a mathematical subject - just as physics, engineering, and computer science are mathematical. But this does not make it a branch of mathematics any more than it makes these other subjects branches of mathematics. At least partly because of this perception, statistics (and statisticians) have been slow to follow up promising new developments. That is, there has been an emphasis on mathematical rigour, a (perfectly reasonable) desire to establish that something is sensible on theoretical grounds before testing it in practice.

In contrast, the machine learning community has its origins very much in computer practice (and not really even in computer science, in general). This has led to a practical orientation, a willingness to test something out to see how well it performs, without waiting for a formal proof of effectiveness.

It goes without saying that both strategies can be very effective. Indeed, ideally one would apply both strategies - establish by experiment that something does seem to work and demonstrate by mathematics when and under what circumstances it does so. Thus, in principle at least, there is a great potential for synergy between the two areas. Although, in general, I think this potential has yet to be realised, one area where it has been realised is in artificial neural network technology.

If the place given to mathematics is one of the major differences between statistics and machine learning, another is in the relative emphasis they give to models and to algorithms.

Modern statistics is almost entirely driven by the notion of a model. This is a postulated structure, or an approximation to a structure, which could have led to the data. Many different types of models exist. Indeed, recalling the comment above that the tools of data analysis have complex interrelationships, and are not simply a collection of isolated techniques, it will come as no surprise if I say that many different families of models exist. There are a few exceptional examples within statistics of schools or philosophies which are not model driven, but they are notable for their scarcity and general isolation. (The Dutch Gifi school of statistics [211], which seeks data transformations to optimise some external criterion, is an example of an algorithm driven statistical school. By applying an algorithm to a variety of data sets, an understanding emerges of the sort of behaviour one can expect when new data sets are explored.) These exceptions have more in common with approaches to data analysis developed in the machine learning community than in traditional statistics (which here is intended to include Bayesian statistics). In place of the statistical emphasis on models, machine learning tends to emphasise algorithms. This is hardly surprising - the very word "learning" contains the notion of process, an implicit algorithm.

The term "model" is very widely used - and, as often occurs when words are widely used, admits a variety of subtly different interpretations. This is perhaps rather unfortunate in data analysis since different types of models are used in different ways and different tools are used for constructing different kinds of models. Several authors have noted the distinction between empirical and mechanistic models (see, for example [81, 129, 239]). The former seek to model relationships without basing them on any underlying theory, while the latter are constructed on the basis of some supposed mechanism underlying the data generating process. Thus, for example, we could build a regression model to relate one variable to several potential explanatory variables, and perhaps obtain a very accurate predictive model, without having any claim or belief that the model in any way represented the causal mechanism; or we might believe that our model described an "underlying reality", in which increasing one variable led to an increase in another.

We can also distinguish models designed for prediction from models designed to help understanding. Sometimes a model which is known to be a poor representation of the underlying processes (and therefore useless for "understanding") may do better in predicting future values than one which is a good representation. For example, the so-called Independence Bayes model, a model for assigning objects to classes based on the usually false assumption of independence between the variables on which the prediction is to be based, often performs well. This sort of behaviour is now understood, but has caused confusion in the past.

Finally here (although doubtless other distinctions can be made) we can distinguish between models and patterns (and as with the distinctions above, there is, inevitably, overlap). This is an important distinction in data mining, where tools for both kinds of structure are often needed (see [244]). A model is a "large scale" structure, perhaps summarising relationships over many cases, whereas a pattern is a local structure, satisfied by a few cases or in a small region of the data space in some sense. A Box-Jenkins analysis of a time series will yield a model, but a local waveform, occurring only occasionally and irregularly, would be a pattern. Both are clearly of potential importance: we would like to detect seasonality, trend, and correlation structures in data, as well as the occasional anomaly which indicates that something peculiar has happened or is about to happen. (It is also worth noting here that the word "pattern", as used in the phrase "pattern recognition", has a rather different meaning. There it refers to the vector of measurements characterising a particular object - a "point" in the language of multivariate statistics.)

I commented above that the modern computer-aided model fitting process is essentially effortless. This means that a huge model space can be searched in order to find a well-fitting model. This is not without its disadvantages. The larger the set of possible models examined in order to fit a given set of data, the better a fit one is likely to obtain. This is fine if we are seeking simply to summarise the available data, but not so fine if the objective is inference. In this case we are really aiming to generalise beyond the data, essentially to other data which could have arisen by the same process (although this generalisation may be via parameters of distributions which are postulated to (approximately) underlie the data generating mechanism). When we are seeking to generalise, the data on which the model must be based will have arisen as a combination of the underlying process and the chance events which led to that particular data set being generated and chosen (sampling variability, measurement error, time variation, and so on). If the chosen model fits the data too well, then it will not merely be fitting the underlying process but will also be fitting the chance factors. Since future data will have different values for the chance factors, this will mean that our inference about future values will be poor. This phenomenon - in which the model goes too far in fitting the data - is termed overfitting. Various strategies have been developed in attempts to overcome it. Some are formal - a probability model for the chance factors is developed - while others are more ad hoc - for example, penalising the measure of how well the model fits the data, so that more complex models are penalised more heavily, or shrinking a well-fitting model. Examples are given in later chapters.

As model fitting techniques have become more refined (and quick to carry out, even on large data sets), and as massive amounts of data have accumulated, so other issues have come to light which generally, in the past, did not cause problems. In particular, subtle aspects of the underlying process can now often be detected, aspects which are so small as to be irrelevant in practice, even though they highly statistically significant are and almost certainly real. The decision as to how complex a model to choose must be based on the size of the effects one regards as important, and not merely on the fact that a feature of the data is very unlikely to be a chance variation.

Moreover, if the data analysis is being undertaken for a colleague, or in collaboration, there are other practical bounds on how sophisticated a model should be used. The model must be comprehensible and interpretable in the terms of the discipline from which it arose.

If a significant chunk of intelligent data analysis is concerned with finding a model for data or the structures which led to the data, then another significant chunk is concerned with algorithms. These are the computational facilitators which enable us to analyse data at all. Although some basic forms of algorithm have been around since before the dawn of the computer age, others have only been developed - could only be imagined - since computers became sufficiently powerful. Computer intensive methods such as resampling, and Bayesian methods based on avoiding integration by generating random samples from arbitrary distributions, have revolutionised modern data analysis. However, to every fundamental idea, every idea which opens up a wealth of new possibilities, there are published descriptions of a hundred (Why be conservative? A thousand.) algorithms which lead to little significant progress. In fact, I would claim that there are too many algorithms being developed without any critical assessment, without any theoretical base, and without any comparison with existing methods. Often they are developed in the abstract, without any real problem in mind. I recall a suggestion being made as far back as twenty years ago, only partly tongue-in-cheek, that a moratorium should be declared on the development of new cluster analysis algorithms until a better understanding had been found for those (many) that had already been developed. This has not happened. Since then work has continued at an even more breakneck pace.

The adverse implication of this is that work is being duplicated, that effort and resources are being wasted, and that sub-optimal methods are being used. I would like to make an appeal for more critical assessment, more evaluation, more meta-analysis and synthesis of the different algorithms and methods, and more effort to place the methods in an overall context by means of their properties. I have appealed elsewhere (for example [240]) for more teaching of higher level courses in data analysis, with the emphasis being placed on the concepts and properties of the methods, rather than on the mechanical details of how to apply them. Obtaining a deeper understanding of the methods, how they behave, and why they behave the way they do, is another side of this same coin. Some institutions now give courses on data analytic consultancy work - aspects of the job other than the mechanics of how to carry out a regression, etc. - but there is still a long way to go.

Of course, because of the lack of this synthesis, what happens in practice at present is that, despite the huge wealth of methods available, the standard methods - those that are readily available in widespread data analytic packages - are the ones that get used. The others are simply ignored, even if a critical assessment might have established that they had some valuable properties, or that they were "best" in some sense under some circumstances.

Although in this section we have focused on the relationship between data analysis and the two disciplines of statistics and machine learning, we should note in passing that those two disciplines also cover other areas. This is one reason why they are not merely subdisciplines of "intelligent data analysis". Statistics, for example, subsumes efficient design methodologies, such as experimental design and survey design, described briefly in the next section, and machine learning also covers syntactic approaches to learning. Likewise, we should also note that there are other influences on modern data analysis. The impact of the huge data sets which are being collected is one example. Modern electronics facilitates automatic data acquisition (e.g. in supermarket point of sale systems, in electronic measurement systems, in satellite photography, and so on) and some vast databases have been compiled. The new discipline of data mining has developed especially to extract valuable information from such huge data sets (see [244] for detailed discussion of such databases and ways to cope with them).

As data sets have grown in size and complexity, so there has been an inevitable shift away from direct hands-on data analysis towards indirect data analysis in which the analyst works via more complex and sophisticated tools. In a sense this is automatic data analysis. An early illustration is the use of variable selection techniques in regression. Given a clearly defined criterion (sum of squared errors, for example), one can let the computer conduct a much larger search than could have been conducted by hand. The program has become a key part of the analysis and has moved the analyst's capabilities into realms which would be impossible unaided. Modern intelligent data analysis relies heavily on such distanced analysis. By virtue of the power it provides, it extends the analyst's capabilities substantially (by orders of magnitude, one might say). The perspective that the analyst instructs a program to go and do the work is essentially a machine learning perspective.

1.3. The Nature of Data

This book is primarily concerned with numerical data, but other kinds exist. Examples include text data and image data. In text data the basic symbols are words rather than numbers, and they can be combined in more ways than can numbers. Two of the major challenges with text data are search and matching. These have become especially important with the advent of the World Wide Web. Note that the objects of textual data analysis are the blocks of text themselves, but the objects of numerical data analysis are really the things which have given rise to the numbers. The numbers are the result of a mapping, by means of measuring instruments, from the world being studied (be it physical, psychological, or whatever), to a convenient representation. The numerical representation is convenient because we can manipulate the numbers easily and relatively effortlessly. Directly manipulating the world which is the objective of the study is generally less convenient. (For example, to discover which of two groups of men is heavier, we could have them all stand on the pans of a giant weighing scales and see which way the scales tipped. Or we could simply add up their weights and compare the two resulting numbers.) The gradual development of a quantitative view of the world which took place around the fourteenth and fifteenth centuries (one can argue about timescales) is what underpinned the scientific revolution and what led ultimately to our current view of the world. The development of the computer, and what it implies about data analysis means that this process is continuing.

The chapters in this book present rather idealised views on data analysis. All data, perhaps especially modern data sets which are often large and many of which relate to human beings, has the potential for being messy. A priori one should expect to find (or rather, not to find) missing values, distortions, misrecording, inadequate sampling and so on. Raw data which do not appear to show any of these problems should immediately arouse suspicion. A very real possibility is that the presented data have been cleaned up before the analyst sees them. This has all sorts of implications. Here are some illustrations.

Data may be missing for a huge variety of reasons. In particular, however, data may be missing for reasons connected with the values they would have had, had they been recorded. For example, in pain research, it would be entirely reasonable to suppose that those patients who would have had the most severe pain are precisely those who have taken an analgesic and dropped out of the study. Imagine the mistakes which would result from studying only the apparent pain scores. Clearly, to cope with this, a larger data analysis is required. Somehow one must model not only the scores which have been presented, but also the mechanism by which the missing ones went missing.

Data may be misrecorded. I can recall one incident in which the most significant digit was missed from a column of numbers because it went over the edge of the printing space. (Fortunately, the results in this case were so counter-intuitive that they prompted a detailed search for an explanation.)

Data may not be from the population they are supposed to be from. In clinical trials, for example, the patients are typically not a random sample from some well-defined population, but are typically those who happened to attend a given clinic and who also satisfied a complex of inclusion/exclusion criteria. Outliers are also a classic example here, requiring careful thought about whether they should be dropped from the analysis as anomalous, or included as genuine, if unusual, examples from the population under study.

Given that all data are contaminated, special problems can arise if one is seeking small structures in large data sets. In such cases, the distortions due to contamination may be just as large, and just as statistically significant, as the effects being sought.

In view of all this, it is very important to examine the data thoroughly before undertaking any formal analysis. Traditionally, data analysts have been taught to "familiarise themselves with their data" before beginning to model it or test it against algorithms. However, with the large size of modern data sets this is less feasible (or even entirely impossible in many cases). Here we must rely on computer programs to check the data for us. There is scope here for much research: anomalous data, or data with hidden peculiarities, can only be shown to be such if we can tell the computer what to search for. Peculiarities which we have not imagined will slip through the net and could have all sorts of implications for the value of the conclusions one draws.

In many, perhaps most, cases a "large" data set is one which has many cases or records. Sometimes, however, the word "large" can refer to the number of variables describing each record. In bank or supermarket records, for example, details of each transaction may be retained, while in high resolution nuclear magnetic resonance spectroscopy there may be several hundred thousand variables. When there are more variables than cases, problems can arise: covariance matrices become singular, so that inversion is impossible. Even if things are not so extreme, strong correlations between variables can induce instability in parameter estimates. And even if the data are well-behaved, large numbers of variables mean that the curse of dimensionality really begins to bite. (This refers to the exponentially increasing size of sample required to obtain accurate estimates of probabilities as the number of variables increases. It manifests itself in such counterintuitive effects as the fact that most of the data in a high dimensional hypercube with uniformly distributed data will lie in a thin shell around the edge, and that the nearest sample point to any given point in a high dimensional space will be far from the given point on average.)

At the end of the previous section we commented about the role of design in collecting data. Adequate design can make the difference between a productive and an unproductive analysis. Adequate design can permit intelligent data analysis - whereas data which has been collected with little thought for how it might be analysed may not succumb to even the most intelligent of analyses. This, of course, poses problems for those concerned with secondary data analysis - the analysis of data which have been collected for some purpose other than that being addressed by the current analysis.

In scientific circles the word "experiment" describes an investigation in which (some of) the potentially influential variables can be controlled. So, for example, we might have an experiment in which we control the diet subjects receive, the distance vehicles travel, the temperature at which a reaction occurs, or the proportion of people over 40 who receive each treatment. We can then study how other variables differ between different values of those which have been controlled. The hope is that by such means one can unequivocally attribute changes to those variables which have been controlled - that one can identify causal relationships between variables.

Of course, there are many subtleties. Typically it is not possible to control all the potentially influential variables. To overcome this, subjects (or objects) are randomly assigned to the classes defined by those variables which one wishes to control. Note that this is a rather subtle point. The random assignment does not guarantee that the different groups are balanced in terms of the uncontrolled variables - it is entirely possible that a higher proportion of men (or whatever) will fall in one group than another, simply by random fluctuations. What the random assignment does do, however, is permit us to argue about the average outcomes over the class of similar experiments, also carried out by such random allocations.

Apart from its use in eliminating bias and other distortions, so that the question of interest is really being answered, experimental design also enables one to choose an efficient data collection strategy - to find the most accurate answer to the question for given resources or the least resources required to achieve a specified accuracy. To illustrate: an obvious way to control for six factors is to use each of them at several levels (say three, for the purposes of this illustration) - but this produces 729 groups of subjects. The numbers soon mount up. Often, however, one can decide a priori that certain high order effects are unlikely to occur - perhaps the way that the effect of treatment changes according age and sex is unlikely to be affected by weight, for example (even though weight itself influences the treatment effect, and so on). In such cases it is possible to collect information on only a subset of the 729 (or however many) groups and still answer the questions of interest.

Experiments are fundamentally manipulative - by definition they require that one can control values of variables or choose objects which have particular values of variables. In contrast, surveys are fundamentally observational. We study an existing population to try to work out what is related to what. To find out how a particular population (not necessarily of people, though surveys are used very often to study groups of people) behaves one could measure every individual within it. Alternatively, one could take measurements merely on a sample. The Law of Large Numbers of statistics tells us that if we repeatedly draw samples of a given size from a population, then for larger samples the variance of the mean of the samples is less. So if we only draw a small sample from our population we might obtain only an inaccurate estimate of the mean value in which we are interested. But we can obtain an accurate estimate, to any degree of accuracy we like, by choosing a large enough sample. Interestingly enough, it is essentially the size of the sample which is of relevance here. A sample of 1000 from a population of 100,000 will have the same accuracy as one from a population of a million (if the two populations have the same distribution shape).

The way in which the sample is drawn is fundamental to survey work. If one wanted to draw conclusions about population of New York one would not merely interview people who worked in delicatessens. There is a classic example of a survey going wildly wrong in predicting people's Presidential voting intentions in the US: the survey was carried out by phone, and failed to take account of the fact that the less well-off sections of the population did not have phones. Such problems are avoided by drawing up a "sampling frame" - a list of the entire population of interest - and ensuring that the sample is randomly selected from this frame.

The idea of a simple random sample, implicit in the preceding paragraph, underlies survey work. However, more sophisticated sampling schemes have been developed - again with the objective of achieving maximum efficiency, as with experimental design. For example, in stratified sampling, the sampling frame is divided into strata according to the value of a (known) variable which is thought to be well correlated with the target variable. Separate estimates are then obtained within each stratum, and these are combined to yield an overall estimate.

The subdisciplines of experimental and survey design have developed over the years and are now very sophisticated, with recent developments involving high level mathematics. They provide good illustrations of the effort which is necessary to ensure good and accurate data so that effective answers can be obtained in data analysis. Without such tools, if the data are distorted or have been obtained by an unknown process, then no matter how powerful one's computers and data analysis tools, one will obtain results of dubious value. The familiar computer adage "garbage in, garbage out" is particularly relevant.

1.4. Modern Data Analytic Tools

The chapters in this book illustrate the range of tools available to the modern data analyst. The opening chapters adopt a mainly statistical perspective, illustrating the modelling orientation.

Chapter 2 describes basic statistical concepts, covering such things as what 'probability means', the notions of sampling and estimates based on samples, elements of inference, as well as more recently developed tools of intelligent data analysis such as cross-validation and bootstrapping.

Chapter 3 describes some of the more important statistical model structures. Most intelligent data analysis involves issues of how variables are related, and this chapter describes such multivariate models, illustrating some of them with simple examples. The discussion includes the wide range of generalised linear models, which are a key tool in the data analyst's armoury.

Up until recently, almost all statistical practice was carried out in the 'frequentist' tradition. This is based on an objective interpretation of probability, regarding it as a real property of events. Chapter 3 assumes this approach. Recently, however, thanks to advances in computer power, an alternative approach, based on a subjective interpretation of probability as a degree of belief, has become feasible. This is the Bayesian approach. Chapter 4 provides an introductory overview of Bayesian methods.

A classical approach to supervised classification methods was to combine and transform the raw measured variables to produce 'features', defining a new data space in which the classes were linearly separable. This basic principle has been developed very substantially in the notion of support vector machines, which use some clever mathematics to permit the use of an effectively infinite number of features. Early experience suggests that methods based on these ideas produce highly effective classification algorithms. The ideas are described in Chapter 5.

Time series occupy a special place in data analysis because they are so ubiquitous. As a result of their importance, a wide variety of methods has been developed. Chapter 6 describes some of these approaches

I remarked above, that statistics and machine learning, the two legs on which modern intelligent data analysis stands, have differences in emphasis. One of these differences is the importance given to the interpretability of a model. For example, in both domains, recursive partitioning or tree methods have been developed. These are essentially predictive models which seek to predict the value of a response variable from one or more explanatory variables. They do this by partitioning the space of explanatory variables into discrete regions, such that a unique predicted value of the response variable is associated with each region. While there is overlap, the statistical development has been more concerned with predictive accuracy, and the machine learning development with interpretability. Tree models are closely related to (indeed, some might say are a form of) methods for rule induction, which is the subject of Chapter 7. A rule is a substructure of a model which recognises a specific pattern in the database and takes some action. From this perspective, such tools for data analysis are very much machine learning tools.

It is unlikely that anyone reading this book will not have heard the phrase "neural network". In the context of this book an artificial neural network is a structure of simple processors with parameterised interconnections. By choosing the values of the parameters appropriately, one can use the network as a very flexible function estimator. This makes such networks powerful tools - essentially providing a very good fit to a data set and then being shrunk to avoid overfitting problems. Their flexibility means that they are less subject to the bias problems intrinsic to methods which assume a particular model form to start with (e.g. the linear form of classical regression). Artificial neural networks are important because of their power as models, but they may turn out to be just as important because of the impetus they are giving to enhanced understanding of inference and the nature of induction. Chapter 8 discusses such tools for intelligent data analysis.

Probability, and the theories of inferential statistics built on it, are the most widely accepted and used tool for handling uncertainty. However, uncertainty comes in many shapes and forms. There is, for example, stochastic uncertainty arising from the basic mechanism leading to the data, but there is also uncertainty about the values of measurements or the meaning of terms. While many - especially Bayesians - feel that this second kind can also be handled by probabilistic arguments, not everyone agrees, and other approaches have been developed. One such is the school of fuzzy reasoning and fuzzy logic. Essentially, this replaces the (acknowledged false) notion that classes are precisely known by a membership function, which allows an object to belong to more than one class, but with differing degrees. The details are given in Chapter 9, which also describes fuzzy numbers and how they may be manipulated.

One of the most exciting developments which has resulted from the growth of computer power has been the probabilistic solution of previously intractable methods by means of stochastic search and optimisation methods, such as simulated annealing and genetic algorithms. These sort of strategies are the subject of Chapter 10.

Methods for graphical display of data are as old as data itself. However, modern computational facilities have extended the scope considerably: with such facilities, dynamic and interactive graphics are possible. Coupled with the increasingly tough demands of modern data analysis, arising from such things as huge data sets and time-dependent data sets, the field of graphical displays -or data visualization, as it is now called - has blossomed. Such developments are described in Chapter 11.

Chapters 12 and Appendix A round off the book. Chapter 12 presents some examples of real applications of the ideas in the book, ranging from relatively standard statistical applications to novel machine learning applications such as the "No hands across America" experiment. Appendix A lists and describes some of the many tools available for intelligent data analysis which now abound. The variety and range of origins of these tools indicates the interdisciplinary nature of intelligent data analysis.

1.5. Conclusion

This book is about intelligent data analysis. But if data analysis can be intelligent, then it can also be unintelligent. Unfortunately, as with good health, departures from intelligent data analysis can be in many directions. Distorted data, incorrect choice of questions, misapplication of data analytic tools, overfitting, too idealised a model, a model which goes beyond the various sources of uncertainty and ambiguity in the data, and so on, all represent possibilities for unintelligent data analysis. Because of this, it is less easy to characterise unintelligent data analysis than it is to characterise intelligent data analysis. Often only in retrospect can we see that an analysis was not such a good idea after all. This is one of the reasons why the domain of intelligent data analysis is so interesting. It is very much not a case of simply applying a directory of tools to a given problem, but rather one of critical assessment, exploration, testing, and evaluation. It is a domain which requires intelligence and care, as well as the application of knowledge and expertise about the data. It is a challenging and demanding discipline. Moreover, it is a fundamentally interdisciplinary, taking ideas from several fields of endeavour. And it is a discipline which is continuing to evolve. People sometimes speak of "new technology" as if it were a change which would happen and then finish - like the transition to decimal coinage from the old pounds, shillings, and pence in the UK in the early 1970s. But so-called "new technology" is not like that. It is really a state of permanent change. And riding on the back of it is the development of new methods of data analysis. Of course, if the problems of data analysis remain the same, then the impact of new technology will be limited - we will simply be able to do things faster and bigger. But as technology advances so the possibilities change - the frontiers of what can be achieved, what can be imagined, move back. The current state of data analysis illustrates this. Neural networks, stochastic search methods, practical Bayesian tools, all illustrate possibilities which were inconceivable not so many years ago. Moreover, new application areas present new challenges, posing new problems and requiring new solutions. Examples include financial applications and biometrics (in the sense of person identification through retina and voice prints). Beyond this, the meaning of data is shifting: we have commented above about the growing importance of non-numeric data such as text and image data. But data about data - metadata - is also attracting growing nterest (in the face of such problems as how to merge or fuse data sets which define their basic units in different ways, for example). One thing is clear, data analysis is in the most exciting period of its history. And the evidence is that the possibilities and excitement will continue to grow for many years to come.

Chapter 2
Statistical Concepts

Ad J. Feelders
Utrecht University, The Netherlands

2.1. Introduction

Statistics is the science of collecting, organizing and drawing conclusions from data. How to properly produce and collect data is studied in experimental design and sampling theory. Organisation and description of data is the subject area of descriptive statistics, and how to draw conclusions from data is the subject of statistical inference. In this chapter the emphasis is on the basic concepts of statistical inference, and the other topics are discussed only inasfar as they are required to understand these basic concepts.

In Section 2.2 we discuss the basic ideas of probability theory, because it is the primary tool of statistical inference. Important concepts such as random experiment, probability, random variable and probability distribution are explained in this section.

In Section 2.3 we discuss a particularly important kind of random experiment, namely random sampling, and a particularly important kind of probability distribution, namely the sampling distribution of a sample statistic. Random sampling and sampling distributions provide the link between probability theory and drawing conclusions from data, i.e. statistical inference.

The basic ideas of statistical inference are discussed in Section 2.4. Inference procedures such as point estimation, interval estimation (confidence intervals) and hypothesis testing are explained in this section. Next to the frequentist approach to inference we also provide a short discussion of likelihood inference and the Bayesian approach to statistical inference. The interest in the latter approach seems to be increasing rapidly, particularly in the scientific community. Therefore a separate chapter of this volume is entirely dedicated to this topic (see Chapter 4).

In Section 2.5 we turn to the topic of prediction. Once a model has been estimated from the available data, it is often used to predict the value of some variable of interest. We look at the different sources of error in prediction in order to gain an understanding of why particular statistical methods tend to work well on one type of dataset (in terms of the dimensions of the dataset, i.e. the number of observations and number of variables) but less so on others. The emphasis in this section is on the decomposition of total prediction error into an irreducible and reducible part, and in turn the decomposition of the reducible part into a bias and variance component. Flexible techniques such as classification and regression trees, and neural networks tend to have low bias and high variance whereas the more inflexible "conventional" statitical methods such as linear regression and linear discriminant analysis tend to have more bias and less variance than their "modern" counterparts. The well-known danger of overfitting, and ideas of model averaging presented in Section 2.6, are rather obvious once the bias/variance decomposition is understood.

In Section 2.6, we address computer-intensive statistical methods based on resampling. We discuss important techniques such as cross-validation and bootstrapping. We conclude this section with two model averaging techniques based on resampling the available data, called bagging and arcing. Their well-documented success in reducing prediction error is primarily due to reduction of the variance component of error.

We close off this chapter with some concluding remarks.

2.2. Probability

The most important tool in statistical inference is probability theory. This section provides a short review of the important concepts.

2.2.1 Random Experiments

A *random experiment* is an experiment that satisfies the following conditions

1. all possible distinct outcomes are known in advance,
2. in any particular trial, the outcome is not known in advance, and
3. the experiment can be repeated under identical conditions.

The *outcome space* Ω of an experiment is the set of all possible outcomes of the experiment.

Example 2.1. Tossing a coin is a random experiment with outcome space $\Omega = \{H,T\}$

Example 2.2. Rolling a die is a random experiment with outcome space $\Omega = \{1,2,3,4,5,6\}$

Something that might or might not happen, depending on the outcome of the experiment, is called an *event*. Examples of events are "coin lands heads" or "die shows an odd number". An event A is represented by a subset of the outcome space. For the above examples we have $A = \{H\}$ and $A = \{1,3,5\}$ respectively. Elements of the outcome space are called elementary events.

2.2.2 Classical Definition of Probability

If all outcomes in Ω are equally likely, the probability of A is the number of outcomes in A, which we denote by $M(A)$ divided by the total number of outcomes M

$$P(A) = \frac{M(A)}{M}$$

If all outcomes are equally likely, the probability of $\{H\}$ in the coin tossing experiment is $\frac{1}{2}$, and the probability of $\{5,6\}$ in the die rolling experiment is $\frac{1}{3}$. The assumption of equally likely outcomes limits the application of the concept of probability: what if the coin or die is not 'fair'? Nevertheless there are random experiments where this definition of probability is applicable, most importantly in the experiment of random selection of a unit from a population. This special and important kind of experiment is discussed in the Section 2.3.

2.2.3 Frequency Definition of Probability

Recall that a random experiment may be repeated under identical conditions. When the number of trials of an experiment is increased indefinitely, the relative frequency of the occurrence of an event approaches a constant number. We denote the number of trials by m, and the number of times A occurs by $m(A)$. The frequency definition of probability states that

$$P(A) = \lim_{m \to \infty} \frac{m(A)}{m}$$

The law of large numbers states that this limit does indeed exist. For a small number of trials, the relative frequencies may show strong fluctuation as the number of trials varies. The fluctuations tend to decrease as the number of trials increases.

Figure 2.1 shows the relative frequencies of heads in a sequence of 1000 coin tosses as the sequence progresses. In the beginning there is quite some fluctuation, but as the sequence progresses, the relative frequency of heads settles around 0.5.

2.2.4 Subjective Definition of Probability

Because of the demand of repetition under identical circumstances, the frequency definition of probability is not applicable to every event. According to the subjective definition, the probability of an event is a measure of the *degree of belief*

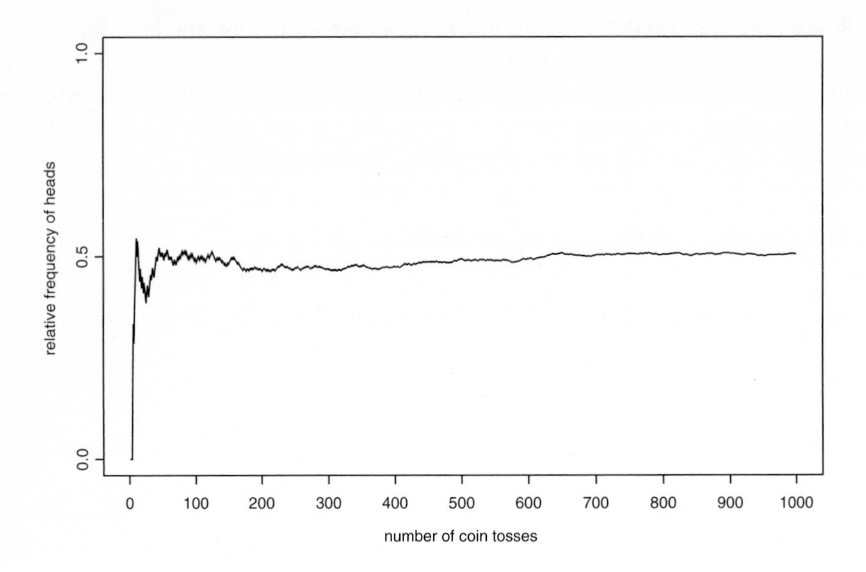

Fig. 2.1. Relative frequency of heads in a sequence of 1000 coin tosses.

that the event will occur (or has occured). Degree of belief depends on the person who has the belief, so my probability for event A may be different from yours.

Consider the statement: "There is extra-terrestrial life". The degree of belief in this statement could be expressed by a number between 0 and 1. According to the subjectivist definition we may interpret this number as the probability that there is extra-terrestrial life.

The subjective view allows the expression of all uncertainty through probability. This view has important implications for statistical inference (see Section 2.4.3 and Chapter 4 of this book).

2.2.5 Probability Axioms

Probability is defined as a function from subsets of Ω to the real line $\mathbb{R}$, that satisfies the following axioms

1. Non-negativity: $P(A) \geq 0$
2. Additivity: If $A \cap B = \emptyset$ then $P(A \cup B) = P(A) + P(B)$
3. $P(\Omega) = 1$

The classical, frequency and subjective definitions of probability all satisfy these axioms. Therefore every property that may be deduced from these axioms holds for all three interpretations of probability.

2.2.6 Conditional Probability and Independence

The probability that event A occurs may be influenced by information concerning the occurrence of event B. The probability of event A, given that B will occur or has occurred, is called the *conditional probability* of A given B, denoted by $P(A \mid B)$. It follows from the axioms of probability that

$$P(A \mid B) = \frac{P(A \cap B)}{P(B)}$$

for $P(B) > 0$. Intuitively we can appreciate this equality by considering that B effectively becomes the new outcome space. The events A and B are called *independent* if the occurrence of one event does not influence the probability of occurrence of the other event, i.e.

$$P(A \mid B) = P(A) \text{ , and consequently } P(B \mid A) = P(B)$$

Since independence of two events is always mutual, it is more concisely expressed by the product rule

$$P(A \cap B) = P(A)\, P(B)$$

2.2.7 Random Variables

A random variable X is a *function* from the outcome space Ω to the real line

$$X : \Omega \rightarrow \mathbb{R}$$

Example 2.3. Consider the random experiment of tossing a coin twice, and observing the faces turning up. The outcome space is

$$\Omega = \{(H,T), (T,H), (H,H), (T,T)\}$$

The number of heads turning up is a random variable defined as follows

$$X((H,T)) = X((T,H)) = 1\,,\; X((H,H)) = 2\,,\; X((T,T)) = 0$$

2.2.8 Probability Distribution

A probability function p assigns to each possible realisation x of a discrete random variable X the probability $p(x)$, i.e. $P(X = x)$. From the axioms of probability it follows that $p(x) \geq 0$, and $\sum_x p(x) = 1$.

Example 2.4. The number of heads turning up in two tosses of a *fair* coin is a random variable with the following probability function: $p(1) = 1/2$, $p(0) = 1/4$, $p(2) = 1/4$.

Since for continuous random variables, $P(X = x) = 0$, the concept of a probability function is useless. The probability distribution is now specified by representing probabilities as areas under a curve. The function $f : \mathbb{R} \to \mathbb{R}^+$ is called the probability density of X if for each pair $a \leq b$,

$$P(a < X \leq b) = \int_a^b f(x)\, dx$$

It follows from the probability axioms that $f(x) \geq 0$ and $\int_{-\infty}^{\infty} f(x)\, dx = 1$.

Example 2.5. Consider the random variable X with the following density function

$$f(x) = \begin{cases} \frac{1}{2} \text{ for } 0 \leq x \leq 2 \\ 0 \text{ otherwise} \end{cases}$$

It follows that

$$P(1/2 < X \leq 5/4) = \int_{1/2}^{5/4} 1/2 dx = 1/2x|_{1/2}^{5/4} = 3/4$$

The *distribution function* is defined for both discrete and continuous random variables as the function F which gives for each $x \in \mathbb{R}$ the probability of an outcome of X at most equal to x:

$$F(x) = P(X \leq x), \quad \text{for } x \in \mathbb{R}$$

2.2.9 Entropy

The entropy of a random variable is the average amount of information generated by observing its value. The information provided by observing realisation $X = x$ is

$$\mathrm{H}(X = x) = \ln \frac{1}{p(x)} = -\ln p(x)$$

Example 2.6. Consider the random experiment of tossing a coin with probability of heads equal to 0.9, and random variable X with $X(H) = 1$ and $X(T) = 0$. What is the information generated by observing $x = 1$? $\mathrm{H}(x = 1) = -\ln 0.9 = 0.105$. The information generated by observing $x = 0$ is $\mathrm{H}(x = 0) = -\ln 0.1 = 2.303$.

Intuitively, one can appreciate that observing the outcome "heads" provides little information, since the probability of heads is 0.9, i.e. heads is almost certain to come up. Observing "tails" on the other hand provides much information, since its probability is low.

If we were to repeat this experiment very many times, how much information would be generated on average? In general

$$\mathrm{H}(X) = -\sum_i p(x_i) \ln p(x_i)$$

Example 2.7. The average amount of information or *entropy* generated by the previous experiment is: $-(0.9 \ln 0.9 + 0.1 \ln 0.1) = 0.325$. The entropy of tossing a fair coin is: $-(0.5 \ln 0.5 + 0.5 \ln 0.5) = 0.693$.

The information provided by the individual outcomes is weighted by their respective probabilities. Tossing a biased coin generates less information on average than tossing a fair coin, because for a biased coin, the realisation that generates much information (tails coming up in example 2.6) occurs less frequently.

2.2.10 Expectation

For a discrete random variable, the *expected value* or mean is defined as

$$\mathrm{E}(X) = \sum_x x\, p(x) \;, \text{ and } \mathrm{E}[h(X)] = \sum_x h(x)\, p(x)$$

for arbitrary function $h : \mathbb{R} \rightarrow \mathbb{R}$.

Example 2.8. Consider once more the coin tossing experiment of example 2.4 and corresponding probability distribution. The expected value or mean of X is

$$\mathrm{E}(X) = 1/2 \cdot 1 + 1/4 \cdot 2 + 1/4 \cdot 0 = 1$$

The definition of expectation for a continuous random variable is analogous, with summation replaced by integration.

$$\mathrm{E}(X) = \int_{-\infty}^{\infty} x\, f(x)\, dx \;, \text{ and } \mathrm{E}[h(X)] = \int_{-\infty}^{\infty} h(x)\; f(x)\, dx$$

Example 2.9. (Continuation of example 2.5) The mean or expected value of the random variable with probability density given in example 2.5 is

$$\mathrm{E}(X) = \int_0^2 \frac{1}{2}\, dx = \frac{1}{2} x \bigg|_0^2 = \frac{1}{2} \cdot 2 - \frac{1}{2} \cdot 0 = 1$$

The expected value $\mathrm{E}(X)$ of a random variable is usually denoted by μ. The variance σ^2 of a random variable is a measure of spread around the mean obtained by averaging the squared differences $(x - \mu)^2$, i.e.

$$\sigma^2 = \mathrm{V}(X) = \mathrm{E}(X - \mu)^2$$

The standard deviation $\sigma = \sqrt{\sigma^2}$ has the advantage that it has the same dimension as X.

Table 2.1. Conditional probability function $p(x \mid C)$.

x	4	6	8	10	12
$p(x \mid C)$	1/9	2/9	1/3	2/9	1/9

2.2.11 Conditional Probability Distributions and Expectation

For a discrete random variable X we define a conditional probability function as follows

$$p(x \mid C) = P(X = x \mid C) = \frac{P(\{X = x\} \cap C)}{P(C)}$$

Example 2.10. Two fair dice are rolled, and the numbers on the top face are noted. We define the random variable X as the sum of the numbers showing. For example $X((3,2)) = 5$. Consider now the event C : both dice show an even number. We have $P(C) = \frac{1}{4}$ and $P(\{X = 6\} \cap C) = \frac{1}{18}$ since

$$\begin{aligned} C &= \{(2,2),(2,4),(2,6),(4,2),(4,4),(4,6),(6,2),(6,4),(6,6)\} \\ \{X = 6\} \cap C &= \{(2,4),(4,2)\} \end{aligned}$$

The probability of $\{X = 6\}$ given C therefore is

$$P(X = 6 \mid C) = \frac{P(\{X = 6\} \cap C)}{P(C)} = \frac{1/18}{1/4} = \frac{2}{9}$$

The conditional probability function of X is shown in Table 2.1. The conditional expectation of X given C is: $\mathrm{E}(X \mid C) = \sum_x x\, p(x \mid C) = 8$.

For continuous random variable X, the conditional density $f(x \mid C)$ of X given C is

$$f(x \mid C) = \begin{cases} f(x)/P(C) & \text{for } x \in C \\ 0 & \text{otherwise} \end{cases}$$

2.2.12 Joint Probability Distributions and Independence

The joint probability distribution of a pair of discrete random variables (X, Y) is uniquely determined by their joint probability function $p : \mathbb{R}^2 \to \mathbb{R}$

$$p(x, y) = P((X, Y) = (x, y)) = P(X = x, Y = y)$$

From the axioms of probability it follows that $p(x, y) \geq 0$ and $\sum_x \sum_y p(x, y) = 1$.

The *marginal* probability function $p_X(x)$ is easily derived from the joint distribution

$$p_X(x) = p(X = x) = \sum_y P(X = x, Y = y) = \sum_y p(x, y)$$

The conditional probability function of X given $Y = y$

$$p(x \mid y) = \frac{P(X = x, Y = y)}{P(Y = y)} = \frac{p(x, y)}{p_Y(y)}$$

Definitions for continuous random variables are analogous with summation replaced by integration. The function $f : \mathbb{R}^2 \to \mathbb{R}$ is the probability density of the pair of random variables (X, Y) if for all $a \leq b$ and $c \leq d$

$$P(a < X \leq b, c < Y \leq d) = \int_a^b \int_c^d f(x, y)\, dx\, dy$$

From the probability axioms it follows that

1. $f(x, y) \geq 0$
2. $\int_{-\infty}^{\infty} \int_{-\infty}^{\infty} f(x, y)\, dx\, dy = 1$

The marginal distribution of X is obtained from the joint distribution

$$f_X(x) = \int_{-\infty}^{\infty} f(x, y)\, dy$$

and the conditional density of X given $\{Y = y\}$ is

$$f(x \mid y) = \frac{f(x, y)}{f_Y(y)}$$

According to the product rule discussed in Section 2.2.6, the events $\{X = x\}$ and $\{Y = y\}$ are independent iff

$$P(X = x, Y = y) = P(X = x)P(Y = y)$$

We now generalize the concept of independence to pairs of random variables. Discrete random variables X and Y are independent iff

$$p(x, y) = p_X(x)p_Y(y) \text{ for all } (x, y),$$

and as a consequence $p(x \mid y) = p_X(x)$, and $p(y \mid x) = p_Y(y)$. Definitions are completely analogous for continuous random variables, with probability functions replaced by probability densities.

2.2.13 The Law of Total Probability

In some cases the (unconditional) probability of an event may not be calculated directly, but can be determined as a weighted average of various conditional probabilities.

Let $B_1, B_2, \ldots, B_s$ be a partition of Ω, that is $B_i \cap B_j = \emptyset$ for all $i \neq j$ and $\bigcup_{i=1}^{s} B_i = \Omega$. It follows from the axioms of probability that

$$P(A) = \sum_{i=1}^{s} P(A|B_i)P(B_i)$$

Table 2.2. Performance of diagnostic test.

	T^+	T^-
D	0.95	0.05
$\bar{D}$	0.02	0.98

Example 2.11. Consider a box containing three white balls and one red ball. First we draw a ball at random, i.e. all balls are equally likely to be drawn from the box. Then a second ball is drawn at random (the first ball has not been replaced in the box). What is the probability that the second draw yields a red ball? This is most easily calculated by averaging conditional probabilities.

$$P(R_2) = P(R_2|W_1)P(W_1) + P(R_2|R_1)P(R_1) = 1/3 \cdot 3/4 + 0 \cdot 1/4 = 1/4,$$

where R_i stands for "a red ball is drawn on i-th draw" and W_i for "a white ball is drawn on i-th draw".

2.2.14 Bayes' Rule

Bayes' rule shows how probabilities change in the light of evidence. It is a very important tool in Bayesian statistical inference (see Section 2.4.3). Let $B_1, B_2, \ldots, B_s$ again be a partition of Ω. Bayes' rule follows from the axioms of probability

$$P(B_i|A) = \frac{P(A|B_i)P(B_i)}{\sum_j P(A|B_j)P(B_j)}$$

Example 2.12. Consider a physician's diagnostic test for the presence or absence of some rare disease D, that only occurs in 0.1% of the population, i.e. $P(D) = .001$. It follows that $P(\bar{D}) = .999$, where $\bar{D}$ indicates that a person does not have the disease. The probability of an event before the evaluation of evidence through Bayes' rule is often called the prior probability. The prior probability that someone picked at random from the population has the disease is therefore $P(D) = .001$.

Furthermore we denote a positive test result by T^+, and a negative test result by T^-. The performance of the test is summarized in Table 2.2.

What is the probability that a patient has the disease, if the test result is positive? First, notice that $D, \bar{D}$ is a partition of the outcome space. We apply Bayes' rule to obtain

$$P(D|T^+) = \frac{P(T^+|D)P(D)}{P(T^+|D)P(D) + P(T^+|\bar{D})P(\bar{D})} = \frac{.95 \cdot .001}{.95 \cdot .001 + .02 \cdot .999} = .045.$$

Only 4.5% of the people with a positive test result actually have the disease. On the other hand, the posterior probability (i.e. the probability after evaluation of evidence) is 45 times as high as the prior probability.

2.2.15 Some Named Discrete Distributions

A random experiment that only distinguishes between two possible outcomes is called a *Bernoulli* experiment. The outcomes are usually referred to as *success* and *failure* respectively. We define a random variable X that denotes the number of successes in a Bernoulli experiment; X consequently has possible values 0 and 1. The probability distribution of X is completely determined by the probability of success, which we denote by π, and is: $p(X=0) = 1-\pi$ and $p(X=1) = \pi$. It easily follows that $\mathrm{E}(X) = \mu = \pi$ and $\sigma^2 = \pi(1-\pi)$.

A number of *independent, identical* repetitions of a Bernoulli experiment is called a *binomial* experiment. We denote the number of successes in a binomial experiment by Y which has possible values $0, 1, \ldots, m$ (where m is the number of repetitions). Any particular sequence with y successes has probability

$$\pi^y(1-\pi)^{m-y}$$

since the trials are independent. The number of distinct ways y successes may occur in a sequence of m is

$$\binom{m}{y} = \frac{m!}{y!(m-y)!}$$

so the probability distribution of Y is

$$p(y) = \binom{m}{y}\pi^y(1-\pi)^{m-y} \quad \text{for } y = 0, 1, \ldots, m.$$

We indicate that Y has binomial distribution with parameters m and π by writing $Y \sim B(m, \pi)$ ($\sim$ should be read "has distribution"). We can derive easily that $\mathrm{E}(Y) = \mu = m\pi$ and $\sigma^2 = m\pi(1-\pi)$.

The multinomial distribution is a generalization of the binomial distribution to random experiments with $n \geq 2$ possible outcomes or categories. Let y_i denote the number of results in category i, and let π_i denote the probability of a result in the $i\,th$ category on each trial (with $\sum_{i=1}^{n} \pi_i = 1$). The joint probability distribution of $Y_1, Y_2, \ldots, Y_n$ for a sequence of m trials is

$$P(Y_1 = y_1, Y_2 = y_2, \ldots, Y_n = y_n) = \frac{m!}{y_1!\,y_2!\ldots y_n!}\pi_1^{y_1}\pi_2^{y_2}\ldots\pi_n^{y_n}$$

The product of powers of the π_i represents the probability of any particular sequence with y_i results in category i for each $1 \leq i \leq n$, and the ratio of factorials indicates the number distinct sequences with y_i results in category i for each $1 \leq i \leq n$.

A random variable Y has Poisson distribution with parameter μ if it has probability function

$$p(y) = \frac{\mu^y}{y!}e^{-\mu} \text{ for } y = 0, 1, 2, \ldots$$

where the single parameter μ is a positive real number. One can easily show that $\mathrm{E}(Y) = \mathrm{V}(Y) = \mu$. We write $Y \sim \mathrm{Po}(\mu)$. Use of the Poisson distribution as an approximation to the binomial distribution is discussed in Section 2.3.

2.2.16 Some Named Continuous Distributions

Continuous distributions of type

$$f(y) = \begin{cases} \frac{1}{\beta-\alpha} & \text{for } \alpha \le y \le \beta \\ 0 & \text{otherwise} \end{cases}$$

are called uniform distributions, denoted $U(\alpha, \beta)$. Mean and variance are respectively

$$\mu = \frac{\alpha+\beta}{2}, \text{ and } \sigma^2 = \frac{(\beta-\alpha)^2}{12}$$

Continuous distributions of type

$$f(y) = \frac{e^{-(y-\mu)^2/(2\sigma^2)}}{\sigma\sqrt{2\pi}} \quad \text{for } y \in \mathbb{R}$$

with $\sigma > 0$ are called *normal* or *Gaussian* distributions. Mean μ and variance σ^2 are the two parameters of the normal distribution, which we denote by $\mathcal{N}(\mu, \sigma^2)$. The special case with $\mu = 0$ and $\sigma^2 = 1$, is called the standardnormal distribution. A random variable of this type is often denoted by Z, i.e. $Z \sim \mathcal{N}(0,1)$. If the distribution of a random variable is determined by many small independent influences, it tends to be normally distributed. In the next section we discuss why the normal distribution is so important in statistical inference.

The binormal distribution is a generelization of the normal distribution to the joint distribution of pairs (X, Y) of random variables. Its parameters are μ_x, μ_y, σ_x^2, σ_y^2, and correlation coefficient ρ, with σ_x^2, $\sigma_y^2 > 0$ and $-1 \le \rho \le 1$. We write

$$(X, Y) \sim \mathcal{N}^2(\mu_x, \mu_y, \sigma_x^2, \sigma_y^2, \rho)$$

The parameter ρ is a measure for the linear dependence between X and Y (for further explanation of this parameter, the reader is referred to Section 2.6.3). Further generelization to the joint distribution of $n \ge 2$ random variables $Y_1, Y_2, \ldots, Y_n$ yields the multivariate normal distribution. For convenience we switch to matrix notation for the parameters

$$(Y_1, Y_2, \ldots, Y_n) \sim \mathcal{N}^n(\mu, \Sigma)$$

where $\mu = (\mu_1, \mu_2, \ldots, \mu_n)$ is the vector of means and Σ is an $n \times n$ covariance matrix. The diagonal elements of Σ contain the variances $(\sigma_1^2, \sigma_2^2, \ldots, \sigma_n^2)$ and element (i, j) with $i \ne j$ contains the covariance between Y_i and Y_j (for an explanation of covariance, the reader is again referred to Section 2.6.3).

A random variable T has *exponential* distribution with rate λ ($\lambda > 0$) if T has probability density

$$f(t) = \lambda e^{-\lambda t} \quad (t \ge 0)$$

We may think of T as a random time of some kind, such as a time to failure for artifacts, or survival times for organisms. With T we associate a survival function

$$P(T > s) = \int_s^\infty f(t)dt = e^{-\lambda s}$$

representing the probability of surviving past time s. Characteristic for the exponential distribution is that it is *memoryless*, i.e.

$$P(T > t + s \,|\, T > t) = P(T > s) \quad (t \geq 0, s \geq 0)$$

Given survival to time t, the chance of surviving a further time s is the same as surviving to time s in the first place. This is obviously not a good model for survival times of systems with aging such as humans. It is however a plausible model for time to failure of some artifacts that do not wear out gradually but stop functioning suddenly and unpredictably.

A random variable Y has a Beta distribution with parameters $l > 0$ and $m > 0$ if it has probability density

$$f(y) = \frac{y^{l-1}(1-y)^{m-1}}{\int_0^1 y^{l-1}(1-y)^{m-1}dy} \quad (0 \leq y \leq 1)$$

For the special case that $l = m = 1$ this reduces to a uniform distribution over the interval $[0, 1]$. The Beta distribution is particularly useful in Bayesian inference concerning unknown probabilities, which is discussed in Section 2.4.3.

2.3. Sampling and Sampling Distributions

The objective of sampling is to draw a sample that permits us to make inferences about a population of interest. We may for example draw a sample from the population of Dutch men of 18 years and older to learn something about the joint distribution of height and weight in this population.

Because we cannot draw conclusions about the population from a sample without error, it is important to know how large these errors may be, and how often incorrect conclusions may occur. An objective assessment of these errors is only possible for a *probability sample*. For a probability sample, the probability of inclusion in the sample is *known* and *positive* for each unit in the population. Drawing a probability sample of size m from a population consisting of M units, may be a quite complex random experiment. The experiment is simplified considerably by subdividing it into m experiments, consisting of drawing the m consecutive units. In a *simple random sample* the m consecutive units are drawn with equal probabilities from the units concerned. In random sampling *with replacement* the subexperiments (drawing of one unit) are all identical and independent: m times a unit is randomly selected from the entire population. We will see that this property simplifies the ensuing analysis considerably.

For units in the sample we observe one or more population variables. For probability samples, each draw is a random experiment. Every observation may therefore be viewed as a random variable. The observation of a population variable $\mathcal{X}$ from the unit drawn in the ith trial, yields a random variable X_i. Observation of the complete sample yields m random variables $X_1, ..., X_m$. Likewise, if we observe for each unit the pair of population variables $(\mathcal{X}, \mathcal{Y})$, we obtain pairs

Table 2.3. A small population.

Unit	1	2	3	4	5	6
$\mathcal{X}$	1	1	2	2	2	3

Table 2.4. Probability distribution of X_1 and X_2.

x	1	2	3
$p_1(x) = p_2(x)$	1/3	1/2	1/6

of random variables (X_i, Y_i) with outcomes (x_i, y_i). Consider the population of size $M = 6$, displayed in Table 2.3.

A random sample of size $m = 2$ is drawn *with replacement* from this population. For each unit drawn we observe the value of $\mathcal{X}$. This yields two random variables X_1 and X_2, with identical probability distribution as displayed in Table 2.4. Furthermore X_1 and X_2 are independent, so their joint distribution equals the product of their individual distributions,

$$p(x_1, x_2) = \prod_{i=1}^{2} p_i(x_i) = [p(x)]^2$$

The distribution of the sample is displayed in the Table 2.5.

Usually we are not really interested in the individual outcomes of the sample, but rather in some sample statistic. A statistic is a function of the sample observations $X_1, ..., X_m$, and therefore is itself also a random variable. Some important sample statistics are the sample mean $\bar{X} = \frac{1}{m}\sum_{i=1}^{m} X_i$, sample variance $S^2 = \frac{1}{m-1}\sum_{i=1}^{m}(X_i - \bar{X})^2$, and sample fraction $Fr = \frac{1}{m}\sum_{i=1}^{m} X_i$ (for 0-1 vari-

Table 2.5. Probability distribution of sample of size $m = 2$ by sampling with replacement from the population in Table 2.3

(x_1, x_2)	$p(x_1, x_2)$	$\bar{x}$	s^2
(1,1)	1/9	1	0
(2,2)	1/4	2	0
(3,3)	1/36	3	0
(1,2)	1/6	1.5	0.5
(1,3)	1/18	2	2
(2,1)	1/6	1.5	0.5
(2,3)	1/12	2.5	0.5
(3,1)	1/18	2	2
(3,2)	1/12	2.5	0.5

Table 2.6. Sampling distribution of $\bar{X}$.

$\bar{x}$	$p(\bar{x})$
1	1/9
1.5	1/3
2	13/36
2.5	1/6
3	1/36

Table 2.7. Sampling distribution of S^2.

s^2	$p(s^2)$
0	14/36
0.5	1/2
2	1/9

able $\mathcal{X}$). In Table 2.5 we listed the values of sample statistics $\bar{x}$ and s^2, for all possible samples of size 2.

The probability distribution of a sample statistic is called its *sampling distribution*. The sampling distribution of $\bar{X}$ and S^2 is calculated easily from Table 2.5; they are displayed in tables 2.6 and 2.7 respectively.

Note that $E(\bar{X}) = \frac{11}{6} = \mu$, where μ denotes the population mean, and $E(S^2) = \frac{17}{36} = \sigma^2$, where σ^2 denotes the population variance.

In the above example, we were able to determine the probability distribution of the sample, and sample statistics, by complete enumeration of all possible samples. This was feasible only because the sample size and the number of distinct values of $\mathcal{X}$ was very small. When the sample is of realistic size, and $\mathcal{X}$ takes on many distinct values, complete enumeration is not possible. Nevertheless, we would like to be able to infer something about the shape of the sampling distribution of a sample statistic, from knowledge of the distribution of X. We consider here two options to make such inferences.

1. The distribution of X has some standard form that allows the mathematical derivation of the exact sampling distribution.
2. We use a limiting distribution to approximate the sampling distribution of interest. The limiting distribution may be derived from some characteristics of the distribution of X.

The exact sampling distribution of a sample statistic is often hard to derive analytically, even if the population distribution of $\mathcal{X}$ is known. As an example we consider the sample statistic $\bar{X}$. The mean and variance of the sampling distribution of $\bar{X}$ are $E(\bar{X}) = \mu$ and $V(\bar{X}) = \sigma^2/m$, but its exact shape can only be derived in a few special cases. For example, if the distribution of $\mathcal{X}$ is

$\mathcal{N}(\mu, \sigma^2)$ then the distribution of $\bar{X}$ is $\mathcal{N}(\mu, \sigma^2/m)$. Of more practical interest is the exact sampling distribution of the sample statistic Fr, i.e. the fraction of successes in the sample, with $\mathcal{X}$ a 0-1 population variable. The number of successes in the sample has distribution $Y \sim B(m, \pi)$ where m is the sample size and π the fraction of successes in the population. We have $\mu_y = m\pi$ and $\sigma_y^2 = m\pi(1-\pi)$. Since $Fr = Y/m$, it follows that $\mu_{fr} = \pi$ and $\sigma_{fr}^2 = \pi(1-\pi)/m$. Since $P(Fr = fr) = P(Y = mfr)$, its sampling distribution is immediately derived from the sampling distribution of Y.

Example 2.13. Consider a sample of size 10 from a population with fraction of successes $\pi = 0.8$. What is the sampling distribution of Fr, the sample fraction of successes? The distribution is immediately derived from the distribution of the number of successes $Y \sim B(10, 0.8)$.

In practice, we often have to rely on approximations of the sampling distribution based on so called *asymptotic* results. To understand the basic idea, we have to introduce some definitions concerning the convergence of sequences of random variables. For present purposes we distinguish between convergence in probability (to a constant) and convergence in distribution (weak convergence) of a sequence of random variables. The limiting arguments below are all with respect to sample size m.

Definition 2.1. *A sequence $\{X_m\}$ of random variables converges in probability to a constant c if, for every positive number ε and η, there exists a positive integer $m_0 = m_0(\varepsilon, \eta)$ such that*

$$P(|X_m - c| > \varepsilon) < \eta, \; m \geq m_0$$

Example 2.14. Consider the sequence of random variables $\{X_m\}$ with probability distributions $P(x_m = 0) = 1 - 1/m$ and $P(x_m = m) = 1/m$. Then $\{X_m\}$ converges in probability to 0.

Definition 2.2. *A sequence $\{X_m\}$ of random variables converges in distribution to a random variable X with distribution function $F(X)$ if for every $\varepsilon > 0$, there exists an integer $m_0 = m_0(\varepsilon)$, such that at every point where $F(X)$ is continuous*

$$|F_m(x) - F(x)| < \varepsilon, \; m \geq m_0$$

where $F_m(x)$ denotes the distribution function of x_m.

Example 2.15. Consider a sequence of random variables $\{X_m\}$ with probability distributions $P(x_m = 1) = 1/2 + 1/(m+1)$ and $P(x_m = 2) = 1/2 - 1/(m+1)$, $m = 1, 2, \ldots$. As m increases without bound, the two probabilities converge to 1/2, and $P(X = 1) = 1/2$, $P(X = 2) = 1/2$ is called the *limiting distribution* of $\{X_m\}$.

Convergence in distribution is a particularly important concept in statistical inference, because the limiting distributions of sample statistics may be used as an approximation in case the exact sampling distribution cannot be (or is prohibitively cumbersome) to derive. A crucial result in this respect is the *central limit theorem*: If $(x_1, ..., x_m)$ is a random sample from any probability distribution with finite mean μ and finite variance σ^2, and $\bar{x} = 1/m \sum x_i$ then

$$\frac{(\bar{x} - \mu)}{\sigma/\sqrt{m}} \xrightarrow{D} \mathcal{N}(0, 1)$$

regardless of the form of the parent distribution. In this expression, $\xrightarrow{D}$ denotes convergence in distribution. This property explains the importance of the normal distribution in statistical inference. Note that this theorem doesn't say anything however about the rate of convergence to the normal distribution. In general, the more the population distribution resembles a normal distribution, the faster the convergence. For extremely skewed distributions $m = 100$ may be required for the normal approximation to be acceptable.

A well-known application of the central limit theorem is the approximation of the distribution of the sample proportion of successes Fr by a normal distribution. Since a success is coded as 1, and failure as 0, the fraction of successes is indeed a mean. This means the central limit theorem is applicable and as a rule of thumb $Fr \approx \mathcal{N}(\pi, \pi(1-\pi)/m)$ if $m\pi \geq 5$ and $m(1-\pi) \geq 5$. Even though the exact sampling distribution can be determined in this case, as m becomes large it becomes prohibitively time-consuming to actually calculate this distribution.

If π is close to 0 or 1, quite a large sample is required for the normal approximation to be acceptable. In that case we may use the following covergence property of the binomial distribution

$$\binom{m}{y} \pi^y (1-\pi)^{m-y} \xrightarrow{D} \frac{(m\pi)^y}{y!} e^{-m\pi}$$

In words, the binomial distribution with parameters m and π converges to a Poisson distribution with parameter $\mu = m\pi$ as m gets larger and larger. Moreover, it can be shown that this approximation is quite good for $\pi \leq 0.1$, regardless of the value of m. This explains the use of the Poisson rather than the normal approximation to the binomial distribution when π is close to 0 or 1.

2.4. Statistical Inference

The relation between sample data and population may be used for reasoning in two directions: from known population to yet to be observed sample data (as discussed in Section 2.3), and from observed data to (partially) unknown population. This last direction of reasoning is of inductive nature and is addressed in statistical inference. It is the form of reasoning most relevant to data analysis, since one typically has available one set of sample data from which one intends to draw conclusions about the unknown population.

2.4.1 Frequentist Inference

According to frequentists, inference procedures should be interpreted and evaluated in terms of their behavior in hypothetical repetitions under the same conditions. To quote David S. Moore, the frequentist consistently asks "What would happen if we did this many times?"[386]. To answer this question, the sampling distribution of a statistic is of crucial importance. The two basic types of frequentist inference are estimation and testing. In estimation one wants to come up with a plausible value or range of plausible values for an unknown population parameter. In testing one wants to decide whether a hypothesis concerning the value of an unknown population parameter should be accepted or rejected in the light of sample data.

Point Estimation In point estimation one tries to provide an estimate for an unknown population parameter, denoted by θ, with *one number*: the point estimate. If G denotes the estimator of θ, then the estimation error is a random variable $G - \theta$, which should preferably be close to zero.

An important quality measure from a frequentist point of view is the bias of an estimator

$$\mathrm{B}_\theta = \mathrm{E}_\theta(G - \theta) = \mathrm{E}_\theta(G) - \theta,$$

where expectation is taken with respect to repeated samples from the population. If $\mathrm{E}_\theta(G) = \theta$, i.e. the expected value of the estimator is equal to the value of the population parameter, then the estimator G is called *unbiased*.

Example 2.16. If π is the proportion of successes in some population and Fr is the proportion of successes in a random sample from this population, then $\mathrm{E}_\pi(Fr) = \pi$, so Fr is an unbiased estimator of π.

Another important quality measure of an estimator is its variance

$$\mathrm{V}_\theta(G) = \mathrm{E}_\theta(G - \mathrm{E}_\theta(G))^2$$

which measures how much individual estimates g tend to differ from $\mathrm{E}_\theta(G)$, the average value of g over a large number of samples.

An overall quality measure that combines bias and variance is the *mean squared error*

$$\mathrm{M}_\theta(G) = \mathrm{E}_\theta(G - \theta)^2$$

where low values indicate a good estimator. After some algebraic manipulation, we can decompose mean squared error into

$$\mathrm{M}_\theta(G) = \mathrm{B}_\theta^2(G) + \mathrm{V}_\theta(G)$$

that is mean squared error equals squared bias plus variance. It follows that if an estimator is unbiased, then its mean squared error equals its variance.

Example 2.17. For the unbiased estimator Fr of π we have $\mathrm{M}_\pi(Fr) = \mathrm{V}_\pi(Fr) = \pi(1 - \pi)/m$.

The so-called "plug-in" principle provides a simple and intuitively plausible method of constructing estimators. The plug-in estimate of a parameter $\theta = t(F)$ is defined to be $\hat{\theta} = t(\hat{F})$. Here F denotes the population distribution function and $\hat{F}$ its estimate, based on the sample. For example, to estimate the population mean μ use its sample analogue $\bar{x} = 1/m \sum x_i$, and to estimate population variance σ^2 use its sample analogue $s^2 = 1/m \sum (x_i - \bar{x})^2$. Another well-known method for finding point estimates is the method of least squares. The least squares estimate of population mean μ is the number g for which the sum of squared errors $(x_i - g)^2$ is at a minimum. If we take the derivative of this sum with respect to g, we obtain

$$\frac{\partial}{\partial g} \sum_{i=1}^{m} (x_i - g)^2 = \sum_{i=1}^{m} (x_i - g)(-2) = -2m(\bar{x} - g)$$

When we equate this expression to zero, and solve for g we obtain $g = \bar{x}$. So $\bar{x}$ is the least squares estimate of μ. A third important method of estimation is *maximum likelihood estimation*, which is discussed in Section 2.4.2.

Interval Estimation An interval estimator for population parameter θ is an interval of type (G_L, G_U). Two important quality measures for interval estimates are:

$$\mathrm{E}_\theta(G_U - G_L),$$

i.e. the expected width of the interval, and

$$P_\theta(G_L < \theta < G_U),$$

i.e. the probability that the interval contains the true parameter value. Clearly there is a trade-off between these quality measures. If we require a high probability that the interval contains the true parameter value, the interval itself has to become wider. It is customary to choose a confidence level $(1 - \alpha)$ and use an interval estimator such that

$$P_\theta(G_L < \theta < G_U) \geq 1 - \alpha$$

for all possible values of θ. A realisation (g_L, g_U) of such an interval estimator is called a $100(1 - \alpha)\%$ *confidence interval.*

The form of reasoning used in confidence intervals is most clearly reflected in the estimation of the mean of a normal population with variance σ^2 known, i.e. $X \sim \mathcal{N}(\mu, \sigma^2)$. The distribution of the sample mean for random samples of size m from this population is known to be $\bar{X} \sim \mathcal{N}(\mu, \sigma^2/m)$. First $\bar{X}$ is standardized to obtain

$$\frac{\bar{X} - \mu}{\sigma/\sqrt{m}} \sim \mathcal{N}(0, 1)$$

which allows us to use a table for the standardnormal distribution $Z \sim \mathcal{N}(0, 1)$ to find the relevant probabilities. The probability that $\bar{X}$ is more than one standard error (standard deviation of the sampling distribution) larger than unknown μ is

$$P(\bar{X} > \mu + \frac{\sigma}{\sqrt{m}}) = P(\frac{\bar{X} - \mu}{\sigma/\sqrt{m}} > 1) = P(Z > 1) = 0.1587$$

But we can *reverse* this reasoning by observing that

$$P(\bar{X} - \frac{\sigma}{\sqrt{m}} < \mu) = 1 - 0.1587 = 0.8413$$

because $\bar{X} - \frac{\sigma}{\sqrt{m}} < \mu$ holds unless $\bar{X} > \mu + \frac{\sigma}{\sqrt{m}}$. Therefore, the probability that the interval $(\bar{X} - \sigma/\sqrt{m}, \infty)$ will contain the true value of μ equals 0.8413. This is called a left-sided confidence interval because it only states a lower bound for μ. In general a $100(1-\alpha)\%$ left-sided confidence interval for μ reads $(\bar{x} - z_\alpha \frac{\sigma}{\sqrt{m}}, \infty)$, where $P(Z > z_\alpha) = \alpha$. Likewise, we may construct a right-sided confidence interval $(-\infty, \bar{x} + z_\alpha \frac{\sigma}{\sqrt{m}})$ and a two-sided confidence interval

$$(\bar{x} - z_{\alpha/2} \frac{\sigma}{\sqrt{m}}, \bar{x} + z_{\alpha/2} \frac{\sigma}{\sqrt{m}}).$$

If the distribution of X is unknown, i.e. $X \sim \mu, \sigma^2$, then for sufficiently large m we may invoke the central limit theorem and use $\bar{X} \approx \mathcal{N}(\mu, \sigma^2/m)$, and proceed as above.

In most practical estimation problems we don't know the value of σ^2, and we have to estimate it from the data as well. A rather obvious estimator is the sample variance

$$S^2 = \frac{1}{m-1} \sum_{i=1}^{m} (x_i - \bar{x})$$

Now we may use

$$\frac{\bar{X} - \mu}{S/\sqrt{m}} \sim t_{m-1}$$

where t_{m-1} denotes the t-distribution with $m - 1$ degrees of freedom. This distribution has a higher variance than the standardnormal distribution, leading to somewhat wider confidence intervals. This is the price we pay for the fact that we don't know the value of σ^2, but have to estimate it from the data. On the other hand we have $t_\nu \approx \mathcal{N}(0, 1)$ for $\nu \geq 100$, so if m is large enough we may use the standardnormal distribution for all practical purposes.

Hypothesis Testing A test is a statistical procedure to make a choice between two hypotheses concerning the value of a population parameter θ. One of these, called the *null hypothesis* and denoted by H_0, gets the "benefit of the doubt". The two possible conclusions are to reject or not to reject H_0. H_0 is only rejected if the sample data contains strong evidence that it is not true. The null hypothesis is rejected iff realisation g of test statistic G is in the *critical region* denoted by C. In doing so we can make two kinds of errors:

Type I Error: Reject H_0 when it is true.
Type II Error: Accept H_0 when it is false.

Type I errors are considered to be more serious than Type II errors. Test statistic G is usually a point estimator for θ, e.g. if we test a hypothesis concerning the value of population mean μ, then $\bar{X}$ is an obvious choice of test statistic. As an example we look at hypothesis test

$$H_0 : \theta \geq \theta_0 \; , \; H_a : \theta < \theta_0$$

The highest value of G that leads to the rejection of H_0 is called the critical value c_u, it is the upper bound of the so-called critical region $C = (-\infty, c_u]$. All values of G to the left of c_u lead to the rejection of H_0, so this is called a left one-sided test. An overall quality measure for a test is its power β

$$\beta(\theta) = P_\theta(\text{Reject } H_0) = P_\theta(G \in C)$$

Because we would like a low probability of Type I and Type II errors, we like to have $\beta(\theta)$ small for $\theta \in H_0$ and $\beta(\theta)$ large for $\theta \in H_a$. It is common practice in hypothesis testing to restrict the probability of a Type I error to a maximum called the *significance level* α of the test, i.e.

$$\max_{\theta \in H_0} \beta(\theta) \leq \alpha$$

Since the maximum is reached for $\theta = \theta_0$ this reduces to the restriction $\beta(\theta_0) \leq \alpha$. If possible the test is performed in such a way that $\beta(\theta_0) = \alpha$ (This may not be possible for discrete sampling distributions). Common levels for α are 0.1, 0.05 and 0.01. If in a specific application of the test, the conclusion is that H_0 should be rejected, then the result is called *significant*.

Consider a left one-sided test on population mean μ with $X \sim \mathcal{N}(\mu, \sigma^2)$ and the value of σ^2 known. That is

$$H_0 : \mu \geq \mu_0 \; , \; H_a : \mu < \mu_0$$

We determine the sampling distribution of the test statistic $\bar{X}$ under the assumption that the $\mu = \mu_0$, i.e. $\bar{X} \sim \mathcal{N}(\mu_0, \sigma^2/m)$. Now

$$\alpha = P_{\mu_0}(\bar{X} \leq c_u) = P(\frac{\bar{X} - \mu_0}{\sigma/\sqrt{m}} \leq \frac{c_u - \mu_0}{\sigma/\sqrt{m}}) = P(Z \leq \frac{c_u - \mu_0}{\sigma/\sqrt{m}})$$

and since $P(Z \leq -z_\alpha) = \alpha$, we obtain

$$\frac{c_u - \mu_0}{\sigma/\sqrt{m}} = -z_\alpha, \text{ and therefore } c_u = \mu_0 - z_\alpha \frac{\sigma}{\sqrt{m}}$$

Example 2.18. Consider a random sample of size $m = 25$ from a normal population with known $\sigma = 5.4$ and unknown mean μ. The observed sample mean is $\bar{x} = 128$. We want to test the hypothesis

$$H_0 : \mu \geq 130, \text{ against } \; H_a : \mu < 130$$

i.e. $\mu_0 = 130$. The significance level of the test is set to $\alpha = 0.05$. We compute the critical value

$$c_u = \mu_0 - z_{0.05}\frac{\sigma}{\sqrt{m}} = 130 - 1.645\frac{5.4}{\sqrt{25}} = 128.22$$

where $z_{0.05} = 1.645$ was determined using a statistical package (many books on statistics contain tables that can be used to determine the value of z_α). So the critical region is $(-\infty, 128.22]$ and since $\bar{x} = 128$ is in the critical region, we reject H_0.

Similarly, if

$$H_0 : \theta \leq \theta_0 \ , \ H_a : \theta > \theta_0$$

the critical region is $[c_l, \infty)$, and for a two-sided test

$$H_0 : \theta = \theta_0 \ , \ H_a : \theta \neq \theta_0$$

it has the form $(-\infty, c_u] \cup [c_l, \infty)$.

As with the construction of a confidence interval for the mean, for a hypothesis test concerning the mean we may invoke the central limit theorem if $X \sim \mu, \sigma^2$ and m is large. Furthermore, if σ^2 is unknown, we have to estimate it from the data and use a t_{m-1} distribution rather than the standardnormal distribution to determine the critical region.

Sometimes one doesn't want to specify the significance level α of the test in advance. In that case it us customary to report so-called p-values, indicating the *observed significance*.

Example 2.19. Consider the test of example 2.18. The p-value of the observed outcome $\bar{x} = 128$ is

$$P_{\mu_0}(\bar{X} \leq 128) = P(Z \leq \frac{128 - \mu_0}{\sigma/\sqrt{m}}) = P(Z \leq -1.852) = 0.0322$$

Since the p-value is 0.0322, we would reject H_0 at $\alpha = 0.05$, but we would accept H_0 at $\alpha = 0.01$.

2.4.2 Likelihood

The deductive nature of probability theory versus the inductive nature of statistical inference is perhaps most clearly reflected in the "dual" concepts of (joint) probability distribution and likelihood.

Given a particular probability model and corresponding parameter values, we may calculate the probability of observing different samples. Consider the experiment of 10 coin flips with probability of heads $\pi = 0.6$. The probability distribution of random variable "number of times heads comes up" is now the following function of the data

$$P(y) = \binom{10}{y} 0.6^y \, 0.4^{10-y}$$

Table 2.8. Probability distributions (columns) and likelihood functions (rows) for $Y \sim B(10, \pi)$.

y	π								
	0.1	0.2	0.3	0.4	0.5	0.6	0.7	0.8	0.9
0	.349	.107	.028	.006	.001				
1	.387	.269	.121	.04	.01	.002			
2	.194	.302	.234	.121	.044	.01	.002		
3	.057	.201	.267	.215	.117	.043	.009	.001	
4	.011	.088	.2	.251	.205	.111	.036	.005	
5	.002	.027	.103	.201	.246	.201	.103	.027	.002
6		.005	.036	.111	.205	.251	.2	.088	.011
7		.001	.009	.043	.117	.215	.267	.201	.057
8			.002	.01	.044	.121	.234	.302	.194
9				.002	.01	.04	.121	.269	.387
10					.001	.006	.028	.107	.349
	1	1	1	1	1	1	1	1	1

We may for example compute that the probability of observing $y = 7$ is

$$\binom{10}{7} 0.6^7 0.4^3 \approx 0.215$$

In statistical inference however, we typically have one data set and want to say something about the relative likelihood of different values of some population parameter. Say we observed 7 heads in a sequence of ten coin flips. The likelihood is now a function of the unknown parameter π

$$L(\pi \mid y = 7) = \binom{10}{7} \pi^7 (1 - \pi)^3$$

where the constant term is actually arbitrary, since we are not interested in absolute values of the likelihood, but rather in ratios of likelihoods for different values of π.

In Table 2.8, each column specifies the probability distribution of Y for a different value of π. Each column sums to 1, since it represents a probability distribution. Each row, on the other hand, specifies a likelihood function, or rather: it specifies the value of the likelihood function for 9 values of π. So for example, in the third row we can read off the probability of observing 2 successes in a sequence of 10 coin flips for different values of π.

In general, if $\mathbf{y} = (y_1, \ldots, y_m)$ are independent observations from a probability density $f(y \mid \theta)$, where θ is the parameter vector we wish to estimate, then

$$L(\theta \mid \mathbf{y}) \propto \prod_{i=1}^{m} f(y_i \mid \theta)$$

The *likelihood function* then measures the relative likelihood that different θ have given rise to the observed $\mathbf{y}$. We can thus try to find that particular $\hat{\theta}$

which maximizes L, i.e. that $\hat{\theta}$ such that the observed $\mathbf{y}$ are more likely to have come from $f(y \mid \hat{\theta})$ than from $f(y \mid \theta)$ for any other value of θ.

For many parameter estimation problems one can tackle this maximization by differentiating L with respect to the components of θ and equating the derivatives to zero to give the *normal equations*

$$\frac{\partial L}{\partial \theta_j} = 0$$

These are then solved for the θ_j and the second order derivatives are examined to verify that it is indeed a maximum which has been achieved and not some other stationary point.

Maximizing the likelihood function L is equivalent to maximizing the (natural) log of L, which is computationally easier. Taking the natural log, we obtain the log-likelihood function

$$l(\theta \mid \mathbf{y}) = \ln(L(\theta \mid \mathbf{y})) = \ln(\prod_{i=1}^{m} f(y_i \mid \theta)) = \sum_{i=1}^{m} \ln f(y_i \mid \theta)$$

since $\ln ab = \ln a + \ln b$.

Example 2.20. In a coin flipping experiment we define the random variable Y with $y = 1$ if heads comes up, and $y = 0$ when tails comes up. Then we have the following probability distribution for one coin flip

$$f(y) = \pi^y (1 - \pi)^{1-y}$$

For a sequence of m coin flips, we obtain the joint probability distribution

$$f(\mathbf{y}) = f(y_1, y_2, ..., y_m) = \prod_{i=1}^{m} \pi^{y_i} (1 - \pi)^{1-y_i}$$

which defines the likelihood when viewed as a function of π. The log-likelihood consequently becomes

$$l(\pi \mid \mathbf{y}) = \sum_{i=1}^{m} y_i \ln(\pi) + (1 - y_i) \ln(1 - \pi)$$

In a sequence of 10 coin flips with seven times heads coming up, we obtain

$$l(\pi) = \ln(\pi^7 (1 - \pi)^3) = 7 \ln \pi + 3 \ln(1 - \pi)$$

To determine the maximum we take the derivative

$$\frac{dl}{d\pi} = \frac{7}{\pi} - \frac{3}{1 - \pi} = 0$$

which yields maximum likelihood estimate $\hat{\pi} = 0.7$.

The reader may notice that the maximum likelihood estimate in this case is simply the fraction of heads coming up in the sample, and we could have spared ourselves the trouble of maximizing the likelihood function to obtain the required estimate. Matters become more interesting (and complicated) however, when we make π a function of data *and* parameters. Suppose that for each y_i in our sample, we observe a corresponding measure x_i which we assume is a continuous variable. We could write $\pi_i = g(x_i)$, where g is some function. In so-called Probit analysis [227] we assume

$$\pi_i = \Phi(\alpha + \beta x_i)$$

where Φ denotes the standard normal distribution function. The parameters of the model are now α and β, and we can write the log-likelihood function as

$$l(\alpha, \beta) = \sum_{i=1}^{m} y_i \ln(\Phi(\alpha + \beta x_i)) + (1 - y_i) \ln(1 - \Phi(\alpha + \beta x_i))$$

This is the expression of the log-likelihood for the Probit model. By maximizing with respect to α and β, we obtain maximum likelihood estimates for these parameters.

Example 2.21. Consider a random sample $\mathbf{y} = (y_1, ..., y_m)$ from a normal distribution with unknown mean μ and variance σ^2. Then we have likelihood

$$L((\mu, \sigma^2)' \,|\, \mathbf{y}) = \prod_{i=1}^{m} \frac{e^{-(y_i - \mu)^2/(2\sigma^2)}}{\sigma\sqrt{2\pi}} = \frac{1}{\sigma^m (2\pi)^{m/2}} \exp\left[-\frac{1}{2}\sum_{i=1}^{m}\left(\frac{y_i - \mu}{\sigma}\right)^2\right]$$

The natural log of this expression is

$$l = \ln(L) = -m \ln \sigma - \left(\frac{m}{2}\right) \ln 2\pi - \frac{1}{2\sigma^2}\sum_{i=1}^{m}(y_i - \mu)^2$$

To determine the maximum likelihood estimates of μ and σ, we take the partial derivative of l with respect to these parameters, and equate them to zero

$$\frac{\partial l}{\partial \mu} = \frac{m}{\sigma^2}(\bar{y} - \mu) = 0$$

$$\frac{\partial l}{\partial \sigma^2} = -\frac{m}{2\sigma^2} + \frac{m}{2\sigma^4}(s^2 + (\bar{y} - \mu)^2) = 0$$

Solving these equations for μ and σ, we obtain maximum likelihood estimates $\hat{\mu} = \bar{y}$ and $\hat{\sigma}^2 = s^2$, where $s^2 = 1/m \sum(y_i - \hat{\mu})^2$.

Another important aspect of the log-likelihood function is its shape in the region near the maximum. If it is rather flat, one could say that the likelihood contains little information in the sense that there are many values of θ with log-likelihood near that of $\hat{\theta}$. If, on the other hand, it is rather steep, one could say that the log-likelihood contains much information about $\hat{\theta}$. The log-likelihood of any other value of θ is approximately given by the Taylor expansion

$$l(\theta) = l(\hat{\theta}) + (\theta - \hat{\theta})\frac{dl}{d\theta} + \frac{1}{2}(\theta - \hat{\theta})^2\frac{d^2 l}{d\theta^2} + \ldots$$

where the differential coefficients are evaluated at $\theta = \hat{\theta}$. At this point, $\frac{dl}{d\theta}$ is zero, so approximately

$$l(\theta) = l(\hat{\theta}) + \frac{1}{2}(\theta - \hat{\theta})^2\frac{d^2 l}{d\theta^2}.$$

Minus the second derivative of the log-likelihood function is known as the (Fisher) *information*. When evaluated at $\hat{\theta}$ (the maximum likelihood estimate of θ) it is called the *observed information.*

Note: This concept of information should not be confused with the one discussed in Section 2.2.9.

Some authors take the view that all statistical inference should be based on the likelihood function rather than the sampling distribution used in frequentist inference (see [164, 454]). In this sense likelihood inference differs from frequentist inference.

Example 2.22. Figure 2.2 displays the likelihood function for π corresponding to 7 successes in a series of 10 coin flips. The horizontal line indicates the range of values of π for which the ratio of $L(\pi)$ to the maximum $L(0.7)$ is greater than 1/8. The 1/8 likelihood interval is approximately $(0.38, 0.92)$. Such an interval is similar in spirit to a confidence interval in the sense that it intends to provide a range of "plausible values" for π based on the sample data. A confidence interval for π is based however on the sampling distribution of some sample statistic (the sample proportion of successes is the most obvious choice) whereas a likelihood interval is based, as the name suggests, on the likelihood function.

On the other hand, maximum likelihood estimation may be used and evaluated from a frequentist perspective. This motivates the study of the sampling distribution of maximum likelihood estimates. If we know the true value of $\theta = \theta^*$, we can determine the *expected* log-likelihood, i.e. the mean value of the log-likelihood conditional on $\theta = \theta^*$ (still expressed as a function of θ). The expected log-likelihood has a maximum at $\theta = \theta^*$. Minus the second derivative of the expected log-likelihood evaluated at $\theta = \theta^*$, is called the *expected information.* Assuming parameter vector θ with several components the expected information matrix is defined as

$$I(\theta) = -\left\{E\left(\frac{\partial^2 l}{\partial\theta_j \partial\theta_k}\right)_{\theta^*}\right\}$$

In large samples, the maximum likelihood estimate $\hat{\theta}$ is approximately normally distributed with mean θ^*, and covariance matrix $I(\theta)^{-1}$. Unfortunately, we

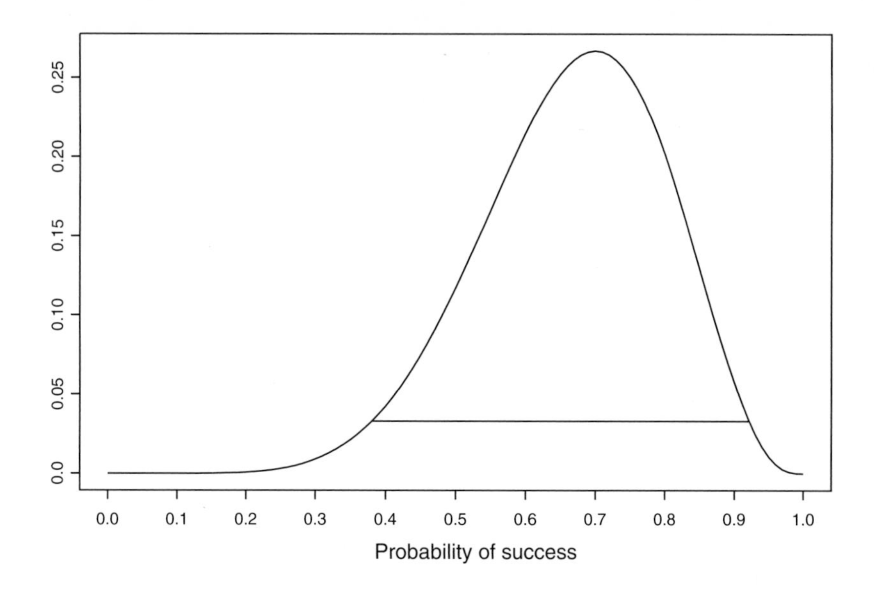

Fig. 2.2. Likelihood function $L(\pi \,|\, y = 7) = 120\pi^7(1-\pi)^3$.

cannot in practice determine $I(\theta)$, since θ^* is unknown. It is therefore set equal to $\hat{\theta}$ so that $I(\theta)$ can be calculated. An alternative estimate for the covariance matrix is the observed information matrix

$$-\left(\frac{\partial^2 l}{\partial\theta_j \partial\theta_k}\right)_{\hat{\theta}}$$

which is easier to compute since it does not involve an expectation. For the exponential family of distributions these two estimates are equivalent.

Example 2.23. Consider a sequence of m coin tosses, with heads coming up y times. We are interested in the probability of heads π. We have seen that

$$l(\pi) = y\ln(\pi) + (m-y)\ln(1-\pi)$$

Setting the first derivative to zero and solving for π yields $\hat{\pi} = y/m$. The information is

$$-\frac{d^2 l}{d\pi^2} = \frac{y}{\pi^2} + \frac{(m-y)}{(1-\pi)^2}$$

Evaluating this expression at $\hat{\pi} = y/m$ we obtain the observed information

$$\frac{m}{\hat{\pi}(1-\hat{\pi})}.$$

LIVERPOOL
JOHN MOORES UNIVERSITY
AVRIL ROBARTS LRC

In large samples, $\hat{\pi}$ is approximately normally distributed with mean π^* and variance $\pi^*(1-\pi^*)/m$, i.e. the reciprocal of the expected information. The estimated variance of $\hat{\pi}$ is equal to the reciprocal of the observed information, i.e. $\hat{\pi}(1-\hat{\pi})/m$.

2.4.3 Bayesian Inference

In this section we briefly consider the principal idea of Bayesian inference. In Chapter 4 of this volume and in [59, 208], Bayesian inference is discussed in greater detail.

Consider again the coin tossing experiment. We stated that the probability of heads, denoted by π, is a fixed yet unknown quantity. From a relative frequency viewpoint, it makes no sense to talk about the probability distribution of π since it is not a random variable. In Bayesian inference one departs from this strict interpretation of probability. We may express prior, yet incomplete, knowledge concerning the value of π through the construction of a *prior distribution*. This prior distribution is then combined with sample data (using Bayes rule, see Section 2.2.14) to obtain a posterior distribution. The posterior distribution expresses the new state of knowledge, in light of the sample data. We reproduce Bayes' rule using symbols that are more indicative for the way it is used in Bayesian inference:

$$P(M_i|D) = \frac{P(D|M_i)P(M_i)}{\sum_j P(D|M_j)P(M_j)}$$

where the M_i specify different models for the data, i.e. hypotheses concerning the parameter value(s) of the probability distribution from which the data were drawn. Note that in doing so, we actually assume that this probability distribution is known up to a fixed number of parameter values.

Example 2.24. Consider the somewhat artificial situation where two hypotheses concerning the probability of heads of a particular coin are entertained, namely M_1: $\pi = 0.8$ and M_2: $\pi = 0.4$ (see Table 2.9). Prior knowledge concerning these models is expressed through a prior distribution as specified in the first column of Table 2.9. Next we observe 5 times heads in a sequence of 10 coin flips, i.e. $y = 5$. The likelihood of this outcome under the different models is specified in the second column of Table 2.9 (the reader can also find them in Table 2.8). The posterior distribution is obtained via Bayes' rule, and is specified in the last column of Table 2.9. Since the data are more likely to occur under M_2, the posterior distribution has clearly shifted towards this model.

In general, the probability distribution of interest is indexed by a number of continuous valued parameters, which we denote by parameter vector θ. Replacing probabilities by probability densities and summation by integration, we obtain the probability density version of Bayes' rule

$$f(\theta\,|\,\mathbf{y}) = \frac{f(\mathbf{y}\,|\,\theta)\;f(\theta)}{\int_\Omega f(\mathbf{y}\,|\,\theta)\;f(\theta)\;d\theta}$$

Table 2.9. Prior and posterior probabilities of M_1 and M_2.

	Prior $P(M_i)$	Likelihood $P(y=5 \mid M_i)$	Posterior $P(M_i \mid y=5)$
M_1: $\pi = 0.8$	0.7	0.027	0.239
M_2: $\pi = 0.4$	0.3	0.201	0.761

where $\mathbf{y}$ denotes the observed data and Ω denotes the parameter space, i.e. the space of possible values of θ.

Consider the case where we have no prior knowledge whatsoever concerning the probability of heads π. How should this be reflected in the prior distribution? One way of reasoning is to say that all values of π are considered equally likely, which can be expressed by a uniform distribution over $\Omega = [0, 1]$: the range of possible values of π. Let's consider the form of the posterior distribution in this special case.

$$f(\pi \mid \mathbf{y}) = \frac{f(\mathbf{y} \mid \pi) f(\pi)}{\int_0^1 f(\mathbf{y} \mid \pi) f(\pi) d\pi}$$

If we observe once again 7 times heads in a sequence of 10 coin flips, then $f(\mathbf{y} \mid \pi) = \pi^7(1-\pi)^3$. Since $f(\pi) = 1$, the denominator of the above fraction becomes

$$\int_0^1 \pi^7(1-\pi)^3 d\pi = \frac{1}{1320}$$

and so the posterior density becomes

$$f(\pi \mid \mathbf{y}) = 1320\,\pi^7(1-\pi)^3$$

It is reassuring to see that in case of prior ignorance the posterior distribution is proportional to the likelihood function of the observed sample. Note that the constant of proportionality merely acts to make the integral of the expression in the numerator equal to one, as we would expect of a probability density!

In general, the computationally most difficult part of obtaining the posterior distribution is the evaluation of the (multiple) integral in the denominator of the expression. For this reason, a particular class of priors, called conjugate priors, have received special attention in Bayesian statistics. Assume our prior knowledge concerning the value of π may be expressed by a Beta(4,6) distribution (see Section 2.2.16), i.e.

$$f(\pi) = \frac{\pi^3(1-\pi)^5}{\int_0^1 \pi^3(1-\pi)^5 d\pi}$$

Since $\int_0^1 \pi^3(1-\pi)^5 d\pi = \frac{1}{504}$, we get $f(\pi) = 504\,\pi^3(1-\pi)^5$. Multiplied with the likelihood this results in $504\,\pi^3(1-\pi)^5\pi^7(1-\pi)^3 = 504\pi^{10}(1-\pi)^8$, so the denominator becomes

$$\int_0^1 504\,\pi^{10}(1-\pi)^8 = \frac{28}{46189}$$

and the posterior density becomes

$$f(\pi \mid \mathbf{y}) = 831402\,\pi^{10}(1-\pi)^8$$

which is a Beta(11,9) distribution. In general, when we have a binomial sample of size m with r successes, and we combine that with a Beta(l, k) prior distribution, then the posterior distribution is Beta($l+r, k+m-r$). Loosely speaking, conjugate priors allow for simple rules to update the prior with sample data to arrive at the posterior distribution. Furthermore, the posterior distribution belongs to the same family as the prior distribution. Since the uniform distribution over the interval $[0, 1]$ is the same as a Beta(1,1) distribution (see Section 2.2.16), we could have used this simple update rule in the "prior ignorance" case as well: combining a Beta(1,1) prior with a binomial sample of size 10 with 7 successes yields a Beta(8,4) posterior distribution.

Once we have calculated the posterior distribution, we can extract all kinds of information from it. We may for example determine the mode of the posterior distribution which represents the value of π for which the posterior density is maximal. When asked to give a point estimate for π, it makes sense to report this value. When asked for a range of plausible values for π we may use the posterior distribution to determine a so-called $100(1-\alpha)\%$ probability interval, which is an interval $[g_l, g_u]$ such that $P(\pi < g_l) = \alpha/2$ and $P(\pi > g_u) = \alpha/2$ where the relevant probabilities are based on the posterior distribution for π.

2.5. Prediction and Prediction Error

The value of a random variable Y depends on the outcome of a random experiment (see Section 2.2.7). Before the experiment is performed, the value of the random variable is unknown. In many cases, we would like to *predict* the value of a yet to be observed random variable. The usual assumption is that the distribution of Y depends in some way on some random vector $\mathbf{X} = (X_1, ..., X_n)$. For example, in the simple linear regression model, the assumption is that $Y_i \sim \mathcal{N}(\beta_0 + \beta_1 x_i, \sigma_\epsilon^2)$, i.e. Y is normally distributed with mean a linear function of x, and constant variance σ_ϵ^2. If Y is a 0-1 variable, a common assumption is $Y_i \sim B(1, \Phi(\beta_0 + \beta_1 x_i))$, i.e. the Y_i are Bernoulli random variables. The probability of success is the area under the standardnormal distribution to the left of a value that depends on x_i: the so-called Probit model [227]. Replacing the standardnormal distribution by the logistic distribution leads to the very popular logistic regression model [370].

Often, the goal of data analysis is to gain information from a training sample $T = \{(\mathbf{x}_1, y_1), (\mathbf{x}_2, y_2), ..., (\mathbf{x}_m, y_m)\}$ in order to estimate the parameters of such a model. Once the model is estimated, we can use it to make predictions about random variable Y. If Y is numeric, we speak about *regression*, and if Y takes its values from a discrete unordered set we speak about *classification*.

A fundamental choice is which assumptions are made concerning *how* the distribution of Y (or some aspect of it: typically the expected value of Y), depends on $\mathbf{x}$. Such assumptions are often called *inductive bias* in the machine learning literature [382] and are part of what is called *model specification* in the econometrics literature [227]. The assumptions of the linear regression model, for example, are quite restrictive. If the true relationship is not linear, the estimated model will produce some prediction error due to the unwarranted assumption of linearity. One might therefore argue that no assumptions at all should be made, but let the data "speak entirely for itself". However if no assumptions whatsoever are made, there is no rational basis to generalize beyond the data observed [382]. If we make very mild assumptions, basically allowing the relation between $\mathrm{E}(Y)$ and $\mathbf{x}$ to be chosen from a very large class of functions, the estimated function is capable to adapt very well to the data. In fact so well that it will also model the peculiarities that are due to random variations. The estimate becomes highly sensitive to the particular sample drawn, that is upon repeated sampling it will exhibit high variance.

We consider a simple regression example to illustrate these ideas. Suppose that $Y_i \sim \mathcal{N}(\mu = 2.0 + 0.5x_i, \sigma_\varepsilon^2 = 1)$, i.e. the true relation between $\mathrm{E}(Y)$ and x is

$$\mathrm{E}(Y) = 2.0 + 0.5x.$$

We have a sample T of ten (x, y) observations, which is displayed in the scatterplot of Fig. 2.3(a). Note that x is not a random variable but its values are chosen by us to be $1, 2, \ldots, 10$. Although $\mathrm{E}(Y)$ is a linear function of x, the observations do not lie on a straight line due to the inherent variability of Y. We pretend we don't know the relation between x and y, but only have T at our disposal, as would be the case in most data analysis settings. We consider three classes of models to describe the relation between x and y

Linear Model: $\mathrm{E}(Y) = f_1(x) = \beta_0 + \beta_1 x$
Quadratic Model: $\mathrm{E}(Y) = f_2(x) = \beta_0 + \beta_1 x + \beta_2 x^2$
Cubic Model: $\mathrm{E}(Y) = f_3(x) = \beta_0 + \beta_1 x + \beta_2 x^2 + \beta_3 x^3$

Note: In this section, the symbol f does not denote a probability density. Probability densities and probability functions are henceforth both denoted by the symbol p.

Note that (2) encompasses (1) in the sense that if $\beta_2 = 0$, (2) reduces to the linear function (1). Likewise, (3) encompasses (2), and consequently also (1). The β_j are the parameters of the model, whose estimates are chosen in such a way that the sum of squared *vertical* distances from the points (x_i, y_i) to the fitted equation is minimized. For example, for the simple linear regression model we choose the estimates of β_0 and β_1 such that

$$\sum_{i=1}^{m} [y_i - (\hat{\beta}_0 + \hat{\beta}_1 \, x_i)]^2$$

is minimal. The expression $\hat{\beta}_0 + \hat{\beta}_1 \, x_i$ denotes the predicted value for y_i, so one effectively minimizes the sum of squared differences between predicted values and realisations of y. The estimates $\hat{\beta}_j$ of the β_j thus obtained are called the *least squares* estimates. The choice for minimizing the squared differences rather than using some other criterion (e.g. minimize the absolute value of the differences) was historically for a large part justified by analytical convenience. Under the usual assumption of the linear regression model concerning the normal distribution of Y (conditional on x), the least squares estimates and maximum likelihood estimates coincide.

The equations obtained by least squares estimation for the respective models are displayed in Fig. 2.3 (b) to (d). Without performing the actual calculations, one can easily see that the linear model gives the worst fit, even though the true (population) relation is linear. The quadratic model gives a somewhat better fit, and the cubic model gives the best fit of the three. In general, the more parameters the model has, the better it is able to adjust to the data in T. Does this mean that the cubic model gives better predictions than the linear model? It does on T, but how about on data that were not used to fit the equation? We drew a second random sample, denoted by T', and looked at the fit of the equations to T' (see Fig. 2.4). The fit of the cubic model is clearly worse than that of the linear model. The reason is that the cubic model has adjusted itself to the random variations in T, leading on average to bad predictive performance on new samples. This phenomenon is called *overfitting*.

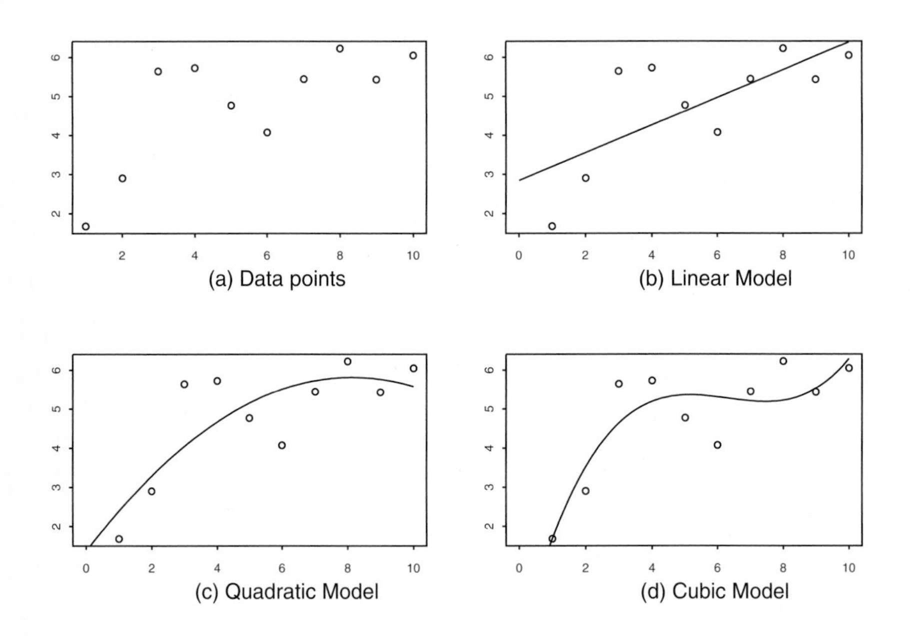

Fig. 2.3. Equations fitted by least squares to the data in T.

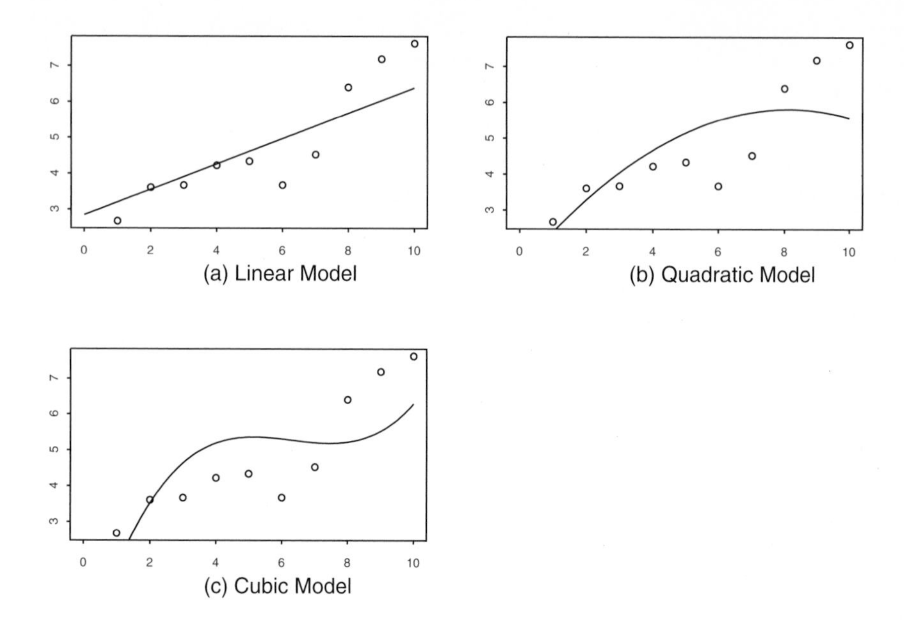

Fig. 2.4. Fit of equations to new sample T'.

In the next section we discuss the decomposition of prediction error into its components to gain a further understanding of the phenomenon illustrated by the above example.

2.5.1 Prediction Error in Regression

Once we have obtained estimates $\hat{\beta}_j$ by estimation from some training set T, we may use the resulting function to make predictions of y when we know the corresponding value of $\mathbf{x}$. Henceforth we denote this prediction by $\hat{f}(\mathbf{x})$. The difference between prediction $\hat{f}(\mathbf{x})$ and realisation y is called prediction error. It should preferrably take values close to zero. A natural quality measure of $\hat{f}$ as a predictor of Y is the mean squared error. For fixed T and $\mathbf{x}$

$$\mathrm{E}\left[(Y - \hat{f}(\mathbf{x} \,|\, T))^2\right]$$

where the expectation is taken with respect to $p(Y \,|\, \mathbf{x})$, the probability distribution of Y at $\mathbf{x}$. We may decompose this overall error into a *reducible* part, and an *irreducible* part that is due to the variability of Y at $\mathbf{x}$, as follows

$$\mathrm{E}\left[(Y - \hat{f}(\mathbf{x} \,|\, T))^2\right] = [f(\mathbf{x}) - \hat{f}(\mathbf{x} \,|\, T)]^2 + \mathrm{E}[(y - f(\mathbf{x}))^2]$$

where $f(\mathbf{x}) \equiv \mathrm{E}[Y \,|\, \mathbf{x}]$. The last term in this expression is the mean square error of the best possible (in the mean squared error sense) prediction $\mathrm{E}[Y \,|\, \mathbf{x}]$. Since we can't do much about it, we focus our attention on the other source of error $[f(\mathbf{x}) - \hat{f}(\mathbf{x}\,|\,T)]^2$. This tells us something about the quality of the estimate $\hat{f}(\mathbf{x}\,|\,T)$ for a particular realisation of T. To say something about the quality of the estimator $\hat{f}$, we take its expectation over all possible training samples (of fixed size) and decompose it into its bias and variance components as discussed in Section 2.4.1:

$$\mathrm{E}_T[(f(\mathbf{x}) - \hat{f}(\mathbf{x}\,|\,T))^2] = (f(\mathbf{x}) - \mathrm{E}_T[\hat{f}(\mathbf{x}\,|\,T)])^2 + \mathrm{E}_T[(\hat{f}(\mathbf{x}\,|\,T) - \mathrm{E}_T[\hat{f}(\mathbf{x}\,|\,T)])^2]$$

The first component is the squared bias, where bias is the difference between the best prediction $f(\mathbf{x})$ and its average estimate over all possible samples of fixed size. The second component, variance, is the expected squared difference between an estimate obtained for a single training sample and the average estimate obtained over all possible samples.

We illustrate these concepts by a simple simulation study using the models introduced in the previous section. The expectations in the above decomposition are taken over *all possible* training samples, but this is a little bit to much to compute. Instead we use the computer to draw a large number of random samples to obtain an estimate of the desired quantities. In the simulation we sampled 1000 times from

$$Y_i \sim \mathcal{N}(\mu = 2 + 0.5x_i, \sigma_\varepsilon^2 = 1)$$

with $x_i = 1, 2, \ldots, 10$. In other words we generated 1000 random samples, $T_1, T_2, \ldots, T_{1000}$ each consisting of 10 (x, y) pairs. For each T_i, the least squares parameter estimates for the three models were computed. Using the estimated models we computed the predicted values $\hat{f}(x)$. From the 1000 predicted values we computed the mean to estimate the expected value $\mathrm{E}(\hat{f}(x))$ and variance to estimate $\mathrm{V}(\hat{f}(x))$. The results of this simulation study are summarized in Table 2.10. Consider the fourth row of this table for the moment. It contains the simulation results of the predictions of the different models for $x = 4$. The expected value is $f(4) = \mathrm{E}(Y|x = 4)$ is $2 + 0.5 \cdot 4 = 4$. From the first three columns we conclude that all models have no or negligable bias; in fact we can prove mathematically they are unbiased since all three models encompass the correct model. But now look at the last three columns of Table 2.10. We see that the linear model has lowest variance, the cubic model has highest variance, and the quadratic model is somewhere inbetween. This is also illustrated by the histograms displayed in Fig. 2.5. We clearly see the larger spread of the cubic model compared to the linear model. Although all three models yield unbiased estimates, the linear model tends to have a lower prediction error because its variance is smaller than that of the quadratic and cubic model.

The so-called bias/variance dilemma lies in the fact that there is a trade-off between the bias and variance components of error. Incorrect models lead to high bias, but highly flexible models suffer from high variance. For a fixed bias, the variance tends to decrease when the training sample gets larger and larger. Consequently, for very large training samples, bias tends to be the most important source of prediction error.

Table 2.10. Expected value and variance of $\hat{f}_j$.

x	$f(x)$	$\mathrm{E}(\hat{f}_1)$	$\mathrm{E}(\hat{f}_2)$	$\mathrm{E}(\hat{f}_3)$	$\mathrm{V}(\hat{f}_1)$	$\mathrm{V}(\hat{f}_2)$	$\mathrm{V}(\hat{f}_3)$
1	2.50	2.48	2.48	2.49	0.34	0.61	0.84
2	3.00	2.99	2.98	2.98	0.25	0.27	0.29
3	3.50	3.49	3.49	3.48	0.18	0.18	0.33
4	4.00	3.99	4.00	3.99	0.13	0.20	0.32
5	4.50	4.50	4.50	4.50	0.10	0.23	0.25
6	5.00	5.00	5.00	5.01	0.10	0.22	0.23
7	5.50	5.50	5.51	5.52	0.13	0.19	0.28
8	6.00	6.01	6.01	6.02	0.17	0.18	0.31
9	6.50	6.51	6.51	6.51	0.24	0.28	0.30
10	7.00	7.01	7.01	7.00	0.33	0.62	0.80

Table 2.11. Bias, variance and mean squared estimation error for different sample sizes.

	Squared bias			Variance			Mean square error		
m	10	100	1000	10	100	1000	10	100	1000
Linear (f_1)	.021	.022	.022	.197	.022	.002	.218	.043	.024
Quadratic (f_2)	.000	.000	.000	.299	.037	.004	.299	.037	.004
Cubic (f_3)	.001	.000	.000	.401	.054	.006	.401	.054	.006

This phenomenon is illustrated by a second simulation. We generated the training sets by drawing from

$$Y_i \sim \mathcal{N}(\mu = 2 + 0.5x_i + 0.02x_i^2, \sigma_\varepsilon^2 = 1).$$

The true model is quadratic, so the linear model is biased whereas the quadratic and cubic model are unbiased. We generated 1000 training samples of size 10, 100 and 1000 respectively. The first three columns of Table 2.11 summarize the estimated squared bias for the different models and sample sizes.

The results confirm that the linear model is biased, and furthermore indicate that the bias component of error does not decrease with sample size. Now consider the variance estimates shown in the middle three columns of Table 2.11. Looking at the rows, we observe that variance does decrease with the size of the sample. Taking these two phenomena together results in the summary of mean square error given in the last three columns of Table 2.11. The linear model outperforms the other models for small sample size, despite its bias. Because variance is a substantial component of overall error the linear model profits from its smaller variance. As the sample size gets larger, variance becomes only a small part of total error, and the linear model becomes worse due to its bias.

These phenomena explain the recent interest in increasingly flexible techniques that reduce estimation bias, since large data sets are quite commonly analyzed in business nowadays, using so-called *data mining* techniques.

LIVERPOOL JOHN MOORES UNIVERSITY
LEARNING & INFORMATION SERVICES

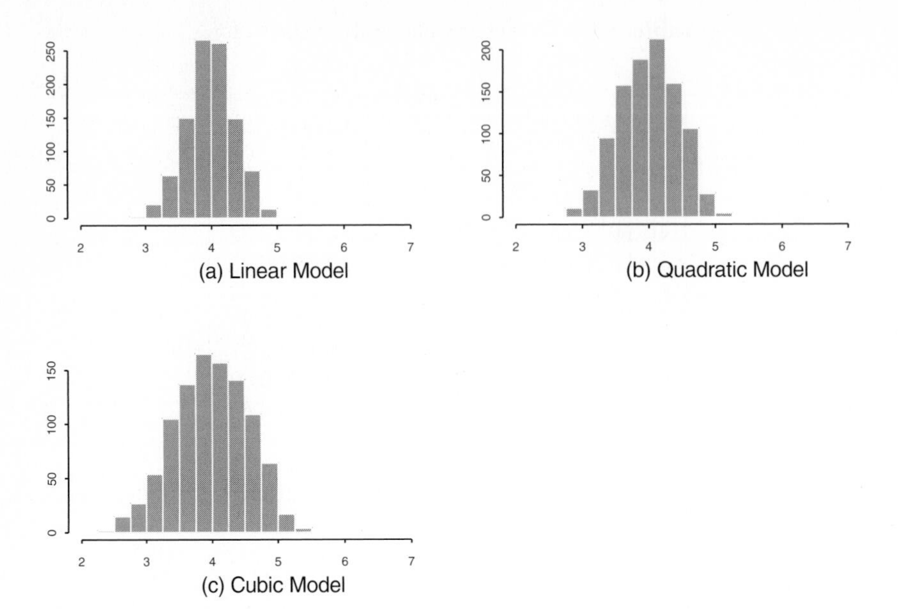

Fig. 2.5. Histograms of $\hat{f}_j(4)$ based on 1000 samples.

2.5.2 Prediction Error in Classification

In classification y assumes values on an unordered discrete set $y \in \{y_1, ..., y_L\}$. For ease of exposition we assume $L = 2$, and $y \in \{0, 1\}$. We now have

$$\mathrm{E}[y \,|\, \mathbf{x}] \equiv f(\mathbf{x}) = P(y = 1 | \, \mathbf{x}) = 1 - P(y = 0 \,|\, \mathbf{x})$$

The role of a classification procedure is to construct a rule that makes a prediction $\hat{y}(\mathbf{x}) \in \{0, 1\}$ for the class label y at every input point $\mathbf{x}$.

For classification it is customary to measure prediction error in terms of the error rate $P(\hat{y} \neq y)$ [241, 370]. The best possible allocation rule, in terms of error rate, is the so called Bayes rule:

$$y_B(\mathbf{x}) = I(f(\mathbf{x}) \geq 1/2)$$

where $I(\cdot)$ is an indicator function of the truth of its argument. This rule simply states that $\mathbf{x}$ should be allocated to the class that has highest probability at $\mathbf{x}$. The probability of error $P(y_B \neq y)$ of the Bayes rule represents the irreducible error rate, analogous to $\mathrm{E}[(y - f(\mathbf{x}))^2]$ in regression.

Training data T are again used to form an estimate $\hat{f}(\mathbf{x} \,|\, T)$ of $f(\mathbf{x})$ to construct an allocation rule:

$$\hat{y}(\mathbf{x} \,|\, T) = I(\hat{f}(\mathbf{x} \,|\, T) \geq 1/2)$$

We discuss two alternative decompositions of the prediction error of a classification method. The first one is due to Friedman [199], and aims to keep a close analogy to the decomposition for regression problems. The second one is an additive decomposition due to Breiman [89], and is aimed at explaining the success of model aggregation techniques such as bagging (see Section 2.6.4) and arcing (see Section 2.6.5).

Friedman's Bias-Variance Decomposition for Classifiers Friedman [199] proposes the following decomposition of the error rate of an allocation rule into reducible and irreducible error

$$P(\hat{y}(\mathbf{x}) \neq y(\mathbf{x})) = |2f(\mathbf{x}) - 1| P(\hat{y}(\mathbf{x}) \neq y_B(\mathbf{x})) + P(y_B(\mathbf{x}) \neq y(\mathbf{x}))$$

The reader should note that the reducible error is expressed as a *product* of two terms, rather than a *sum*, in the above equation. The first term of the product denotes the increase in error if $\hat{f}$ is on the wrong side of $1/2$, and the second term of the product denotes the probability of this event. We consider a simple example to illustrate this decomposition.

Example 2.25. Suppose $P(y = 1|\mathbf{x}) = f(\mathbf{x}) = 0.8$, and $\mathrm{E}(\hat{f}(\mathbf{x})) = 0.78$. Since $f(\mathbf{x}) \geq 1/2$, Bayes rule allocates $\mathbf{x}$ to class 1, that is $y_B(\mathbf{x}) = 1$. The error rate of Bayes rule is $P(y = 0|\mathbf{x}) = 1 - 0.8 = 0.2$. Now $\hat{y}(\mathbf{x})$ differs from Bayes rule if it is below $1/2$. Assume the distribution $p(\hat{f}(\mathbf{x}))$ of $\hat{f}(\mathbf{x})$ is such that $P(\hat{f}(\mathbf{x}) < 1/2) = 0.1$, and consequently $P(\hat{y}(\mathbf{x}) \neq y_B(\mathbf{x})) = 0.1$. We can now compute the error rate as follows

$$\begin{aligned} P(\hat{y}(\mathbf{x}) \neq y(\mathbf{x})) &= P(\hat{f}(\mathbf{x}) \geq 1/2) P(y = 0|\mathbf{x}) + P(\hat{f}(\mathbf{x}) < 1/2) P(y = 1|\mathbf{x}) \\ &= 0.9 \cdot 0.2 + 0.1 \cdot 0.8 = 0.26 \end{aligned}$$

The decomposition of Friedman shows how this error rate may be split into reducible and irreducible parts as follows. We saw the irreducibe error $P(y_B(\mathbf{x}) \neq y(\mathbf{x})) = 0.2$. We assumed $P(\hat{y}(\mathbf{x}) \neq y_B(\mathbf{x})) = 0.1$. If $\hat{f}(\mathbf{x}) < 1/2$ the error rate increases from 0.2 to 0.8, an increase of 0.6 that is represented by the term

$$|2f(\mathbf{x}) - 1| = 2 \cdot 0.8 - 1 = 0.6$$

Summarizing Friedman's decomposition gives

$$P(\hat{y}(\mathbf{x}) \neq y(\mathbf{x})) = 0.6 \cdot 0.1 + 0.2 = 0.26.$$

The important difference with regression is that a wrong estimate $\hat{f}(\mathbf{x})$ of $f(\mathbf{x})$ does not necessarily lead to a different allocation as Bayes rule, as long as they are on the same side of $1/2$ [199]. In Fig. 2.6 we show a simple representation of the decision boundary when $\mathbf{x}$ only has two components, i.e. $\mathbf{x} = (x_1, x_2)$. The picture shows the optimal decision boundary corresponding to the Bayes allocation rule as the curved solid line $f(x_1, x_2) = 1/2$. The dotted line denotes the average decision boundary $\mathrm{E}[\hat{f}(x_1, x_2)]$ for some unspecified classifier. The

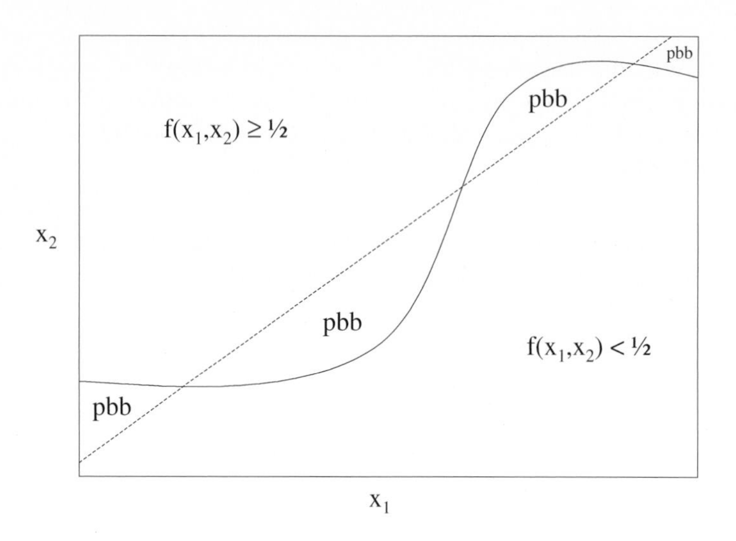

Fig. 2.6. Example regions of positive boundary bias: solid line denotes $f(x_1, x_2) = 1/2$ and dotted line denotes $\mathrm{E}[\hat{f}(x_1, x_2)] = 1/2$.

areas in Fig. 2.6 labeled "pbb" correspond to the regions where the classifier is on average on the wrong side of 1/2; these regions are called regions of positive boundary bias (hence "pbb") by Friedman [199]. In the remaining regions the classifier is either unbiased, or its bias is of the "harmless" kind in the sense that the classifier is on average on the right side of 1/2. In the latter case we speak of negative boundary bias.

When $\mathrm{E}[\hat{f}]$ and f are on the same side of 1/2 (negative boundary bias), then

1. the classification error decreases with increasing $|\,\mathrm{E}[\hat{f}] - 1/2\,|$ irrespective of the amount of bias $f - \mathrm{E}[\hat{f}]$, and
2. one can reduce the classification error towards its minimal value by reducing the variance of $\hat{f}$ alone.

Consider the distributions of $\hat{f}(\mathbf{x})$ displayed in Fig. 2.7 to get an intuitive appreciation for these phenomena. Assume $f(\mathbf{x}) = 0.6$, so in the top row $\mathrm{E}[\hat{f}(\mathbf{x})]$ is on the correct side of 1/2 (negative boundary bias). The shaded areas indicate the probability that $\hat{y}(\mathbf{x})$ differs from $y_B(\mathbf{x})$. In Fig. 2.7 (a) $\mathrm{E}[\hat{f}(\mathbf{x})] = 0.6$, so $\hat{f}(\mathbf{x})$ is unbiased in the mean squared error sense. In Fig. 2.7 (b) $\mathrm{E}[\hat{f}(\mathbf{x})] = 0.65$, so $\hat{f}(\mathbf{x})$ is biased in the mean squared error sense; $f - \mathrm{E}[\hat{f}] = -0.05$. Since the distributions have the same variance, the biased estimator actually has a lower error rate! An increase in bias (with constant variance) actually leads to a decrease in error rate, something that could never occur with mean squared

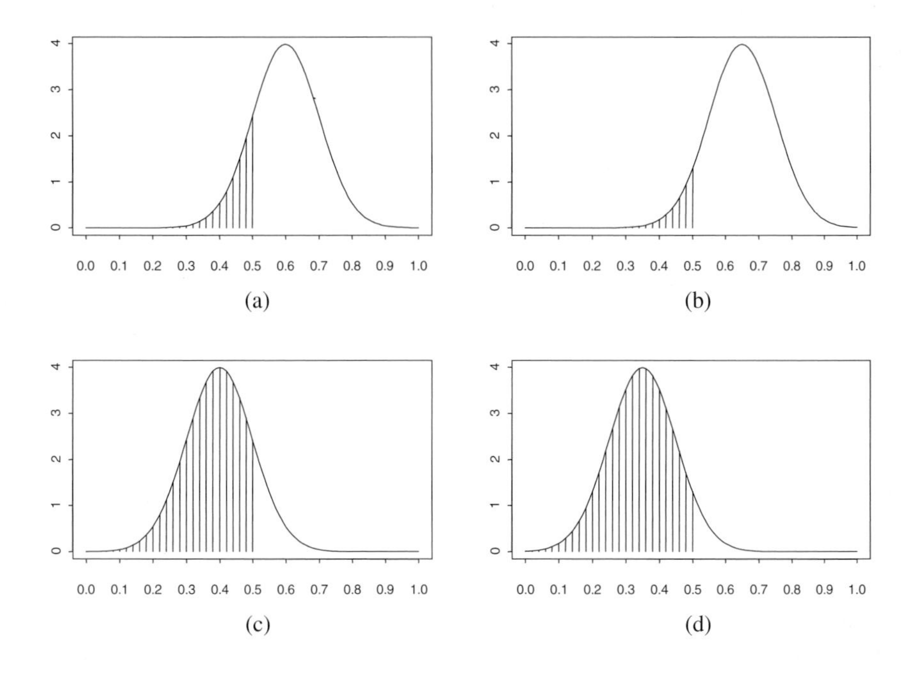

Fig. 2.7. Distributions $p(\hat{f}(\mathbf{x}))$; $f(\mathbf{x}) = 0.6$.

error since that is simply the sum of squared bias and variance. In the bottom row of Fig. 2.7 $\mathrm{E}[\hat{f}(\mathbf{x})]$ is on the wrong side of 1/2 (positive boundary bias). In this case an increase in bias increases the error rate, as indicated by the size of the shaded areas in panel (c) and (d).

Next consider the influence of variance on the error rate, assuming constant bias. Again the qualitative behaviour differs, depending on whether there is negative boundary bias (top row of Fig. 2.8) or positive boundary bias (bottom row of Fig. 2.8). In the first case an increase in variance leads to an increase of the error rate, as witnessed by the increase of the shaded area. In the latter case however, an increase in variance leads to a decrease in the error rate.

We may summarize these findings as follows. The effect of bias and variance (as defined for regression models) on prediction error in classification involves a rather complex interaction, unlike the nice additive decomposition obtained for prediction error in regression. This interaction explains the success of some obviously biased classification procedures. As long as boundary bias is predominantly negative the error rate can be made arbitrarily small by reducing variance. As indicated in [199] classification procedures such as naive Bayes and k-nearest neighbour tend to produce negative boundary bias. This explains their relative

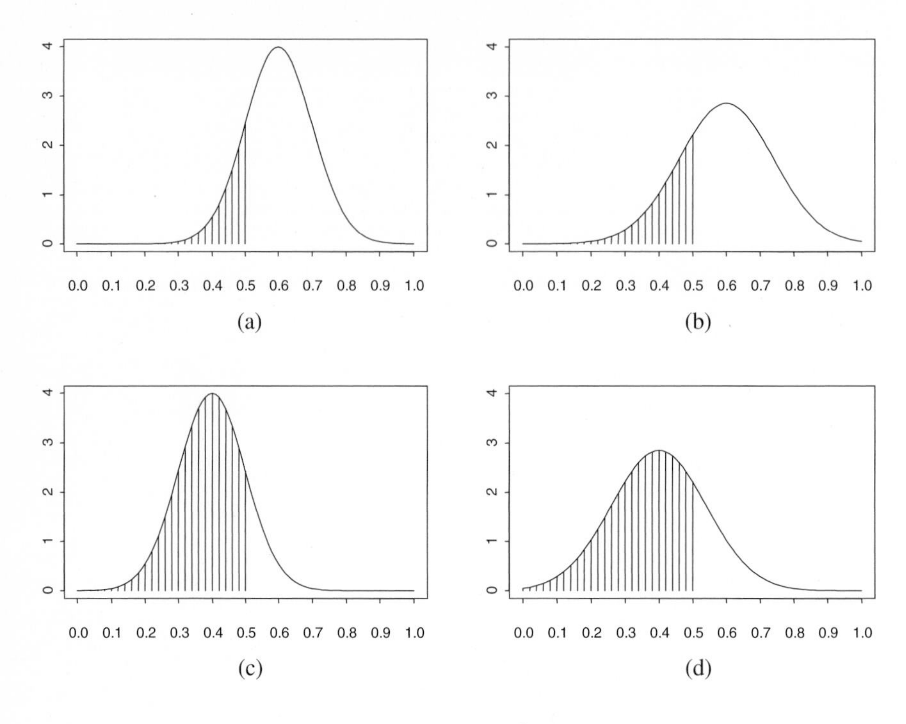

Fig. 2.8. Distributions $p(\hat{f}(\mathbf{x}))$; $f(\mathbf{x}) = 0.6.$.

success in comparison to more flexible classifiers, even on data sets of moderate size [269].

Breiman's Bias-Variance Decomposition for Classifiers In [89] Breiman proposes an additive decomposition of the prediction error of classifiers aimed at explaining the success of particular aggregation techniques such as bagging and arcing (see sections 2.6.4 and 2.6.5). To this end an aggregate classifier is definded as follows

$$y_A(\mathbf{x}) = I(\mathrm{E}\hat{f}(\mathbf{x}) \geq 1/2)$$

where the subscript A denotes aggregation.

Note: In fact, Breiman defines the aggregate classifier somewhat differently as follows

$$f_A(\mathbf{x}) = \mathrm{E}(I(\hat{f}(\mathbf{x}) \geq 1/2))$$

and then

$$y_A(\mathbf{x}) = I(f_A(\mathbf{x}) \geq 1/2)$$

In the first definition the class probabilities of the individual classifiers are averaged and the aggregate classifier assigns $\mathbf{x}$ to the class with largest average probability. In the second definition the aggregate classifier assigns $\mathbf{x}$ to the class to which the individual classifiers allocate $\mathbf{x}$ most often (majority voting). If $p(\hat{f})$ is symmetric and unimodal, the two definitions are equivalent. Henceforth we stick to the first definition.

We already saw that the reducible error at $\mathbf{x}$ is

$$r(\mathbf{x}) = P(\hat{y}(\mathbf{x}) \neq y(\mathbf{x})) - P(y_B(\mathbf{x}) \neq y(\mathbf{x}))$$

The bias at $\mathbf{x}$ is now defined as follows

$$\text{bias}(\mathbf{x}) = I(y_A(\mathbf{x}) \neq y_B(\mathbf{x}))\, r(\mathbf{x})$$

and likewise

$$\text{var}(\mathbf{x}) = I(y_A(\mathbf{x}) = y_B(\mathbf{x}))\, r(\mathbf{x})$$

By definition $r(\mathbf{x}) = \text{bias}(\mathbf{x}) + \text{var}(\mathbf{x})$, so

$$P(\hat{y}(\mathbf{x}) \neq y(\mathbf{x})) = \text{bias}(\mathbf{x}) + \text{var}(\mathbf{x}) + P(y_B(\mathbf{x}) \neq y(\mathbf{x}))$$

Example 2.26. (Continuation of example 2.25) Recall that $\mathrm{E}[\hat{f}(\mathbf{x})] = 0.78$. The aggregate classifier allocates $\mathbf{x}$ to class

$$y_A(\mathbf{x}) = I(\mathrm{E}[\hat{f}(\mathbf{x})] \geq 1/2) = I(0.78 \geq 0.5) = 1$$

The reducible error $r(\mathbf{x}) = 0.26 - 0.2 = 0.06$. Since the aggregate classifier allocates $\mathbf{x}$ to the same class as Bayes rule, i.e. $y_A(\mathbf{x}) = y_B(\mathbf{x})$, we have

$$\text{bias}(\mathbf{x}) = I(y_A(\mathbf{x}) \neq y_B(\mathbf{x}))\, r(\mathbf{x}) = 0 \cdot 0.06 = 0$$

and

$$\text{var}(\mathbf{x}) = I(y_A(\mathbf{x}) = y_B(\mathbf{x}))\, r(\mathbf{x}) = 1 \cdot 0.06 = 0.06$$

In comparison to Friedman's decomposition, the reducible error is called bias in regions of positive boundary bias, and variance in regions of negative boundary bias. At any given point $\mathbf{x}$, the classifier has either bias or variance, depending upon whether or not the aggregated classifier disagrees with the Bayes rule there.

2.6. Resampling

2.6.1 Introduction

Resampling techniques are computationally expensive techniques that reuse the available sample to make statistical inferences. Because of their computational requirements these techniques were infeasible at the time that most of "classical" statistics was developed. With the availability of ever faster and cheaper computers, their popularity has grown very fast in the last decade. In this section we provide a brief introduction to some important resampling techniques.

2.6.2 Cross-Validation

Cross-Validation is a resampling technique that is often used for model selection and estimation of the prediction error of a classification- or regression function. We have seen already that squared error is a natural measure of prediction error for regression functions:

$$\mathrm{PE} = \mathrm{E}(y - \hat{f})^2$$

Estimating prediction error on the same data used for model estimation tends to give downward biased estimates, because the parameter estimates are "fine-tuned" to the peculiarities of the sample. For very flexible methods, e.g. neural networks or tree-based models, the error on the training sample can usually be made close to zero. The true error of such a model will usually be much higher however: the model has been "overfitted" to the training sample.

An alternative is to divide the available data into a training sample and a test sample, and to estimate the prediction error on the test sample. If the available sample is rather small, this method is not preferred because the test sample may not be used for model estimation in this scenario. Cross-validation accomplishes that all data points are used for training as well as testing. The general K-fold cross-validation procedure works as follows

1. Split the data into K roughly equal-sized parts.
2. For the kth part, estimate the model on the other $K-1$ parts, and calculate its prediction error on the kth part of the data.
3. Do the above for $k = 1, 2, \ldots, K$ and combine the K estimates of prediction error.

If $K = m$, we have the so-called *leave-one-out* cross-validation: one observation is left out at a time, and $\hat{f}$ is computed on the remaining $m-1$ observations.

Now let $k(i)$ be the part containing observation i. Denote by $\hat{f}_i^{-k(i)}$ the value predicted for observation i by the model estimated from the data with the $k(i)$th part removed. The cross-validation estimate of mean squared error is now

$$\widehat{\mathrm{PE}}_{cv} = \frac{1}{m}\sum_{i=1}^{m}(y_i - \hat{f}_i^{-k(i)})^2$$

We consider a simple application of model selection using cross-validation, involving the linear, quadratic and cubic model introduced in Section 2.5. In a simulation study we draw 50 (x, y) observations from the probability distributions

$$X \sim U(0, 10) \quad \text{and} \quad Y \sim \mathcal{N}(\mu = 2 + 3x + 1.5x^2, \sigma_\varepsilon = 5),$$

i.e. $\mathrm{E}(Y)$ is a quadratic function of x. For the purposes of this example, we pretend we don't know the true relation between x and y, as would usually be the case in a practical data analysis setting. We consider a linear, quadratic and cubic model as the possible candidates to be selected as the model with lowest prediction error, and we use leave-one-out cross validation to compare the three candidates.

Table 2.12. Mean square error of candidate models: in-sample and leave-one-out.

	in-sample	leave-one-out
linear	150.72	167.63
quadratic	16.98	19.89
cubic	16.66	20.66

The first column of Table 2.12 contains the "in-sample" estimate of the mean square error of all three models. Based on the in-sample comparison one would select the cubic model as the best model since it has the lowest prediction error. We already noted however that this estimate tends to be too optimistic, and the more flexible the model the more severe the optimism tends to be. In the second column the cross-validation estimates of prediction error are listed. As one would expect they are higher than their in-sample counterparts. Furthermore, we see that the quadratic model (the true model) has the lowest cross-validation prediction error of the three. The lower in-sample prediction error of the cubic was apparently due to a modest amount of overfitting.

2.6.3 Bootstrapping

In Section 2.3 we saw that in some special cases we can derive mathematically the exact distribution of a sample statistic, and in some other cases we can rely on limiting distributions as an approximation to the sampling distribution for a finite sample. For many statistics that may be of interest to the analyst, such exact or limiting distributions cannot be derived analytically. In yet other cases the asymptotic approximation may not provide a good fit for a finite sample. In such cases an alternative approximation to the sampling distribution of a statistic $t(\mathbf{x})$ may be obtained using just the data at hand, by a technique called *bootstrapping* [166, 145]. To explain the basic idea of the *non-parametric* bootstrap, we first introduce the *empirical distribution function*

$$\hat{F}(z) = \frac{1}{m} \sum_{i=1}^{m} I\left(x_i \leq z\right) \quad -\infty < z < \infty$$

where I denotes the indicator function and $\mathbf{x} = (x_1, x_2, ..., x_m)$ is a random sample from population distribution function F. We now approximate the sampling distribution of $t(\mathbf{x})$ by repeated sampling from $\hat{F}$. This is achieved by drawing samples $\mathbf{x}^{(r)}$ of size m by sampling independently *with replacement* from $(x_1, x_2, ..., x_m)$. If all observations are distinct, there are $\binom{2m-1}{m}$ distinct samples in

$$\mathcal{B} = \left\{ \mathbf{x}^{(r)}, r = 1, ..., \binom{2m-1}{m} \right\}$$

LIVERPOOL
JOHN MOORES UNIVERSITY
AVRIL ROBARTS LRC
TEL. 0151 231 4022

with respective multinomial probabilities (see Section 2.2.15)

$$P(\mathbf{x}^{(r)}) = \frac{m!}{j_1^{(r)}!\, j_2^{(r)}! \ldots j_m^{(r)}!}\left(\frac{1}{m}\right)^m$$

where $j_i^{(r)}$ is the number of copies of x_i in $\mathbf{x}^{(r)}$. The bootstrap distribution of $t(\mathbf{x})$ is derived by calculating the realisation $t(\mathbf{x}^{(r)})$ for each of the resamples and assigning each one probability $P(\mathbf{x}^{(r)})$. As $m \to \infty$, the empirical distribution $\hat{F}$ converges to the underlying distribution F, so it is intuitively plausible that the bootstrap distribution is an asymptotically valid approximation to the sampling distribution of a statistic.

We can in principle compute all $\binom{2m-1}{m}$ values of the statistic to obtain its "ideal" bootstrap distribution, but this is computationally infeasible even for moderate m. For $m = 15$ there are already 77558760 distinct samples. The usual alternative is to use Monte Carlo simulation, by drawing a number B of samples and using them to approximate the bootstrap distribution.

If a *parametric* form is adopted for the underlying distribution, where θ denotes the vector of unknown parameters, then the parametric bootstrap uses an estimate $\hat{\theta}$ formed from $\mathbf{x}$ in place of θ. If we write F_θ to signify its dependence on θ, then bootstrap samples are generated from $\hat{F} = F_{\hat{\theta}}$.

The non-parametric bootstrap makes it unneccesary to make parametric assumptions about the form of the underlying distribution. The parametric bootstrap may still provide more accurate answers than those provided by limiting distributions, and makes inference possible when no exact or limiting distributions can be derived for a sample statistic.

We present an elementary example to illustrate the parametric and non-parametric bootstrap. The population parameter of interest is the correlation coefficient, denoted by ρ. We first discuss this parameter before we show how to use bootstrapping to make inferences about it.

The linear dependence between population variables $\mathcal{X}$ and $\mathcal{Y}$ is measured by the covariance

$$\sigma_{xy} = \frac{1}{M}\sum_{i=1}^{M}(x_i - \mu_x)(y_i - \mu_y)$$

A term $(x_i - \mu_x)(y_i - \mu_y)$ from this sum is positive if both factors are positive or both are negative, i.e. if x_i and y_i are both above or both below their mean. Such a term is negative if x_i and y_i are on opposite sides of their mean. The dimension of σ_{xy} is the product of the dimensions of X and Y; division by both σ_x and σ_y yields a dimensionless number called the correlation coefficient, i.e.

$$\rho_{xy} = \frac{\sigma_{xy}}{\sigma_x \sigma_y}$$

Evidently ρ has the same sign as σ_{xy}, and always lies between -1 and $+1$. If $\rho = 0$ there is no linear dependence: the two variables are uncorrelated. The linear dependence increases as $|\rho|$ gets closer to 1. If all pairs (x, y) are on a straight line with positive slope, then $\rho = 1$; if all pairs are on a straight line with negative slope then $\rho = -1$.

To make inferences about ρ we use the sample correlation coefficient

$$r_{xy} = \frac{s_{xy}}{s_x s_y} = \frac{\sum_{i=1}^{m}(x_i - \bar{x})(y_i - \bar{y})}{\sqrt{\sum_{i=1}^{m}(x_i - \bar{x})^2 \sum_{i=1}^{m}(y_i - \bar{y})^2}}$$

The sampling distribution of this statistic can't be mathematically derived in general, in fact there is no general expression for the expected value of r_{xy}. Therefore it makes sense to use the bootstrap to make inferences concerning ρ.

In our study, we draw 30 (x, y) pairs from a standard binormal distribution with $\rho = 0.7$, i.e.

$$(X, Y) \sim \mathcal{N}^2(\mu_x = 0, \mu_y = 0, \sigma_x^2 = 1, \sigma_y^2 = 1, \rho = 0.7)$$

Based on this dataset, bootstrapping proceeds as follows

Non-parametric: Draw samples of 30 (x, y) pairs (with replacement) from the data. For each bootstrap sample, compute r, to obtain an empirical sampling distribution.

Parametric: Make appropriate assumptions about the joint distribution of X and Y. In our study we assume

$$(X, Y) \sim \mathcal{N}^2(\mu_x, \mu_y, \sigma_x^2, \sigma_y^2, \rho)$$

which happens to be correct. In a practical data analysis situation we would evidently not know that, and it would usually be hard to ascertain that our assumptions are appropriate. We build an empirical sampling distribution by drawing samples of size 30 from

$$\mathcal{N}^2(\bar{x}, \bar{y}, s_x^2, s_y^2, r)$$

In both cases we draw 1000 samples to generate the empirical sampling distribution of r. To construct $100(1 - \alpha)\%$ confidence intervals for ρ, we simply take the $100(\alpha/2)$ and $100(1 - \alpha/2)$ percentiles of this distribution.

In order to determine whether the bootstrap provides reliable confidence intervals with the right coverage, we repeated the following procedure 100 times

1. Draw a sample of size 30 from the population.
2. Build a bootstrap distribution for r, and construct 90% confidence intervals for ρ. (both parametric and non-parametric)
3. Determine whether the true value of ρ is inside the confidence interval.

Like any conventional method for constructing confidence intervals, the bootstrap will sometimes miss the true value of the population parameter. This happens when the data is not representative for the population. For example, in 1 of the 100 samples the sample correlation coefficient was 0.36. This is highly unlikely to occur when sampling from a population with $\rho = 0.7$ but it will

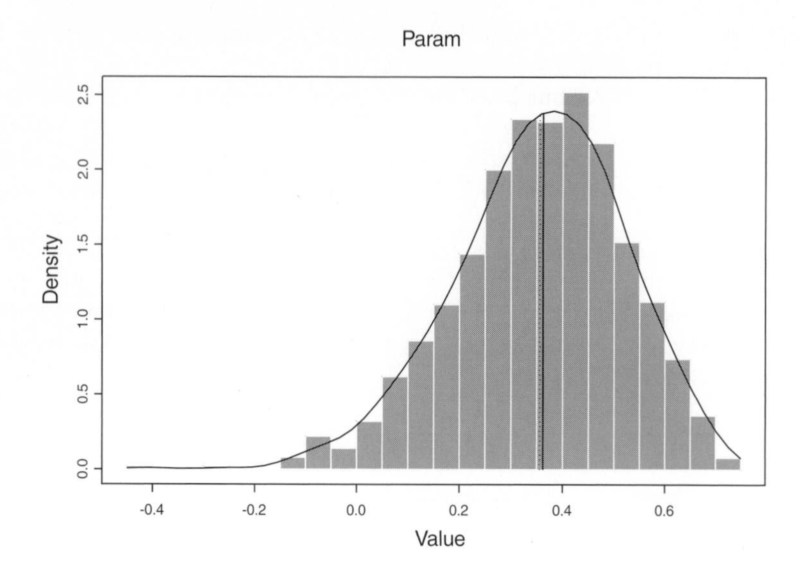

Fig. 2.9. Bootstrap distribution for r. Observed value of r is 0.36.

occur occasionally. In such a case the bootstrap distribution of r is bound to be way off as well. In Fig. 2.9 the non-parametric bootstrap distribution for this particular sample is displayed. The 90% confidence interval computed from this distribution is $(0.064, 0.610)$. Not surprisingly it does not contain the true value of ρ.

On average, one would expect a 90% confidence interval to miss the true value in 10% of the cases; that's why it's called a 90% confidence interval. Furthermore the narrower the confidence intervals, the more informative they are. Both the parametric and non-parametric bootstrap missed the true value of ρ 13 times out of 100, where one would expect 10 misses. Now we may test whether the bootstrap confidence intervals have the right coverage:

$$H_0 : \alpha = 0.1 \quad \text{against} \quad H_a : \alpha \neq 0.1$$

We observed 13 misses out of 100, so the observed value of our test statistic is $a = 0.13$. The distribution of $\hat{\alpha}$ under H_0 (the null-hypothesis) may be approximated by

$$\hat{\alpha} \approx_{H_0} \mathcal{N}(\mu = \alpha, \sigma^2 = \alpha(1-\alpha)/m)$$

which yields $\hat{\alpha} \approx \mathcal{N}(0.1, 0.0009)$. We may now compute the p-value of the observed value under the null-hypothesis as follows

$$P_{H_0}(\hat{\alpha} > a) = P_{H_0}(\hat{\alpha} > 0.13) = P(Z > \frac{0.13 - 0.1}{\sqrt{0.0009}}) = P(Z > 1) = 0.1587$$

where the value 0.1587 was looked-up in a table for the standardnormal distribution. Since we are performing a two-sided test this probability should be doubled, so we obtain a p-value of $2 \times 0.1587 = 0.3174$. This means we would not reject the null-hypothesis under any conventional significance level. The probability under the null-hypothesis of obtaining a result at least as far from $\alpha_0 = 0.1$ (to either side) as the one we observed is "pretty high".

The mean length of the confidence intervals is 0.31 for the non-parametric bootstrap, and 0.32 for the parametric bootstrap. Even though the assumptions of the parametric bootstrap were correct it did not give shorter confidence intervals on average.

2.6.4 Bagging Predictors

Bagging [88] is an acronym for **b**ootstrap **agg**regat**ing**, and is used to construct aggregated predictors with the intention of reducing the variance component of prediction error (see Section 2.5). The rationale of this idea uses a form of "reasoning by analogy" from population to sample that is characteristic for bootstrap procedures.

If we were to have access to the entire population of interest we could in principle compute aggregated regression and classification functions as follows. If y is numerical, we replace $\hat{f}(\mathbf{x} \,|\, T)$ by the average of $\hat{f}(\mathbf{x} \,|\, T)$ over all possible samples T, that is by

$$f_A(\mathbf{x}) = \mathrm{E}_T \, \hat{f}(\mathbf{x} \,|\, T)$$

where the A in f_A denotes aggregation. This aggregation reduces mean prediction error, since the bias of $f_A(\mathbf{x})$ is the same as that of $\hat{f}(\mathbf{x} \,|\, T)$, but its variance is 0. The degree of this reduction depends on the relative importance of variance $V(\hat{f}(\mathbf{x} \,|\, T))$ as compared to squared bias $B^2(\hat{f}(\mathbf{x} \,|\, T))$. In general high variance predictors such as neural networks and regression trees tend to benefit substantially from aggregation.

If y is a class label, say $y \in \{0, 1\}$, recall from Section 2.5.2 that the aggregated classifier may defined as

$$y_A(\mathbf{x}) = I(f_A \geq 1/2) = I(\mathrm{E}_T \hat{f}(\mathbf{x} \,|\, T) \geq 1/2)$$

where again the A in y_A denotes aggregation.

Now of course in practice we only have a single sample, so we cannot compute f_A. Here's where the bootstrap "reasoning by analogy" comes in. Instead of sampling repeatedly from the population, we *resample* the single sample T we have, and estimate f_A by

$$f_B(\mathbf{x}) = \hat{f}_A(\mathbf{x}) = \mathrm{E}_{T^{(r)}} \hat{f}(\mathbf{x} \,|\, T^{(r)})$$

where $T^{(r)}$ are bootstrap samples from T. Now there are two factors influencing the performance of f_B. If the predictor has high variance , it can give improvement through aggregation. If the predictor has low variance, then the unaggregated predictor $\hat{f}(\mathbf{x} \,|\, T)$ tends to be close to $\mathrm{E}_T \, \hat{f}(\mathbf{x} \,|\, T)$, and so $f_B = \mathrm{E}_{T^{(r)}} \hat{f}(\mathbf{x} \,|\, T^{(r)})$ may be less accurate then $\hat{f}(\mathbf{x} \,|\, T)$.

Table 2.13. Bias and variance of single and aggregated neural network.

												unaggregated			bagged		
	x_1	x_2	x_3	x_4	x_5	x_6	x_7	x_8	x_9	x_{10}	$f(\mathbf{x})$	$E(\hat{f})$	B^2	V	$E(\hat{f})$	B^2	V
1	.17	.29	.33	.11	.70	.69	.86	.83	.11	.78	6.79	5.62	1.36	98.58	6.75	0.00	23.59
2	.67	.25	.37	.51	.73	.99	.94	.35	.64	.36	14.14	14.27	0.02	1.81	14.50	0.13	0.55
3	.91	.34	.58	.17	.91	.26	.73	.40	.31	.72	14.60	14.31	0.08	2.72	13.93	0.45	5.29
4	.80	.09	.25	.08	.40	.70	.46	.36	.35	.09	6.31	7.85	2.37	77.48	7.80	2.20	1.28
5	.88	.43	.70	.83	.36	.55	.30	.90	.68	.05	20.19	19.80	0.15	1.29	19.24	0.90	1.24
6	.91	.53	.12	.04	.42	.88	.81	.87	.35	.33	15.36	15.07	0.09	83.13	13.42	3.77	2.03
7	.01	.53	.98	.09	.32	.91	.26	.06	.09	.67	7.22	5.99	1.52	105.79	5.83	1.94	36.42
8	.54	.33	.20	.67	.93	.42	.47	.14	.60	.06	18.53	18.36	0.03	14.62	18.35	0.03	1.57
9	.32	.42	.64	.07	.04	.12	.39	.14	.75	.07	5.40	5.38	0.00	1.70	5.77	0.14	7.17
10	.54	.93	.71	.79	.84	.18	.31	.33	.84	.06	23.05	22.04	1.01	12.86	22.12	0.85	2.64

A simple simulation example serves to illustrate the idea and potential benefit of bagging. In this example we use a neural network as the regression method. We do not consider the details of neural networks here, as this is not required to understand the idea of bagging. Neural networks can be used as flexible regression methods and can approximate complex non-linear functions. In general, neural networks tend to have a low bias and high variance, so they may benefit from aggregation methods such as bagging.

The following steps were performed in the simulation study. We generated a simulated data set of size 200 with ten independent predictor variables $x_1, \ldots, x_{10}$ with $x_j \sim U(0,1)$, $j = 1, \ldots, 10$. The response is given by

$$y = 10 \sin(\pi x_1 x_2) + 20(x_3 - 0.5)^2 + 10x_4 + 5x_5 + \varepsilon$$

with $\varepsilon \sim \mathcal{N}(0,1)$. This data generating mechanism was used by Friedman in a paper on a flexible regression technique called MARS [198]. On this dataset we performed the following computations. First, we estimated ("trained") a single neural network on the dataset. This represented the "unaggregated" regression function. We computed the predictions of this network on a test sample of 1000 observations not used in training. Next, we drew 25 bootstrap samples from the data set at hand, and trained a neural network on each bootstrap sample. The predictions of the 25 networks were averaged to obtain averaged predictions for the test sample.

This whole procedure was repeated 100 times to obtain 100 predictions for each point in the test sample. We may now decompose the mean square estimation error into its bias and variance components. We would expect the aggregated neural network to show smaller variance and comparable bias, on average leading to smaller prediction error than the unaggregated version.

Averaged over the 1000 test observations, the variance of the unaggregated neural network is 47.64, whereas the variance of its aggregated counterpart is about 10 times smaller, namely 4.66. The bias components of error are about equal: 1.51 for the unaggregated neural network against 1.97 for the aggregated version. This example illustrates that bagging is a variance reduction technique; the bias component of error is not reduced by bagging. Table 2.13 displays the bias-variance decomposition for the unaggregated and bagged neural network for 10 of the 1000 observations in the test set.

In [88] and [89], elaborate experiments with bagging are presented on simulated as well as "real-life" data sets. The overall conclusion is that bagging tends to lead to a substantial reduction in prediction error for regression as well as classification methods. Since it is a variance-reduction technique, it tends to work well for methods with high variance (such as neural networks and tree-based methods) but does not improve the performance of methods with low variance such as linear regression and linear discriminant analysis.

2.6.5 Arcing Classifiers

Arcing [89] is an acronym for **a**daptive **r**esampling and **c**ombin**ing**, and has proved to be a succesfull technique in reducing the prediction error of classifiers. Like bagging, arcing is an aggregation method based on resampling the available data. The main difference with bagging is the *adaptive* resampling scheme. In bagging bootstrap samples are generated by sampling the original data where each data point has probability $1/m$ (where m is the number of elements in the original sample). In the arcing algorithm these inclusion probabilities may change.

At the start of each construction, there is a probability distribution $\{p(i)\}, i = 1, ..., m$ on the cases in the training set. A bootstrap sample $T^{(r)}$ is constructed by sampling m times (with replacement) from this distribution. Then the probabilities are updated in such a way that the probability of point $\mathbf{x}$ is increased if it is misclassified by the classifier constructed from $T^{(r)}$ and remains the same otherwise (of course the revised "probabilities" are rescaled to sum to one). After R classifiers have been constructed in this manner, a (possibly weighted) voting for the class allocation is performed.

The intuitive idea of arcing is that the points most likely to be selected for the replicate data sets are those most likely to be misclassified. Since these are the troublesome points, focussing on them using the adaptive resampling scheme of arcing may do better than the "neutral" bagging approach.

We perform a simple simulation study to illustrate the idea. First we specify the exact arcing algorithm used in this simulation study. It is essentially the arcx4 algorithm proposed by Breiman [89] except that we don't aggregate the classifiers by voting, but rather by averaging the predicted class probabilities (see Section 2.5.2). The algorithm works as follows

1) At the rth step, sample with replacement from T to obtain training set $T^{(r)}$, using inclusion probabilities $\{p^{(r)}(i)\}$. Use $T^{(r)}$ to construct (estimate) $\hat{f}_r$.

2) Predict the class of each point in T using $\hat{f}_r$, i.e. determine $\hat{y}_r = I(\hat{f}_r \geq 1/2)$ for each point in T. Now let $k(i)$ be the number of misclassifications of the ith point by $\hat{y}_1, \ldots, \hat{y}_r$, the classifiers constructed so far.

3) The new $r+1$ step inclusion probabilities are defined by

$$p^{(r+1)}(i) = (1 + k(i)^4) / \sum_{j=1}^{m} (1 + k(j)^4)$$

4) Finally, after R steps the aggregated classifier is determined

$$\hat{y}_A = I(\hat{f}_A \geq 1/2), \quad \text{with} \quad \hat{f}_A = 1/R \sum \hat{f}_r$$

For bagging we use the same manner of aggregating classifiers, but the inclusion probabilities are simply $1/m$ at each step.

The simulation study may now be summarized as follows. Repeat the following steps 100 times

1. Generate a training set T containing 150 observations from class 0 and 150 observations from class 1.
2. Estimate the unaggregated function $\hat{f}$ and classifier $\hat{y}$ using T.
3. Apply arcx4 to T to obtain an aggregated function $\hat{f}_a$ and classifier $\hat{y}_a$, and apply bagging to T to construct an aggregated function $\hat{f}_b$ and classifier $\hat{y}_b$ (the subscript "a" stands for arcing and "b" for bagging). We chose $R = 25$ so the aggregated classifiers are based on averages of 25 functions.
4. Use the classifiers so obtained to predict the class of 1000 observations in an independent test.

After these computations have been performed we have 100 predictions for each observation in the test set for the arced classifier as well as the bagged classifier and the unaggregated classifier. Furthermore we know the bayes error for each point in the test set since we specified the distribution from which the class 0 points were drawn, the distribution from which the class 1 points were drawn, and the prior probabilities of both classes. This allows us the compute the probability of class 1 at point $\mathbf{x}$ using Bayes rule as follows

$$P(y = 1 \mid \mathbf{x}) = \frac{p(\mathbf{x} \mid y = 1)P(y = 1)}{p(\mathbf{x} \mid y = 1)P(y = 1) + p(\mathbf{x} \mid y = 0)P(y = 0)}$$

where p denotes a probability density.

The class 0 observations were drawn from a 10-dimensional normal distribution with mean 0 and covariance matrix 5 times the identity. The class 1 observations were drawn from a 10-dimensional normal distribution with mean 0.5 and identity covariance matrix.

$$\mathbf{x} \mid y = 0 \sim \mathcal{N}^{10}(\boldsymbol{\mu}_0, \Sigma_0), \quad \mathbf{x} \mid y = 1 \sim \mathcal{N}^{10}(\boldsymbol{\mu}_1, \Sigma_1)$$

Since the class-conditional distributions are normal with distinct covariance matrices, the optimal decision boundary is a quadratic function [370]. Again we used a simple neural network as the classifier in the simulation study. We don't discuss the details of neural networks here but simply state that it may be

Table 2.14. Prediction error of arcing and bagging and its bias-variance decomposition.

	Breiman				Squared error		
	$\bar{e}$	e_B	bias	var	$E(f-\hat{f})^2$	$(f-E\hat{f})^2$	$E(\hat{f}-E\hat{f})^2$
unaggregated	21.6%	3.7%	4.4%	13.6%	.138	.068	.070
arced	10.4%	3.7%	1.8%	4.9%	.095	.089	.006
bagged	14.1%	3.7%	6.1%	4.3%	.082	.072	.010

used to estimate a classification function from the training set (see Chapter 8 of this volume for an introduction to neural networks and their interpretation). Since neural networks tend to have substantial variance, one would expect that averaging through either bagging or arcing will reduce overall prediction error through variance reduction.

The results of the simulation study are summarized in Table 2.14. Column 1 contains the average error rate (denoted by $\bar{e}$) of the three methods on the test set. We conclude from this column that both arcing and bagging result in a substantial reduction of the error rate compared to the "conventional" unaggregated classifier. Furthermore arcing seems to have a slight edge over bagging. These conclusions are in line with more elaborate studies to be found in [89]. Column 2 contains the Bayes error rate on the test set which we were only able to calculate because we specified the data generating mechanism ourselves. In any practical data analysis setting the Bayes error rate is of course unknown, although methods have been suggested for its estimation (see [448]). The third and fourth column contain the decomposition of reducible error into its bias and variance components, using the additive decomposition of Breiman (see Section 2.5.2). We see that both aggregation methods lead to a substantial variance reduction. Furthermore, arcing leads to a reduction in bias as well, but bagging leads to some increase in bias in this study. More elaborate studies suggest that both arcing and bagging tend to reduce bias somewhat, but the major improvement in overall performance is caused by a substantial reduction in variance.

The last three columns of Table 2.14 contain the decomposition of mean square estimation error into its bias and variance components. This decompostition provides some additional insight. Again both aggregation methods lead to an improvement of overall mean square estimation error. Furthermore we see that this improvement is entirely due to a substantial decrease in the variance component of error. Note that arcing leads to some increase in squared bias but to a decrease in bias according to Breiman's definition. This suggest that the bias induced by arcing is of the "harmless" kind, i.e. negative boundary bias in Friedman's terminology. Since boundary bias is predominantly negative, the strong reduction in variance (in the mean squared error sense) leads to a substantial reduction of the error rate.

2.7. Conclusion

Probability theory is the primary tool of statistical inference. In the probability calculus we reason deductively from known population to possible samples and their respective probabilities under a particular sampling model (we only discussed simple random sampling). In statistical inference we reason inductively from the observed sample to unknown (usually: up to a fixed number of parameters) population. The three main paradigms of statistical inference are frequentist inference, likelihood inference and Bayesian inference. Simplifying matters somewhat, one may say that frequentist inference is based on the sampling distribution of a sample statistic, likelihood inference is based on the likelihood function, and Bayesian inference on the posterior distribution of the parameter(s) of interest.

One of the goals of statistical data analysis is to estimate functions $y = f(\mathbf{x})$ from sample data, in order to predict realisations of random variable Y when $\mathbf{x}$ is known. We may evaluate methods to estimate such functions by comparing their expected prediction error under repeated sampling from some fixed population (distinct samples gives rise to a different estimates $\hat{f} \mid T$). Decomposition of reducible prediction error into its bias and variance components explains to a large extent the relative performance of different methods (e.g. linear regression, neural networks and regression trees) on different problems. Well-known phenomena such as overfitting and variance reduction techniques such as model averaging (e.g. bagging and arcing) follow naturally from the decomposition into bias and variance.

With the increased computer power, computationally expensive inference methods such as bootstrapping are becoming widely used in practical data analysis. From a frequentist perspective, the bootstrap constructs an empirical sampling distribution of the statistic of interest by substituting "sample" for "population" and "resample" for "sample", and mimicing the repeated sampling process on the computer. In many cases this frees the analyst from "heroic" assumptions required to obtain analytical results.

We have attempted to provide the reader with an overview of statistical concepts particularly relevant to intelligent data analysis. In doing so, we emphasized and illustrated the basic ideas, rather than to provide a mathematically rigorous in-depth treatment. Furthermore, this chapter serves as a basis for the more advanced chapters of this volume, most notably Chapters 3 and 4 that give comprehensive overviews of topics only touched upon here.

Chapter 3
Statistical Methods

Paul Taylor
University of Hertfordshire, United Kingdom

3.1. Introduction

This chapter describes a collection of statistical methods. The emphasis is upon what the methods are used to do and upon how to interpret the results. There is little on how to calculate the results, because the algorithms required have already been included in statistical packages for computers and this is how the calculations are performed in practice.

Examples, including computer output and my interpretation of what the output means, are given for some of the more widely used techniques. Other techniques are simply described, in terms of what they can be used to do, along with references to more detailed descriptions of these methods. Presenting examples for all the techniques would have made the chapter far too long.

Section 3.2 describes the most widely used statistical technique, namely regression analysis. Regression analysis is widely used, because there are so many statistical problems that can be presented as finding out how to predict the value of a variable from the values of other variables.

The techniques presented in Section 3.3 are regression analysis techniques for use in specific situations, which arise in practice and are not easy extensions of methods in Section 3.2. In fact, the techniques in Section 3.3.2 and Section 3.3.3 are part of an area of current research in statistical methodology.

Despite being around quite a long time (since the late 1970s or earlier, in most cases) the multivariate analysis techniques of Section 3.4 do not seem to be used as much as they might. When they are used, they are often used inappropriately. It seems likely that these techniques will start to be used more and more, because they are useful and the misconceptions about their use will gradually be eliminated. In particular, with the increase in automatic data collection, the multivariate methods which aim to reduce the number of variables in a data set, discarding uninformative variables, ought to become more important.

3.2. Generalized Linear Models

The fitting of generalized linear models is currently the most frequently applied statistical technique. Generalized linear models are used to describe the relationship between the mean, sometimes called the *trend*, of one variable and the values taken by several other variables. Modelling this type of relationship is sometimes called *regression*. Regression, including alternatives to generalized linear modelling, is described in Section 3.2.1.

Fitting models is not the whole story in modelling; having fitted several plausible models to a set of data, we often want to select one of these models as being the most appropriate. An objective method for choosing between different models is called *analysis of variance*. Analysis of variance is presented in Section 3.2.2.

Within the generalized linear models, there is a subset of models called *linear models*. Sections 3.2.1 and 3.2.2 concentrate on linear models, because these are the most commonly used of the generalized linear models. Log-linear models and logistic regression models are two other heavily used types of generalized linear model. Section 3.2.3 describes log-linear modelling; Section 3.2.4 describes logistic regression.

Section 3.2.5 is about the analysis of survival data. The models used in the analysis of survival data are not generalized linear models. The reason that they are included here is that the techniques used for fitting generalized linear models can also be applied to the fitting of models in the analysis of survival data.

3.2.1 Regression

Regression analysis is the process of determining how a variable, y, is related to one, or more, other variables, $x_1, x_2, \ldots, x_n$. The y variable is usually called the *response* by statisticians; the x_i's are usually called the *regressors* or simply the *explanatory variables*. Some people call the response the *dependent variable* and the regressors the *independent variables*. People with an electrical-engineering background often refer to the response as the *output* and the regressors as the *inputs*. Here, we will use the terms output and inputs. Common reasons for doing a regression analysis include:

- the output is expensive to measure, but the inputs are not, and so cheap predictions of the output are sought;
- the values of the inputs are known earlier than the output is, and a working prediction of the output is required;
- we can control the values of the inputs, we believe there is a causal link between the inputs and the output, and so we want to know what values of the inputs should be chosen to obtain a particular target value for the output;

- it is believed that there is a causal link between some of the inputs and the output, and we wish to identify which inputs are related to the output.

The most widely used form of regression model is the *(general) linear model.* The linear model is usually written as

$$y_j = \beta_0 + \beta_1 x_{1j} + \beta_2 x_{2j} + \cdots + \beta_n x_{nj} + \varepsilon_j \quad j = 1, 2, \ldots, m \tag{3.1}$$

where the ε_j's are independently and identically distributed as $\mathcal{N}(0, \sigma^2)$ and m is the number of data points.

The linear model is called *linear*, because the expected value of y_j is a linear function, or weighted sum, of the β's. In other words, we can write the expected value of y_j as

$$\mathrm{E}(y_j) = \beta_0 + \sum_{i=1}^{n} \beta_i x_{ij}. \tag{3.2}$$

The main reasons for the use of the widespread use of the linear model are given below.

- The maximum likelihood estimators of the β's are the same as the least squares estimators; see Section 2.4 of Chapter 2.
- There are explicit formulae for the least squares estimators of the β's. This means that the estimates of the β's can be obtained without electronic computers, so the use of the linear model was common in agricultural research in the 1920s and the chemical industry in the 1930s (in the UK, at least). There are also rapid and reliable numerical methods for obtaining these estimates using a computer; see chapter 3 of [511].
- There is a large class of regression problems that can be converted into the form of the general linear model. For example, a polynomial relationship, such as
$$\mathrm{E}(y) = \beta_0 + \beta_1 x_1 + \beta_2 x_2 + \beta_3 x_1 x_2 + \beta_4 x_1^2 + \beta_5 x_2^2, \tag{3.3}$$
can be converted to the form in (3.2) by setting $x_3 = x_1 x_2$, $x_4 = x_1^2$ and $x_5 = x_2^2$. (This conversion would not usually be explicitly carried out.) For a wide range of examples of the use of the linear model see, for example [157].
- Even when the linear model is not strictly appropriate, there is often a way to transform the output and/or the inputs, so that a linear model can provide useful information.

Another major attraction of the linear model is a technical point: it is possible to determine the statistical properties of virtually any object that we might derive from the linear model. In other words, we can perform statistical inference for the linear model. In particular, we can use hypothesis testing to compare different linear models (see Section 3.2.2) and we can obtain interval estimates for predictions and for the β's.

The linear model is the main type of model used in regression. We will return to the linear model in Section 3.2.2, but now we will look at other types of regression models.

Non-linear Regression A non-linear regression model differs from a linear model in that the expected output in not a weighted sum of the inputs. An example of a non-linear model is the *allometric* model,

$$y_j = \beta_0 x_{1j}^{\beta_1} + \varepsilon_j \quad j = 1, 2, \ldots, m \tag{3.4}$$

where the ε's and m are as in (3.1). This model is known to represent the relationship between the weight of part of a plant, such as its root system, and the weight of the whole plant. Another example is the *Mitscherlich* model for relating crop yield to the amount of fertiliser applied to the crop,

$$y_j = \beta_0 \left(1 - e^{-\beta_1 (x_{1j} + \beta_2)}\right) + \varepsilon_j \quad j = 1, 2, \ldots, m. \tag{3.5}$$

Here, y_j is the crop yield for the jth plot in a field, x_{1j} is the amount of fertiliser applied to the jth plot in the field, β_0 represents an upper limit on crop yield, β_2 represents the amount of fertiliser already available in the soil, while β_1 is related to how quickly the upper limit for yield is reached.

The drawbacks to using non-linear regression boil down to the fact that there is no general algebraic solution to finding the least squares estimators of the β's. (As for the linear model, least squares estimation will produce the maximum likelihood estimators, because the ε's have identical, independent normal distributions.) The problems that arise because of the lack of a general algebraic solution are as follows.

1. Estimation is carried out using iterative methods which require good choices of starting values, might not converge, might converge to a local optimum rather than the global optimum, and will require human intervention to overcome these difficulties.
2. The statistical properties of the estimates and predictions from the model are not known, so we cannot perform statistical inference for non-linear regression.
   (This is not strictly true; we could perform statistical inference using techniques such as the bootstrap, which is explained in Section 2.6.3 of Chapter 2.)

Non-linear models are usually used when the form of the relationship between the output and the inputs is known and the primary aim of the study is determination of the β's. In this case, the β's represent fundamental biological, chemical or physical constants and are of interest for their own sake. An example of this is β_0 in the Mitscherlich model at (3.5). If prediction of the output is the primary aim, then it is usually possible to use a linear model to make quite good predictions for some range of interest.

An introduction to non-linear modelling can be found in chapter 10 of [157].

Generalized Linear Models The linear model is very flexible and can be used in many situations, even when it is not strictly valid. Despite this, there is a stage where the linear model cannot do what we need. The non-linear model above is one way to expand the possibilities of regression. The *generalized linear model* offers a much greater expansion of the problems that can be addressed by regression. The definitive reference for this topic is [368].

The generalization is in two parts.

1. The distribution of the output does not have to be the normal, but can be any of the distributions in the exponential family; see, for example [368, p. 28].
2. Instead of the expected value of the output being a linear function of the β's, we have

$$g\left(\mathrm{E}(y_j)\right) = \beta_0 + \sum_{i=1}^{n} \beta_i x_{ij} \tag{3.6}$$

   where $g(\cdot)$ is a monotone differentiable function. The function $g(\cdot)$ is called the *link* function.

So, using this generalization we can branch out to consider output distributions such as the binomial or the Poisson (see Chapter 2), or the gamma [1], as well as the normal. It is also possible to restrict the range of values predicted for the expected output, to capture a feature such as an asymptote. Some non-linear models can be framed as generalized linear models. The allometric model at (3.4) can, by choosing a log link function (and a normal output distribution).

A major step forward made by [368] was the development of a general algorithm for fitting generalized linear models. The algorithm used by [368] is iterative, but has natural starting values and is based on repeated application of least squares estimation. The generalized linear model also gives us a common method for making statistical inferences, which we will use in Sections 3.2.3 and 3.2.4.

Examples of widely used generalized linear models are given in Sections 3.2.3, 3.2.4 and 3.2.5.

Generalized Additive Models Generalized additive models are a generalization of generalized linear models. The generalization is that $g\left(\mathrm{E}(y_j)\right)$ need not be a linear function of a set of β's, but has the form

$$g\left(\mathrm{E}(y_j)\right) = \beta_0 + \sum_{i=1}^{n} s_i(x_{ij}) \tag{3.7}$$

where the s_i's are arbitrary, usually smooth, functions. In particular, the scatterplot smoothers that have been one focus of statistical research for the last fifteen

[1] The gamma distribution is a continuous distribution that is not symmetric about its population mean, and only takes positive values. The exponential distribution of Section 2.2.16 in Chapter 2 is a special case of the gamma distribution. The gamma distribution with parameters n and λ, written $\Gamma(n, \lambda)$, is the distribution of the sum of n independent exponential random variables, all with rate parameter λ. So, the $\Gamma(1, \lambda)$ distribution is the same as the exponential distribution with parameter λ.

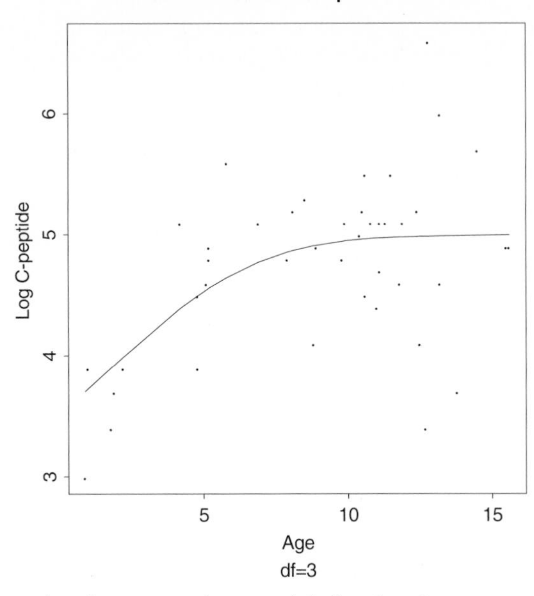

Fig. 3.1. An example of a regression model fitted using one type of scatterplot smoother; the form of the relationship between the output and the input is allowed to vary over the range of the input. The specific type of smoother used is called a *cubic B-spline* with 3 degrees of freedom.

years can be used. An example of the model produced using a type of scatterplot smoother is shown in Figure 3.1. All that has been specified about the shape of the fitted curve in Figure 3.1 is that it has to be smooth (has a continuous second derivative) and must not change direction frequently, that is it must not contain high frequency oscillations (it must not be 'wiggly'). This means that we do not have to specify the functional form of the relationship between the output and the inputs, but can use local regression methods to identify the functional form.

Methods for fitting generalized additive models exist and are generally reliable. The main drawback is that the framework of statistical inference that is available for generalized linear models has not yet been developed for generalized additive models. Despite this drawback, generalized additive models can be fitted by several of the major statistical packages already. The definitive reference for this topic is [249].

3.2.2 Analysis of Variance

The *analysis of variance*, or ANOVA, is primarily a method of identifying which of the β's in a linear model are non-zero. This technique was developed for the analysis of agricultural field experiments, but is now used quite generally.

Table 3.1. Yields of turnips (in kg) from an experiment carried out at Sonning on a farm owned by The University of Reading. The experiment had a *randomized block* design, with four blocks, each of size sixteen. There are sixteen different treatment combinations, labelled *A–R*, each appearing once in each block. Each treatment is a combination of turnip variety (Barkant or Marco), sowing date (21/8/90 or 28/8/90) and seed density (1, 2, 4 or 8 kg/ha).

Treatments				Blocks			
Variety	Date	Density	Label	I	II	III	IV
Barkant	21/8/90	1 kg/ha	A	2.7	1.4	1.2	3.8
		2 kg/ha	B	7.3	3.8	3.0	1.2
		4 kg/ha	C	6.5	4.6	4.7	0.8
		8 kg/ha	D	8.2	4.0	6.0	2.5
	28/8/90	1 kg/ha	E	4.4	0.4	6.5	3.1
		2 kg/ha	F	2.6	7.1	7.0	3.2
		4 kg/ha	G	24.0	14.9	14.6	2.6
		8 kg/ha	H	12.2	18.9	15.6	9.9
Marco	21/8/90	1 kg/ha	J	1.2	1.3	1.5	1.0
		2 kg/ha	K	2.2	2.0	2.1	2.5
		4 kg/ha	L	2.2	6.2	5.7	0.6
		8 kg/ha	M	4.0	2.8	10.8	3.1
	28/8/90	1 kg/ha	N	2.5	1.6	1.3	0.3
		2 kg/ha	P	5.5	1.2	2.0	0.9
		4 kg/ha	Q	4.7	13.2	9.0	2.9
		8 kg/ha	R	14.9	13.3	9.3	3.6

Example 3.1 (Turnips for Winter Fodder). The data in Table 3.1 are from an experiment to investigate the growth of turnips. These types of turnip would be grown to provide food for farm animals in winter. The turnips were harvested and weighed by staff and students of the Departments of Agriculture and Applied Statistics of The University of Reading, in October, 1990.

The blocks correspond to four different pieces of land. These pieces of land have different characteristics and so these four areas are expected to produce different amounts of turnips.

The four areas of land are all the same size and have each been split into sixteen identical plots; this gives a total of sixty-four plots. Each plot has a treatment applied to it; there are sixteen treatments. Each of the sixteen treatments is a combination of three properties: the variety (type) of turnip sown (*Barkant* or *Marco*); when the seed was sown (*21/8/90* or *28/8/90*); how much seed was sown (*1*, *2*, *4* or *8* kg/ha).

The following linear model

$$\begin{aligned} y_j = \beta_0 &+ \beta_B x_{Bj} + \beta_C x_{Cj} + \cdots + \beta_R x_{Rj} \\ &+ \beta_{\mathrm{II}} x_{\mathrm{II},j} + \beta_{\mathrm{III}} x_{\mathrm{III},j} + \beta_{\mathrm{IV}} x_{\mathrm{IV},j} + \varepsilon_j \quad j = 1, 2, \ldots, 64 \end{aligned} \tag{3.8}$$

or an equivalent one could be fitted to these data. The inputs take the values 0 or 1 and are usually called *dummy* or *indicator* variables. The value of x_{Cj}

would be 1 if treatment C (Barkant, sown on 21/8/90 at 4 kg/ha) were applied to plot j, but zero otherwise. Similarly, the value of $x_{\mathrm{III},j}$ would be 1 if plot j were in block III, but zero otherwise.

On first sight, (3.8) should also include a β_A and a β_{I}. Indeed, many statisticians would write this model with terms in β_A and β_{I}. If β_A and β_{I} are included, then the model is over-parameterised. Over-parameterised means that there are more parameters (β's) than are necessary to describe the structure in the expected values. For the turnip experiment we need a total of nineteen parameters: a baseline value; fifteen parameters to specify the **differences** between the expected yields for the sixteen treatments; three parameters to specify the **differences** between the expected yields for the four blocks. If β_A and β_{I} are included, then there are twenty-one parameters, which is two more than necessary.

The choice of model specified by (3.8) implies the following interpretations for the β's. The baseline value, β_0, is the expected turnip yield for a plot in block I that has received treatment A. The expected yield for a plot in block I that has received treatment B is $\beta_0 + \beta_B$, that is, β_B is the difference between the expected yields for treatment B and treatment A; similarly for treatments C to R. The expected yield for a plot in block II that has received treatment A is $\beta_0 + \beta_{\mathrm{II}}$, so β_{II} is the difference in expected yields for block II and block I; similarly for blocks III and IV. Finally, the expected yield for a plot that is neither in block I nor in receipt of treatment A, such as block III with treatment G, would be the baseline, plus the expected increase for using treatment G rather than treatment A, plus the increase for being in block III rather than block I. Algebraically, this is $\beta_0 + \beta_G + \beta_{\mathrm{III}}$.

The first question that we would try to answer about these data is

Does a change in treatment produce a change in the turnip yield?

which is equivalent to asking

Are any of $\beta_B, \beta_C, \ldots, \beta_R$ *non-zero?*

which is the sort of question that can be answered using ANOVA.

This is how the ANOVA works. Recall, the general linear model of (3.1),

$$y_j = \beta_0 + \beta_1 x_{1j} + \beta_2 x_{2j} + \cdots + \beta_n x_{nj} + \varepsilon_j \quad j = 1, 2, \ldots, m$$

where the ε_j's are independently and identically distributed as $\mathcal{N}(0, \sigma^2)$ and m is the number of data points. Estimates of the β's are found by least squares estimation, which happens to be the same as maximum likelihood estimation for this model. Writing the estimate of β_i as $\hat{\beta}_i$, we can define the *fitted values*

$$\hat{y}_j = \hat{\beta}_0 + \sum_{i=1}^{n} \hat{\beta}_i x_{ij}. \tag{3.9}$$

The fitted values are what the fitted model would predict the output to be for the inputs in the set of data. The residuals are the differences between the observed output values and the fitted values,

$$r_j = y_j - \hat{y}_j. \tag{3.10}$$

The size of the residuals is related to the size of σ^2, the variance of the ε_j's. It turns out that we can estimate σ^2 by

$$S^2 = \frac{\sum_{j=1}^{m}(y_j - \hat{y}_j)^2}{m - (n+1)} \tag{3.11}$$

assuming that the model is not over-parameterised. The numerator of the right-hand side of (3.11) is called the *residual sum of squares*, while the denominator is called the *residual degrees of freedom*.

The key facts about S^2 that allow us to compare different linear models are:

- if the fitted model is adequate ('the right one'), then S^2 is a good estimate of σ^2;
- if the fitted model includes redundant terms (that is includes some β's that are really zero), then S^2 is still a good estimate of σ^2;
- if the fitted model does not include one or more inputs that it ought to, then S^2 will tend to be larger than the true value of σ^2.

So if we omit a useful input from our model, the estimate of σ^2 will shoot up, whereas if we omit a redundant input from our model, the estimate of σ^2 should not change much. The use of this concept is described in detail below. Note that omitting one of the inputs from the model is equivalent to forcing the corresponding β to be zero.

Suppose we have fitted a particular linear model, Ω_1, to a set of data. We can estimate the variance, σ^2, of the ε_j's, as the ratio

$$S_1^2 = \frac{RSS_1}{\nu_1}, \tag{3.12}$$

where RSS_1 and ν_1 are the residual sum of squares and the residual degrees of freedom, respectively, for model Ω_1.

As long as Ω_1 includes the inputs that are associated with the output, S_1^2 will be a reasonable estimate of σ^2. In particular, it is a valid estimate even if some (or all) of the β's are zero.

If we want to investigate whether certain β's are zero, then we can fit another model, Ω_0, in which those β's and the corresponding inputs are absent. We will be able to obtain another estimate of the variance

$$S_0^2 = \frac{RSS_0}{\nu_0}, \tag{3.13}$$

where RSS_0 and ν_0 are the residual sum of squares and the residual degrees of freedom, respectively, for model Ω_0.

Now, if the β's that are absent from Ω_0 really are zero, then S_0^2 and S_1^2 are both valid estimates of σ^2 and should be approximately equal. On the other hand, if some of the β's that are absent from Ω_0 are not really zero, then S_1^2 is still a valid estimate of σ^2, but S_0^2 will tend to be bigger than σ^2. The further the absent β's are from zero, the bigger S_0^2 is.

In principle then, we can compare S_0^2 with S_1^2. If S_0^2 is big relative to S_1^2, then some of the absent β's are non-zero and Ω_0 is an inadequate model. If S_0^2 and S_1^2 are similar, then this suggests that Ω_0 is a satisfactory model.

In fact we can make a much better comparison by forming a third estimate of σ^2. We calculate the extra sum of squares

$$ESS = RSS_0 - RSS_1, \tag{3.14}$$

the extra degrees of freedom

$$\nu_E = \nu_0 - \nu_1 \tag{3.15}$$

and our third estimate of σ^2 as

$$S_E^2 = \frac{ESS}{\nu_E}. \tag{3.16}$$

This third estimate, S_E^2, shares some of the properties of S_0^2 in that it is an estimate of σ^2 if Ω_0 is an adequate model, but will be higher than σ^2 if some of the absent β's are non-zero. The advantage of S_E^2 over S_0^2 is that it is much more sensitive to the absent β's being non-zero than S_0^2 is. (In addition, S_E^2 and S_1^2 are statistically independent, which simplifies the statistical theory required.)

We can carry out a hypothesis test of whether Ω_0 is adequate (versus the alternative that it is not, which implies that we would prefer to use Ω_1) by forming the following test statistic

$$F = \frac{S_E^2}{S_1^2} = \frac{ESS/\nu_E}{RSS_1/\nu_1} \tag{3.17}$$

which has various names, including the 'F-statistic', the 'F-ratio' and the 'variance ratio'.

If Ω_0 is an adequate model for the data, then F should be approximately 1 and will have an F-distribution[2] with parameters, ν_E and ν_1; if Ω_0 is not adequate then F should be large.

So we can carry out a formal hypothesis test of the null hypothesis that Ω_0 is the correct model versus the alternative hypothesis that Ω_1 is the correct model, by forming the F-statistic and comparing it with the appropriate F-distribution; we reject the null hypothesis if the F-statistic appears to be too big to come from the appropriate F-distribution. A good rule-of-thumb is that an F-statistic of over 4 will lead to rejection of Ω_0 at the 5% level of significance.

---

[2] The F-distribution is a well-known continuous distribution. Most statistical packages can generate the associated probabilities, which are tabulated in many elementary statistics books. If $X_1, X_2, \ldots, X_{\nu_1}$ and $W_1, W_2, \ldots, W_{\nu_2}$ are statistically independent standard normal random variables, then the ratio

$$\frac{(X_1^2 + X_2^2 + \cdots + X_{\nu_1}^2)/\nu_1}{(W_1^2 + W_2^2 + \cdots + W_{\nu_2}^2)/\nu_2}$$

will have an F-distribution with parameters (*degrees of freedom*) ν_1 and ν_2.

Table 3.2. ANOVA table for fitting the model at (3.18) to the data in Table 3.1.

	Df	Sum of Sq	Mean Sq	F Value	Pr(F)
block	3	163.737	54.57891	2.278016	0.08867543
Residuals	60	1437.538	23.95897		

Table 3.3. ANOVA table for fitting the model at (3.8) to the data in Table 3.1.

	Df	Sum of Sq	Mean Sq	F Value	Pr(F)
block	3	163.737	54.57891	5.690430	0.002163810
treat	15	1005.927	67.06182	6.991906	0.000000171
Residuals	45	431.611	9.59135		

Example 3.2 (Turnips for Winter Fodder continued). To make this idea more concrete, we can take Ω_1 to be the model at (3.8) on page 75, and Ω_0 to be the following model

$$y_j = \beta_0 + \beta_{II} x_{II,j} + \beta_{III} x_{III,j} + \beta_{IV} x_{IV,j} + \varepsilon_j \quad j = 1, 2, \ldots, 64. \tag{3.18}$$

So here, Ω_0 is the special case of Ω_1 in which all of β_B, β_C, ..., β_R are zero.

The ANOVA tables for models Ω_0 and Ω_1 are shown in Tables 3.2 and 3.3. These tables are exactly how they appear in the output from a statistical package. The column labelled 'Df' contains the degrees of freedom, 'Sum of Sq' is the sum of squares column. The entries in the 'Mean Sq' column are obtained by dividing the entries in the sum of squares column by their corresponding degrees of freedom. The label, 'Mean Sq', is an abbreviation of *mean square.* We come back to the meanings of the other columns below.

The numbers that we are interested in are in the row labelled 'Residuals' in each table. We can read the 'Residuals' rows of these two tables, to obtain

$$RSS_0 = 1437.5, \quad \nu_0 = 60 \quad \text{and} \quad s_0^2 = \frac{1437.5}{60} = 23.96$$

(from Table 3.2) and

$$RSS_1 = 431.6, \quad \nu_1 = 45 \quad \text{and} \quad s_1^2 = \frac{431.6}{45} = 9.59$$

(from Table 3.3). These estimates of σ^2 are quite different, which suggests that one or more of the inputs that were omitted from Ω_0 were necessary after all. The formal statistical hypothesis testing procedure to determine whether Ω_0 is adequate is completed below.

From here, we could find the extra sum of squares and extra degrees of freedom, then calculate the F-statistic and compare it with the relevant F-distribution. We do not need to do this calculation ourselves, as the computer does it for us in producing Table 3.3. The extra sum of squares and degrees of freedom are presented in the row labelled 'treat' in Table 3.3. We can read the 'treat' row of Table 3.3 to get

$$ESS = RSS_0 - RSS_1 = 1437.5 - 431.6 = 1005.9,$$

$$\nu_E = \nu_0 - \nu_1 = 60 - 45 = 15$$

and

$$s_E^2 = \frac{1005.9}{15} = 67.06.$$

Notice that the extra degrees of freedom, ν_E, is the same as the number of additional inputs in Ω_1 compared to Ω_0.

The procedure for obtaining the F-statistic is quite automatic; the column labelled 'F Value' is simply the 'Mean Sq' column divided by the residual mean square. So, we find the F-statistic is

$$\frac{s_E^2}{s_1^2} = \frac{67.06}{9.59} = 6.99$$

and the significance probability ("p-value") can be read from the final column, labelled 'Pr(F)'. The p-value is less than $\frac{1}{10{,}000}$ of one percent.

Recall (from Section 2.4.1 of Chapter 2) that, the p-value is the (conditional) probability of the test-statistic being so extreme, if the null hypothesis is true. Our null hypothesis is that Ω_0 is an adequate description of the variation in the turnip yields. So, the p-value that we have obtained tells us that **if** Ω_0 is an adequate model, **then** the probability of getting an F-statistic as high as 6.99 is less than $\frac{1}{10{,}000}$%; in other words, getting an F-statistic of 6.99 is amazing if Ω_0 is the right model. On the other hand, if Ω_0 is inadequate, then we would expect the F-statistic to be inflated. The rational choice between these alternatives is to decide that Ω_0 is not a good model for the turnip yields, because the observed F-statistic is nothing like what it ought to be if Ω_0 were adequate. This tells me that model Ω_0 is inadequate, so I should use model Ω_1.

Having chosen Ω_1, in preference to Ω_0, I conclude that the different treatments produce systematic differences in the turnip yield. I draw this conclusion because the difference between Ω_0 and Ω_1 is that Ω_1 allows for treatment to alter yield, but Ω_0 does not.

Some might wonder where the 'block' row in the ANOVA tables has come from. The 'block' row in Table 3.3 comes from working out the extra sums of squares and degrees of freedom for comparing the model at (3.18) with the model at (3.19).

$$y_j = \beta_0 + \varepsilon_j \quad j = 1, 2, \ldots, 64. \tag{3.19}$$

The test for 'blocks' indicates that the yield is systematically different for different blocks, but we knew this already, so the result of this test is of no interest.

It might be thought that that is the end of the analysis for the turnip data. In some ways it is, as we have established that the different treatments result in

Table 3.4. ANOVA table that would be produced in the real-world analysis of the data in Table 3.1.

	Df	Sum of Sq	Mean Sq	F Value	Pr(F)
block	3	163.7367	54.5789	5.69043	0.0021638
variety	1	83.9514	83.9514	8.75282	0.0049136
sowing	1	233.7077	233.7077	24.36650	0.0000114
density	3	470.3780	156.7927	16.34730	0.0000003
variety:sowing	1	36.4514	36.4514	3.80045	0.0574875
variety:density	3	8.6467	2.8822	0.30050	0.8248459
sowing:density	3	154.7930	51.5977	5.37960	0.0029884
variety:sowing:density	3	17.9992	5.9997	0.62554	0.6022439
Residuals	45	431.6108	9.5914		

different yields. If I had not been illustrating the technique of ANOVA, I would have approached the analysis differently. I would be trying to find out what it was about the treatments that made a difference to the yield. I would investigate this by looking at the ANOVA for a sequence of nested models.

Table 3.4 shows the ANOVA that would usually be produced for the turnip data. Notice that the 'block' and 'Residuals' rows are the same as in Table 3.3. The basic difference between Tables 3.3 and 3.4 is that the treatment information is broken down into its constituent parts in Table 3.4.

The 'variety' row in Table 3.4 corresponds to the extra sum of squares and degrees of freedom for comparing the model at (3.18) with

$$\begin{aligned} y_j = &\beta_0 + \beta_{\mathrm{marco}} x_{\mathrm{marco},j} \\ &+ \beta_{\mathrm{II}} x_{\mathrm{II},j} + \beta_{\mathrm{III}} x_{\mathrm{III},j} + \beta_{\mathrm{IV}} x_{\mathrm{IV},j} + \varepsilon_j \quad j = 1, 2, \ldots, 64 \end{aligned} \tag{3.20}$$

which allows for differences in yield between the two varieties. The variable x_{marco} is an indicator variable, like those used in (3.8). Similarly, the 'sowing' row corresponds to the extra sum of squares and degrees of freedom for comparing the model at (3.20) with one that allows for difference between sowing dates too. The remaining rows in Table 3.4 all correspond to the introduction of new inputs to make the model more complex. All the F-statistics are worked out using the estimate of variance obtained from the 'Residuals' row of the ANOVA.

The rows with labels such as 'variety:sowing' are the extra sums of squares and degrees of freedom corresponding to including inputs that are products of other inputs. This sort of term is called an *interaction*. As an example, take the inputs corresponding to *variety*, *sowing* and *variety:sowing*. These are, respectively:

- x_{marco}, the indicator variable that *Marco* turnips have been used;
- x_{28}, the indicator variable that sowing date was 28/8/90;
- x_{m28}, the indicator variable that the turnips were of type *Marco* and were sown on 28/8/90—$x_{\mathrm{m28},j} = x_{\mathrm{marco},j} \times x_{28,j}$.

Table 3.5. Variety means for the data in Table 3.1.

Barkant	Marco
6.522	4.231

Table 3.6. Sowing date by density means for the data in Table 3.1.

	Density			
Date	1kg/ha	2kg/ha	4kg/ha	8kg/ha
21/8/90	1.7625	3.0125	3.9125	5.1750
28/8/90	2.5125	3.6875	10.7375	12.2125

The reasons for using this sort of input are illustrated in the discussion of Table 3.6, below. Note that various other statistical packages use '.', '*' or '×', instead of ':' to denote interactions.

Based on the significance probabilities (the column labelled 'Pr(F)') in Table 3.4 most statisticians would identify *variety*, *sowing*, *density* and *sowing:density* as the terms that make a difference to turnip yield. The presence of the *sowing:density* interaction in the model means that we cannot consider *sowing* or *density* in isolation. Instead, we have to look at the combinations of *sowing* and *density*. There are no interactions involving *variety* in our set of terms related to turnip yield, so we can think of it in isolation.

Tables 3.5 and 3.6 show the relevant patterns: *Barkant* produces about 2.3 kg more per plot than *Marco* does regardless of sowing date and density; for the two lowest sowing densities, delaying the sowing by one week adds about 0.7 kg to the yield per plot, but for the two highest sowing densities it adds about 7 kg per plot. For the statistician this is close to the end of the analysis. The fact that the increase in yield due to delaying the sowing by a week is different for different sowing densities is the reason for the inclusion of the *sowing:density* interaction. This interaction could only be omitted from the model if the increase due to delaying sowing was the same regardless of the sowing density. There are three inputs corresponding to the *sowing:density* interaction. These inputs are calculated by forming the products of each of the three inputs corresponding to *density* with the input corresponding to *sowing*.

In this section, we have seen how the ANOVA can be used to identify which inputs are related to the output and which are not. The basic idea is to produce different estimates of the variance, σ^2, based on different assumptions, and to compare these estimates in order to decide which assumptions are true. The ANOVA procedure is only valid if the models being compared are nested, in other words, one model is a special case of the other.

3.2.3 Log-linear Models

Log-linear modelling is a way of investigating the relationships between categorical variables. Categorical variables are non-numeric variables and are sometimes called *qualitative* variables; the set of values that a categorical variable can take are called its *categories* or *levels*. There are two types of categorical variables: *nominal* variables have no ordering to their categories; the categories of an *ordinal* variable have a natural ordering, such as *none–mild–moderate–severe*. The data shown in Table 3.7 show the sort of problem attacked by log-linear modelling. There are five categorical variables displayed in Table 3.7:

centre is one of three health centres for the treatment of breast cancer;
age is the age of the patient when her breast cancer was diagnosed;
survived is whether the patient survived for at least three years from the date of diagnosis;
appear the visual appearance of the patient's tumour—either *malignant* or *benign*;
inflam the amount of inflammation of the tumour—either *minimal* or *greater*.

The combination of *appear* and *inflam* represents the state of the tumour, or the disease progression. The entries in the tables are numbers of patients in each group. For these data, the output is the number of patients in each cell.

A log-linear model is a type of generalized linear model; the output, Y_i, is assumed to have a Poisson distribution with expected value μ_i. The (natural) logarithm of μ_i is assumed to be linear in the β's. So the model is

$$y_j \sim \text{Pois}(\mu_j) \quad \text{and} \quad \log(\mu_j) = \beta_0 + \beta_1 x_{1j} + \beta_2 x_{2j} + \cdots + \beta_n x_{nj}. \tag{3.21}$$

Since all the variables of interest are categorical, we need to use indicator variables as inputs in the same way as in (3.8).

The aim in log-linear modelling is to identify associations between the categorical variables. Associations correspond to interaction terms in the model. So, our problem becomes that of finding out which β's are zero; this is the same problem that we used the ANOVA to solve for linear models in Section 3.2.2. For the log-linear model, and indeed all generalized linear models, we consider a quantity called *deviance* when we are trying to compare two models. Deviance is analogous to the sums of squares in an ANOVA. (In fact, if the generalized linear model happens to be a linear model, the deviance and the sum of squares are the same thing.) We can identify which β's are zero by considering an analysis of deviance table, such as that in Table 3.8. As in Section 3.2.2, this table is in exactly the form that a statistical package produced.

The analysis of deviance works in a similar way to the ANOVA. Here we will consider how to interpret the results of analysis of deviance. No attempt will be made to explain the underlying statistical theory; the relevant theory and derivations can be found in [368, chapter 6].

The change in deviance for adding the interaction between *inflam* and *appear* is 95.4381 (from Table 3.8). The change in deviance is the counterpart of the extra

Table 3.7. Incidence of breast cancer at three different health centres. The data are taken from [391].

			State of Tumour			
			Minimal Inflammation		Greater Inflammation	
			Malignant	Benign	Malignant	Benign
Centre	Age	Survived	Appearance	Appearance	Appearance	Appearance
Tokyo	Under 50	No	9	7	4	3
		Yes	26	68	25	9
	50–69	No	9	9	11	2
		Yes	20	46	18	5
	70 or over	No	2	3	1	0
		Yes	1	6	5	1
Boston	Under 50	No	6	7	6	0
		Yes	11	24	4	0
	50–69	No	8	20	3	2
		Yes	18	58	10	3
	70 or over	No	9	18	3	0
		Yes	15	26	1	1
Glamorgan	Under 50	No	16	7	3	0
		Yes	16	20	8	1
	50–69	No	14	12	3	0
		Yes	27	39	10	4
	70 or over	No	3	7	3	0
		Yes	12	11	4	1

sum of squares. For the ANOVA, we take the extra sum of squares and divide by its degrees of freedom to obtain an estimate of the variance, which we compare with another estimate of variance (by forming the F-statistic and finding its p-value). For the analysis of deviance we do not need to find an estimate of variance, because the approximate distribution of the change in deviance is known.

If the *inflam:appear* interaction is superfluous, then the change in deviance will have a distribution that is approximately chi-squared[3], with one degree of freedom. The number of degrees of freedom is read from the same row of Table 3.8 as the change in deviance. So, by comparing 95.4381 with the χ_1^2 distribution and noting that 95.4381 is incredibly high for an observation from that distribution, we can say that the change in deviance is massively higher than it should be if the

[3] The chi-squared distribution is a well-known continuous distribution. Most statistical packages can generate the associated probabilities, which are tabulated in many elementary statistics books. If $X_1, X_2, \ldots, X_\nu$ are statistically independent standard normal random variables, then the sum

$$(X_1^2 + X_2^2 + \cdots + X_\nu^2)$$

will have an chi-squared distribution with ν degrees of freedom. The chi-squared distribution with ν degrees of freedom is often written χ_ν^2.

Table 3.8. An analysis of deviance table for fitting log-linear models to the data in Table 3.7.

Terms added sequentially (first to last)

	Df	Deviance	Resid. Df	Resid. Dev	Pr(Chi)
NULL			71	860.0076	
centre	2	9.3619	69	850.6457	0.0092701
age	2	105.5350	67	745.1107	0.0000000
survived	1	160.6009	66	584.5097	0.0000000
inflam	1	291.1986	65	293.3111	0.0000000
appear	1	7.5727	64	285.7384	0.0059258
centre:age	4	76.9628	60	208.7756	0.0000000
centre:survived	2	11.2698	58	197.5058	0.0035711
centre:inflam	2	23.2484	56	174.2574	0.0000089
centre:appear	2	13.3323	54	160.9251	0.0012733
age:survived	2	3.5257	52	157.3995	0.1715588
age:inflam	2	0.2930	50	157.1065	0.8637359
age:appear	2	1.2082	48	155.8983	0.5465675
survived:inflam	1	0.9645	47	154.9338	0.3260609
survived:appear	1	9.6709	46	145.2629	0.0018721
inflam:appear	1	95.4381	45	49.8248	0.0000000

inflam:appear interaction is redundant, therefore we can say that the interaction between *inflam* and *appear* ought to be included in the model. (The significance probability is presented in the final column of the table and turns out to be less than $5 \times 10^{-6}\%$, that is, zero to seven decimal places.)

Applying this procedure to each row of the analysis of deviance table leads me to conclude that the following interactions can be omitted from the model: `age:survived`, `age:inflam`, `age:appear` and `survived:inflam`. Strictly, we should be more careful about this, as the changes in deviance vary with the order in which the terms are included. It turns out in this particular case that this quick method leads to the answer that I would have got by a more careful approach.

The model that I would select contains all five main effects, all the two-way interactions involving *centre*, the interaction between *appear* and *inflam*, and the interaction between *appear* and *survived*. This means that: all the categorical variables are associated with *centre*; *appear* and *inflam* are associated; *appear* and *survived* are associated. To summarise this model, I would construct its conditional independence graph and present tables corresponding to the interactions.

The conditional independence graph is shown in Figure 3.2. There is an arc linking each pair of variables that appear in an interaction in the chosen model. Using the graph in Figure 3.2, I can say things like

> *'If I know the values of* centre *and* appear, *then knowing the values of* age *and* inflam *tells me nothing more about* survived.*'*

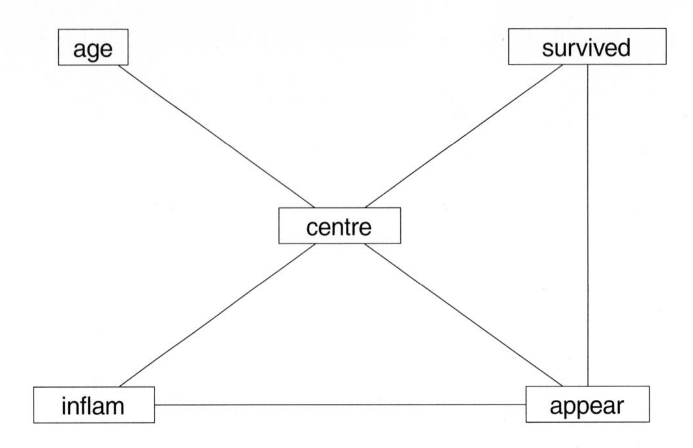

Fig. 3.2. The conditional indepence graph of the model selected for the data in Table 3.7.

because the routes from *age* and *inflam* to *survived* all pass through either *centre* or *appear*.

The frequency tables corresponding to interactions in the model are given in Tables 3.9, 3.10, 3.11, 3.12, 3.13 and 3.14. From these tables we can make statements such as

> *'The three-year survival rate in Tokyo is much bigger than in Boston or Glamorgan.'*

based on Table 3.12. By only looking at the tables corresponding to interactions that are in the selected model, we can be confident that the patterns we see in these tables are statistically significant and not simply due to chance.

The log-linear model gives us a systematic way to investigate the relationships between categorical variables. If there is an interaction in a log-linear model, this implies that the variables involved in the interaction are not independent, in other words they are related. In the particular example considered here, one of the variables, *survived*, can be considered to be an output, so we are primarily interested in associations with that variable. There is no requirement for one of the categorical variables to be considered an output. On the other hand, if one of the variables is an output, then we might prefer to model it directly; if the output is binary (has exactly two categories), as *survived* is, we can use *logistic regression* to model it directly. We take up logistic regression in Section 3.2.4. A more detailed discussion of log-linear models can be found in [368, chapter 6].

3.2.4 Logistic Regression

An alternative to using a log-linear model to analyse the breast cancer data in Table 3.7 (on page 84) would be to use logistic regression.

Table 3.9. Frequency table corresponding to the *age* by *centre* interaction for the data in Table 3.7.

	Tokyo	Boston	Glamorgan
Under 50	151	58	71
50–69	120	122	109
70 or over	19	73	41

Table 3.10. Frequency table corresponding to the *inflam* by *centre* interaction for the data in Table 3.7.

	Tokyo	Boston	Glamorgan
Minimal	206	220	184
Greater	84	33	37

Table 3.11. Frequency table corresponding to the *appear* by *centre* interaction for the data in Table 3.7.

	Tokyo	Boston	Glamorgan
Malignant	131	94	119
Benign	159	159	102

Table 3.12. Frequency table corresponding to the *survived* by *centre* interaction for the data in Table 3.7.

	Tokyo	Boston	Glamorgan
No	60	82	68
Yes	230	171	153

Table 3.13. Frequency table corresponding to the *survived* by *appear* interaction for the data in Table 3.7.

	Malignant	Benign
No	113	97
Yes	231	323

Table 3.14. Frequency table corresponding to the *inflam* by *appear* interaction for the data in Table 3.7.

	Malignant	Benign
Minimal	222	388
Greater	122	32

In logistic regression, the output is the number of successes out of a number of trials, each trial resulting in either a success or failure. For the breast cancer data, we can regard each patient as a 'trial', with success corresponding to the patient surviving for three years.

Instead of the 72 frequencies given in Table 3.7, we would have 36 pairs, each pair being the number surviving to three years and the total (surviving plus dying) number of patients. The first pair would be the 26 survivors out of the 35 patients who were in Tokyo, under 50 years old, and had a malignant tumour with minimal inflammation. Alternatively, the output could simply be given as the number of successes, either 0 or 1, for each of the 764 patients involved in the study. This is a simpler approach and is easy to do in the statistical package that I have available, so we will do this. This approach is also more stable numerically.

The model that we will fit is $P(y_j = 0) = 1 - p_j$, $P(y_j = 1) = p_j$ and

$$\log\left(\frac{p_j}{1 - p_j}\right) = \beta_0 + \beta_1 x_{1j} + \beta_2 x_{2j} + \cdots + \beta_n x_{nj}. \tag{3.22}$$

Again, the inputs here will be indicators for the breast cancer data, but this is not generally true; there is no reason why any of the inputs should not be quantitative. The function $\log(p/(1-p))$ is often written $\text{logit}(p)$ by statisticians and referred to as the *logit* (of p). There are various reasons for modelling the logit, rather than the probability of success. Most of these reasons are mathematical, the simplest being that:

- using the logit prevents us from predicting success probabilities that are not between 0 and 1;
- if success has a higher probability than failure, then the logit is positive, otherwise it is negative;
- swapping the labels between success and failure simply changes the sign of the logit.

There are deeper mathematical reasons for using the logits; there are also statistical justifications based on the idea that some patients will be more robust than others—the technical name for this is the *tolerance distribution* of the patients. These deeper mathematical and statistical reasons are discussed by [123] and [368].

For a binary output, like this, the expected value is just the probability of the output being 1; as a formula this is $p_j = \mu_j$. Thus, (3.22) does conform to the form of the generalized linear model given at (3.6).

To investigate the relationship between three-year survival and the other variables we fit a logistic regression model that has 36 β's (as there are 36 possible combinations of *centre*, *age*, *inflam* and *appear*) and try to identify those that could be assumed to be zero. As with log-linear modelling, this decision is made using an analysis of deviance table. The relevant table is shown in Table 3.15.

Looking at the significance probabilities in Table 3.15 tells us that we do not need any interaction terms at all, nor do we need *inflam*, as long as we have the three main effects, *centre*, *age* and *appear*. This can be deduced because

Table 3.15. An analysis of deviance table for fitting a logistic regression to the data in Table 3.7.

	Df	Deviance	Resid. Df	Resid. Dev	Pr(Chi)
NULL			763	898.5279	
centre	2	11.26979	761	887.2582	0.0035711
age	2	3.52566	759	883.7325	0.1715588
appear	1	9.69100	758	874.0415	0.0018517
inflam	1	0.00653	757	874.0350	0.9356046
centre:age	4	7.42101	753	866.6140	0.1152433
centre:appear	2	1.08077	751	865.5332	0.5825254
centre:inflam	2	3.39128	749	862.1419	0.1834814
age:appear	2	2.33029	747	859.8116	0.3118773
age:inflam	2	0.06318	745	859.7484	0.9689052
appear:inflam	1	0.24812	744	859.5003	0.6184041
centre:age:appear	4	2.04635	740	857.4540	0.7272344
centre:age:inflam	4	7.04411	736	850.4099	0.1335756
centre:appear:inflam	2	5.07840	734	845.3315	0.0789294
age:appear:inflam	2	4.34374	732	840.9877	0.1139642
centre:age:appear:inflam	3	0.01535	729	840.9724	0.9994964

all the significance probabilities from *inflam* to the bottom of the table are relatively large. Therefore, the changes in deviance produced by adding these terms conform to what we would expect to see if these terms have no impact upon the output. Hence, we discard these terms, because we make the assumption that each term has no impact upon the output, unless we obtain evidence that it does. The evidence that we seek is a change in deviance that is not consistent with that term having no impact on the output; Table 3.15 indicates that adding *appear* to the set of inputs produces such a change in deviance—the significance probability for *appear* is small.

From here, we should check whether we need to keep *age* in the model, if *appear* and *centre* are in the model, and whether we need to keep *centre*, if *appear* and *age* are in the model. (We know that *appear* is needed, because the order of fitting in Table 3.15 adds *appear* to the model after *centre* and *age*.) Not surprisingly, it turns out that *age* is superfluous. Notice how the selected model matches the conditional independence graph in Figure 3.2 on page 86; the node for *survived* is linked to two other nodes, *centre* and *appear*.

The fitted model is simple enough in this case for the parameter estimates to be included here; they are shown in the form that a statistical package would present them in Table 3.16. Using the estimates given in Table 3.16, the fitted model is

$$\mathrm{logit}(p_j) = 1.080257 - 0.6589141x_{Bj} - 0.4944846x_{Gj} + 0.5157151x_{aj}. \quad (3.23)$$

Here x_{Bj} is 1 if patient j was treated at the Boston centre, and 0 otherwise, that is, x_{Bj} is the indicator variable for *centre* being Boston (which is the second centre, hence the labelling in Table 3.16). Similarly, x_{Gj} is the indicator for *centre*

Table 3.16. Parameter estimates for the logistic regression model selected for the data in Table 3.7, in the form that a statistical package would present them.

```
Coefficients:
 (Intercept)    centre2    centre3    appear
    1.080257 -0.6589141 -0.4944846 0.5157151
```

being Glamorgan and x_{aj} is the indicator for *appear* being Benign. So, for a patient in Tokyo with a tumour that appears benign, the estimated logit of the probability of survival to three years is

$$1.080257 - 0 - 0 + 0.5157151 = 1.595972$$

so, solving $\log\left(\frac{p}{1-p}\right) = 1.595972$ for p, the estimated three-year survival probability is

$$\frac{1}{1 + e^{-1.595972}} = 0.8314546.$$

As we have two different ways to analyse the data in Table 3.7, how should we decide whether to use log-linear modelling or logistic regression? The choice really depends on our aims. If the overriding interest is in modelling patient survival, then the logistic regression is more natural. On the other hand the logistic regression does not tell us that *centre* and *age* are related, because this does not help in the prediction of survival. The log-linear model tells us about relationships between all the variables, not just which ones are related to *survived.*

Other differences that determine whether to use logistic regression or log-linear modelling include:

- the output has to be binary for a logistic regression, but could take, for example, one of the four values *none*, *mild*, *moderate* or *severe*, in a log-linear model;
- the variables in a log-linear model have to be categorical, but need not be in a logistic regression, so the patient's age in the breast cancer data had to be converted from the actual age in years to the three categories for use in the log-linear model.

For an extensive description of logistic regression and other ways of analysing binary data see [123].

3.2.5 Analysis of Survival Data

Analysis of survival data is not strictly within the framework of generalized linear models. This sort of analysis can be carried out using the same numerical techniques as generalized linear models, so this is a natural place to describe it.

Survival data are data concerning how long it takes for a particular event to happen. In many medical applications the event is death of a patient with an illness, and so we are analysing the patient's survival time. In industrial applications the event is often failure of a component in a machine.

The output in this sort of problem is the survival time. As with all the other problems that we have seen in this section, the task is to fit a regression model to describe the relationship between the output and some inputs. In the medical context, the inputs are usually qualities of the patient, such as age and sex, or are determined by the treatment given to the patient.

There are two main characteristics of survival data that make them different from the data in the regression problems that we have already seen.

Censoring For many studies, the event has not necessarily happened by the end of the study period. So, for some patients in a medical trial, we might know that the patient was still alive after five years, say, but do not know when the patient died. This sort of observation would be called a *censored* observation. So, if the output is censored, we do not know the value of the output, but we do have some information about it.

Time Dependent Inputs Since collecting survival data entails waiting until the event happens, it is possible for the inputs to change in value during the wait. This could happen in a study of heart disease in which it is likely that the patient has been given medical advice to, for example, change diet or stop smoking. If the patient takes the advice, then things such as the patient's weight and blood cholesterol level are likely to change; so there would be a difficulty in deciding which value of weight to use if we wanted to model survival time using weight as an input.

These difficulties lead us to concentrate upon the *hazard* function and the *survivor* function for this sort of data. The survivor function is the probability of survival time being greater than time t. So, if the output is y, then the survivor function is

$$S(t) = P(y \geq t) = 1 - F(t)$$

where $F(t)$ is the cumulative distribution function of the output. The hazard function is the probability density of the output at time t conditional upon survival to time t, that is,

$$h(t) = f(t)/S(t),$$

where $f(t)$ is the (marginal) probability density of the output.

The hazard function indicates how likely failure is at time t, given that failure has not occurred before time t. So, for example, suppose the output is the time at which your car breaks down, measured from its last service. I would expect the hazard function to be high for the first few days after servicing (break-down due to a mistake during servicing), then it would drop, as the car (having survived the first few days) would be in good condition; after that I would anticipate a gradual rise in hazard as the condition of the car deteriorates over time, assuming no maintenance is carried out whilst the car is still running unless a service is scheduled. Notice that the hazard of break-down at, say, two years after servicing would be very high for most types of car, but the probability density would be low, because most cars would not survive to two years; the hazard is only looking at the cars that have survived to time t and indicating the probability of their imminent failure. Most modelling of survival data is done using a *proportional-hazards model.* A proportional-hazards model assumes that the hazard function is of the form

$$h(t) = \lambda(t) \exp\{G(\boldsymbol{x}, \boldsymbol{\beta})\}, \tag{3.24}$$

where $\boldsymbol{\beta}$ is a vector of parameters to be estimated, G is a function of arbitrary, but known, form, and $\lambda(t)$ is a hazard function in its own right. The multiplier of $\lambda(t)$ is written in exponential form, because the multiplier must be greater than zero. The $\lambda(t)$ hazard function is called the *baseline hazard.* This is called a *proportional-hazard* model, because the hazard functions for two different patients have a constant ratio for different values of t (as long as the inputs are not time-dependent).

The conventional choice for G, for patient j, is

$$G(\boldsymbol{x}_j, \boldsymbol{\beta}) = \beta_0 + \sum_{i=1}^{n} \beta_i x_{ij}. \tag{3.25}$$

This choice of G means that the effect (on hazard) of each input is multiplicative. So, for example, if x_{1j} were an indicator for whether patient j smoked, and $\beta_1 = 1$, then we would be able to say that smoking increased the hazard by a factor of $\exp(1) = 2.718$, that is, the hazard for a smoker would be 2.718 times that for a non-smoker. The simple interpretation of the β's in this model and the plausibility (to statisticians) of the effects being multiplicative has made the fitting of proportional-hazard models the primary method of analysing survival data.

The choice of G in (3.25) has another massive advantage. If we know, or are prepared to assume, the baseline hazard, $\lambda(t)$, then it turns out that the log likelihood can be converted to the same form as that of a generalized linear model, with a Poisson output and a log link function; this is the same form as the log-linear model in (3.21) on page 83. The log-linear model implied by this conversion of the log likelihood has the same inputs and β's as in (3.25), but the output is the indicator w_j, which is 1 if the jth survival time is not censored, and 0 if it is. This means that there is a method for estimating the β's, which is available in any proprietary statistical package that can fit generalized linear models; all the major statistical packages can do this.

As an example of this fitting, we could assume that the baseline hazard was

$$\lambda(t) = \alpha t^{\alpha - 1},$$

which would mean that the survival times had a distribution called the *Weibull* distribution. So, we would start by assuming $\alpha = 1$ and then construct our artificial log-linear model. We would then fit the log-linear model and obtain estimates of the β's based on the assumption that $\alpha = 1$. We then assume that the estimates of the β's are in fact the true values, and find the maximum likelihood estimate of α under this assumption. We then assume the estimate of α is the true value of α, and construct another log-linear model to obtain updated estimates of the β's, which we then use to update the estimate of α. This process continues until convergence of the estimates of α and the β's (which will happen and will be at the global optimum; see, for example [368, chapter 13]).

In the industrial context, it is reasonable to believe that the baseline hazard is known. Experiments to determine its form could be carried out in most industrial contexts, because the values of the inputs are likely to be controllable. In the medical context, it would be neither ethical nor practical to carry out an experiment to determine the form of the baseline hazard function. To overcome this problem, we can use *Cox's proportional-hazards model*, which was introduced by [128].

Cox's proportional-hazards model allows the baseline hazard to be arbitrary. The key idea for this model is to base the estimation of the β's upon the likelihood derived from knowing the order in which patients died (or were censored), rather than the likelihood derived from the actual survival (or censoring) times. This discards information, but the information it discards is related to the baseline hazard, and the resulting likelihood does not depend upon the baseline hazard. We have to accept this loss of information, if we do not know and are unwilling to assume what form the baseline hazard takes. Again, it turns out that the likelihood for Cox's proportional-hazards model can be converted into that of an artificial log-linear model (which is not the same as the one referred to already). Fitting this artificial log-linear model produces estimates for the β's amongst other things. Cox's proportional-hazards model is used in virtually all medical analyses of survival data, sometimes regardless of whether it is appropriate.

Details of the likelihoods and references to even more detailed discussions of this sort of analysis can be found in [368, chapter 13].

3.3. Special Topics in Regression Modelling

The topics in this section are special in the sense that they are extensions to the basic idea of regression modelling. The techniques have been developed in response to methods of data collection in which the usual assumptions of regression modelling are not justified.

Multivariate analysis of variance concerns problems in which the output is a vector, rather than a single number. One way to analyse this sort of data would be to model each element of the output separately, but this ignores the possible relationship between different elements of the output vector. In other words, this analysis would be based on the assumption that the different elements of the output vector are not related to one another. Multivariate analysis of variance is a form of analysis that does allow for different elements of the output vector to be correlated to one another. This technique is outlined in Section 3.3.1.

Repeated measures data arise when there is a wish to model an output over time. A natural way to perform such a study is to measure the output and the inputs for the same set of individuals at several different times. This is called taking repeated measurements. It is unrealistic to assume (with no evidence) that the output measurements taken on a particular individual at different times are unrelated. Various ways of coping with the relationship between output recorded on the same individual are presented in Section 3.3.2.

Random effects modelling is described in Section 3.3.3 and also as one approach to analysing repeated measures data in Section 3.3.2. Amongst other things, this technique allows (the coefficients of) the fitted regression model to vary between different individuals.

3.3.1 Multivariate Analysis of Variance

In Section 3.2.2, we saw how to test whether particular β's of the linear model of (3.1) on page 71 were zero. The technique used was called analysis of variance, or ANOVA.

Sometimes the output in a data set is a vector of variables rather than a single variable. We will see a special case of this in Section 3.3.2, where the vector of variables consists of the same quantity, blood pressure say, at several different times, recorded on the same individual. The output variables need not be the same quantity; they could be something like heights and weights for a set of children.

Given this sort of data, we might be able to analyse it using a multivariate linear model, which is

$$\underset{(c\times 1)}{\boldsymbol{y}_j} = \boldsymbol{\beta}_0 + \boldsymbol{\beta}_1 x_{1j} + \boldsymbol{\beta}_2 x_{2j} + \cdots + \boldsymbol{\beta}_n x_{nj} + \boldsymbol{\varepsilon}_j \quad j = 1, 2, \ldots, m \tag{3.26}$$

where the $\boldsymbol{\varepsilon}_j$'s are independently and identically distributed as[4] $\mathcal{N}^c(\mathbf{0}, \Sigma)$ and m is the number of data points. The '$(c\times 1)$' under '$\boldsymbol{y}_j$' indicates the dimensions of the vector, in this case c rows and 1 column; the $\boldsymbol{\beta}$'s are also $(c \times 1)$ vectors.

This model can be fitted in exactly the same way as a linear model (by least squares estimation). One way to do this fitting would be to fit a linear model to each of the c dimensions of the output, one-at-a-time.

Having fitted the model, we can obtain fitted values

$$\hat{\boldsymbol{y}}_j = \hat{\boldsymbol{\beta}}_0 + \sum_{i=1}^{n} \hat{\boldsymbol{\beta}}_i x_{ij} \quad j = 1, 2, \ldots, m$$

and hence residuals

$$\boldsymbol{y}_j - \hat{\boldsymbol{y}}_j \quad j = 1, 2, \ldots, m.$$

The analogue of the residual sum of squares from the (univariate) linear model is the matrix of residual sums of squares and products for the multivariate linear

[4] The notation $\mathcal{N}^c(\boldsymbol{\mu}, \Sigma)$ denotes the c-dimensional normal distribution. If $\boldsymbol{w} \sim \mathcal{N}^c(\boldsymbol{\mu}, \Sigma)$, then $\boldsymbol{w}$ is a $(c \times 1)$ vector, whose elements are each normally distributed. The expected values of the elements of $\boldsymbol{w}$ are the corresponding elements of $\boldsymbol{\mu}$, which is also a $(c \times 1)$ vector. The variance of element i of $\boldsymbol{w}$ is element (i, i) of the $(c \times c)$ variance matrix, Σ; the covariance between elements i and k of $\boldsymbol{w}$ is element (i, k) of Σ.

model. This matrix is defined to be

$$R = \sum_{j=1}^{m} (\boldsymbol{y}_j - \hat{\boldsymbol{y}}_j)(\boldsymbol{y}_j - \hat{\boldsymbol{y}}_j)^\top .$$

The R matrix has the residual sum of squares for each of the c dimensions stored on its leading diagonal. The off-diagonal elements are the residual sums of cross-products for pairs of dimensions.

The residual degrees of freedom is exactly the same as it would be for a univariate linear model with the same set of inputs, namely m minus the number of linearly independent inputs. If we call the residual degrees of freedom ν, then we can estimate Σ by $(1/\nu)R$.

If we wish to compare two nested linear models, to determine whether certain $\boldsymbol{\beta}$'s are equal to the zero vector, $\mathbf{0}$, then we can construct an extra sums of squares and products matrix in the same way as we constructed the extra sum of squares in Section 3.2.2 on page 78. In other words, if we have model Ω_0 which is a special case of model Ω_1, and the residual sums of squares and products matrices for these models are R_0 and R_1, respectively, then the extra sums of squares and products matrix is

$$R_E = R_0 - R_1 .$$

Thus, in the same way that we can form an ANOVA table, which shows the effect of introducing new inputs into the model, we can build a multivariate ANOVA, or *MANOVA*. Instead of the sums of squares of the ANOVA table, the MANOVA table contains sums of squares and products matrices.

The only sensible choices for the test statistic to compare models Ω_0 and Ω_1 are functions of the eigenvalues of the matrix

$$R_1^{-1} R_E ,$$

where R_1^{-1} is the matrix inverse of R_1. By 'sensible', we mean that the test statistic does not change if the co-ordinate origin (for the output vectors) is changed, nor if the output variables are rescaled; this is demonstrated in [21]. There are four commonly used test statistics. If the dimensionality of the output is $c = 1$, then these statistics are all the same as the F-statistic of the ANOVA. Further, they are all equivalent to one another if the extra degrees of freedom is 1.

The names of the four commonly used test statistics are: *Roy's greatest root*; the *Lawley-Hotelling trace*; the *Pillai trace*; *Wilks' lambda*. Discussions of which one to use under what circumstances can be found in [326, section 13.2] and [112, section 8.3.3]. Most statistical packages that can perform MANOVA will produce these four statistics and the corresponding significance probabilities.

Having performed a MANOVA and discovered some $\boldsymbol{\beta}$'s are not equal to $\mathbf{0}$, we usually want to say which of the inputs is related to which element of the output, and how. If the input is a quantitative variable, then it is straightforward to identify the relationship, simply by considering the corresponding $\boldsymbol{\beta}$. If the input is one of a set of indicator variables corresponding to a factor, then we need to consider the set of β's corresponding to the factor. This is harder, because there are too many different contributions in the β's for someone to grasp. In this case, we would use a technique called *canonical variates analysis*, to display the mean output for each level of the factor. Canonical variates analysis will not be described here, but descriptions of this technique can be found in many textbooks on multivariate analysis, for example [326] and [112]. One way to think of canonical variates analysis is that it is the same as principal components analysis (see Section 3.4.1) applied to the mean outputs for each distinct level of the factor. Though this is not strictly true, it conveys the basic idea, which is that of trying to focus on the differences between the means.

3.3.2 Repeated Measures Data

Repeated measures data are generated when the output variable is observed at several points in time, on the same individuals. Usually, the covariates are also observed at the same time points as the output; so the inputs are time-dependent too. Thus, as in Section 3.3.1 the output is a vector of measurements. In principle, we can simply apply the techniques of Section 3.3.1 to analyse repeated measures data. Instead, we usually try to use the fact that we have the same set of variables (output and inputs) at several times, rather than a collection of different variables making up a vector output.

Repeated measures data are often called *longitudinal data*, especially in the social sciences. The term *cross-sectional* is often used to mean 'not longitudinal'.

There are several ways to analyse this sort of data. As well as using the MANOVA, another straightforward approach is to regress the outputs against time for each individual and analyse particular parameter estimates, the gradient say, as though they were outputs. If the output variable were measured just twice, then we could analyse the difference between the two measurements, using the techniques that we saw in Section 3.2.

The three most common ways to analyse longitudinal data, at least in the statistical research literature, are *marginal* modelling, *subject-specific* modelling and *transition* modelling.

Marginal modelling is what we have been considering in the regression modelling that we have already seen. We have a model, such as (3.26) on page 94. The primary complication is that it is unrealistic to assume that there is no correlation between outputs from the same individual. We could make no assumptions about the correlation structure, as we did for the MANOVA, but usually there is reason to believe that the variance matrix has a particular form. For example, if the individuals are (adult) patients, then we might expect the correlation between the outputs in weeks 1 and 2 to be the same as that between outputs in weeks 5 and 6 (because the pairs have the same time lapse between them), with the correlation between outputs for weeks 2 and 5 being smaller.

Subject-specific modelling is regression modelling in which we allow some of the parameters to vary between individuals, or groups of individuals. An example of this sort of model is

$$y_{jt} = \beta_0 + U_j + \beta_1 x_{1jt} + \beta_2 x_{2jt} + \cdots + \beta_n x_{njt} + \varepsilon_{jt} \quad j = 1, 2, \ldots, m, \tag{3.27}$$

where t indexes the observation time, the ε_{jt}'s are independently and identically distributed as $\mathcal{N}(0, \sigma^2)$, the U_j's are independently and identically distributed as $\mathcal{N}(0, \sigma_U^2)$ and m is the number of individuals.
The model in (3.27) is called a *random-intercept* model, because it is equivalent to a standard model with the intercept (β_0) replaced by an intercept for each individual ($\beta_0 + U_j$). This random-intercept model does not have a correlation structure to specify, as was required for the marginal model. Instead, the relationship between outputs from the same individual (at different times) is assumed to be represented by the U_j's being constant across time, but different between individuals. If U_j happens to be high, then all the outputs for individual j will be expected to be higher than the average across all individuals.
Other subject-specific models allow the other β's, not just the intercept, to vary between individuals, or allow things like a random intercept for, say, the clinic an individual attends, or the district an individual lives or works in.
Subject-specific models are a type of model known as random effects models; see Section 3.3.3.

Transition models allow the output to depend on outputs at previous time points. Thus, the relationship between outputs from the same individual is allowed for by allowing outputs from the past to be used as inputs for the output in the present. A simple example of a transition model is

$$y_{jt} = \alpha y_{j(t-1)} + \beta_0 + \beta_1 x_{1jt} + \beta_2 x_{2jt} + \cdots + \beta_n x_{njt} + \varepsilon_{jt} \quad j = 1, 2, \ldots, m, \tag{3.28}$$

where t indexes the observation time, the ε_{jt}'s are independently and identically distributed as $\mathcal{N}(0, \sigma^2)$ and m is the number of individuals.
As an example, suppose y_{jt} is a measurement of the progression of a disease. The model at (3.28) implies that the progression of the disease at time t is related to the progression at the previous time point $(t-1)$ and to the values of a set of inputs at time t.
The dependence on $y_{j(t-1)}$ can be far more complicated than in (3.28). For example, a much more general dependence would be to allow the β's to depend on $y_{j(t-1)}$. This would be difficult if the output were a continuous quantity, but might even be considered natural if the output were binary, as in the models in Section 3.2.4.

Descriptions of these three types of models, along with details on how to fit them using maximum likelihood estimation, practical examples of the their use and bibliographic notes, can be found in [155].

The choice of which of these types of models to use boils down to what the aims of the study are and what is believed about the underlying mechanism of change in the output. The implications of these choices, and hence reasons for choosing one type of model rather than another are also given in [155], but, as an example, a marginal model is appropriate when you want to make public health statements like

Reduce your blood pressure by eating less salt.

which apply on average. This sort of public health advice is not necessarily appropriate to everybody in a population, but if everyone followed the advice then the *average* blood pressure across the population would be reduced; for some people eating less salt might make no difference to their blood pressure and there might even be individuals whose blood pressure actually increases. If you wanted to give health advice on an individual basis, then you would want to use a subject-specific model and estimate random effects, such as the U_j's in (3.27), for each individual. This would allow you give different advice to different people, each person getting advice appropriate to themselves; the advice would vary because people vary.

3.3.3 Random Effects Models

We have seen the use of random effects for longitudinal data in the subject-specific model described in Section 3.3.2. The random effects in the subject-specific model are used as a way to cope with the fact that outputs from one individual are likely to be more similar to one another than outputs from different individuals are.

There are several other situations in which random effects could be included in a model. Some of the main uses of random effects in modelling are given below.

Overdispersion Overdispersion is the phenomenon of the observed variability of the output being greater than the fitted model predicts. This can be detected for some generalized linear models; for example, overdispersion can be detected for models in which the output has a binomial distribution. The reason why overdispersion can be detected for a binomial output is that the variance is a function of the expected value for the binomial distribution. Thus, the model happens to predict the variance as well as the expected value, so we can compare the sample variance with that predicted by the model.

There are two main ways in which overdispersion might arise, assuming that we have identified the correct output distribution. One way is that we failed to record a variable that ought to have been included as an input. The reasons for not recording the missing input might include: time or expense of recording it; an incorrect belief that it was not relevant to the study; we do not actually know how to measure it. The second way is that the expected output for an individual is not entirely determined by the inputs, and that there is variation in the expected output between different individuals who have the same values for the inputs.

In either case, the solution is to allow a random effect into the linear part of the model. In a logistic regression we might replace (3.22) from page 88 with

$$\text{logit}(p_j) = \beta_0 + \beta_1 x_{1j} + \beta_2 x_{2j} + \cdots + \beta_n x_{nj} + U_j, \tag{3.29}$$

where the U_j's are independently and identically distributed as $\mathcal{N}(0, \sigma_U^2)$. We can think of U_j as representing either the effect of the missing input on p_j or simply as random variation in the success probabilities for individuals that have the same values for the input variables.

The topic of overdispersion is discussed in detail by [123, chapter 6].

Hierarchical Models We have seen a hierarchical model already. Equation (3.8) on page 75 has three parameters, β_{II}, β_{III} and β_{IV}, that concern differences between groups of individuals, in this case the suitability of different parts of a field for growing turnips. The turnip experiment of Example 3.1 features four groups of plots. Plots in the same group are physically close to one another, so they are assumed to have similar characteristics to one another. The different groups correspond to different pieces of land, so we allow for the possibility that plots in different groups might have different characteristics (in terms of suitability for growing turnips).

This allocation of individuals into groups, so as to put similar individuals in the same group as one another, is called *blocking*. The groups are referred to as *blocks*.

The hierarchy which gives its name to hierarchical models is, for the turnip experiment, made up of two levels, the upper level being the four blocks and the lower level being the plots within the blocks. It takes little imagination to envisage the extension of the hierarchy to a third level, different farms say, so that there are groups of blocks corresponding to different farms and groups of plots corresponding to blocks. To move away from the agricultural context, a medical hierarchy could be patients within wards within hospitals within health trusts; an educational hierarchy might be student within class within school within education authority.

In many cases the levels in the hierarchy consist of samples of groups taken from populations of groups. In the turnip experiment, the growth of the turnips is affected by the different blocks, but the effects (the β's) for each block are likely to be different in different years. So we could think of the β's for each block as coming from a population of β's for blocks. If we did this, then we could replace the model in (3.8) on page 75 with

$$\begin{aligned} y_j = \beta_0 &+ \beta_B x_{Bj} + \beta_C x_{Cj} + \cdots + \beta_R x_{Rj} \\ &+ b_{\text{I}} x_{\text{I},j} + b_{\text{II}} x_{\text{II},j} + b_{\text{III}} x_{\text{III},j} + b_{\text{IV}} x_{\text{IV},j} + \varepsilon_j \quad j = 1, 2, \ldots, 64 \end{aligned} \tag{3.30}$$

where b_{I}, b_{II}, b_{III} and b_{IV} are independently and identically distributed, each being $\mathcal{N}(0, \sigma_b^2)$. This is how random effects, like the b's can be used in hierarchical models. We can do the same thing in both the medical and educational scenarios given above.

We would not use random effects to replace β_B to β_R, because there is an implicit belief in this experiment that these are constants that will be the same in the future, that is, they are not random. In addition, they can be attributed to specific actions by the person growing the turnips, whereas the block effects are attributable to random fluctuations in the quality of the land, the weather, whether the field was flooded by the river, and so on.

For the turnip data using a random effects model like (3.30) makes no difference to our conclusions about the effects of the different treatments. This is because the turnip data were generated using a carefully designed experiment; one of the goals of the design used was to eliminate any dependence of the treatment comparisons upon σ_b^2. In other studies, a random effects model might be essential. There are two common reasons for requiring a random effects model. First, we actually want to estimate variances such as σ_b^2, either for their own sake, or because a treatment has been applied to the groups in a particular level, rather than to individuals in the lowest level. Secondly, the data may have been generated by a study in which we had less control over the values taken by the inputs than we had for the turnip experiment. This can lead to treatment effects whose estimates do depend on the variance of random effects, which we would want to take into account, or it might lead to situations in which the β's cannot be estimated without including random effects in the model. Examples of both these possibilities are given by [368, chapter 14].

3.4. Classical Multivariate Analysis

The techniques in this section are used for analysing samples that are in the form of observations that are vectors. They all have their roots in linear algebra and geometry, which is perhaps why these diverse techniques are grouped together under the heading of *multivariate analysis*.

Principal components analysis is a method of transforming the vector observations into a new set of vector observations. The goal of this transformation is to concentrate the information about the differences between observations into a small number of dimensions. This allows most of the structure of the sample to be seen by plotting a small number of the new variables against one another. A fairly detailed description of principal components analysis is given in Section 3.4.1.

The relationships between the rows and columns of a table of counts can be highlighted using correspondence analysis, which is described in Section 3.4.2. Despite having different aims from principal components analysis, it turns out that the mathematical derivation of correspondence analysis is very closely related to that of principal components analysis. Many of the techniques regarded as *multivariate analysis* are related to one another mathematically, which is another reason for techniques with disparate aims being grouped together under this heading.

There are situations in which the information available is in the form of the dissimilarities (or distances) between pairs of individuals, rather than the vectors corresponding to each individual. Multidimensional scaling is a set of techniques for constructing the set of vectors from the set of dissimilarities. Multidimensional scaling is described, along with a contrived example, in Section 3.4.3. The number of situations in which this technique is useful in practice is surprisingly high.

Cluster analysis is a collection of techniques for creating groups of objects. The groups that are created are called clusters. The individuals within a cluster are similar in some sense. An outline of the various different types of cluster analysis is given in Section 3.4.4, along with a short description of *mixture decomposition*. Mixture decomposition is similar to cluster analysis, but not the same. The difference is given in Section 3.4.4.

The relationships between the elements of the observation vectors can be investigated by the techniques in Section 3.4.5. This is a different emphasis from the other multivariate techniques presented here, as those techniques are trying to achieve goals in spite of the relationships between elements of the observation vectors. In other words, the structure being studied in Section 3.4.5 is regarded as a nuisance in most other multivariate analysis.

The main omission from this set of multivariate techniques is anything for the analysis of data in which we know that there is a group structure (and we know what that structure is, as well as knowing it exists). Discrimination, or *pattern recognition*, or *supervised classification*, is a way to model this sort of data, and is addressed in (parts of) Chapters 7, 8 and 9; for a statistical view of this topic see, for example [241]. Canonical variates analysis is a way to obtain a diagram highlighting the groups structure; see, for example [326] and [112].

3.4.1 Principal Components Analysis

Principal components analysis is a way of transforming a set of c-dimensional vector observations, $\boldsymbol{x}_1, \boldsymbol{x}_2, \ldots, \boldsymbol{x}_m$, into another set of c-dimensional vectors, $\boldsymbol{y}_1, \boldsymbol{y}_2, \ldots, \boldsymbol{y}_m$. The $\boldsymbol{y}$'s have the property that most of their information content is stored in the first few dimensions (features).

The idea is that this will allow us to reduce the data to a smaller number of dimensions, with low information loss, simply by discarding some of the elements of the $\boldsymbol{y}$'s. Activities that become possible after the dimensionality reduction include:

- obtaining (informative) graphical displays of the data in 2-D;
- carrying out computer intensive methods on reduced data;
- gaining insight into the structure of the data, which was not apparent in c dimensions.

One way that people display c-dimensional data, with the hope of identifying structure is the pairwise scatterplot. We will look at this method and see what its weaknesses are before going on to look at how principal components analysis overcomes these problems.

Figure 3.3 shows a pairwise scatterplot for a data set with $c = 4$ dimensions and $m = 150$ individuals. The individuals are iris flowers; the four dimensions

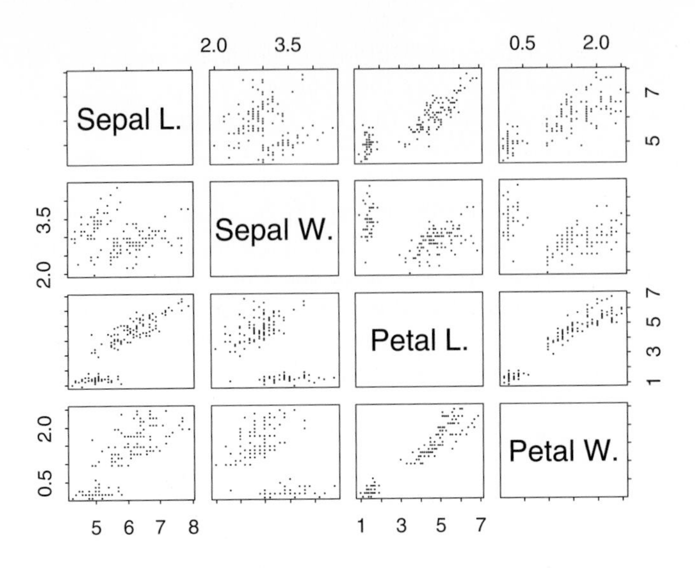

Fig. 3.3. Pairwise scatterplots for Fisher's Iris Data.

correspond to the sepal lengths and widths, and the petal lengths and widths of the flowers. It is known that there are three different species of iris represented in this set of data, but we are not using this information in the analysis. These data are very famous and are usually known as *Fisher's Iris Data*, even though they were collected by Anderson and described in [20], because they were used by Fisher in [180] (the first paper on linear discriminant analysis).

The pairwise scatterplot is constructed by plotting each variable against every other variable. The idea is to have all information available simultaneously. The main structure that can be identified in Figure 3.3 is that there is a distinct group of individuals with low values for both *Petal W.* and *Petal L.* (petal width and length).

The drawbacks with pairwise scatterplots are as follows.

- A high number of dimensions leads to a very complicated picture.
- It is often difficult to locate the same observation in several different panels (plots).
- There is an implicit assumption that the structure within the data is related to pairs of features. There might be other 2-D projections that are clearer.

There are ways to improve upon pairwise scatterplots using dynamic graphics to look at 3-D projections (this is called *spinning*) or to allow highlighting of individuals in all the panels simultaneously (this is called *brushing*). It is also possible to use symbols or *glyphs*, such as *Chernoff's faces* introduced by [115], as described in [326, pages 35–40]. Even though they are improvements, none of these techniques is particularly easy to use when searching for structure and they are all time consuming, requiring a human 'expert' to interpret them.

We need a method that will regularly reveal structure in a multivariate data set. Principal components analysis is such a method.

The main idea behind principal components analysis is that high information corresponds to high variance. So, if we wanted to reduce the $\boldsymbol{x}$'s to a single dimension we would transform $\boldsymbol{x}$ to $y = \boldsymbol{a}^\top \boldsymbol{x}$, choosing $\boldsymbol{a}$ so that y has the largest variance possible. (The length of $\boldsymbol{a}$ has to be constrained to be a constant, as otherwise the variance of y could be increased simply by increasing the length of $\boldsymbol{a}$ rather than changing its direction.) An alternative, but equivalent, approach is to find the line that is closest to the $\boldsymbol{x}$'s; here we measure the distance from the $\boldsymbol{x}$'s to a line by finding the squared perpendicular distance from each $\boldsymbol{x}_j$ to the line, then summing these squared distances for all the $\boldsymbol{x}_j$'s. Both approaches give the same answer; we will use the variance approach because this makes it simpler to write down the procedure for carrying out principal components analysis.

It is possible to show that the direction of maximum variance is parallel to the eigenvector corresponding to the largest eigenvalue of the variance (covariance) matrix of $\boldsymbol{x}$, Σ. It is also possible to show that of all the directions orthogonal to the direction of highest variance, the (second) highest variance is in the direction parallel to the eigenvector of the second largest eigenvalue of Σ. These results extend all the way to c dimensions. The eigenvectors of Σ can be used to construct a new set of axes, which correspond to a rotation of the original axes. The variance parallel to the first axis will be highest, the variance parallel to the second axis will be second highest, and so on.

In practice, we do not know Σ, so we use the sample variance (covariance), S, which is defined to be

$$\underset{(c \times c)}{S} = \frac{1}{m-1} \sum_{j=1}^{m} (\boldsymbol{x}_j - \overline{\boldsymbol{x}})(\boldsymbol{x}_j - \overline{\boldsymbol{x}})^\top, \tag{3.31}$$

where $\overline{\boldsymbol{x}} = \frac{1}{m} \sum_j \boldsymbol{x}_j$.

To specify the principal components, we need some more notation.

- The eigenvalues of S are

$$\lambda_1 \geq \lambda_2 \geq \cdots \geq \lambda_c \geq 0.$$

- The eigenvectors of S corresponding to $\lambda_1, \lambda_2, \ldots, \lambda_c$ are $\boldsymbol{e}_1, \boldsymbol{e}_2, \ldots, \boldsymbol{e}_c$, respectively.
  The vectors $\boldsymbol{e}_1, \boldsymbol{e}_2, \ldots, \boldsymbol{e}_c$ are called the *principal axes*. ($\boldsymbol{e}_1$ is the first principal axis, etc.)
- The $(c \times c)$ matrix whose ith column is $\boldsymbol{e}_i$ will be denoted as E.

The principal axes (can be and) are chosen so that they are of length 1 and are orthogonal (perpendicular). Algebraically, this means that

$$\boldsymbol{e}_i^\top \boldsymbol{e}_{i'} = \begin{cases} 1 & \text{if } i = i' \\ 0 & \text{if } i \neq i' \end{cases}. \tag{3.32}$$

The vector $\boldsymbol{y}$ defined as,

$$\underset{(c\times 1)}{\boldsymbol{y}} = \underset{(c\times c)}{\begin{bmatrix} \boldsymbol{e}_1^\top \\ \boldsymbol{e}_2^\top \\ \vdots \\ \boldsymbol{e}_c^\top \end{bmatrix}} \underset{(c\times 1)}{\boldsymbol{x}} = E^\top \boldsymbol{x}$$

is called the vector of *principal component scores* of $\boldsymbol{x}$. The ith principal component score of $\boldsymbol{x}$ is $y_i = \boldsymbol{e}_i^\top \boldsymbol{x}$; sometimes the principal component scores are referred to as the principal components.

The principal component scores have two interesting properties.

1. The elements of $\boldsymbol{y}$ are uncorrelated and the sample variance of the ith principal component score is λ_i. In other words the sample variance matrix of $\boldsymbol{y}$ is

$$\underset{(c\times c)}{\begin{bmatrix} \lambda_1 & & & 0 \\ & \lambda_2 & & \\ & 0 & \ddots & \\ & & & \lambda_c \end{bmatrix}}.$$

So an alternative way to view principal components analysis is as a method for obtaining a set of uncorrelated variables.

2. The sum of the sample variances for the principal components is equal to the sum of the sample variances for the elements of $\boldsymbol{x}$. That is,

$$\sum_{i=1}^{c} \lambda_i = \sum_{i=1}^{c} s_i^2,$$

where s_i^2 is the sample variance of x_i.

The principal components scores are plotted in Figure 3.4. Compare Figure 3.4 with Figure 3.3. In Figure 3.3 every panel exhibits two clusters and all four variables carry information about the two clusters. In Figure 3.4, on the other hand most of the information about these two clusters is concentrated in the first principal component, y_1. The other three principal components do not appear to carry any structure. One might wish to argue that the plots of y_3 and y_4 against y_2 suggest two overlapping clusters separated by the line $y_2 = -5.25$; this indicates that y_2 carries some structural information. The clear message of Figure 3.4 is that the structural information for these data is stored in at most two principal components, so we can achieve a dimensionality reduction from four dimensions down to two dimensions, with no important loss of information.

There are two more issues to consider before we finish our look at principal components analysis.

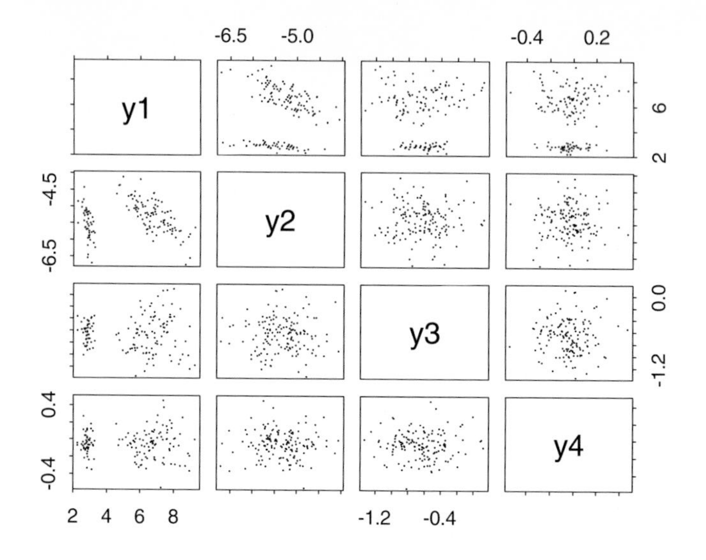

Fig. 3.4. Plot of the principal components for Fisher's Iris Data.

- How many of the principal components are needed to get a good representation of the data? That is, what is the effective dimensionality of the data?
- Should we normalise the data before carrying out principal components analysis?

Effective Dimensionality There are three main ways of determining how many principal components are required to obtain an adequate representation of the data.

1. **The proportion of variance accounted for** Take the first r principal components and add up their variances. Divide by the sum of all the variances, to give
$$\frac{\sum_{i=1}^{r} \lambda_i}{\sum_{i=1}^{c} \lambda_i}$$
which is called the *proportion of variance accounted for by the first r principal components.*
Usually, projections accounting for over 75% of the total variance are considered to be good. Thus, a 2-D picture will be considered a reasonable representation if
$$\frac{\lambda_1 + \lambda_2}{\sum_{i=1}^{c} \lambda_i} > 0.75.$$
2. **The size of *important* variance** The idea here is to consider the variance if all directions were equally important. In this case the variances would be

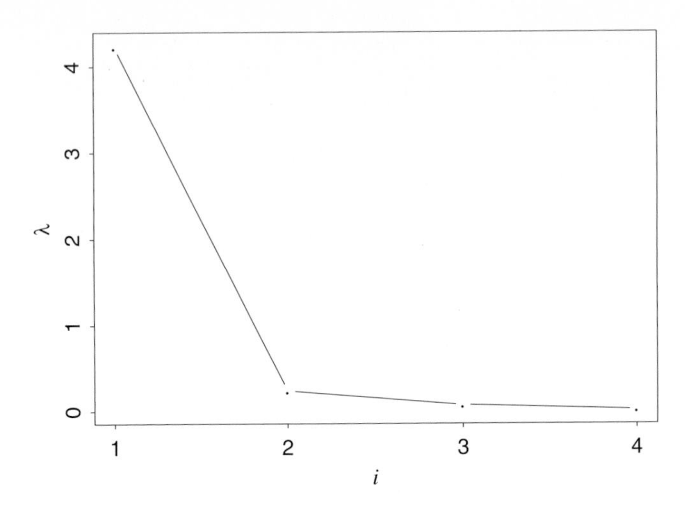

Fig. 3.5. A scree diagram for Fisher's Iris Data.

approximately

$$\overline{\lambda} = \frac{1}{c}\sum_{i=1}^{c} \lambda_i.$$

The argument runs

If $\lambda_i < \overline{\lambda}$, then the ith principal direction is less interesting than average.

and this leads us to discard principal components that have sample variances below $\overline{\lambda}$.

3. **Scree diagram** A scree diagram is an index plot of the principal component variances. In other words it is a plot of λ_i against i. An example of a scree diagram, for the Iris Data, is shown in Figure 3.5.
   The idea is that where the curve flattens out is where the variation is just random fluctuation with no structure. So we look for the elbow; in this case we only need the first component.

These methods are all ad-hoc, rely on judgement and have no theoretical justification for their use. The first two methods have the convenient trait of being precisely defined, so that these methods could be performed by computer. Though they are not supported by any theory, these methods work well in practice. My personal preference is to use the scree diagram and the proportion of variance methods, in tandem.

Normalising Consider the two scatterplots in Figure 3.6. The scatterplot on the left shows two of the variables from Fisher's iris data. The scatterplot on the

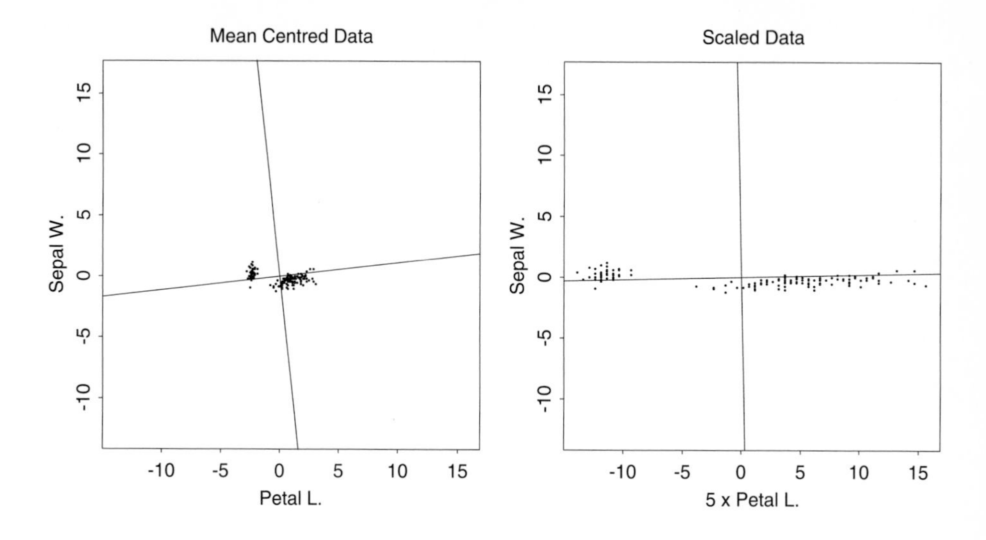

Fig. 3.6. The effect of non-uniform dilation on the principal axes.

right shows the same two variables but with one of them, *Petal L.*, multiplied by 5. The variables have been centred (had the mean value subtracted from each observation), so that the point being illustrated will be easier to see in the diagrams. Principal components analysis was carried out separately for each pair of variables. The (directions of) the principal axes are shown, as two lines passing through $(0, 0)$. The effect of multiplying *Petal L.* by 5 can be seen in the change of direction of the principal axes; the first principal axis moves closer to the *Petal L.* axis. As we increase the dilation in the *Petal L.* direction the first principal axis tends to become parallel to the axis of the first variable.

This is important, because it means the results of principal components analysis depend upon the measurement units (centimetres, inches, feet, stones, pounds, kilograms, etc.). Usually we want to identify structure, regardless of the particular choice of measurement scale, so we normalise the data.

The data can be normalised by carrying out the following steps.

- Centre each variable. In other words, subtract the mean of each variable to give

$$\mathring{\boldsymbol{x}}_j = \boldsymbol{x}_j - \overline{\boldsymbol{x}}.$$

- Divide each element of $\mathring{\boldsymbol{x}}_j$ by its standard deviation; as a formula this means calculate

$$z_{ij} = \frac{\mathring{x}_{ij}}{s_i},$$

where s_i is the sample standard deviation of x_i.

The result is to produce a $\boldsymbol{z}_j$ for each $\boldsymbol{x}_j$, with the property that the sample variance matrix of $\boldsymbol{z}$ is the same as the correlation matrix for $\boldsymbol{x}$.

The only reason that I can think of for not normalising is if the measurements of each element of $\boldsymbol{x}$ are in the same units, **and** that the scale is relevant; spatial data would be an example of a type of data for which normalising would not usually be appropriate.

So, in general, many statisticians would recommend normalising and applying principal components analysis to the $\boldsymbol{z}_j$'s rather than the $\boldsymbol{x}_j$'s matrix.

Interpretation The final part of a principal components analysis is to inspect the eigenvectors in the hope of identifying a meaning for the (important) principal components. For the normalised *Iris Data* the eigenvector matrix (E) is

$$\begin{pmatrix} 0.521 & 0.377 & 0.720 & 0.261 \\ -0.269 & 0.923 & -0.244 & -0.124 \\ 0.580 & 0.024 & -0.142 & -0.801 \\ 0.565 & 0.067 & -0.634 & 0.524 \end{pmatrix}$$

and the eigenvalues (the λ's) are

$$(2.92, 0.91, 0.15, 0.02)$$

so, for the normalised data, the first two principal components should be adequate, with the third and fourth components containing very little variation.

The first eigenvector (column 1 of E) suggests that the first principal component is large for large values of variables 1, 3, and 4, and for small values of variable 2, using the original (unnormalised) scale. (On the normalised scale variable 2 takes negative values when the first principal component is large; negative values on the normalised scale correspond to values that are below the mean on the unnormalised scale.) The sort of meaningful interpretation that can be made here is that y_1 is large for irises with long, thin sepals and big petals.

The second principal component essentially measures the size of the sepals (variables 1 and 2); it is dominated by variable 2 (sepal width); there is essentially no contribution from the petal dimensions.

Interpretation of the other two principal components is pointless, as they contain very little information.

Principal Components and Neural Networks In Chapter 8, it is stated that principal components analysis can be performed using neural network and Hebbian learning. Here we will demonstrate how principal components analysis is related to Hebbian learning.

In the neural network context, the network produces an output, o, from a set of inputs, $\boldsymbol{x}$, using a set of weights $\boldsymbol{w}$. The relationship between these three components is

$$o = \boldsymbol{w}^\top \boldsymbol{x}.$$

Earlier in this chapter, we considered the problem of finding $\boldsymbol{a}$ so as to maximise the variance of y, where

$$y = \boldsymbol{a}^\top \boldsymbol{x}.$$

Clearly, if we set $\boldsymbol{w} = \boldsymbol{a}$ then the neural network output, o, will be the first principal component score of $\boldsymbol{x}$, namely y.

We can, without any loss in generality, assume that the (input) variables have been centred (see 'Normalising' above). In this case, the (sample) variance of y will be

$$\frac{1}{m-1}\sum_{j=1}^{m} y_j^2.$$

So, for the neural network to perform principal components analysis, it must have the goal of maximising

$$\sum_{j=1}^{m} o_j^2 = \sum_{j=1}^{m} (\boldsymbol{w}^\top \boldsymbol{x}_j)^2.$$

It is apparent that we can increase the size of this function by making the elements of $\boldsymbol{w}$ bigger, and we can make it as large as we like. Therefore, we need a constraint on $\boldsymbol{w}$; the standard constraint is $\boldsymbol{w}^\top \boldsymbol{w} = 1$.

Putting all these bits together, we conclude that a neural network that learns how to maximise

$$\sum_{j=1}^{m} o_j^2$$

subject to

$$\boldsymbol{w}^\top \boldsymbol{w} = 1$$

will produce the first principal component score on its output.

To solve this problem we form the Lagrangian

$$L = \sum_{j=1}^{m} o_j^2 + \lambda(1 - \boldsymbol{w}^\top \boldsymbol{w}),$$

differentiate it to give

$$\frac{\partial L}{\partial \boldsymbol{w}} = \sum_{j=1}^{m} 2 o_j \boldsymbol{x}_j - 2\lambda \boldsymbol{w}$$

and then try to determine what λ is when $\frac{\partial L}{\partial \boldsymbol{w}} = 0$.

If $\frac{\partial L}{\partial \boldsymbol{w}} = 0$, then

$$\lambda \boldsymbol{w} = \sum_{j=1}^{m} o_j \boldsymbol{x}$$

so

$$\lambda \boldsymbol{w}^\top \boldsymbol{w} = \sum_{j=1}^{m} o_j \boldsymbol{w}^\top \boldsymbol{x}_j = \sum_{j=1}^{m} o_j^2$$

and hence (because we have the constraint $\boldsymbol{w}^\top \boldsymbol{w} = 1$)

$$\lambda = \sum_{j=1}^{m} o_j^2.$$

We can now substitute this value into the expression for $\frac{\partial L}{\partial \boldsymbol{w}}$, giving

$$\frac{1}{2}\frac{\partial L}{\partial \boldsymbol{w}} = \sum_{j=1}^{m} o_j \boldsymbol{x}_j - \sum_{j=1}^{m} o_j^2 \boldsymbol{w} = \sum_{j=1}^{m} o_j(\boldsymbol{x}_j - o_j \boldsymbol{w}).$$

Comparing this with the modified version of the plain Hebbian rule given in Chapter 8

$$\Delta w_k^q = \eta o^q (x_k^q - o^q w_k^q)$$

we can see that the modified Hebbian rule is performing a gradient descent for the problem of finding the first principal component.

3.4.2 Correspondence Analysis

Correspondence analysis is a way to represent the structure within *incidence matrices*. Incidence matrices are also called *two-way contingency tables*. The definitive text on this subject in English is [225], but there has been a great deal of work in this area by French data theorists. An example of a (5×4) incidence matrix, with marginal totals, is shown in Table 3.17. The aim is to produce a picture that shows which groups of staff have which smoking habits. There is an implicit assumption that smoking category and staff group are related.

The first step towards this goal is to transform the incidence matrix into something that is related more directly to association between the two variables. Having seen that log-linear modelling in Section 3.2.3 is concerned with association of categorical variables, we might attempt to find some quantity based on the log-linear model for our transformation of the incidence matrix. We do not do this, but we do something very similar. We convert the incidence matrix to a different measure of distance (chi-squared distance), which is based on the

Table 3.17. Incidence of smoking amongst five different types of staff. The data are taken from [225, page 55].

	Smoking Category				
Staff Group	None	Light	Medium	Heavy	Total
Senior Managers	4	2	3	2	11
Junior Managers	4	3	7	4	18
Senior Employees	25	10	12	4	51
Junior Employees	18	24	33	13	88
Secretaries	10	6	7	2	25
Total	61	45	62	25	193

traditional test for association between two categorical variables, the chi-squared test of association. This is the test that was used before log-linear models were developed and is still used if either you do not have a computer handy or you do not want to introduce your clients/students to log-linear modelling.

Notation Denote the incidence matrix as $\underset{(m \times n)}{X}$.

The row totals are

$$X_{j+} = \sum_{i=1}^{n} X_{ji} \quad \text{for } j = 1, \ldots, m,$$

the column totals are

$$X_{+i} = \sum_{j=1}^{m} X_{ji} \quad \text{for } i = 1, \ldots, n$$

and the grand total is

$$X_{++} = \sum_{j=1}^{m} X_{j+} = \sum_{i=1}^{n} X_{+i}.$$

From these totals we can calculate the *expected values* under the assumption that there is no association between the row variable and the column variable—smoking and staff group in the example. The expected values are

$$E_{ji} = \frac{X_{j+} X_{+i}}{X_{++}}.$$

So in the smoking example $E_{23} = \frac{18 \times 62}{193}$.

Chi-squared test of association For the smoking example, we can perform a statistical test of the hypothesis

H_0: P(Being in smoking category i) is the same for each staff group.

This boils down to saying that each row is generated by the same smoking distribution, or that rows (and columns) are proportional to one another. This is the same as saying smoking category and staff group are unrelated. A test statistic for this test is Pearson's chi-squared statistic

$$\chi^2 = \sum_{j=1}^{m} \sum_{i=1}^{n} \left\{ \frac{(X_{ji} - E_{ji})^2}{E_{ji}} \right\}. \tag{3.33}$$

The bigger χ^2, the greater the evidence against H_0.

Chi-squared distance It turns out that χ^2 can be decomposed in a way that will allow us to highlight the patterns in the incidence matrix. This decomposition arises from consideration of row profiles.

The *row profiles* of an incidence matrix are defined to be the vectors

$$\boldsymbol{p}_j = (p_{j1}, p_{j2}, \ldots, p_{jn})^\top \quad \text{for } j = 1, \ldots, m,$$

where $p_{ji} = X_{ji}/X_{j+}$.
The *mean row profile* of an incidence matrix is the vector

$$\overline{\boldsymbol{p}} = (\overline{p}_1, \overline{p}_2, \ldots, \overline{p}_n)^\top,$$

where $\overline{p}_i = X_{+i}/X_{++}$.
The main structure of interest in the incidence matrix is captured by how the row profiles differ from the mean row profile. For the data in Table 3.17, these differences correspond to the ways in which the proportion of each type of smoker for a particular type of worker vary relative to the average proportions.
We could measure how a row profile differed from the mean row profile by forming the squared (euclidean) distance between them, namely

$$(\boldsymbol{p}_j - \overline{\boldsymbol{p}})^\top (\boldsymbol{p}_j - \overline{\boldsymbol{p}}).$$

In correspondence analysis, we used chi-squared distance instead of the squared distance. The chi-squared distance is

$$d_j^2 = (\boldsymbol{p}_j - \overline{\boldsymbol{p}})^\top D_{\overline{\boldsymbol{p}}}^{-1} (\boldsymbol{p}_j - \overline{\boldsymbol{p}}), \tag{3.34}$$

where $D_{\overline{\boldsymbol{p}}}$ is a diagonal matrix with the elements of $\overline{\boldsymbol{p}}$ along the leading diagonal. In other words,

$$D_{\overline{\boldsymbol{p}}} = \begin{bmatrix} \overline{p}_1 & 0 & \cdots & 0 \\ 0 & \overline{p}_2 & \cdots & 0 \\ \vdots & \vdots & \ddots & \vdots \\ 0 & 0 & \cdots & \overline{p}_n \end{bmatrix}$$

and (consequently)

$$D_{\overline{\boldsymbol{p}}}^{-1} = \begin{bmatrix} 1/\overline{p}_1 & 0 & \cdots & 0 \\ 0 & 1/\overline{p}_2 & \cdots & 0 \\ \vdots & \vdots & \ddots & \vdots \\ 0 & 0 & \cdots & 1/\overline{p}_n \end{bmatrix}.$$

The chi-squared distance is related to the chi-squared statistic through the formula

$$\chi^2 = \sum_{j=1}^{m} X_{j+} d_j^2.$$

The quantity $\chi_j^2 = X_{j+} d_j^2$ is called the *weighted chi-squared distance* of $\boldsymbol{p}_j$ from $\overline{\boldsymbol{p}}$.

Given the relationship between the chi-squared statistic and the weighted chi-squared distances of the row profiles, we can transform the incidence matrix, X, into a new matrix, Z say, which has the following properties.

Table 3.18. The Z matrix for the incidence matrix given in Table 3.17.

Staff Group	Smoking Category			
	None	Light	Medium	Heavy
Senior Managers	0.020	-0.025	-0.020	0.035
Junior Managers	-0.051	-0.042	0.036	0.079
Senior Employees	0.159	-0.039	-0.078	-0.073
Junior Employees	-0.134	0.055	0.064	0.034
Secretaries	0.054	0.005	-0.026	-0.050

- The squared length of the vector made from the jth row of Z is χ_j^2/X_{++}.
- The direction of this vector indicates how the jth row profile differs from the mean row profile.
- If a particular row profile happens to be the same as the mean row profile, then the vector made from the jth row of Z will consist of zeroes. (In other words, the mean profile corresponds to the co-ordinate origin.)

The (j, i)th element of the Z matrix is

$$Z_{ji} = \sqrt{(X_{j+}/X_{++})}(p_{ji} - \overline{p}_i)\sqrt{\frac{1}{\overline{p}_i}}. \tag{3.35}$$

An alternative, but equivalent, formulation of Z_{ji}, which is more obviously related to the chi-squared statistic of (3.33), is

$$Z_{ji} = \frac{1}{\sqrt{X_{++}}} \times \frac{(X_{ji} - E_{ji})}{\sqrt{E_{ji}}}. \tag{3.36}$$

The formulation in (3.36) also illustrates that we could consider column profiles instead of row profiles, and we would arrive at the same Z matrix. The Z matrix for the data in Table 3.17 is shown in Table 3.18.

We could try to interpret the Z matrix directly, by looking for entries that are large in absolute value. For these data the chi-squared test of association indicates that observed association between type of worker and amount of smoking is not statistically significant (the significance probability, or p-value, is 17%). As a result most of the values in the Z matrix are quite small. There are two relatively high values, 0.159 and -0.134 in the first column. These indicate that the *senior employees* group contains more non-smokers than average and that the *junior employees* group contains fewer.

If the Z matrix were more complicated, then we would seek to draw a diagram to display the structure of the Z matrix. We could apply the same methods as we used for principal components analysis; for principal components analysis we searched for directions of high variance, but for correspondence analysis we seek directions of high weighted chi-squared distance. The matrix

$$Z^\top Z$$

is analogous to the sample variance matrix in principal components analysis. The diagonal elements of $Z^\top Z$ sum to χ^2/X_{++} and indicate how much of the chi-squared statistic is related to each of the four smoking categories; these elements are the analogues of the sample variances for the untransformed variables in principal components analysis. The off-diagonal elements are analogues of the sample covariances.

Having met principal components analysis, we know that the direction of highest weighted chi-squared distance is found by performing an eigen decomposition of $Z^\top Z$. The eigenvectors of $Z^\top Z$ for the data in Table 3.17 are

$$\begin{pmatrix} -0.8087001 & 0.17127755 & 0.0246170 & 0.5621941 \\ 0.1756411 & -0.68056865 & -0.5223178 & 0.4828671 \\ 0.4069601 & -0.04167443 & 0.7151246 & 0.5667835 \\ 0.3867013 & 0.71116353 & -0.4638695 & 0.3599079 \end{pmatrix}$$

the corresponding eigenvalues are

$$(0.0748, 0.0100, 0.0004, 0.0000)$$

and so we can see that most of the chi-squared statistic is accounted for by the first eigenvector. (The eigenvalues sum to the sum of the diagonal elements of $Z^\top Z$, in the same way that the eigenvalues in principal components analysis sum to the sum of the diagonal elements of the sample variance matrix. Here, the sum of the eigenvalues is χ^2/X_{++}.) The first two eigenvectors account for virtually all of the chi-squared statistic.

To convert Z to scores corresponding to the eigenvectors (in the same way that we converted to the principal component scores in principal components analysis) we simply multiply Z by the eigenvector matrix, V say, to get the matrix of row scores, G say. That is we find

$$G = ZV.$$

For the incidence matrix in Table 3.17 we find that G is

$$\begin{pmatrix} -0.01570127 & 0.046251973 & -0.016945718 & 2.775558 \times 10^{-17} \\ 0.07908382 & 0.074303261 & 0.010293294 & -6.245005 \times 10^{-17} \\ -0.19564528 & 0.005479739 & 0.002650324 & -5.204170 \times 10^{-17} \\ 0.15730008 & -0.038991402 & -0.002231942 & -1.387779 \times 10^{-17} \\ -0.07237356 & -0.028400773 & 0.002908443 & -1.387779 \times 10^{-17} \end{pmatrix}.$$

As we discovered that the first two components accounted for most of the chi-squared statistic, we can display the relevant structure of Table 3.17 by plotting the first two columns of G against each other. In addition, we plot the first two columns of V on the same diagram. The resulting plot is shown in Figure 3.7. The technical name for this sort of diagram is a *biplot*; see [222] for uses, interpretation and generalisations of this sort of diagram.

For our purposes, the interpretation of Figure 3.7 is fairly easy. The different categories of staff are all close to the origin; this indicates that their row profiles

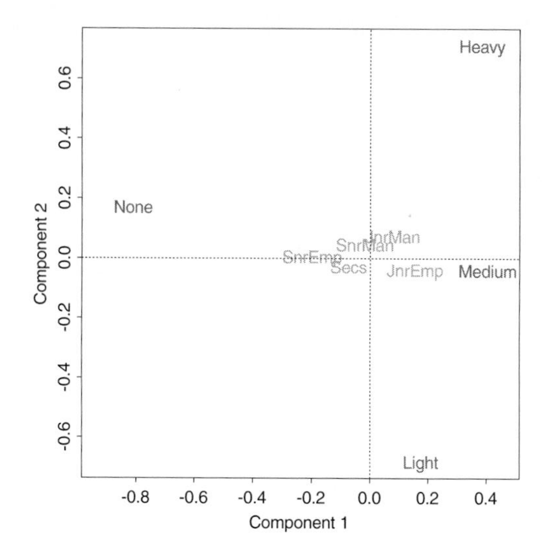

Fig. 3.7. Biplot of the first two components produced by applying correspondence analysis to the data in Table 3.17.

are all similar to the mean row profile, in other words, the differences in smoking habits are not large between different types of staff. We know that the patterns that we see in this diagram are not statistically significant, so perhaps we should leave the interpretation as it stands; for illustration of how to interpret the diagram, we will over interpret it. The *senior employees* group is displaced from the origin in the direction of the *none* category for smoking. This indicates that the proportion of non-smokers is relatively high for senior employees. The *junior employees* group, on the other hand, is shown as being in the opposite direction from the *none* category, indicating a lower proportion of non-smokers than the mean profile. The shortfall in non-smokers for the *junior employees* is taken up by an excess of *medium* smokers, which can be seen from the fact that *junior employees* and *medium* are plotted in the same direction as one another (relative to the origin).

3.4.3 Multidimensional Scaling

Multidimensional scaling is the process of converting a set of pairwise dissimilarities for a set of points, into a set of co-ordinates for the points. The key problem here is that the locations of the points in space, or the *configuration* of the points, is unknown. The aim is to find a configuration in which the pairwise distances between points approximate the dissimilarities as closely as possible.

At first sight, it is difficult to see how the pairwise dissimilarities might be known if the configuration is not known. Examples of dissimilarities could be: the price of an airline ticket between pairs of cities; the difference between the aggregate scores of pairs of teams in a sports league, the aggregation being across the matches between that pair of teams; road distances between towns (as opposed to straight-line distances); a coefficient indicating how different the artefacts found in pairs of tombs within a graveyard are. For the airline ticket example, the idea would be to construct a map in which the distances correspond to cost of travel between the cities; we might wish to compare this map with a standard map showing the geographical locations of the cities, but we are not trying to reconstruct the geographical map. Similarly, with road distances, we would be trying to construct a map in which distance corresponds to how far one has to drive a car to travel between the towns.

Two types of multidimensional scaling will be described here. *Classical scaling* assumes that the dissimilarities are euclidean distances (that is, distances like those that we meet in everyday life, with the same geometric properties). *Ordinal scaling* assumes merely that the dissimilarities are in the same order as the distances between points in the configuration; if the dissimilarity between two points, A and B, is 0.5 and the dissimilarity between A and a third point, C, is 1, then this says that the distance between A and C in the configuration should be bigger than the distance from A to B, but not necessarily twice as big.

Classical Scaling Classical scaling is also known as *metric scaling* and as *principal co-ordinates analysis.* The name 'metric' scaling is used because the dissimilarities are assumed to be distances—or in mathematical terms the measure of dissimilarity is the *euclidean metric.* The name 'principal co-ordinates analysis' is used because there is a link between this technique and principal components analysis. The name 'classical' is used because it was the first widely used method of multidimensional scaling, and pre-dates the availability of electronic computers.

To derive a configuration, we can start off by imagining that we knew the configuration and working out what the pairwise distances would be, and then trying to reverse the process. So, let X be the matrix whose rows specify the positions of the m objects in the configuration. That is

$$\underset{(m\times c)}{X} = \begin{bmatrix} \boldsymbol{x}_1^\top \\ \boldsymbol{x}_2^\top \\ \vdots \\ \boldsymbol{x}_m^\top \end{bmatrix},$$

where $\boldsymbol{x}_j$ is the $(c \times 1)$ vector representing the position of the jth object in c-dimensional space; if $c = 2$ then all the objects lie on a plane, if $c = 3$ then they occupy a 3-D space, just as we do.

This configuration would lead to a matrix of pairwise distances, D say,

$$D = \begin{bmatrix} d_{11} & d_{12} & \cdots & d_{1m} \\ d_{21} & d_{22} & \cdots & d_{2m} \\ \vdots & \vdots & \ddots & \vdots \\ d_{m1} & d_{m2} & \cdots & d_{mm} \end{bmatrix},$$

where $d_{k\ell}$ is the distance between objects k and ℓ. The elements of D are obtained from X using the following relationship:

$$d_{k\ell}^2 = \boldsymbol{x}_k^\top \boldsymbol{x}_k + \boldsymbol{x}_\ell^\top \boldsymbol{x}_\ell - 2\boldsymbol{x}_k^\top \boldsymbol{x}_\ell. \tag{3.37}$$

What we want to do is go from D to X. A step on the way is the matrix

$$B = XX^\top = \begin{bmatrix} b_{11} & b_{12} & \cdots & b_{1m} \\ b_{21} & b_{22} & \cdots & b_{2m} \\ \vdots & \vdots & \ddots & \vdots \\ b_{m1} & b_{m2} & \cdots & b_{mm} \end{bmatrix}$$

because it is easy to construct an appropriate configuration given B, and we can rewrite (3.37) in terms of the elements of B as in

$$d_{k\ell}^2 = b_{kk} + b_{\ell\ell} - 2b_{k\ell}. \tag{3.38}$$

If we make an assumption, then (3.38) can be rearranged to give the $b_{k\ell}$ in terms of the $d_{k\ell}$'s. The assumption that we need to make is that the origin of the co-ordinate system is at the mean of the configuration; this is equivalent to assuming that the columns of X have been centred. Specifying where the co-ordinate origin is cannot alter the pairwise distances, so we can make this assumption, without loss of generality. The point of this assumption is that it implies that

$$\mathbf{1}_m^\top X = \mathbf{0}_c^\top,$$

where $\mathbf{1}_m$ denotes the $(m \times 1)$ vector whose elements are all equal to 1, and $\mathbf{0}_c$ is a similar $(c \times 1)$ vector of zeroes. So,

$$\mathbf{1}_m^\top B = \mathbf{0}_c^\top X^\top = \mathbf{0}_m^\top,$$

and

$$B\mathbf{1}_m = \mathbf{0}_m.$$

This is equivalent to saying that the rows and columns of B sum to zero. In other words,

$$\sum_{k=1}^m b_{k\ell} = \sum_{\ell=1}^m b_{k\ell} = 0.$$

This result allows us to sum (3.38) over ℓ to get

$$\begin{aligned}\sum_{\ell=1}^m d_{k\ell}^2 &= mb_{kk} + \sum_{\ell=1}^m b_{\ell\ell} - (2 \times 0) \\ &= mb_{kk} + \sum_{\ell=1}^m b_{\ell\ell},\end{aligned}$$

which can in turn be summed over k, to give

$$\sum_{k=1}^m \sum_{\ell=1}^m d_{k\ell}^2 = \sum_{k=1}^m mb_{kk} + m\sum_{\ell=1}^m b_{\ell\ell} = 2m\sum_{k=1}^m b_{kk}.$$

Thus, we have a formula for $\sum_{k=1}^{m} b_{kk}$ (or, equivalently, $\sum_{\ell=1}^{m} b_{\ell\ell}$) in terms of the $d_{k\ell}$'s, which we can use to obtain a formula for b_{kk} in terms of the $d_{k\ell}$'s, which we can substitute into (3.38) to give a formula for $b_{k\ell}$ in terms of the $d_{k\ell}$'s.

If we obtain all these formulae, we find that the route from D to B is to calculate the matrix

$$D^* = \begin{bmatrix} d_{11}^2 & d_{12}^2 & \cdots & d_{1m}^2 \\ d_{21}^2 & d_{22}^2 & \cdots & d_{2m}^2 \\ \vdots & \vdots & \ddots & \vdots \\ d_{m1}^2 & d_{m2}^2 & \cdots & d_{mm}^2 \end{bmatrix},$$

then use the formula

$$B = -\frac{1}{2}\left(I_m - \frac{1}{m}\mathbf{1}_m\mathbf{1}_m^\top\right) D^* \left(I_m - \frac{1}{m}\mathbf{1}_m\mathbf{1}_m^\top\right)$$

to find B.

To get a configuration that has the same pairwise distances as in D, we need to find a matrix, X, which satisfies

$$XX^\top = B$$

and any such matrix will do. The choice made in classical scaling is the matrix

$$X^* = \underset{(m\times c)}{E} \begin{bmatrix} \sqrt{\lambda_1} & 0 & \cdots & 0 \\ 0 & \sqrt{\lambda_2} & \cdots & 0 \\ \vdots & \vdots & \ddots & \vdots \\ 0 & 0 & \cdots & \sqrt{\lambda_c} \end{bmatrix}, \tag{3.39}$$

where $\lambda_1 \geq \lambda_2 \geq \cdots \geq \lambda_c > 0$ are the non-zero eigenvalues of B and the ith column of E is $\boldsymbol{e}_i$, the eigenvector corresponding to λ_i. These eigenvectors can be, and are, chosen to satisfy (3.32) on page 103, that is, to be *orthonormal.*

This method of choosing X from B is called the *spectral decomposition* of B. The eigenvalues of B will all be greater than or equal to zero, as long as the distances in D are euclidean. Here, we can regard euclidean as meaning just like distances in the physical world that we live in. If any eigenvalue turns out to be negative, then we can either just ignore the problem, or use ordinal scaling (see below) instead.

The main advantage of using the spectral decomposition of B to find a configuration is that X^* turns out to be the matrix containing the principal components scores that would be obtained if we knew X and performed principal components analysis upon it. This is why classical scaling is sometimes called principal co-ordinates analysis. This means that if we want, for example, a 2-D map in which the distances between points in the configuration most closely approximate the distances in D, then all we need do is plot the first two columns of X^* against one another. This is equivalent to plotting the first two principal components against one another.

The results of applying classical scaling to British road distances are shown in Figure 3.8. These road distances correspond to the routes recommended by the

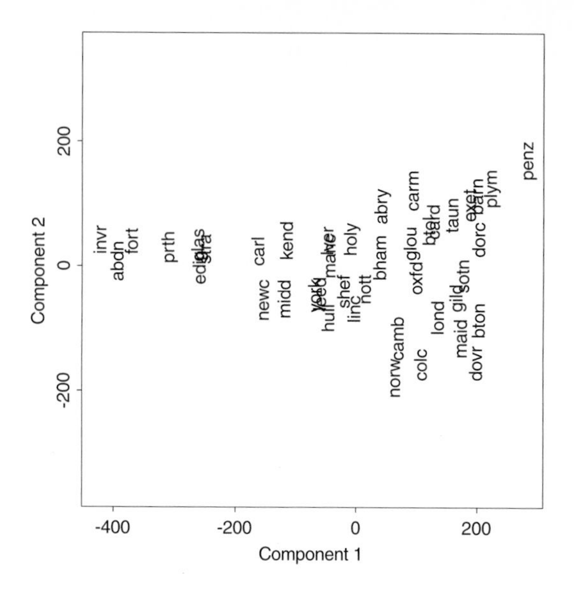

Fig. 3.8. A 2-D map of the island of Great Britain, in which the distances are approximations to the road distances between the towns and cities labelled. The labels are rotated so as to reduce the amount of overlap between labels.

Automobile Association; these recommended routes are intended to give the minimum travelling time, not the the minimum journey distance. An effect of this, that is visible in Figure 3.8 is that the towns and cities have lined up in positions related to the motorway (high-speed road) network. I can see the places on the M6 motorway, Birmingham (*bham*), Manchester (*manc*), Kendal (*kend*), Carlise (*carl*), and its continuation the A74 which goes to Glasgow (*glas*), all appear along a straight line in the configuration. Similarly, the M1 and A1 routes up the eastern side of England can be seen by the alignment of Nottingham (*nott*), Sheffield (*shef*), Leeds (*leed*), York (*york*), Middlesborough (*midd*) and Newcastle (*newc*); the line from Gloucester (*glou*) to Plymouth (*plym*) corresponds to the M5. The map also features distortions from the geographical map such as the position of Holyhead (*holy*), which appears to be much closer to Liverpool (*lver*) and Manchester than it really is, and the position of Cornish peninsula (the part ending at Penzance, *penz*) is further from Carmarthen (*carm*) than it is physically. These distortions are due to features such as large water bodies (between Carmarthen and Penzance) making the road distance much bigger than the straight-line distance, or because a place (Holyhead) is isolated. The isolation of Holyhead results in there being just one fast road to it, and most routes from the South of England (places with high component 1 values) to Holyhead go through Birmingham; this means that most of the distances to Holyhead are the same as the distance to Birmingham plus a constant amount (because most of the places are in the South of England). The best compromise in terms of getting most distances approximately right is to place Holyhead north of Birmingham, that is, in the same area as Manchester and Liverpool.

Ordinal Scaling Ordinal scaling is used for the same purposes at classical scaling, but for dissimilarities that are not metric, that is, they are not what we would think of as distances. Ordinal scaling is sometimes called *non-metric scaling*, because the dissimilarities are not metric. Some people call it *Shepard-Kruskal scaling*, because Shepard and Kruskal are the names of two pioneers of ordinal scaling.

In ordinal scaling, we seek a configuration in which the pairwise distances between points have the same rank order as the corresponding dissimilarities. So, if $\delta_{k\ell}$ is the dissimilarity between points k and ℓ, and $d_{k\ell}$ is the distance between the same points in the derived configuration, then we seek a configuration in which

$$d_{k\ell} \leq d_{ab}$$

if

$$\delta_{k\ell} \leq \delta_{ab}.$$

If we have such a configuration to hand, then all is well and the problem is solved. If we do not have such a configuration, or no such configuration exists, then we need a way to measure how close a configuration is to having this property. The quantity used to measure how close a configuration is to having this property is called *STRESS*, which is zero if the configuration has this property and greater than zero if it does not.

The definition of STRESS is omitted in the interests of brevity; see [326, page 117] for the definition. Basically, the STRESS measures the size of the smallest change in the $d_{k\ell}$'s required to obtained distances with the correct ordering. A set of fitted distances, $\hat{d}_{k\ell}$'s, say, that are in the correct order is created. The STRESS is

$$\sum_k \sum_\ell \left(d_{k\ell} - \hat{d}_{k\ell} \right)^2 .$$

The part of the definition that is omitted here is how to obtain the $\hat{d}_{k\ell}$'s, which are themselves the result of an optimisation procedure. Whilst the $d_{k\ell}$'s are distances between points in a configuration, the $\hat{d}_{k\ell}$'s used in calculating the STRESS do not generally correspond to any valid configuration; the STRESS is purely a measure of how far the distances are from being in the correct order.

The principle of using STRESS was introduced by [324]. Given that we have STRESS, we can think of it as a function of the co-ordinates of the points in the configuration. We can then start from some arbitrary configuration and iteratively improve it. A practical steepest descent algorithm is given by [325]; obtaining the derivative of STRESS is non-trivial.

Armed with STRESS and the algorithm for minimising it, ordinal scaling boils down to:

– choose a dimensionality, c say;
– choose a c-dimensional configuration;
– update the configuration iteratively, so as to minimise the value of STRESS.

Often classical scaling is applied directly to the dissimilarities, or some transformation of them, to obtain a starting configuration. This is believed to both shorten computing time and increase the chance of finding the globally optimal configuration. Sometimes a higher dimensionality than is required, $c+1$ say, is used during the minimisation. The aim of this is to allow the configuration to change in ways that would not be allowed in c dimensions. For example, in a 2-D configuration a point might be on the right of the configuration, when its optimal position is on the left. In this situation it is possible that moving the point to the left is prevented because small movements to the left increase the STRESS. If the same configuration is then embedded in a 3-D space, then it is sometimes possible for the move from the right to the left to be made by using a route that comes out of the 2-D plane. These strategies generally work well, but they are not guaranteed to produce the global optimum; if it is vital that the globally optimal configuration be found, then the techniques described in Chapter 10 might be used.

Choice of c is sometimes part of the problem. If this is so, then a range of values of c are tried. The minimised STRESS values are then plotted against c and this gives a diagram like a scree diagram (see Figure 3.5), in which the 'elbow' is sought. In this procedure, it is usual to start with the highest value of c under consideration. The optimum configuration for c dimensions would be used to give a starting configuration for $c-1$ dimensions, either by simply discarding the cth dimension from the co-ordinates of the configuration, or by some transformation such as principal components analysis (see Section 3.4.1).

3.4.4 Cluster Analysis and Mixture Decomposition

Cluster analysis and mixture decomposition are both techniques to do with identification of concentrations of individuals in a space.

Cluster Analysis Cluster analysis is used to identify groups of individuals in a sample. The groups are not pre-defined, nor, usually, is the number of groups. The groups that are identified are referred to as *clusters*. There are two major types of clustering, *hierarchical* and *non-hierarchical*; within hierarchical clustering, there are two main approaches, *agglomerative* and *divisive*.

In hierarchical clustering, we generate m different partitions of the sample into clusters, where m is the number of individuals in the sample. One of these partitions corresponds to a single cluster made up of all m individuals in the sample, while at the opposite extreme there is a partition corresponding to m clusters, each made up of just one individual. Between these extremes there is a partition with 2 clusters, one with 3 clusters, and so on up to a partition with $m-1$ clusters. The key characteristic of these partitions, which makes them hierarchical, is that the partition with r clusters can be used to produce the partition with $r-1$ clusters by merging two clusters, and it can also be used to produce the partition with $r+1$ clusters by splitting a cluster into two.

As we know both the top layer and the bottom layer of the hierarchy, there are two natural approaches to finding the intervening layers. We could start with m clusters, each containing one individual, and merge a pair of clusters to get $m-1$ clusters, and continue successively merging pairs of clusters; this approach is called agglomerative clustering. Alternatively, we could start with a single cluster, split it into two, then split one of the new clusters to give a total of three clusters, and so on; this approach is called divisive clustering.

Agglomerative clustering has been preferred traditionally, because the number of partitions considered in building the hierarchy is much smaller than for divisive clustering; the number of partitions considered is cubic in m for agglomerative, but exponential in m for divisive.

In order to decide which clusters to merge, or which clusters to split, we need a way to measure distance between clusters. The usual methods for measuring distance between clusters are given below.

- **Minimum distance** or *single-link* defines the distance between two clusters as the minimum distance between an individual in the first cluster and an individual in the second cluster.
- **Maximum distance** or *complete-link* defines the distance between two clusters as the maximum distance between an individual in the first cluster and an individual in the second cluster.
- **Average distance** defines the distance between two clusters as the mean distance between an individual in the first cluster and an individual in the second cluster, taken over all such pairs of individuals.
- **Centroid distance** defines the distance between two clusters as the squared distance between the mean vectors (that is, the centroids) of the two clusters.
- **Sum of squared deviations** defines the distance between two clusters as the sum of the squared distances of individuals from the joint centroid of the the two clusters minus the sum of the squared distances of individuals from their separate cluster means.

These different methods of measuring distance between clusters lead to different characteristics in the clusters. This means that you ought to choose a method that is appropriate, rather than trying as many as you can. The characteristics of the various methods, and hence how to choose between them, are described and discussed in [326, section 3.1] and [235, section 7.3.1].

The results of a hierarchical cluster analysis are almost always presented as a *dendrogram*, such as that in Figure 3.9. The dendrogram in Figure 3.9 is the result of applying complete linkage clustering (agglomerative clustering where the distance between clusters is *maximum distance*, also called *complete-link*, as defined above) to a set of nine individuals, which are labelled 1 to 9. If we consider the partition giving two clusters, we can see that one cluster consists of individuals 1–4 and the other consists of 5–9. We can see this by following the two vertical lines which are linked at the top of the diagram, down to the labels for the individuals at the bottom of the diagram. Similarly, the partition

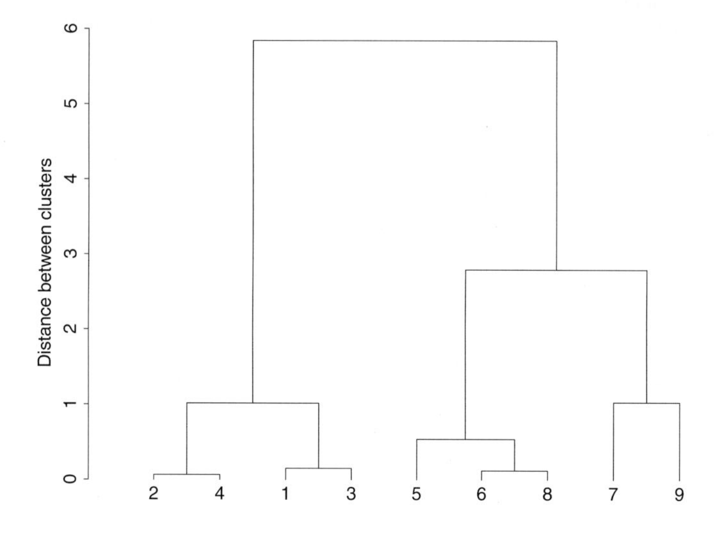

Fig. 3.9. An example of a dendrogram.

corresponding to three clusters gives the clusters as: 1–4; 5, 6 and 8; 7 and 9. We can see this by looking for a height where there are three vertical lines (around about 2 on the vertical scale) and tracing the three vertical lines down to the individuals. The vertical scale indicates the distance between the two clusters being merged. So, for example, in the two-cluster partition, the clusters are just under 6 units apart, while for the three-cluster partition the two closest clusters are just under 3 units apart.

Non-hierarchical clustering is essentially trying to partition the sample so as to optimize some measure of clustering. The choice of measure of clustering is usually based on properties of sums of squares and products matrices, like those met in Section 3.3.1, because the aim in the MANOVA is to measure differences between groups. The main difficulty here is that there are too many different ways to partition the sample for us to try them all, unless the sample is very small (around about $m = 10$ or smaller). Thus our only way, in general, of guaranteeing that the global optimum is achieved is to use a method such as branch-and-bound, as used by [510], for example.

One of the best known non-hierarchical clustering methods is the *k-means* method of [362]. The k-means method starts with k clusters and allows each individual to be moved from its current cluster to another cluster. Individuals are moved between clusters until it becomes impossible to improve the measure of clustering. An algorithm for performing this method of clustering is given by [248], which uses the sum of the squared distances of individuals from their cluster centroids as the measure of clustering (the smaller the better). There is no guarantee that the global optimum will be achieved.

A description of the various criteria that are used and the optimisation algorithms available can be found in [235, section 7.4].

Mixture Decomposition Mixture decomposition is related to cluster analysis in that it is used to identify concentrations of individuals. The basic difference between cluster analysis and mixture decomposition is that there is an underlying statistical model in mixture decomposition, whereas there is no such model in cluster analysis. The probability density that has generated the sample data is assumed to be a mixture of several underlying distributions. So we have

$$f(\boldsymbol{x}) = \sum_{k=1}^{K} w_k f_k(\boldsymbol{x}; \boldsymbol{\theta}_k),$$

where K is the number of underlying distributions, the f_k's are the densities of the underlying distributions, the $\boldsymbol{\theta}_k$'s are the parameters of the underlying distributions, the w_k's are positive and sum to one, and f is the density from which the sample has been generated.

The reason for considering such a model is that we want to summarise the mechanism that generated the sample by stating which probability distribution produced the sample. The problem overcome by using a mixture distribution is that f might not correspond to a standard distribution with well known properties. The f_k's are chosen so that they do correspond to standard distributions, usually multivariate normal distributions.

We can consider this sort of model either because we believe that there are K sub-populations mixed together to give the population from which our sample has been drawn, or simply because it is a pragmatic way of summarising a non-standard distribution.

The $\boldsymbol{\theta}_k$'s and the w_k's can be estimated using iterative methods. Descriptions of these methods and examples of the use of mixture decomposition can be found in [235, chapter 3].

3.4.5 Latent Variable and Covariance Structure Models

A *latent* variable is a variable that is not included in our set of data, either because it cannot be measured directly, or because it was not thought of, but the latent variable is related to the variables that are included in the data set. A latent variable is often of interest because it is believed that the latent variable causes changes in the values of several of the variables that we have observed; this underlying cause of variation induces a covariance structure between the variables that it changes.

The (k, ℓ)th off-diagonal element of the variance-covariance matrix of variables x_1 to x_n contains the covariance between the kth and ℓth variable. The diagonal elements of the matrix contain the variances of the variables. In this section we examine ways to model the structure of such a matrix. There are various reasons for wanting to do this, but an important one is simply that it provides a way to reduce the number of parameters from $n(n+1)/2$ to something more manageable. This can result in a model which is easier to understand and also one which has improved predictive power.

Several such methods for modelling variance-covariance matrices have been developed, with differing complexity and power. All of them have convenient graphical (in the mathematical sense) representations, in which the nodes of a graph represent the variables and edges connecting the nodes represent various kinds of relationships between the variables (the nature of the relationship depending upon the type of graph in question). We shall begin with the simplest case—that of *path analysis*—which goes back to the work of [548, 549].

Path Analysis In path analysis the aim is to decompose the relationships between the observed variables into *causal paths*, and attribute strengths to each of these paths. To start with, we shall assume that

(a) the variables are known to have a *weak causal order*. This means that we can order the variables such that a given variable may (but need not) be a cause of variables later than itself in the list, but cannot be a cause of variables earlier than itself in the list; and
(b) *causal closure* holds. This means that the covariation between two of the measured variables is due to the causal effects of one on the other or to the effects on both of them of some other variable(s) in the model.

Of course, in order to make this rigorous we really ought to define what we mean by *cause*. Since this has occupied philosophers since the dawn of time, we shall content ourselves with the following working definition: *x is a cause of y if and only if the value of y can be changed by manipulating x and x alone.* Note that the effect of x on y may be *direct* or it may be *indirect*. In the latter case, the effect arises because x affects some *intermediate* variable z which in turn affects y.

If a unit change in x directly induces a change of c units in y then we will write $y = cx$. Conventionally, if unstandardised variables are used then the coefficient is called an *effect coefficient*, written as c, while if standardised variables are used the coefficient is called a *path coefficient*, written as p. Note the distinction between these and regression coefficients (the β's): a regression coefficient simply tells us the expected difference in y between two groups that happen to have a unit difference on x. This may be direct causal, indirect causal, or it may be due to some selection or other mechanism. We shall see examples of the difference between path (denoted p) and regression (denoted b) coefficients below. To begin with, however, Figure 3.10 shows four possible patterns of causal relationships between three variables x, y, and z when they have the weak causal order $x > y > z$ (the four examples in Figure 3.10 do not exhaust the possibilities). The arrows indicate the direction of causation.

In Figure 3.10(a), x causes y, which in turn causes z. If y is fixed (by, for example, selecting only those cases which have a given value of y), then x and z are conditionally independent (this follows from assumption (b) above). In

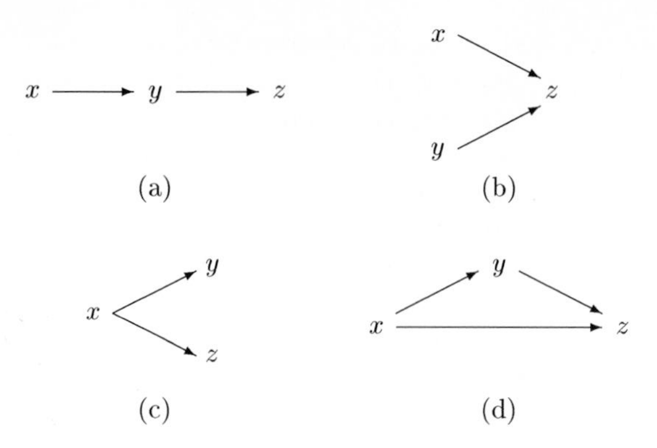

Fig. 3.10. Four of the possible patterns of causal relationships between three variables, x, y and z.

Figure 3.10(b), x and y separately have a causal effect on z, but x and y are marginally independent (there is no arrow linking them). Note, however, that they may not be conditionally independent given values of z. In Figure 3.10(c), x has a causal effect on y and a causal effect on z. Here y and z are conditionally independent given x, but if x is uncontrolled then they may have an induced covariance.

Figure 3.10(d) is more interesting than the others, so we will develop it in a little more detail, with illustrative numbers. First, however, note that it is clear from the description of the first three models that the notion of conditional independence has an important role to play in disentangling relationships between variables—we will have more to say about this below. Secondly, note that, given the two assumptions above, one can estimate the path or effect coefficients between n variables by a set of $(n-1)$ multiple regression analyses. That is, starting with (one of) the variable(s) at the end of the weak causal order, one can construct a regression model to predict it in terms of all the variables which precede it. This is repeated for all variables in turn, working back towards the beginning of the order. Given this modelling process, it is reasonable that the resulting structures should be called *univariate recursive regression models.*

In Figure 3.10(d) the path coefficients for the pairs (y, z) and (x, z) are estimated by regressing z on x and y. That is, the model $z = p_{(z,y)} \cdot y + p_{(z,x)} \cdot x$. Suppose that, if we do this, we obtain the regression (and here also path) coefficients $p_{(z,y)} = -0.2$ and $p_{(z,x)} = -0.3$. Similarly, we can regress y on x to obtain the path coefficient for the pair (x, y)—yielding 0.8, say. Finally, we can also regress z just on x. This regression coefficient will not be a path coefficient, since it will include direct effects of x on z and also indirect effects via y. Decomposing the overall effect of x on z in this way, we find that the regression coefficient is $-0.46 = (-0.3) + (0.8)(-0.2)$. Here the first term is the direct effect and the second term the indirect via y.

Finally, in this model, let us look at the covariation between y and z. This does not simply arise from the regression effect of y on z. There is also a 'spurious' effect arising from the effect of x on both y and z, so that the overall covariation is given by $(-0.2) + (0.8)(-0.3) = -0.44$.

This sort of process yields a decomposition of the overall covariation between any two variables into direct causal relationships, indirect causal relationships, and noncausal relationships (induced by the effect of other variables on the two in question).

The set of variables in a path model can be divided into two groups: those which have no predecessors, called *exogenous* variables, and those which are predicted by others in the model, called *endogenous* variables. Denoting the former by $\boldsymbol{z}$ and the latter by $\boldsymbol{y}$, we can express a path model as

$$\boldsymbol{y} = C\boldsymbol{y} + D\boldsymbol{z} + \boldsymbol{e}$$

where C and D are matrices of regression coefficients and $\boldsymbol{e}$ is a vector of random error terms. Note that assumption (a) means that there are no causal loops.

Up until this point things have been straightforward, but this ease of manipulation and interpretation has been bought by assumptions (a) and (b), which are fairly drastic. Often it is not so easy to assert a causal order between variables (straightforward for *income* and *expenditure*, but not so straightforward for *depression* and *quality of life*). This means that one would sometimes like to relax things—replacing the arrows between the nodes representing the variables by an undirected edge. Likewise, the assumption that the model is complete, that all indirect links between two variables have been included, is often hard to support. To cope with problems such as these, path models have been extended. One very important extension is to include *latent* variables.

Latent Variables A latent variable is an unmeasured (and typically unmeasurable) variable which accounts for and explains the observed correlations between the *manifest* or measured variables. That is, the observed variables are postulated to have high correlations with each other precisely because they have high correlations with the unmeasured latent variable. Latent variables can be considered as convenient mathematical fictions to explain the relationships between the manifest variables, or attempts can be made to interpret them in real terms. Social class, intelligence, and ambition are examples of latent variables which arise in the social and behavioural sciences. One of the earliest types of latent variable models was the *factor analysis model* (originally developed as a model for intelligence in the early decades of the twentieth century, when a 'general' intelligence factor was postulated). In mathematical terms, the manifest variables $\boldsymbol{x}$ are to be explained in terms of a number of latent variables $\boldsymbol{y}$, assumed to follow a distribution $\mathcal{N}(\boldsymbol{0}, I)$, so that we have

$$\boldsymbol{x} = \boldsymbol{\mu} + \Gamma\boldsymbol{y} + \boldsymbol{e}$$

where Γ is a matrix of *factor loadings*, and $\boldsymbol{e}$ is a random vector typically assumed to follow a distribution $\mathcal{N}(\boldsymbol{0}, \Theta)$. If the covariance matrix of the manifest variables is Σ, then we have

$$\Sigma = \Gamma\Gamma^{\top} + \Theta.$$

(We note parenthetically here that factor loading matrices of the form $\Gamma M^{\top}$, where M is a non-singular orthogonal matrix, will also satisfy this expression. This means that it is the space spanned by the factors which is unique, and not the individual factors themselves. This ambiguity is probably one of the reasons that factor analysis acquired a reputation of not being an entirely respectable technique, a reputation which is no longer deserved—the method is now well-understood and can shed valuable light on data structures.)

Nowadays the most common method of estimating the factor loadings is probably via maximum likelihood.

Linear Structural Relational (LISREL) Models In recent decades, the basic factor analysis model has been substantially developed into the general class of linear structural relational (or LISREL) models; see, for example [74, 161]. These often involve multiple latent variables, and either hypothesise or set out to discover relationships between the latent variables themselves, as well as between the latent and the manifest variables. The applicability of such models is immediately seen when one recognises that no measurements are made without error (especially in the human sciences, where these methods were originally developed and continue to be heavily applied). This means that the measurement actually observed is, in fact, merely a manifest version of an unobserved latent variable, so that a more elaborate (and correct) model will consist of a *measurement model* describing the relationships between the measured and latent variables and a *structural model* describing the relationships between the latent variables themselves.

Examples of important kinds of LISREL models are *multitrait multimethod* models and *multiple indicator multiple cause* models. In the former, several hypothesised traits are measured by several methods. The model is used to tease apart effects due to the trait being measured and effects due to the method of measurement used. In the latter, unobserved latent variables are influenced by several causes and the effects are observed on several indicators. This is proposed by [174] as a suitable model for quality of life measurements, where some of the measured variables are indicative of poor quality of life while others cause poor quality of life.

Recent Developments Up until this point, the methods we have been describing have their origins in structural relationships between the variables. In the last decade or so, however, an alternative type of model has attracted considerable interest. Models of this type go under various names—conditional independence graphs, Bayesian belief networks, graphical models, and so on; see, for example [540, 165, 337]. They are described in more detail in Chapter 4. They differ from the models above because they are based on modelling the joint probability distribution of the observed variables in terms of conditional independence relationships. Such models have been developed in greatest depth for multivariate normal data and for categorical data—Figure 3.2 showed a conditional independence graph for a categorical data set. With multivariate normal data, the inverse of the variance-covariance matrix has a particularly useful interpretation: the off-diagonal elements of the matrix give the conditional covariances between the variables. (The conditional covariances are often called *partial covariances.*) In particular, it follows that if a given off-diagonal element is zero then the two variables in question are independent given the other variables in the matrix. In terms of a graphical representation, this corresponds to the two variables not being connected by an edge in the graph. Note that, unlike the structural models introduced above, conditional independence graphs are essentially undirected—though the relationships between undirected graphs and directed graphs is now well understood.

To develop a conditional independence model, two sorts of information are required. The first is the topology of the graph—the edges linking the nodes in the pictorial representation and the conditional independence structure in the mathematical formalism. The second is the numerical detail of the conditional dependencies: if variable A depends on variables B, C, and D, what does the conditional distribution of A look like for each combination of the levels of B, C, and D? If sufficient human expertise is available, then, of course both of these can be obtained by interrogation. Otherwise data analytic methods are required. (In fact, conditional independence models have their origins in the Bayesian school of inference, so that it is common to find prior expertise and data integrated in developing the models.) Extracting the numeric details of the conditional independencies from data is, in principle, straightforward enough. This is not quite so true for constructing the model's basic structure since the space of possible models is often vast. To do this effectively requires a large data set. This is currently a hot research topic.

3.5. Conclusion

The techniques presented in this chapter do not form anything like an exhaustive list of useful statistical methods. These techniques were chosen because they are either widely used or ought to be widely used. The regression techniques are widely used, though there is some reluctance amongst researchers to make the jump from linear models to generalized linear models.

The multivariate analysis techniques ought to be used more than they are. One of the main obstacles to the adoption of these techniques may be that their roots are in linear algebra.

I feel the techniques presented in this chapter, and their extensions, will remain or become the most widely used statistical techniques. This is why they were chosen for this chapter.

Chapter 4
Bayesian Methods

Marco Ramoni[*] and Paola Sebastiani[**]
[*]Harvard University and [**]University of Massachusetts, Amherst, USA

4.1. Introduction

Classical statistics provides methods to analyze data, from simple descriptive measures to complex and sophisticated models. The available data are processed and then conclusions about a hypothetical population — of which the data available are supposed to be a representative sample — are drawn.

It is not hard to imagine situations, however, in which data are not the only available source of information about the population.

Suppose, for example, we need to guess the outcome of an experiment that consists of tossing a coin. How many biased coins have we ever seen? Probably not many, and hence we are ready to believe that the coin is fair and that the outcome of the experiment can be either head or tail with the same probability. On the other hand, imagine that someone would tell us that the coin is forged so that it is more likely to land head. How can we take into account this information in the analysis of our data? This question becomes critical when we are considering data in domains of application for which knowledge *corpora* have been developed. Scientific and medical data are both examples of this situation.

Bayesian methods provide a principled way to incorporate this external information into the data analysis process. To do so, however, Bayesian methods have to change entirely the vision of the data analysis process with respect to the classical approach. In a Bayesian approach, the data analysis process starts already with a given probability distribution. As this distribution is given *before* any data is considered, it is called *prior* distribution. In our previous example, we would represent the fairness of the coin as a uniform prior probability distribution, assigning probability 0.5 of landing to both sides of the coin. On the other hand, if we learn, from some external source of information, that the coin is biased then we can model a prior probability distribution that assigns a higher probability to the event that the coin lands head.

The Bayesian data analysis process consists of using the sample data to update this prior distribution into a *posterior* distribution. The basic tool for this updating is a theorem, proved by Thomas Bayes, an Eighteen century clergyman. The fundamental role of Bayes' theorem in this approach is testified by the fact that the whole approach is named after it.

The next section introduces the basic concepts and the terminology of the Bayesian approach to data analysis. The result of the Bayesian data analysis process is the posterior distribution that represents a revision of the prior distribution on the light of the evidence provided by the data. The fact that we use the posterior distribution to draw conclusions about the phenomenon at hand changes the interpretation of the typical statistical measures that we have seen in the previous chapters. Section 4.3 describes the foundations of Bayesian methods and their applications to estimation, model selection, and reliability assessment, using some simple examples. More complex models are considered in Section 4.4, in which Bayesian methods are applied to the statistical analysis of multiple linear regression models and Generalized Linear Models. Section 4.5 will describe a powerful formalism known as *Bayesian Belief Networks* (BBN) and its applications to prediction, classification and modeling tasks.

4.2. The Bayesian Paradigm

Chapters 2 and 3 have shown that classical statistical methods are usually focused on the distribution $p(\mathbf{y}|\boldsymbol{\theta})$ of data $\boldsymbol{y}$, where $p(\cdot|\boldsymbol{\theta})$ denotes either the probability mass function or the density function of the sample of n cases $\mathbf{y} = (y_1, \ldots, y_n)$ and is known up to a vector of parameters $\boldsymbol{\theta} = (\theta_1, \ldots, \theta_k)$. The information conveyed by the sample is used to refine this probabilistic model by estimating $\boldsymbol{\theta}$, by testing hypotheses on $\boldsymbol{\theta}$ and, in general, by performing statistical inference. However, classical statistical methods do not allow the possibility of incorporating external information about the problem at hand. Consider an experiment that consists of tossing a coin n times. If the results can be regarded as values of independent binary random variables Y_i taking values 1 and 0 — where $\theta = p(Y_i = 1)$ and $Y_i = 1$ corresponds to the event "head in trial i" — the likelihood function $L(\theta) = p(\mathbf{y}|\theta)$ (see Chapter 2) is

$$L(\theta) = \theta^{(\sum_i y_i)}(1-\theta)^{(n-\sum_i y_i)}$$

and the ML estimate of θ is

$$\hat{\theta} = \frac{\sum_i y_i}{n},$$

which is the relative frequency of heads in the sample. This estimate of the probability of head is only a function of the sample information.

Bayesian methods, on the other hand, are characterized by the assumption that it is also meaningful to talk about the conditional distribution of $\boldsymbol{\theta}$, given the information I_0 currently available. Thus, a crucial aspect of Bayesian methods is to regard $\boldsymbol{\theta}$ as a random quantity whose *prior* density $p(\boldsymbol{\theta}|I_0)$ is known before seeing the data. In our previous example, the probability θ of the event "head in trial i" would be regarded as a random variable whose prior probability distribution captures all prior information I_0 about the coin before seeing the experimental results. The prior distribution can arise from data previously observed, or it can be the *subjective* assessment of some domain expert and, as such, it represents the information we have about the problem at hand, that is not conveyed by the sample data. We shall call this information prior, to distinguish it from the data information. For the coin tossing example, we could represent the prior information that the coin is fair by adopting a prior density for θ, defined in [0,1], and with expectation 0.5.

The available information changes as new data $\mathbf{y}$ are observed, and so does the conditional distribution of $\boldsymbol{\theta}$. This operation of revising the conditional distribution of $\boldsymbol{\theta}$ is done by Bayes' Theorem:

$$p(\boldsymbol{\theta}|\mathbf{y}, I_0) = \frac{p(\mathbf{y}|\boldsymbol{\theta}, I_0)p(\boldsymbol{\theta}|I_0)}{p(\mathbf{y}|I_0)}, \tag{4.1}$$

which updates $p(\boldsymbol{\theta}|I_0)$ into the *posterior* density $p(\boldsymbol{\theta}|\mathbf{y}, I_0)$. Hence, regarding $\boldsymbol{\theta}$ as a random quantity gives Bayesian methods the ability to introduce prior information into the inferential process that results in a distribution of $\boldsymbol{\theta}$ conditional on the total information available or posterior information

$$\begin{array}{ccccc} \text{posterior information} & = & \text{prior information} & + & \text{data information} \\ I_1 & = & I_0 & + & \mathbf{y} \end{array}$$

The probability density $p(\mathbf{y}|I_0)$ in (4.1) is computed by using the Total Probability Theorem:

$$p(\mathbf{y}|I_0) = \int p(\mathbf{y}|\boldsymbol{\theta}, I_0)p(\boldsymbol{\theta}|I_0)d\boldsymbol{\theta} \tag{4.2}$$

and it is also called the *marginal* density of data, to stress the fact that it is no longer conditional on $\boldsymbol{\theta}$.

The posterior distribution is the result of Bayesian inference. This distribution is then used to find a point estimate of the parameters, to test a hypothesis or, in general, to find credibility intervals, or to predict future data $\mathbf{y}$, conditional on the posterior information I_1. The latter task is done by computing the *predictive* density:

$$p(\mathbf{y}|I_1) = \int p(\mathbf{y}|\boldsymbol{\theta}, I_1)p(\boldsymbol{\theta}|I_1)d\boldsymbol{\theta}. \tag{4.3}$$

Note that (4.3) is essentially (4.2) with I_0 replaced by I_1 which is now the information currently available. If interest is in a subset of the parameters, e.g. $\boldsymbol{\theta}_1$, then the conditional density of $\boldsymbol{\theta}_1$ given I_1 can be obtained from the posterior density $p(\boldsymbol{\theta}|I_1)$ by integrating out the *nuisance* parameters $\boldsymbol{\theta}_2 = \boldsymbol{\theta} \backslash \boldsymbol{\theta}_1$:

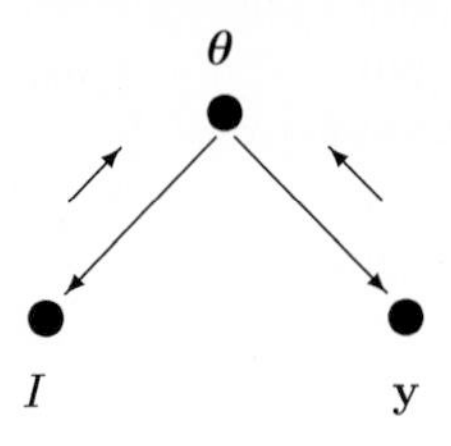

Fig. 4.1. Graphical representation of conditional independence assumptions $i(I, \mathbf{y}|\boldsymbol{\theta})$.

$$p(\boldsymbol{\theta}_1|I_1) = \int p(\boldsymbol{\theta}|I_1) d\boldsymbol{\theta}_2.$$

In particular, inference on a single parameter, say θ_1, is based on its marginal posterior density:

$$p(\theta_1|I_1) = \int p(\boldsymbol{\theta}|I_1) d\theta_2 \ldots d\theta_k.$$

Similarly, inference on any function of the parameters can be performed.

Let now I denote the information currently available and $\mathbf{y}$ be future data. Thus I is either the prior information about a phenomenon or the posterior information resulting from updating of prior information via sample data. We shall see, later on, that this distinction can be relaxed. In some circumstances, it is reasonable to assume that, conditional on $\boldsymbol{\theta}$, knowledge of I is irrelevant to $\mathbf{y}$, and hence

$$p(\mathbf{y}|\boldsymbol{\theta}, I) = p(\mathbf{y}|\boldsymbol{\theta}).$$

In this case, $\mathbf{y}$ and I are said to be *conditionally independent* given $\boldsymbol{\theta}$, and we will write $i(I, \mathbf{y}|\boldsymbol{\theta})$. The conditional independence assumption is reasonable when $\boldsymbol{\theta}$ specifies completely the current state of knowledge, so that I cannot add any relevant information to $\mathbf{y}$, if $\boldsymbol{\theta}$ is known.

The stochastic dependence among I, $\boldsymbol{\theta}$ and $\mathbf{y}$, together with the conditional independence of I and $\mathbf{y}$ given $\boldsymbol{\theta}$ can be graphically represented using the Directed Acyclic Graph (DAG) in Figure 4.1. From a qualitative viewpoint, the two directed links pointing from $\boldsymbol{\theta}$ to I and $\mathbf{y}$ represent the stochastic dependence of I and $\mathbf{y}$ on $\boldsymbol{\theta}$, so that $\boldsymbol{\theta}$ is called a *parent* of I and $\mathbf{y}$, and they are both *children* of $\boldsymbol{\theta}$. There is not a directed link between $\mathbf{y}$ and I, which can "communicate" only via $\boldsymbol{\theta}$. In other words, $\boldsymbol{\theta}$ separates I from $\mathbf{y}$ and this separation can be interpreted as a conditional independence assumption, that is $i(I, \mathbf{y}|\boldsymbol{\theta})$ [415]. The stochastic dependence of I and $\mathbf{y}$ on $\boldsymbol{\theta}$ is quantified by the conditional densities $p(I|\boldsymbol{\theta})$ and $p(\mathbf{y}|\boldsymbol{\theta})$ that are used to sequentially revise the distribution of $\boldsymbol{\theta}$. First, the conditional density $p(\boldsymbol{\theta}|I)$ is computed by processing the information I. This updating operation is represented by the arrow from I

towards $\boldsymbol{\theta}$ that represents the "flow" of information from I to $\boldsymbol{\theta}$ via application of Bayes' Theorem. After this first updating, the probability density associated with $\boldsymbol{\theta}$ is $p(\boldsymbol{\theta}|I)$ and this is going to be the prior density in the next inferencial step. When new data arrive, and their probability density $p(\mathbf{y}|\boldsymbol{\theta}, I) = p(\mathbf{y}|\boldsymbol{\theta})$ is known, the new piece of information is processed locally, along the path $\boldsymbol{\theta} \rightarrow \mathbf{y}$, by applying Bayes' Theorem again, and the conditional density of $\boldsymbol{\theta}$, after this second updating, is $p(\boldsymbol{\theta}|I, \mathbf{y})$. Note that this updating process can be continuously applied so that inference is a continuous, dynamic process in which new data are used to revise the current knowledge. Thus, the posterior information I_1 that is the updating of some prior information and sample data, becomes the current prior information when new data are to be analyzed. Furthermore, this updating process can be iteratively applied to each datum at a time and the inference procedure can process data as a whole (*in batch*), as classical methods do, but it can also process data one at the time (*sequentially*). This incremental nature is a further advantage of Bayesian methods: the statistical analysis of new data does not require to process again data considered so far.

4.3. Bayesian Inference

Suppose we have a sample of n cases $\mathbf{y} = \{y_1, \ldots, y_n\}$, generated from a density function $p(y|\boldsymbol{\theta})$. The density $p(\cdot|\boldsymbol{\theta})$ is known, up to a vector of unknown parameters. We assume that the cases are independent given $\boldsymbol{\theta}$, and hence the joint probability density of the sample is

$$p(\mathbf{y}|\boldsymbol{\theta}) = \prod_{i=1}^{n} p(y_i|\boldsymbol{\theta}).$$

The likelihood function $L(\boldsymbol{\theta}) = p(\mathbf{y}|\boldsymbol{\theta})$ plays a central role in classical methods. For Bayesian methods, the likelihood function is the instrument to pass from the prior density $p(\boldsymbol{\theta}|I_0)$ to the posterior density $p(\boldsymbol{\theta}|I_0, \mathbf{y})$ via Bayes' Theorem. Compared to classical methods, Bayesian methods involve the use of a further element: the prior density of $\boldsymbol{\theta}$. The first step of a Bayesian data analysis is therefore the assessment of this prior density.

4.3.1 Prior Elicitation

The prior density of $\boldsymbol{\theta}$ can arise either as posterior density derived from past data or it can be the *subjective* assessment elicited from some domain expert. Several methods for eliciting prior densities from experts exist, and O'Hagan [407, Ch. 6] reports a comprehensive review.

A common approach is to choose a prior distribution with density function similar to the likelihood function. In doing so, the posterior distribution of $\boldsymbol{\theta}$ will be in the same class and the prior is said to be *conjugate* to the likelihood. Conjugate priors play an important role in Bayesian methods, since their adoption can simplify the integration procedure required by the marginalization in (4.2), because the computation reduces to updating the parameters. A list of standard conjugate distributions is given in [57, Ch. 5].

Example 4.1. Let $y_1, \ldots, y_n|\theta$ be independent variables taking values 0 and 1, and let $\theta = p(Y_i = 1|\theta)$, $\theta \in [0,1]$. The likelihood function is therefore

$$L(\theta) \propto \theta^{\sum_i y_i}(1-\theta)^{n-\sum_i y_i}. \tag{4.4}$$

The parameter θ is univariate, and constrained to be in the interval $[0,1]$. This restriction limits the class of possible priors. A prior density of the form

$$p(\theta|I_0) \propto \theta^{a-1}(1-\theta)^{b-1}, \quad \theta \in [0,1], \quad a, b > 0$$

will be conjugate, since it has the same functional form as the likelihood $\theta^x(1-\theta)^z$, except for the exponents. This distribution is called a *Beta* distribution, with *hyper-parameters* a and b, and it is sometimes denoted by Beta(a,b). The term hyper-parameter is used to distinguish a and b from the parameter θ of the sampling model. Note that, compared to the likelihood function (4.4), the hyper-parameters $a-1$ and $b-1$ of $p(\theta|I_0)$ play the roles of $\sum_i y_i$ and $n - \sum_i y_i$ respectively. Thus, $a-1$ and $b-1$ can be chosen by assuming that the expert has an imaginary sample of 0s and 1s, of size $a+b-2$, and he can distribute the imaginary cases between 0 and 1 as his prior knowledge dictates. The size of this imaginary sample can be used to characterize the subjective confidence of the expert in her/his own assessment. Summaries of this distribution are

$$\begin{aligned}
\mathrm{E}(\theta|I_0) &= \frac{a}{a+b} \\
\text{mode} &= \frac{a-1}{a+b-2} \\
V(\theta|I_0) &= \frac{ab}{(a+b)^2(a+b+1)} = \frac{\mathrm{E}(\theta|I_0)(1-\mathrm{E}(\theta|I_0))}{a+b+1}
\end{aligned}$$

where the mode of a random variable θ with probability density $p(\theta|I_0)$ is defined as the value maximizing the density function. The prior expectation $\mathrm{E}(\theta|I_0)$ corresponds to the marginal probability of Y before seeing any data:

$$\mathrm{E}(\theta|I_0) = \int \theta p(\theta|I_0)d\theta = \int p(Y=1|\theta)p(\theta|I_0)d\theta = p(Y=1|I_0).$$

Since the variance of θ is a decreasing function of $a+b$ for a given mean, the sum of the hyper-parameters $a+b$ is also called the *precision* of the distribution. The posterior density is easily found to be:

$$p(\theta|I_0) \propto \theta^{a+\sum_i y_i - 1}(1-\theta)^{n+b-\sum_i y_i - 1}, \quad \theta \in [0,1]$$

which identifies a Beta distribution with hyper-parameters $a + \sum_i y_i$ and $b + n - \sum_i y_i$. Thus, the posterior precision is the prior precision increased by the sample size n.

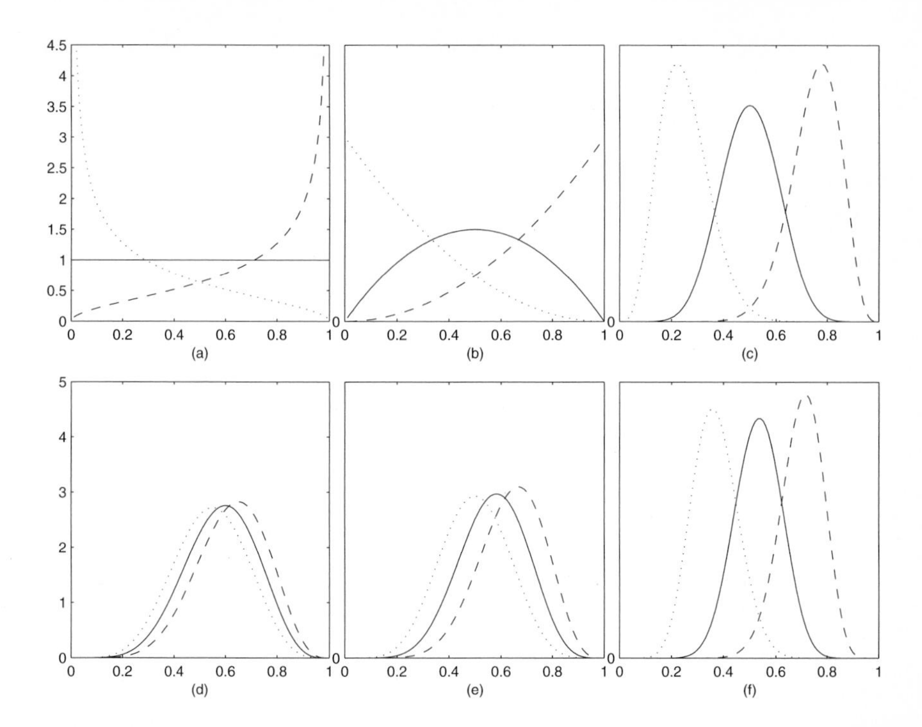

Fig. 4.2. Densities of Beta(a, b) distributions for different choices of the hyper-parameters a and b. Continuous lines report symmetric densities: (a) a=b=1; (b) a=b=2; (c) a=b=10. Dotted lines are positively skewed densities: (a) a=1.5; b=0.5; (b) a=3; b=1; (c) a=15; b=5. Dashed lines represent negatively skewed densities: (a) a=0.5; b=1.5; (b) a=1; b=3; (c) a=5; b=15. Plots (d), (e) and (f) report the corresponding posterior densities derived from a sample of size $n = 10$, and $\sum_i y_i = 6$.

Conjugacy restricts the choice of priors to a limited class of distributions and prior information can only be used to choose the hyper-parameters. However, if the class of distributions is large enough, this limitation is not a serious issue. For instance, in Example 4.1 a choice $a = b = 1$ yields a prior distribution which is uniform in [0,1], and hence uniform prior uncertainty about θ, so that all values are equally likely. Choosing $a = b$ implies that the distribution of θ is symmetrical about the prior mean and mode that are both equal to 0.5. A choice $a < b$ induces negative skew, so that large values of θ are, a priori, more likely. Positive skew can be modeled by chosing $a > b$. Several examples are given in plots (a), (b) and (c) in Figure 4.2. The corresponding posterior densities derived from a sample of size $n = 10$, and $\sum_i y_i = 6$ are given in plots (d), (e) and (f). Figure 4.2 shows that the effect of the prior distribution is small compared to the data when the sample size n is larger than the prior precision (plots (a), (b) versus (d) and (e)): posterior densities are very different from prior densities and exhibit a similar shape. For instance, when the prior hyper-parameters are $a = 0.5$ and $b = 1.5$ (dashed line in plot (a)), the prior distribution assigns larger probability to values of θ larger than 0.5. The impact of the sample information is to concentrate the posterior distribution in the range [0.2,0.9], with a median around 0.6. The difference is less evident in plots (c) and (f), when the prior precision is larger than the sample size: in this case the posterior densities are slightly shifted to take into account the effect of the sample.

An alternative way to assess prior distributions is based on the concept of *Maximum Entropy*: it requires the expert to specify some summaries of the prior distribution, such as the mean or the variance, and it returns the prior distribution having maximum entropy among the class of distributions with the given summaries. This assessment method, due to Jaynes [289, 290], was devised in order to provide probability assessments that are subject only to the available information and the prior returned contains as little information as possible, apart from the summaries specified. In this way, bias due to unintentional subjective components included in the elicitation process is removed, and two experts with the same prior information will return the same prior distribution.

Example 4.2 (Maximum Entropy Priors). When θ is univariate and takes all real values, and the prior mean and variance are specified, the maximum entropy prior is a normal distribution with the specified mean and variance.

An open problem of Bayesian methods is the choice of a prior distribution representing genuine ignorance. When this prior exists, it is called *non-informative*. If a distribution for $\boldsymbol{\theta}$ is non-informative, and we make a parameter transformation $\psi = g(\boldsymbol{\theta})$, then the distribution of ψ must be non-informative. The Jeffreys' rule [291] allows us to find prior distributions that are invariant under reparameterizations. We first need to recall the definition of Fisher information matrix, that was introduced in Chapter 2. If the likelihood function $L(\boldsymbol{\theta})$ is known and we define $l(\boldsymbol{\theta}) = \log L(\boldsymbol{\theta})$, the Fisher information matrix $I(\boldsymbol{\theta}|\mathbf{y})$ is defined as minus the matrix of the second derivatives of the log-likelihood function

$$I(\boldsymbol{\theta}|\mathbf{y}) = -\left\{\frac{\partial^2 \mathbf{l}(\boldsymbol{\theta})}{\partial\theta_\mathbf{i}\partial\theta_\mathbf{j}}\right\}.$$

The expected Fisher information matrix is the matrix $I(\boldsymbol{\theta})$ whose elements are the expectations — over the conditional distribution of the data given $\boldsymbol{\theta}$ — of $I(\boldsymbol{\theta}|\mathbf{y})$ and hence $I(\boldsymbol{\theta}) = \mathrm{E}(I(\boldsymbol{\theta}|\mathbf{y}))$. The matrix $I(\boldsymbol{\theta})$ is used to formulate the Jeffreys prior that has density

$$p(\boldsymbol{\theta}|I_0) \propto \det\{I(\boldsymbol{\theta})\}^{1/2},$$

with $\det(\cdot)$ denoting the determinant of a matrix. If $\psi = g(\boldsymbol{\theta})$, then $p(\psi|I_0) \propto \det\{I(\psi)\}^{1/2}$, and the prior distribution is invariant with respect to reparameterization. In most cases, Jeffreys priors are not technically probability distributions, since their density functions do not have finite integrals over the parameter space, and are therefore termed *improper* priors. It is often the case that Bayesian inference based on improper priors returns proper posterior distributions which then turn out to be numerically equivalent to the results of classical inference [82].

Example 4.3. Let $y_1, \ldots, y_n|\theta$ be independent, normally distributed variates with mean θ and known variance σ^2. Then,

$$p(\mathbf{y}|\theta) \propto \exp\{-n(\bar{y} - \theta)^2/2\sigma^2\}$$

and $I(\theta) = n/\sigma^2$, so that the Jeffreys prior for θ is the (improper) uniform distribution over the real numbers. Nonetheless, the posterior distribution is

$$\theta|\mathbf{y} \sim \mathcal{N}(\bar{y}, \sigma^2/n)$$

which is a proper distribution.

Problems related to the use of improper prior distributions can be overcome by assigning prior distributions that are as uniform as possible but still remain probability distributions. For instance, in Example 4.3, the prior distribution can be normal with a very large variance. This would also be the Maximum Entropy prior, when the prior knowledge is extremely uncertain.

The use of uniform prior distribution to represent uncertainty clearly assumes that equally probable is an adequate representation of lack of information. Recent advances question this assumption and advocate the use of bounds on the set of possible values that $\boldsymbol{\theta}$ can take [221, 530]. Finally, it is worth mentioning that prior distributions elicited from several experts can be combined into a single mixture of different distributions with weights representing the reliability of each expert (see [407] for more references.)

4.3.2 Estimation

Bayesian inference returns the posterior density $p(\boldsymbol{\theta}|I_0, \mathbf{y}) = p(\boldsymbol{\theta}|I_1)$. Marginal inference on parameters of interest is then based on the marginal posterior distribution, and for an individual parameter θ_1 on

$$p(\theta_1|I_1) = \int p(\boldsymbol{\theta}|I_1) d\theta_2 d\theta_3 \ldots d\theta_k.$$

A standard point estimate of θ_1 is the posterior expectation:

$$\mathrm{E}(\theta_1|I_1) = \int \theta_1 p(\theta_1|I_1) d\theta_1. \tag{4.5}$$

An alternative point estimate, based on a principle similar to Maximum Likelihood (see Chapters 2 and 3), is the posterior mode:

$$\hat{\theta}_1 = \arg\max_{\theta} p(\boldsymbol{\theta}|I_1). \tag{4.6}$$

Since $p(\boldsymbol{\theta}|I_1)$ is proportional to $p(\mathbf{y}|\boldsymbol{\theta})p(\boldsymbol{\theta}|I_0)$, the vector of posterior modes $\hat{\boldsymbol{\theta}}$ maximizes the *augmented* likelihood $p(\mathbf{y}|\boldsymbol{\theta})p(\boldsymbol{\theta}|I_0)$, and it is therefore called the Generalized Maximum Likelihood Estimate (GML estimate) of $\boldsymbol{\theta}$. When the parameters $\theta_1, \ldots, \theta_k$ are, a posteriori, independent then

$$p(\boldsymbol{\theta}|I_1) = \prod_i p(\theta_i|I_1)$$

and the posterior mode of individual parameters can be computed from the marginal posterior densities.

Example 4.1 (continued). Since the posterior distribution is still Beta, the posterior mean is

$$\begin{aligned}
\mathrm{E}(\theta|I_1) = \frac{a + \sum_i y_i}{a+b+n} &= \frac{a+b}{a+b+n}\frac{a}{a+b} + \frac{n}{a+b+n}\frac{\sum_i y_i}{n} \\
&= \frac{a+b}{a+b+n}\mathrm{E}(\theta|I_0) + \frac{n}{a+b+n}\bar{y}
\end{aligned}$$

which is a weighted average of the prior mean $\mathrm{E}(\theta|I_0)$ and sample mean $\bar{y}$, with weights depending on the sample size n and the prior precision $a+b$. If $n < a+b$, the prior mean has a larger weight than the posterior mean. As the sample size increases, prior information becomes negligible and the posterior mean approximates the ML estimate $\bar{y}$ of θ. The posterior mode

$$\hat{\boldsymbol{\theta}}_1 = \frac{a + \sum_i y_i - 1}{a+b+n-2}$$

reduces to the standard ML estimate of θ when $a = b = 1$, and hence the prior distribution assumes that all values of θ are equally likely.

Example 4.4. Let $y_1, \ldots, y_n|\theta$ be independent, normally distributed with mean θ and known variance $\sigma^2 \equiv \tau^{-2}$, where τ^2 is called *precision*. Then,

$$p(\mathbf{y}|\theta) \propto \exp\{-n\tau^2(\bar{y}-\theta)^2/2\}$$

and $\theta|I_0 \sim \mathcal{N}(\mu_0, \sigma_0^2)$ is a conjugate prior with hyper-parameters μ_0 and $\sigma_0^2 \equiv \tau_0^{-2}$, (this is also the Maximum Entropy prior for specified mean and variance.) By conjugacy, the posterior distribution is $\mathcal{N}(\mu_1, \sigma_1^2)$ and it can be easily shown, for instance [407, page 7], that the posterior hyper-parameters are:

$$\begin{aligned}
\sigma_1^{-2} = \tau_1^2 &= \tau_0^2 + n\tau^2 \\
\mu_1 &= \frac{\tau_0^2\mu_0 + n\tau^2\bar{y}}{\tau_1^2}.
\end{aligned}$$

Thus, the posterior mean is again a weighted average of prior and sample means, with weights that depend on the sample size n, the *datum precision* τ^2 and the *prior precision* τ_0^2. As in Example 4.1, there is a trade-off between data and prior information, and for small samples the prior mean has a weight larger than the sample mean. As the sample size increases, the prior input becomes negligible, and asymptotically the Bayesian estimate is the sample mean $\bar{y}$. By symmetry, the posterior mode and mean are coincident.

4.3.3 Credibility Intervals

Posterior mean and mode provide simple summaries of the posterior distribution that can be further used to evaluate the probability that $\boldsymbol{\theta}$ is in some given region R, or to find a region R that contains $\boldsymbol{\theta}$ with a specified probability $1-\alpha$. The latter is called a $(1-\alpha)\%$ *credibility region*, i.e.

$$p(\boldsymbol{\theta} \in R|I_1) = 1 - \alpha.$$

When R is the region of smallest volume, it is also called the *Posterior Highest Density* (PHD) region. When $\boldsymbol{\theta}$ is the univariate parameter θ, the PHD region is an interval. If the posterior density of θ is unimodal — i.e. it has a unique mode — the PHD interval is $[l_1, l_2]$ where the two values l_1 and l_2 are such that

$$\begin{aligned} \int_{l_1}^{l_2} p(\theta|I_1)d\theta &= \ 1-\alpha \\ p(l_1|I_1) \quad &= p(l_2|I_1). \end{aligned}$$

Computational methods for finding the PHD region in particular problems can be found in [56, page 140].

Example 4.3 (continued). Given that the posterior distribution of θ is $\mathcal{N}(\bar{y}, \sigma^2/n)$, a $(1-\alpha)\%$ PHD interval is easily found to be:

$$\bar{y} \pm z_{\alpha/2}\frac{\sigma}{\sqrt{n}}; \quad p(Z > z_{\alpha/2}) = \alpha/2; \quad Z \sim \mathcal{N}(0,1)$$

which is identical to the classical $(1-\alpha)\%$ confidence interval for the mean of a normal population when the variance is known (see Chapter 2). However, the meaning of the latter is different. The frequentist interpretation of the $(1-\alpha)\%$ confidence interval is based on the repeatability of the sampling process, so that if we could take, say, 100 samples, we would expect that in $(1-\alpha)\%$ of cases the interval $\bar{y} \pm z_{\alpha/2}\frac{\sigma}{\sqrt{n}}$ *contains* the true value of θ. The $(1-\alpha)\%$ PHD interval returned by the Bayesian method is a credibility statement, conditional on the information I_1 currently available: we believe that, with probability $(1-\alpha)$, θ *belongs* to the interval $\bar{y} \pm z_{\alpha/2}\frac{\sigma}{\sqrt{n}}$.

4.3.4 Hypothesis Testing

The Bayesian approach to hypothesis testing is based on the computation of the conditional probability of a hypothesis H given the information currently available. Thus, when the null hypothesis $H_0 : \boldsymbol{\theta} \in \Theta_0$ and the alternative hypothesis $H_1 : \boldsymbol{\theta} \in \Theta_1$, with $\Theta_0 \cap \Theta_1 = \emptyset$, are formulated, there are prior probabilities on both of them, say $p(H_0|I_0)$ and $p(H_1|I_0)$, with $p(H_0|I_0) + p(H_1|I_0) = 1$. By the Total Probability Theorem (applied to the discrete case), the prior density of $\boldsymbol{\theta}$ is then:

$$p(\boldsymbol{\theta}|I_0) = p(\boldsymbol{\theta}|H_0, I_0)p(H_0|I_0) + p(\boldsymbol{\theta}|H_1, I_0)p(H_1|I_0)$$

where $p(\boldsymbol{\theta}|H_i, I_0)$ are the prior densities of $\boldsymbol{\theta}$, conditional on each hypothesis. The sample information is then used to compute from the *prior odds*

$$\frac{p(H_0|I_0)}{p(H_1|I_0)}$$

the *posterior odds* in favor of H_0 as

$$\frac{p(H_0|I_1)}{p(H_1|I_1)} = \frac{p(\mathbf{y}|H_0)}{p(\mathbf{y}|H_1)} \frac{p(H_0|I_0)}{p(H_1|I_0)},$$

from which the following decision rule is derived:

$$\begin{array}{ll} \text{if } p(H_0|I_1) < p(H_1|I_1) & \text{Reject } H_0 \\ \text{if } p(H_0|I_1) > p(H_1|I_1) & \text{Accept } H_0 \\ \text{if } p(H_0|I_1) = p(H_1|I_1) & \text{Undecidability.} \end{array}$$

Compared to classical methods, in which the sampling variability is taken into account in the definition of the rejection region of the test (see Chapter 2), the Bayesian approach to hypothesis testing is to accept, as true, the hypothesis with the largest posterior probability, since it is the most likely given the information available. The ratio $p(\mathbf{y}|H_0)/p(\mathbf{y}|H_1)$ is called the *Bayes factor*, and when the prior probabilities of H_0 and H_1 are equal, the Bayes factor determines the decision rule. The evaluation of the Bayes factor involves the computation of two quantities:

$$\begin{array}{l} p(\mathbf{y}|H_0) = \int p(\mathbf{y}|H_0, \boldsymbol{\theta}) p(\boldsymbol{\theta}|H_0, I_0) d\boldsymbol{\theta} \\ p(\mathbf{y}|H_1) = \int p(\mathbf{y}|H_1, \boldsymbol{\theta}) p(\boldsymbol{\theta}|H_1, I_0) d\boldsymbol{\theta} \end{array}$$

representing the marginal densities of the data on the two parameter spaces specified in H_0 and H_1 respectively. When the two hypotheses are simple, that is, they specify completely the parameters values as $H_0 : \boldsymbol{\theta} = \boldsymbol{\theta}_0$ and $H_1 : \boldsymbol{\theta} = \boldsymbol{\theta}_1$, then the Bayes factor reduces to the classical likelihood ratio test to discriminate between two simple hypotheses. Further details can be found in [56, Ch. 4].

Example 4.5. Let $y_1, \ldots, y_n|\theta$ be independent, identically distributed Poisson variates with mean θ. Thus

$$p(y_i|\theta) = \frac{\theta^{y_i}}{y_i!} e^{-\theta}; \quad \theta > 0 \quad y_i = 0, 1, 2 \ldots$$

Let $H_0 : \theta = \theta_0$ and $H_1 : \theta = \theta_1$ be two simple hypotheses, with $p(H_0|I_0) = p(H_1|I_0)$. The Bayes factor is

$$\left(\frac{\theta_0}{\theta_1}\right)^{\sum_i y_i} e^{\mathcal{N}(\theta_1 - \theta_0)}$$

and hence, since the prior odds are equal to 1, the decision rule is to accept H_0 if the Bayes factor is greater than 1.

4.4. Bayesian Modeling

The examples discussed in Sections 4.3.1–4.3.4 are "toy" examples used to explain the Bayesian approach. In this section, we will focus on more realistic models in which a response variable Y is a function of some covariates $X_1, \ldots, X_c \in \mathcal{X}$. We begin by considering the multiple linear regression model, in which data have a normal distribution whose expectation is a linear function of the parameters. We then consider Bayesian methods for the analysis of Generalized Linear Models, which provide a general framework for cases in which normality and linearity are not viable assumptions. These cases point out the major computational bottleneck of Bayesian methods: when the assumptions of normality and/or linearity are removed, usually the posterior distribution cannot be computed in closed form. We will then discuss some computational methods to approximate this distribution.

4.4.1 Multiple Linear Regression

We begin by considering the standard multiple linear regression model, in which data are assumed to have the distribution

$$Y_i|(\mu_i, \tau^2) \sim \mathcal{N}(\mu_i, \tau^{-2}) \tag{4.7}$$

conditional on μ_i and τ^2. The expectation μ_i is a linear function of the regression parameters $\boldsymbol{\beta}$, with coefficients that are functions of the regression variables $X_1, \ldots, X_c$:

$$\mu_i = f(\mathbf{x}_i)^T \boldsymbol{\beta} = \beta_0 + \sum_{j=1}^{p} \beta_j f_j(\mathbf{x}_i) \equiv \beta_0 + \sum_{j=1}^{p} \beta_j t_{ij}.$$

The function $f(\cdot)$ is defined in $\mathcal{X}$ and takes values in $\mathcal{T} \subset \mathcal{R}^{p+1}$. This definition allows us to consider general regression models as polynomial regression. For example, with $c = 1$ and $f(x_i) = (1, x_i, x_i^2)^T$ we have a quadratic regression model $\mu_i = \beta_0 + \beta_1 x_i + \beta_2 x_i^2$. The linear model can be written in matrix form as

$$\mathbf{Y}|\boldsymbol{\theta} \sim \mathcal{N}(X\boldsymbol{\beta}, \tau^{-2} I_n)$$

where X is the $n \times (p+1)$ design matrix, whose i-th row contains the coefficients $f(\mathbf{x}_i)$ of the regression parameters, and I_n is the $n \times n$ identity matrix.

Parameter Estimation The parameter vector $\boldsymbol{\theta}$ is given by $(\boldsymbol{\beta}, \tau^2)$. We further suppose $i(Y_1, \ldots, Y_n|\boldsymbol{\theta})$, so that the likelihood function is:

$$L(\boldsymbol{\theta}) \propto \tau^n \exp(-\tau^2 \sum_i (y_i - f(\mathbf{x}_i)^T \boldsymbol{\beta})^2/2).$$

When both $\boldsymbol{\beta}$ and τ^2 are unknown, the simplest analysis is obtained by assuming a conjugate prior for $\boldsymbol{\theta}$, which is specified in two steps:

(i) Conditional on τ^2, we assume

$$\boldsymbol{\beta}|\tau^2, I_0 \sim \mathcal{N}(\boldsymbol{\beta}_0, (\tau^2 R_0)^{-1})$$

where $\tau^2 R_0$ is the prior precision.

(ii) The datum precision τ^2 is assigned a prior distribution:

$$\tau^2|I_0 \sim \chi^2_{\nu_0}/(\nu_0\sigma_0^2)$$

which corresponds to assigning the error variance σ^2 an *Inverse Gamma* distribution [57, page 119] with density function:

$$p(\sigma^2|I_0) \propto (\sigma^2)^{-(\nu_0+2)/2} \exp(-\nu_0\sigma_0^2/(2\sigma^2)).$$

The specification of the hyper-parameters ν_0, σ_0^2, $\boldsymbol{\beta}_0$ and R_0 allows the encoding of the prior information I_0. For instance, the expectation and variance of τ^2 are respectively σ_0^{-2} and $2\sigma_0^{-4}/\nu_0$, so that the expert's information on the variability of the data can be used to define σ_0^{-2}, while the choice of ν_0 may represent the expert's assessment of his ability in terms of size of an imaginary sample used to elicit this prior distribution [407, Ch. 9]. The prior hyper-parameters and the distribution chosen imply that

$$\mathrm{E}(\sigma^2|I_0) = \frac{\nu_0\sigma_0^2}{\nu_0 - 2}, \quad V(\sigma^2|I_0) = \frac{2\nu_0^2\sigma_0^4}{(\nu_0-2)^2(\nu_0-4)}$$

provided that $\nu_0 > 4$. For $2 < \nu_0 \leq 4$ the variance does not exists, and for $\nu_0 \leq 2$ the mean does not exist.

Similarly, $\boldsymbol{\beta}_0$ represents the prior information about the regression model, while R_0 is a measure of the precision of this assessment, for a fixed value of τ^2. In this way, the prior elicitation of the distribution of $\boldsymbol{\beta}$ can be done independently of the sampling variability. The marginal variance of $\boldsymbol{\beta}$ can be easily found to be:

$$V(\boldsymbol{\beta}|I_0) = \mathrm{E}\{V(\boldsymbol{\beta}|I_0, \tau^2)\} + V\{\mathrm{E}(\boldsymbol{\beta}|I_0, \tau^2)\} = \frac{\nu_0\sigma_0^2}{\nu_0 - 2} R_0^{-1}.$$

The joint prior distribution is known as a *Normal-Inverse-Gamma* prior. The elicitation of the prior distribution in two steps and the conditional independence assumptions are represented by the DAG — the Directed Acyclic Graph as defined in Section 4.2 — in Figure 4.3. The link from τ^2 to $\boldsymbol{\beta}$ represents the two-step prior elicitation process described above, so that the distribution of τ^2 depends only on the information I_0 currently available, and then the distribution of $\boldsymbol{\beta}$ is elicited, conditional on τ^2. For consistency, the node τ^2 should be child of a root node I_0 representing the prior information collected from past experience. For simplicity, this node has been removed from the graph. The paths from τ^2 and $\boldsymbol{\beta}$ to $Y_1, \ldots, Y_n$ — via μ_i — specify the sampling model, and hence the stochastic dependence of Y on $\boldsymbol{\beta}$ — via μ — and on τ^2. The conditional independence $i(Y_1, \ldots, Y_n|\boldsymbol{\theta})$ is represented by the lack of directed links among

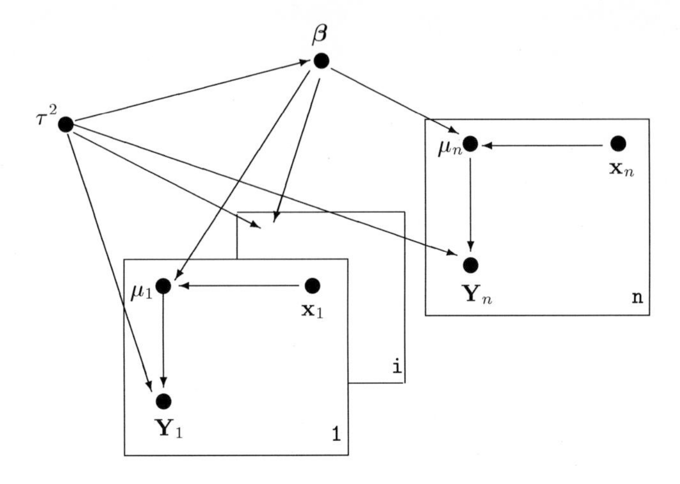

Fig. 4.3. Graphical representation of a regression model with independent observations given $\boldsymbol{\theta} = (\boldsymbol{\beta}, \tau^2)$.

$Y_1, \ldots, Y_n$ that can communicate only via τ^2 and $\boldsymbol{\beta}$. The conditional independence of the observations given $\boldsymbol{\theta}$ is encoded by representing the associations among Y_i, μ_i and $\mathbf{x}_i$ on different plateaux, one for each observation.

The quantification of the dependencies is done by associating the prior distribution $\chi^2_{\nu_0}/(\nu_0 \sigma_0^2)$ to the root node τ^2, and the distribution $\mathcal{N}(\boldsymbol{\beta}_0, (\tau^2 R_0)^{-1})$ to the node $\boldsymbol{\beta}$. The joint information of $\boldsymbol{\beta}$ and values of the covariates are summarized into the nodes $\mu_1, \ldots, \mu_n$ that represent linear functions of $\boldsymbol{\beta}$ with coefficients $f(\mathbf{x}_i)$ and hence they inherit their variability from the "stochastic" parents $\boldsymbol{\beta}$. The sampling models $\mathcal{N}(\mu_i, \tau^{-2})$ are attached to the nodes $Y_1, \ldots, Y_n$.

Data are then processed by applying Bayes' Theorem in an order which is opposite to the order of elicitation of the prior density, so that first there is a flow of information from $Y_1, \ldots, Y_n$ to $\boldsymbol{\beta}$ and the conditional posterior distribution is found to be multivariate normal with updated hyper-parameters:

$$\boldsymbol{\beta}|I_1, \tau^2 \sim \mathcal{N}(\boldsymbol{\beta}_1, (\tau^2 R_1)^{-1}) \tag{4.8}$$

$$R_1 = (R_0 + X^T X) \tag{4.9}$$

$$\boldsymbol{\beta}_1 = R_1^{-1}(R_0 \boldsymbol{\beta}_0 + X^T \mathbf{y}). \tag{4.10}$$

Thus, for fixed τ^2, the posterior precision of $\boldsymbol{\beta}$ is increased by the datum precision $X^T X$, that is the expected Fisher information matrix, and the posterior expectation is an average of prior expectation $\boldsymbol{\beta}_0$ and data $\mathbf{y}$. Thus, a point estimate of $\boldsymbol{\beta}$ (conditional on τ^2) is

$$\mathrm{E}(\boldsymbol{\beta}|I_1, \tau^2) = R_1^{-1} R_0 \boldsymbol{\beta}_0 + R_1^{-1} X^T \mathbf{y} = R_1^{-1} R_0 \boldsymbol{\beta}_0 + (R_0 + X^T X)^{-1} X^T \mathbf{y}$$

and compared to the ML estimate $\hat{\beta} = (X^T X)^{-1} X^T \mathbf{y}$ (see Chapter 3) there is an adjustment to take into account prior information. As the weight of the prior information becomes negligible compared to the sample information, the Bayesian estimate approximates $\hat{\beta}$.

The next step is to find the marginal posterior distribution of τ^2 that, by conjugacy, is

$$\tau^2 | I_1 \sim \chi^2_{\nu_1} / (\nu_1 \sigma_1^2) \tag{4.11}$$

$$\nu_1 = \nu_0 + n \tag{4.12}$$

$$\nu_1 \sigma_1^2 = \nu_0 \sigma_0^2 + \mathbf{y}^T \mathbf{y} + \beta_0^T R_0 \beta_0 - \beta_1^T R_1 \beta_1. \tag{4.13}$$

in which the degrees of freedom are increased by the sample size n. Denote by $\hat{\mathbf{y}} = X\hat{\beta}$ the fitted values of the classical regression model and let RSS denote the residual sum of squares $(\mathbf{y} - \hat{\mathbf{y}})^T (\mathbf{y} - \hat{\mathbf{y}})$, so that $\hat{\sigma}^2 = \text{RSS}/(n-p-1)$ is the classical unbiased estimate of the error variance (see Chapters2 and 3.) Then, it can be shown [407, page 249] that

$$\nu_1 \sigma_1^2 = \nu_0 \sigma_0^2 + \text{RSS} + (\beta_0 - \hat{\beta})^T (R_0^{-1} + (X^T X)^{-1})^{-1} (\beta_0 - \hat{\beta})$$

so that the expected posterior variance is

$$\text{E}(\sigma^2 | I_1) = \frac{\nu_0 - 2}{\nu_0 + n - 2} \text{E}(\sigma^2 | I_0) + \frac{n - p - 1}{\nu_0 + n - 2} \hat{\sigma}^2 + \frac{p+1}{\nu_0 + n - 2} d$$

where $d = (\beta_0 - \hat{\beta})^T (R_0^{-1} + (X^T X)^{-1})^{-1} (\beta_0 - \hat{\beta})/(p+1)$. Therefore, $\text{E}(\sigma^2|I_1)$ provides an estimate of the error variance, which combines prior information $\text{E}(\sigma^2|I_0)$, the standard unbiased estimate of the error variance $\hat{\sigma}^2$ and d, that is a weighted discrepancy between β_0 and $\hat{\beta}$. Thus, a large discrepancy between β_0 and $\hat{\beta}$ represents the fact that the prior information is scarce and hence the expected posterior variance is large.

Inference on the regression parameters is performed by computing the marginal posterior distribution of β:

$$\beta | I_1 = \beta_1 + \frac{\mathcal{N}(\mathbf{0}, \sigma_1^2 R_1^{-1})}{\sqrt{\chi^2_{\nu_1}/\nu_1}}$$

which is a non-central multivariate t-distribution — that is a multivariate t-distribution, with non zero expectation [57, page 139] — and it can be used to provide credibility regions or PHD regions. For instance, with $t_{\alpha/2}$ denoting the upper $\alpha/2$ quantile of a Student's t distribution on ν_1 degrees of freedom, the $(1-\alpha)100\%$ PHD interval for β_i is given by:

$$\beta_{1i} \pm t_{\alpha/2} \sqrt{\sigma_1^2 v_i}$$

where v_i denotes the ith diagonal element of $(R_0 + X^T X)^{-1}$. Further details can be found for instance in [407, Ch. 9]. Prior uncertainty can be modeled by assuming $R_0 \approx O$, $\nu_0 = -(p+1)$ and $\sigma_0^2 = 0$, so that

$$\tau^2|I_1 \sim \chi^2_{n-p-1}/\text{RSS} \tag{4.14}$$

$$\boldsymbol{\beta}|\tau^2, I_1 \sim \mathcal{N}(\hat{\boldsymbol{\beta}}, (\tau X^T X)^{-1}) \tag{4.15}$$

and

$$\begin{aligned}
&\nu_1 = n - p - 1 \\
&R_1 = X^T X \\
&\text{E}(\boldsymbol{\beta}|I_1) = \hat{\boldsymbol{\beta}} \\
&\text{E}(\sigma^2|I_1) = \text{RSS}/(n - p - 1)
\end{aligned}$$

In this way, the Bayesian estimates of σ^2 and $\boldsymbol{\beta}$ reduce to the classical ML estimate.

Example 4.6 (Simple Linear Regression).

Suppose that we are interested in a simple linear regression model for Y on X. We assume that

$$Y|(\boldsymbol{\beta}, \tau) \sim \mathcal{N}(\mu = \beta_0 + \beta_1 x, \tau^{-2})$$

and the parameters are assigned the prior distributions:

$$\tau^2|I_0 \sim \chi^2_{\nu_0}/(\nu_0 \sigma_0^2)$$

and

$$\boldsymbol{\beta} = \begin{pmatrix} \beta_0 \\ \beta_1 \end{pmatrix} |\tau^2, I_0 \sim \mathcal{N}\left(\begin{pmatrix} \beta_{00} \\ \beta_{01} \end{pmatrix}, \left(\tau^2 \begin{pmatrix} r_{00} & r_{01} \\ r_{01} & r_{11} \end{pmatrix}\right)^{-1}\right).$$

With a sample of n independent observations given $\boldsymbol{\theta} = (\tau^2, \beta_0, \beta_1)$, the posterior distribution of the parameters is

$$\tau^2|I_1 \sim \chi^2_{\nu_1}/(\nu_1 \sigma_1^2); \quad \boldsymbol{\beta}|\tau^2, I_1 \sim \mathcal{N}\left(\boldsymbol{\beta}_1, \left(\tau^2 R_1\right)^{-1}\right)$$

with

$$R_1 = \begin{pmatrix} r_{00} + n & r_{01} + n\bar{x} \\ r_{01} + n\bar{x} & r_{11} + \sum_i x_i^2 \end{pmatrix}.$$

A choice $R_0 \approx O$, $\nu_0 = -2$ and $\sigma_0^2 = 0$ yields

$$\tau^2|I_1 \sim \chi^2_{n-2}/\text{RSS}; \quad \boldsymbol{\beta}|\tau^2, I_1 \sim \mathcal{N}(\hat{\boldsymbol{\beta}}, (\tau^2 X^T X)^{-1})$$

and inference on the regression parameters is based on the Student's t distributions:

$$\frac{\beta_0 - \beta_{10}}{\sqrt{\text{RSS} v_1/(n-2)}} \sim t_{n-2}; \qquad \frac{\beta_1 - \beta_{11}}{\sqrt{\text{RSS} v_2/(n-2)}} \sim t_{n-2}$$

where v_1, v_2 are the diagonal elements of $R_1^{-1} = (X^T X)^{-1}$.

The posterior distribution of the parameters can also be used to predict new cases $\tilde{\mathbf{y}}|\boldsymbol{\theta} \sim \mathcal{N}(\tilde{X}\boldsymbol{\beta}, \tau^{-2}I_m)$. The analysis is based on the evaluation of the conditional distribution of $\tilde{\mathbf{y}}|\mathbf{y}$, which again involves the use of a multivariate Students' t distribution. Details are given for instance in [407, Ch. 9], and application to real data-sets are provided by [208]. The generalization of the approach described in this section to models with correlated observations involves the use of "matrix-form" distributions as Wishart and can be found in [57, 82].

Model Selection Our task becomes more challenging when we are interested in discovering the statistical model best fitting the available information. Let $\mathcal{M} = \{M_0, M_1, \ldots, M_m\}$ be the set of models that is believed a priori to contain the true model of dependence of $\mathbf{y}$ on $X_1, \ldots, X_c$. For instance with $c = 1$, the set $\mathcal{M}$ can contain the nested models

$$\begin{aligned} M_0 &: \mu = \beta_0 \\ M_1 &: \mu = \beta_0 + \beta_1 x \\ M_2 &: \mu = \beta_0 + \beta_1 x + \beta_2 x^2. \end{aligned}$$

Each model induces a parameterization $M_i \to \boldsymbol{\theta}^{(i)} = (\boldsymbol{\beta}^{(i)}, \tau^2)$, e.g. $\boldsymbol{\beta}^{(0)} = \beta_0$, $\boldsymbol{\beta}^{(1)} = (\beta_0, \beta_1)^T$, $\boldsymbol{\beta}^{(2)} = (\beta_0, \beta_1, \beta_2)^T$ in the example above. Prior information I_0 allows the elicitation of prior probabilities of $M_0, \ldots, M_m$, and, conditional on M_i, of prior distributions for $\boldsymbol{\theta}^{(i)}$. Graphically, this is equivalent to assume the existence of a further node corresponding to $\mathcal{M}$ in Figure 4.3, with links towards τ^2 and $\boldsymbol{\beta}^{(i)}$, and the parameterization induced by each model is represented by a different plateau, as in Figure 4.4. Note that the parameters $\boldsymbol{\beta}^{(i)}$ are conditionally independent given M and τ^2. Suppose that we wish to use the prior and sample information to select a regression model M_i from $\mathcal{M}$. We can use the sample information to compute the posterior probability of M_i given the data and the prior information I_0

$$p(M_i|I_1) = \frac{p(M_i|I_0)p(\mathbf{y}|M_i)}{p(\mathbf{y}|I_0)},$$

and then we choose the model with the largest posterior probability. Since the denominator is constant, in the comparison between rival models M_i and M_j, M_i is chosen if

$$p(M_i|I_0)p(\mathbf{y}|M_i) > p(M_j|I_0)p(\mathbf{y}|M_j).$$

When $p(M_i|I_0) = p(M_j|I_0)$ the model choice reduces to the evaluation of the Bayes factor

$$\frac{p(\mathbf{y}|M_i)}{p(\mathbf{y}|M_j)}$$

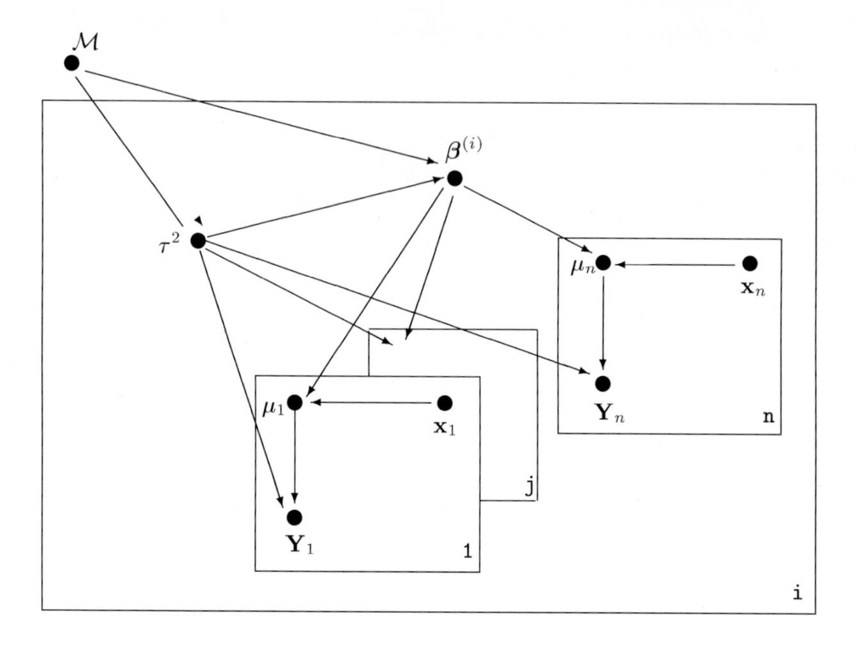

Fig. 4.4. Graphical representation of a regression model with independent observations given $\boldsymbol{\theta}^{(i)} = (\boldsymbol{\beta}, \tau^2)$, conditional on a model.

and M_i is chosen if the Bayes factor is greater than one. The quantity $p(\mathbf{y}|M_i)$ is the *marginal likelihood* (marginal with respect to $\boldsymbol{\theta}^{(i)}$) of the data given the model M_i, which is computed as:

$$p(\mathbf{y}|M_i) = \int p(\mathbf{y}|\boldsymbol{\beta}^{(i)}, \tau^2) p(\boldsymbol{\beta}^{(i)}, \tau^2|I_0) d\boldsymbol{\beta}^{(i)} d\tau^2$$

and the integral has the closed form solution:

$$p(\mathbf{y}|M_i) = \frac{\det(R_0^{(i)})^{1/2} (\nu_0 \sigma_0^2)^{\nu_0/2} \Gamma(\nu_1/2)}{\det(R_1^{(i)})^{1/2} (\nu_1 \sigma_1^2)^{\nu_1/2} \Gamma(\nu_0/2) \pi^{n/2}}$$

where the indices 0 and 1 specify the prior and posterior hyper-parameters of $\boldsymbol{\beta}^{(i)}$ and τ^2. We note that the approach described here gives the same weight to the likelihood and the complexity of a model. More advanced techniques, based on Decision Theory or Information Theory, let us trade off between the complexity and the likelihood of a model. A complete treatment of this problem can be found in [56, Ch. 4] and [407, Ch. 9].

When the inference task is limited to the prediction of future cases, a weighted average of the models in $\mathcal{M}$, with the posterior probability of each model as weights, can be used instead of one single model [258].

4.4.2 Generalized Linear Models

Generalized Linear Models (GLM) provide a unified framework to encompass several situations which are not adequately described by the assumptions of normality of the data and linearity in the parameters. As described in Chapter 3, the features of a GLM are the fact that the distribution of $Y|\boldsymbol{\theta}$ belongs to the exponential family, and that a transformation of the expectation of the data, $g(\mu)$, is a linear function of the parameters $f(\mathbf{x}_i)^T\boldsymbol{\beta}$. The parameter vector is made up of $\boldsymbol{\beta}$ and of the dispersion parameter ϕ. The problem with a Bayesian analysis of GLMs is that, in general, the posterior distribution of $\boldsymbol{\theta}$ cannot be calculated exactly, since the marginal density of the data

$$p(\mathbf{y}|I_0) = \int p(\mathbf{y}|I_0, \boldsymbol{\theta})p(\boldsymbol{\theta}|I_0)d\boldsymbol{\theta}$$

cannot be evaluated in closed form, as the next example shows.

Example 4.7 (Logistic regression). Suppose that, conditional on the vector of parameters $\boldsymbol{\theta} = \boldsymbol{\beta}$, data have a Binomial distribution with $p(Y = 1|\boldsymbol{\theta}) = \mu$. The dispersion parameter is $\phi = 1$. The logit function is the canonical link (see Chapter 3)

$$g(\mu_i) = \log \frac{\mu_i}{1-\mu_i} = \eta_i = f(\mathbf{x}_i)^T\boldsymbol{\beta}.$$

Let $p(\boldsymbol{\beta}|I_0)$ be the prior density. With a sample of n cases — corresponding to n combinations of values of the covariates — and supposed to be conditionally independent given $\boldsymbol{\beta}$, the likelihood function is

$$L(\boldsymbol{\beta}) = \prod_{i=1}^{n} \mu_i^{y_i}(1-\mu_i)^{1-y_i} = \prod_{i=1}^{n} \frac{e^{\eta_i y_i}}{1+e^{\eta_i}}$$

and the marginal density of the data solves the integral

$$\int \prod_{i=1}^{n} \frac{e^{\eta_i y_i}}{1+e^{\eta_i}} p(\boldsymbol{\beta}|I_0)d\boldsymbol{\beta}. \tag{4.16}$$

To date, there are no known prior distributions which lead to a closed form solution of (4.16).

Numerical integration techniques [407] can be exploited to approximate (4.16), from which a numerical approximation of the posterior density of $\boldsymbol{\beta}$ can be found. The basic idea is to select a grid of points $\{\boldsymbol{\beta}_1, \ldots, \boldsymbol{\beta}_g\}$ and replace the value of the integral by their weighted sum:

$$\int \prod_{i=1}^{n} \frac{e^{\eta_i y_i}}{1+e^{\eta_i}} p(\boldsymbol{\beta}|I_0)d\boldsymbol{\beta} \approx \sum_j w_j \prod_{i=1}^{n} \frac{e^{\eta_{ij} y_i}}{1+e^{\eta_{ij}}} p(\boldsymbol{\beta}_j|I_0).$$

However, as the dimension of the parameter space increases, numerical integration becomes infeasible, because it is difficult to select a suitable grid of points and it is hard to evaluate the error of the approximation [172].

4.4.3 Approximate Methods

When numerical integration techniques become infeasible, we are left with two main ways to perform approximate posterior analysis: (i) to provide an asymptotic approximation of the posterior distribution or (ii) to use stochastic methods to generate a sample from the posterior distribution.

Asymptotic Posterior Distributions When the sample size is large enough, posterior analysis can be based on an asymptotic approximation of the posterior distribution to a normal distribution with some mean and variance. This idea generalizes the asymptotic normal distribution of the ML estimates when their exact sampling distribution cannot be derived or it is too difficult to be used (see Chapter 2.)

Berger [56, page 224] mentions four asymptotic normal approximations of the posterior distribution of the parameter vector $\boldsymbol{\theta}$. The approximations are listed below differ in the way the mean and variance of $\boldsymbol{\theta}|I_1$ are computed and are of decreasing accuracy. Recall that the dimension of $\boldsymbol{\theta}$ is k and that $E(\boldsymbol{\theta}|I_1)$ and $V(\boldsymbol{\theta}|I_1)$ denote the exact posterior mean and variance of $\boldsymbol{\theta}$.

1. $\boldsymbol{\theta}|I_1$ is approximately $\mathcal{N}(\mathrm{E}(\boldsymbol{\theta}|I_1), V(\boldsymbol{\theta}|I_1))$.
2. $\boldsymbol{\theta}|I_1$ is approximately $\mathcal{N}(\hat{\boldsymbol{\theta}}_1, (I(\hat{\boldsymbol{\theta}}_1|I_1))^{-1})$, where $\hat{\boldsymbol{\theta}}_1$ is the GML estimate of $\boldsymbol{\theta}$, i. e., $\hat{\boldsymbol{\theta}}_1$ maximizes the augmented likelihood $L(\boldsymbol{\theta})p(\boldsymbol{\theta}|I_0) = p(\mathbf{y}|\boldsymbol{\theta})p(\boldsymbol{\theta}|I_0)$ and $I(\hat{\boldsymbol{\theta}}_1|I_1)$ is the value of the $k \times k$ matrix having element i, j:
$$I(\boldsymbol{\theta}|I_1)_{ij} = -\left(\frac{\partial^2 \log\{p(\mathbf{y}|\boldsymbol{\theta})p(\boldsymbol{\theta}|I_0)\}}{\partial\theta_i\partial\theta_j}\right)$$
evaluated in the GML estimate $\hat{\boldsymbol{\theta}}_1$.
3. $\boldsymbol{\theta}|I_1$ is approximately $\mathcal{N}(\hat{\boldsymbol{\theta}}, (I(\hat{\boldsymbol{\theta}}|\mathbf{y}))^{-1})$, where $\hat{\boldsymbol{\theta}}$ is the ML estimate of $\boldsymbol{\theta}$, i.e. $\hat{\boldsymbol{\theta}}$ maximizes the likelihood $L(\boldsymbol{\theta}) = p(\mathbf{y}|\boldsymbol{\theta})$ and the matrix $I(\hat{\boldsymbol{\theta}}|\mathbf{y})$ is the *observed* Fisher information matrix, that is, the value of the Fisher information matrix
$$I(\boldsymbol{\theta}|\mathbf{y})_{ij} = -\left(\frac{\partial^2 \log p(\mathbf{y}|\boldsymbol{\theta})}{\partial\theta_i\partial\theta_j}\right)$$
evaluated in the ML estimate $\hat{\boldsymbol{\theta}}$.
4. $\boldsymbol{\theta}|I_1$ is approximately $\mathcal{N}(\hat{\boldsymbol{\theta}}, (I(\hat{\boldsymbol{\theta}}))^{-1})$, where the matrix $I(\hat{\boldsymbol{\theta}})$ is the *expected* Fisher information matrix $I(\boldsymbol{\theta})$ evaluated in the ML estimates. Thus, $I(\hat{\boldsymbol{\theta}})$ is the value of the $k \times k$ matrix having element i, j:
$$I(\hat{\boldsymbol{\theta}})_{ij} = -\mathrm{E}\left(\frac{\partial^2 \log p(\mathbf{y}|\boldsymbol{\theta})}{\partial\theta_i\partial\theta_j}\right)$$
evaluated in the ML estimate $\hat{\boldsymbol{\theta}}$ and the expectation is over the conditional distribution of the data given $\boldsymbol{\theta}$.

The first approximation is the most accurate, as it preserves the exact moments of the posterior distribution. However, this approximation relies on the computation of the exact first and second moments [407, Ch. 8]. When exact first and second moments cannot be computed in closed form, we must resort to the second approximation, which replaces them by approximations based on the maximization of the augmented likelihood. Hence, the prior information is taken into account in the calculations of both moments. As the sample size increases and the prior information becomes negligible, the second and third approximation become equivalent. The fourth approximation is the least accurate and relies on the idea that, for exponential family models, expected and observed Fisher information matrices are identical, since the matrix of second derivatives depends on the data only via the ML estimates.

Asymptotic normality of the posterior distribution provides notably computational advantages, since marginal and conditional distributions are still normal, and hence inference on parameters of interest can be easily carried out. However, for relatively small samples, the assumption of asymptotic normality can be inaccurate.

Stochastic Methods For relatively small samples, stochastic methods (or Monte Carlo methods) provide an approximate posterior analysis based on a sample of values generated from the posterior distribution of the parameters. The posterior analysis requires the evaluations of integrals:

$$E(g(\boldsymbol{\theta})|I_1) = \int g(\boldsymbol{\theta})p(\boldsymbol{\theta}|I_1)d\boldsymbol{\theta}.$$

For instance, for $g(\boldsymbol{\theta}) = \boldsymbol{\theta}$, $E(g(\boldsymbol{\theta})|I_1)$ is the posterior expectation. When the exact integration is not possible, Monte Carlo methods replace the exact integral $E(g(\boldsymbol{\theta}|I_1))$ by $\sum_i g(\boldsymbol{\theta}_i)/s$ where $\boldsymbol{\theta}_1, \ldots, \boldsymbol{\theta}_s$ is a random sample of size s generated from $\boldsymbol{\theta}|I_1$. Thus, the task reduces to generating a sample from the posterior distribution of the parameters. Here we will describe Gibbs Sampling (Gs), a special case of Metropolis-Hastings algorithms [213], which is becoming increasingly popular in the statistical community. Gs is an iterative method that produces a Markov Chain, that is a sequence of values $\{\boldsymbol{\theta}^{(0)}, \boldsymbol{\theta}^{(1)}, \boldsymbol{\theta}^{(2)} \ldots\}$ such that $\boldsymbol{\theta}^{(i+i)}$ is sampled from a distribution that depends on the current state i of the chain. The algorithm works as follows.

Let $\boldsymbol{\theta}^{(0)} = \{\theta_1^{(0)}, \cdots, \theta_k^{(0)}\}$ be a vector of initial values of $\boldsymbol{\theta}$ and suppose that the conditional distributions of $\theta_i|(\theta_1, \cdots, \theta_{i-1}, \theta_{i+1}, \cdots, \theta_k, \mathbf{y})$ are known for each i. The first value in the chain is simulated as follows:

$\theta_1^{(1)}$ is sampled from the conditional distribution of $\theta_1|(\theta_2^{(0)}, \cdots, \theta_k^{(0)}, \mathbf{y})$;
$\theta_2^{(1)}$ is sampled from the conditional distribution of $\theta_2|(\theta_1^{(1)}, \theta_3^{(0)}, \cdots, \theta_k^{(0)}, \mathbf{y})$

$\vdots$

$\theta_k^{(1)}$ is sampled from the conditional distribution of $\theta_k|(\theta_1^{(1)}, \theta_2^{(1)}, \cdots, \theta_{k-1}^{(1)}, \mathbf{y})$

Then $\boldsymbol{\theta}^{(0)}$ is replaced by $\boldsymbol{\theta}^{(1)}$ and the simulation is repeated to generate $\boldsymbol{\theta}^{(2)}$, and so forth. In general, the i-th value in the chain is generated by simulating from the distribution of $\boldsymbol{\theta}$ conditional on the value previously generated $\boldsymbol{\theta}^{(i-1)}$. After an initial long chain, called *burn-in*, of say b iterations, the values

$$\{\boldsymbol{\theta}^{(b+1)}, \boldsymbol{\theta}^{(b+2)}, \boldsymbol{\theta}^{(b+3)} \dots\}$$

will be approximately a sample from the posterior distribution of $\boldsymbol{\theta}$, from which empirical estimates of the posterior means and any other function of the parameters can be computed as

$$\frac{1}{s-b} \sum_{i=b+1}^{s} g(\boldsymbol{\theta}^{(i)})$$

where s is the total length of the chain. Critical issues for this method are the choice of the starting value $\boldsymbol{\theta}^{(0)}$, the length of the burn-in and the selection of a stopping rule. The reader is referred to [215] for a discussion of these problems. The program BUGS [512] provides an implementation of Gs suitable for problems in which the likelihood function satisfies certain factorization properties.

4.5. Bayesian Networks

The graphical models that we have used to represent dependencies between data and parameters associated with the sampling model can be further used to describe directed associations among sets of variables. When used in this way, these models are known as Bayesian Belief Networks (BBN) and they result in a powerful knowledge representation formalism, based on probability theory, widely use in Artificial Intelligence. In this section, we will outline the foundations of BBNs and we will describe how to use BBNs to analyze and model data.

4.5.1 Foundations

Formally, a BBN is defined by a set of variables $\mathcal{Y} = \{Y_1, \dots, Y_I\}$ and a DAG defining a model M of conditional dependencies among the elements of $\mathcal{Y}$. We will consider discrete variables and denote by c_i the number of states of Y_i and by y_{ik} a state of Y_i. A conditional dependency links a *child* variable Y_i to a set of *parent* variables Π_i, and it is defined by the conditional probability distributions of Y_i given each *configuration* $\pi_{i1}, \dots, \pi_{iq_i}$ of the parent variables. We term *descendents* of a node Y_i all nodes that can be reached from Y_i through a directed path, that is, following the direction of the arrows. Nodes that are not descendent of Y_i are, obviously, called *non-descendent* of Y_i. The separation of Y_i from its non-descendent $Nd(Y_i)$ given its parents Π_i implies that $i(Y_i, Nd(Y_i)|\Pi_i)$. Hence, the conditional independence assumptions encoded by the directed graphical structure induce a factorization of the joint probability of a set of values $\mathbf{y}_k = \{y_{1k}, \dots, y_{Ik}\}$ of the variables in $\mathcal{Y}$ as

$$p(\mathbf{y}_k) = \prod_{i=1}^{I} p(y_{ik}|\pi_{ij(ik)}), \tag{4.17}$$

where $\pi_{ij(ik)}$ denotes the configuration of states of Π_i in $\mathbf{y}_k$. The index $j(ik)$ is a function of i and k, as the parents configuration $\pi_{ij(ik)}$ in a set of values $\mathbf{y}_k$ is determined by the index i, that specifies the child variable and hence identifies the set of parent variables, and the index k that specifies the states of the parent variables. For notation simplicity, we will denote a parents configuration by π_{ij}.

Example 4.8 (A simple BBN).

Consider the BBN in Figure 4.5, in which the set $\mathcal{Y}$ is $\{Y_1, Y_2, Y_3\}$ and $c_i = 2$ for $i = 1, 2, 3$. The graph encodes the marginal independence of Y_1 and Y_2, which in turn are both parents of Y_3. Thus

$$\Pi_1 = \Pi_2 = \emptyset, \;\; \Pi_3 = (Y_1, Y_2).$$

Note that Π_3 takes four values π_{ij} corresponding to the four combinations of states of Y_1 and Y_2. We will denote these four states as $\pi_{31} = (y_{11}, y_{21})$, $\pi_{32} = (y_{11}, y_{22})$, $\pi_{33} = (y_{12}, y_{21})$ and $\pi_{34} = (y_{12}, y_{22})$. The joint probability of a case $\mathbf{y}_k = \{y_{11}, y_{21}, y_{32}\}$ can then be written as

$$p(\mathbf{y}_k) = p(y_{11})p(y_{21})p(y_{32}|y_{11}, y_{21}) = p(y_{11})p(y_{21})p(y_{32}|\pi_{31}).$$

If Y_3 has a child variable, say Y_4, then the separation of Y_4 from Y_1, Y_2 via Y_3 implies that $i(Y_4, (Y_1, Y_2)|Y_3)$, and the joint probability of $\mathbf{y}_k = \{y_{1k}, y_{2k}, y_{3k}, y_{4k}\}$ factorizes into

$$p(\mathbf{y}_k) = p(y_{1k})p(y_{2k})p(y_{3k}|\pi_{3j})p(y_{4k}|\pi_{4j}),$$

with $\pi_{4j} = y_{3k}$. Thus, the graphical component of a BBN provides a simple way to describe the stochastic dependencies among variables. As mentioned in Chapter 3, the directed links can be given, under some conditions, a causal interpretation, see for instance the review in [258].

The conditional independence assumptions encoded by a BBN have the further advantage of simplifying the computations of conditional probabilities given

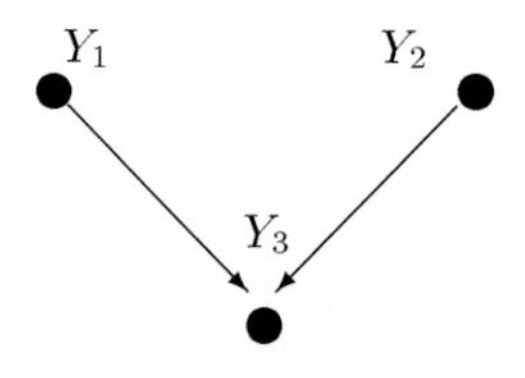

Fig. 4.5. A simple BBN.

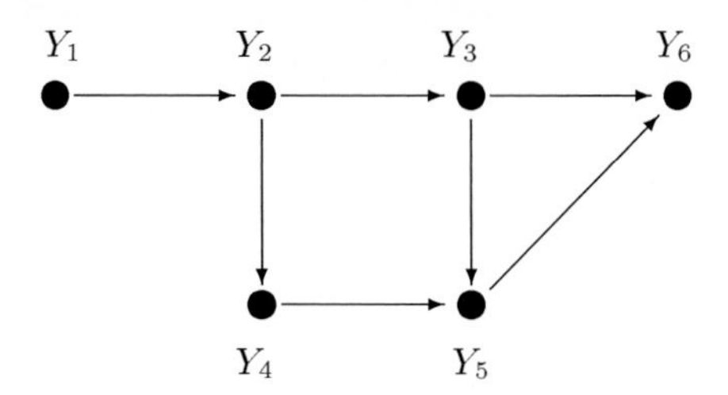

Fig. 4.6. A BBN with 6 binary variables.

some evidence, that is a set of values observed in the network. Thus, tasks as prediction, explanation and classification can be efficiently performed, as shown in the next two examples.

Example 4.9 (Representation). Consider the BBN in Figure 4.6, in which the variables are all binary. Directed links identify parent-child dependencies, and hence:

$$\begin{array}{ll} \Pi_1 = \emptyset & \Pi_2 = Y_1 \\ \Pi_3 = Y_2 & \Pi_4 = Y_2 \\ \Pi_5 = (Y_3, Y_4) & \Pi_6 = (Y_3, Y_5) \end{array}$$

The graph encodes the following conditional independence assumptions:

1. $i(Y_3, Y_1 | Y_2)$
2. $i(Y_4, Y_1 | Y_2)$
3. $i(Y_5, (Y_1, Y_2) | (Y_3, Y_4))$
4. $i(Y_6, (Y_1, Y_2, Y_4) | (Y_3, Y_5))$.

Thus, the joint probability of one case

$$\mathbf{y}_k = (y_{1k}, y_{2k}, y_{3k}, y_{4k}, y_{5k}, y_{6k})$$

can be decomposed into a set of independent parent-child contributions as:

$$\begin{aligned} p(\mathbf{y}_k) &= p(y_{1k})p(y_{2k}|y_{1k})p(y_{3k}|y_{2k})p(y_{4k}|y_{2k})p(y_{5k}|y_{3k}, y_{4k})p(y_{6k}|y_{3k}, y_{5k}) \\ &= p(y_{1k})p(y_{2k}|\pi_{2j})p(y_{3k}|\pi_{3j})p(y_{4k}|\pi_{4j})p(y_{5k}|\pi_{5j})p(y_{6k}|\pi_{6j}) \end{aligned}$$

The advantage of this description is that the joint probability of the 6 variables would require $2^6 - 1 = 63$ independent numbers, that are reduced to $1+2+2+2+4+4 = 15$ when the conditional independence assumptions 1–4 are exploited.[1]

[1] Since the variables are binary, each conditional distribution is identified by one number, and hence $1+2+2+2+4+4$ is the sum of conditional distributions defined by the parent configurations.

Using BBNs, we can easily make predictions about the value of a variable in a given situation by computing the conditional probability distribution of the variable given the values of a set of some other variables in the network. Suppose, for instance, that we are interested in the value of variable Y_6 when the variables Y_3 and Y_4 are observed to be in the states y_{31} and y_{41}. By the Total Probability Theorem

$$\begin{aligned} p(y_{61}|y_{31}, y_{41}) &= \sum_j p(y_{61}, y_{5j}|y_{31}, y_{41}) \\ &= \sum_j p(y_{5j}|y_{31}, y_{41}) p(y_{61}|y_{5j}, y_{31}). \end{aligned}$$

Thus, the conditional probability of interest is expanded to include all variables between the conditioning variables Y_3 and Y_4, and Y_6, and then factorized to account for the conditional independence assumptions encoded by the DAG. Within this framework, the marginal probability $p(y_{61})$ that would be computed by marginalizing the joint probability

$$p(y_{61}) = \sum_{jkmnr} p(y_{61}, y_{5j}, y_{4k}, y_{3m}, y_{2n}, y_{1r})$$

can be computed as:

$$p(y_{61}) = \sum_{jm} p(y_{61}|y_{5j}, y_{3m}) \sum_k p(y_{5j}|y_{3m}, y_{4k}) \sum_n p(y_{3m}|y_{2n}) p(y_{4k}|y_{2n}) \times \\ \sum_r p(y_{2n}|y_{1r}) p(y_{1r})$$

by taking advantage of the locality structure of the parent-child configurations.

A network structure of particular interest for data analysis is displayed in Figure 4.7 and it is known as the Naive Bayesian Classifier.

Example 4.10 (Naive Bayesian Classifier).
The BBN in Figure 4.7 represents a Naive Bayesian Classifier, in which a set of mutually exclusive classes is represented as the root node, and the attributes of the objects to be classified depend on this variable. The simplifying assumptions made by this model, which brings it the name of "naive", are that the classes are mutually exclusive and that the attributes are independent given the class. The meaning of this assumption is that, once we know the class to which an object belongs, the relations between its attributes become irrelevant. In the example depicted in Figure 4.7, the root node Y_1 represent the set of mutually exclusive classes and the leaf nodes Y_2 and Y_3 are the attributes. As the model assumes $i(Y_2, Y_3|Y_1)$, the joint probability distribution is decomposed into

$$p(y_{1k}, y_{2k}, y_{3k}) = p(y_{1k}) p(y_{2k}|y_{1k}) p(y_{3k}|y_{1k}).$$

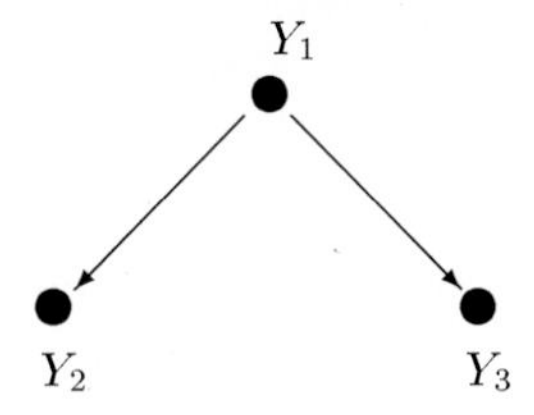

Fig. 4.7. A Naive Bayesian Classifier.

Although the underlying structure is so simple to be termed "naive", this model performs well in classification problems with a large number of attributes [201] in which the task is to classify a unit presenting a combination of attribute values into one of the states of the variable Y_1.

The conditional independence assumptions represented by the model allow the classification to be performed in a very efficient way. Suppose that a unit with attributes y_{2j} and y_{3k} is to be assigned to one of the states of the variable Y_1. The solution is to compute the posterior probability $p(y_{1i}|y_{2j}, y_{3k})$ for all i, and then the unit is assigned to the class with the largest posterior probability. Thus, the problem reduces to computing $p(y_{1i}|y_{2j}, y_{3k})$ which is given by:

$$p(y_{1i}|y_{2j}, y_{3k}) = \frac{p(y_{1i})p(y_{2j}|y_{1i})}{\sum_r p(y_{2j}|y_{1r})p(y_{1r})} \frac{p(y_{3k}|y_{1i})}{\sum_r p(y_{3k}|y_{1r})p(y_{1r}|y_{2j})}.$$

The factorization into terms that depend on the associations Y_1, Y_2, and Y_1, Y_3 is an advantage of Bayesian methods described in Section 4.2. We first have a flow of information from Y_2 to Y_1 by applying Bayes' Theorem:

$$\frac{p(y_{1i})p(y_{2j}|y_{1i})}{\sum_r p(y_{2j}|y_{1r})p(y_{1r})} = p(y_{1i}|y_{2j}).$$

After the first updating, the probability distribution of the class node is $P(Y_1|y_{2j})$, and this is used as prior distribution in the next step, to process the information incoming from node Y_3:

$$\frac{p(y_{3k}|y_{1i})p(y_{1i}|y_{2j})}{\sum_r p(y_{3k}|y_{1r})p(y_{1r}|y_{2j})}.$$

This can be clearly extended to the case of several attributes, and the computation of the conditional probability of Y_1 given a combination of attribute values is found by processing the information incoming from one attribute node at a time.

In this special case, the network structure allows an efficient computation of the posterior distribution of the variable of interest, in time linear with respect to the number of attributes. Unfortunately, this is not the case for any DAG.

Nonetheless, more general (but less efficient) algorithms are available to compute a conditional probability distribution in a generic DAG. The interested reader can find a review of some of these algorithms in [415, 108].

4.5.2 Learning Bayesian Networks From Data

In their original concept, BBNs were supposed to rely on domain experts to supply information about the conditional independence graph and the subjective assessment of the conditional probability distributions that quantify the dependencies. However, the statistical roots of BBNs soon led to the development of learning methods to extract them directly from databases of cases rather than from the insight of human domain experts [126, 100, 259], thus turning BBNs into a powerful tool for the analysis of data.

Suppose we are given a sample of n cases $\mathbf{y} = \{\mathbf{y}_1, \ldots, \mathbf{y}_n\}$ from which we wish to induce a BBN. Note that the sample is now multivariate, since each case $\mathbf{y}_k$ in the sample is a row vector

$$\mathbf{y}_k = (y_{1k}, \ldots, y_{Ik})$$

corresponding to a combination of states of the I variables. Thus, $\mathbf{y}$ is a $n \times I$ matrix. Two components of a BBN can be learned: the graphical structure M, specifying the conditional independence assumptions among the variables in $\mathcal{Y}$, and, given a graph M, the conditional probabilities associated to the remaining dependencies in the graph. We first suppose the graphical structure M given and we focus attention on the second task.

Parameter Estimation Given a DAG M, the conditional probabilities defining the BBN are the parameters $\boldsymbol{\theta} = (\theta_{ijk})$, where $\theta_{ijk} = p(y_{ik}|\pi_{ij}, \boldsymbol{\theta})$, that we wish to estimate from $\mathbf{y}$. We shall denote by $\boldsymbol{\theta}_{ij} = (\theta_{ij1}, \ldots, \theta_{ijc_i})$ the parameter vector associated to the conditional distribution of $Y_i|\pi_{ij}$, to be inferred from $\mathbf{y}$. Graphically, we can expand the BBN by adding, to each variable Y_i, new parent variables representing the parameters that quantify the conditional distribution of $Y_i|\pi_{ij}$.

Example 4.8 (continued). The BBN in Figure 4.5 can be expanded into the BBN in Figure 4.8 by adding the parameters that quantify the dependencies. we show that six parameters $\boldsymbol{\theta} = (\theta_1, \theta_2, \theta_{31}, \theta_{32}, \theta_{33}, \theta_{34})$ are needed. Since the variables Y_1 and Y_2 are binary, their marginal distributions are defined by two parameters: $\theta_1 = p(y_{11}|\boldsymbol{\theta})$ and $\theta_2 = p(y_{21}|\boldsymbol{\theta})$. From the two distributions of Y_1 and Y_2, we can define the joint distribution of the parent variable Π_3, parent of Y_3. Note that Y_1 and Y_2 are marginally independent, so that the distribution of Π_3 is specified by θ_1 and θ_2 as $p(\pi_{31}|\boldsymbol{\theta}) = \theta_1\theta_2$; $p(\pi_{32}|\boldsymbol{\theta}) = \theta_1(1-\theta_2)$; $p(\pi_{33}|\boldsymbol{\theta}) = (1-\theta_1)\theta_2$ and $p(\pi_{34}|\boldsymbol{\theta}) = (1-\theta_1)(1-\theta_2)$. Each parent configuration π_{3j} defines a conditional distribution $Y_3|\pi_{3j}$. The variable Y_3 is binary, and hence each of these conditional distributions is identified by one parameter: $\theta_{3j1} = p(y_{31}|\pi_{3j}, \boldsymbol{\theta})$ for $j = 1, 2, 3, 4$. From these parameters, we obtain the parameter vectors $\boldsymbol{\theta}_1 =$

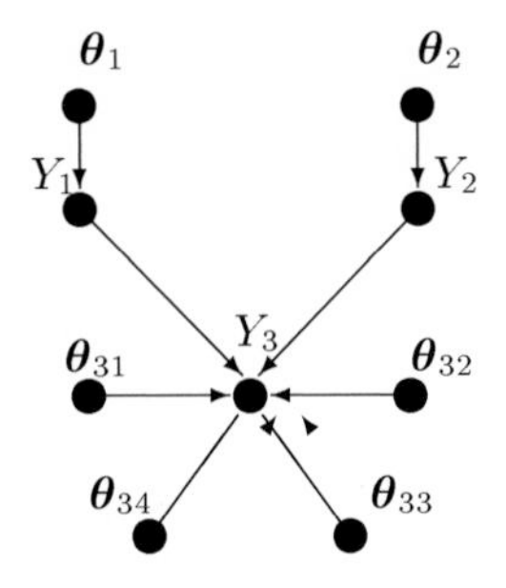

Fig. 4.8. A Simple BBN augmented by parameters.

$(\theta_{11}, 1-\theta_{11})$ associated to the distribution of Y_1, $\boldsymbol{\theta}_2 = (\theta_{21}, 1-\theta_{21})$ associated to the distribution of Y_2, $\boldsymbol{\theta}_{31} = (\theta_{311}, 1-\theta_{311})$ associated to the distribution of $Y_3|\pi_{31}$, $\boldsymbol{\theta}_{32} = (\theta_{321}, 1-\theta_{321})$ associated to the distribution of $Y_3|\pi_{32}$, $\boldsymbol{\theta}_{33} = (\theta_{331}, 1-\theta_{331})$ associated to the distribution of $Y_3|\pi_{33}$ and $\boldsymbol{\theta}_{34} = (\theta_{341}, 1-\theta_{341})$ associated to the distribution of $Y_3|\pi_{34}$.

The standard Bayesian method to estimate $\boldsymbol{\theta}$ uses conjugate analysis. Let $n(y_{ik}|\pi_{ij})$ be the frequency of pairs (y_{ik}, π_{ij}) in the sample, and let $n(\pi_{ij}) = \sum_k n(y_{ik}|\pi_{ij})$ be the frequency of π_{ij}. The joint probability (4.17) of a case $\mathbf{y}_k$ can be written as a function of the unknown θ_{ijk} as

$$p(\mathbf{y}_k|\boldsymbol{\theta}) = \prod_{i=1}^{I} \theta_{ijk},$$

where the index j is uniquely identified by i and k. If cases $\mathbf{y}_k$ are independent, the likelihood function is given by the product of terms $p(\mathbf{y}_k|\theta)$

$$L(\boldsymbol{\theta}) = \prod_{k=1}^{n} p(\mathbf{y}_k|\boldsymbol{\theta}) = \prod_{i=1}^{I} \prod_{j=1}^{q_i} \prod_{k=1}^{c_i} \theta_{ijk}^{n(y_{ik}|\pi_{ij})}.$$

Thus, $p(\mathbf{y}|\boldsymbol{\theta})$ factorizes into a product of local parents-child contributions:

$$\prod_{j=1}^{q_i} \prod_{k=1}^{c_i} \theta_{ijk}^{n(y_{ik}|\pi_{ij})}$$

and each of these terms is itself a product of terms that depend on the parent configurations: $\prod_{k=1}^{c_i} \theta_{ijk}^{n(y_{ik}|\pi_{ij})}$.

A common assumption [496], matching the factorization of the likelihood into parents-child contributions, is that the parameter vectors $\boldsymbol{\theta}_{ij}$ and $\boldsymbol{\theta}_{i'j'}$ associated to different variables Y_i and $Y_{i'}$ are independent for $i \neq i'$ (*global independence*). If the parameters $\boldsymbol{\theta}_{ij}$ and $\boldsymbol{\theta}_{ij'}$ associated to the distributions of Y_i, given different parent configurations π_{ij} and $\pi_{ij'}$ $(j \neq j')$, are further assumed to be independent (*local independence*), the joint prior density $p(\boldsymbol{\theta}|I_0)$ factorizes into the product

$$p(\boldsymbol{\theta}|I_0) = \prod_{i=1}^{I} \prod_{j=1}^{q_i} p(\boldsymbol{\theta}_{ij}|I_0).$$

When the sample **y** is complete, that is there is not entry reported as unknown, local and global independence induce an equivalent factorization of the posterior density of $\boldsymbol{\theta}$:

$$p(\boldsymbol{\theta}|I_1) \propto \prod_{ij} \left\{ p(\boldsymbol{\theta}_{ij}|I_0) \prod_{k=1}^{c_i} \theta_{ijk}^{n(y_{ik}|\pi_{ij})} \right\}$$

and this factorization allows us to independently update the distribution of $\boldsymbol{\theta}_{ij}$, for all i, j, thus reducing the updating process to a local procedure. A further saving in computation is achieved if, for all i and j, the prior distribution of $\boldsymbol{\theta}_{ij}$ is a *Dirichlet* distribution with *hyper-parameters* $\{\alpha_{ij1}, \cdots, \alpha_{ijc_i}\}$, $\alpha_{ijk} > 0$ for all i, j, k. We use the notation

$$\boldsymbol{\theta}_{ij}|I_0 \sim D(\alpha_{ij1}, \ldots, \alpha_{ijc_i}).$$

This distribution generalizes the Beta distribution described in Example 4.1 to vectors of parameters that represent probabilities. In this case, the prior density of $\boldsymbol{\theta}_{ij}$ is, up to a proportionality constant,

$$p(\boldsymbol{\theta}_{ij}|I_0) \propto \prod_{k} \theta_{ijk}^{\alpha_{ijk}-1}$$

which is conjugate to the local likelihood $\prod_{k=1}^{c_i} \theta_{ijk}^{n(y_{ik}|\pi_{ij})}$, since the functional form of this density function matches that of the likelihood. The prior hyper-parameters α_{ijk} encode the observer's prior belief and, since $\alpha_{ijk} - 1$ plays the role of $n(y_{ik}|\pi_{ij})$ in the likelihood, they can be regarded as frequencies of imaginary cases needed to formulate the prior distribution. The quantity $\alpha_{ij} - c_i = \sum_{k=1}^{c_i}(\alpha_{ijk} - 1)$ represents the frequency of imaginary cases observed in the parent configuration π_{ij} and hence α_{ij} is the *local precision.* If the hyper-parameters α_{ijk} are given this interpretation, $\alpha_i = \sum_j \alpha_{ij}$ is the *global precision* on $\boldsymbol{\theta}_i$, that is, the parameter vector associated to the marginal distribution of Y_i. For consistency, we must assume that the imaginary sample has an equal number of observations $\alpha_i = \alpha$ for all the variables Y_i. It can also be shown [207], that this consistency condition is necessary to enforce the local and global independence of the parameters.

The effect of the quantities α_{ijk}s is to specify the marginal probability of $(y_{ik}|\pi_{ij})$ as

$$\mathrm{E}(\theta_{ijk}|I_0) = \frac{\alpha_{ijk}}{\alpha_{ij}} = p(y_{ik}|\pi_{ij}).$$

Furthermore, the prior variance is

$$V(\theta_{ijk}|I_0) = \frac{\mathrm{E}(\theta_{ijk})\{1 - \mathrm{E}(\theta_{ij})\}}{\alpha_{ij} + 1},$$

and, for fixed $E(\theta_{ijk})$, $V(\theta_{ijk})$ is a decreasing function of α_{ij}, so that a small value of α_{ij} will denote great uncertainty. The situation of initial ignorance can be represented by assuming $\alpha_{ijk} = \alpha/(c_i q_i)$ for all i, j and k, so that the prior probability of $(y_{ik}|\pi_{ij})$ is simply $1/c_i$. An important property of the Dirichlet distribution is that it is closed under marginalization, so that if

$$\boldsymbol{\theta}_{ij}|I_0 \sim D(\alpha_{ij1}, \ldots, \alpha_{ijc_i})$$

then any subset of parameters $(\theta_{ij1}, \ldots, \theta_{ijs}, 1 - \sum_{k=1}^{s} \theta_{ijk})$ will have a Dirichlet distribution $D(\alpha_{ij1}, \ldots, \alpha_{ijs}, \alpha_{ij} - \sum_{k=1}^{s} \alpha_{ijk})$. In particular, the parameter θ_{ijk} will have a Beta distribution with hyper-parameters $\alpha_{ijk}, \alpha_{ij} - \alpha_{ijk}$. Thus, marginal inference on parameters of interest can be easily carried out.

Spiegelhalter and Lauritzen [496] show that the assumptions of parameter independence and prior Dirichlet distributions imply that the posterior density of $\boldsymbol{\theta}$ is still a product of Dirichlet densities and

$$\boldsymbol{\theta}_{ij}|I_1 \sim D(\alpha_{ij1} + n(y_{i1}|\pi_{ij}), \ldots, \alpha_{ijc_i} + n(y_{ic_i}|\pi_{ij}))$$

so that local and global independence are retained after the updating. The information conveyed by the sample is therefore captured by simply updating the hyper-parameters of the distribution of $\boldsymbol{\theta}_{ij}$ by increasing them of the frequency of cases (y_{ijk}, π_{ij}) observed in the sample. Thus, the sample information can be summarized into the contingency tables collecting the frequencies of the parents-child dependency. Table 4.1 below provides an example of such a contingency table. The posterior expectation of θ_{ijk} becomes:

$$E(\theta_{ijk}|I_1) = \frac{\alpha_{ijk} + n(y_{ik}|\pi_{ij})}{\alpha_{ij} + n(\pi_{ij})}$$

and the posterior mode is

$$\frac{\alpha_{ijk} + n(y_{ik}|\pi_{ij}) - 1}{\alpha_{ij} + n(\pi_{ij}) - c_i}.$$

Table 4.1. Contingency table collecting the frequencies of cases $(Y_i = y_{ik}, \Pi_i = \pi_{ij})$.

Π_i	Y_i					Row Totals
	y_{i1}	$\ldots$	y_{ik}	$\ldots$	y_{ic_i}	
π_{i1}	$n(y_{i1}\|\pi_{i1})$	$\ldots$	$n(y_{ik}\|\pi_{i1})$	$\ldots$	$n(y_{ic_i}\|\pi_{i1})$	$n(\pi_{i1})$
$\vdots$			$\vdots$			$\vdots$
π_{ij}	$n(y_{i1}\|\pi_{ij})$	$\ldots$	$n(y_{ik}\|\pi_{ij})$	$\ldots$	$n(y_{ic_i}\|\pi_{ij})$	$n(\pi_{ij})$
$\vdots$			$\vdots$			$\vdots$
π_{iq_i}	$n(y_{i1}\|\pi_{iq_i})$	$\ldots$	$n(y_{ik}\|\pi_{iq_i})$	$\ldots$	$n(y_{ic_i}\|\pi_{iq_i})$	$n(\pi_{iq_i})$

Table 4.2. Data from the British General Election Panel Survey. Voting Intention (Y_1), Sex (Y_2), Social Class (Y_3).

Y_2	Y_3	Y_1 = 1	2	3	4
1	1	28	8	7	0
	2	153	114	53	14
	3	20	31	17	1
2	1	1	1	0	1
	2	165	86	54	6
	3	30	57	18	4

The posterior variance is given by:

$$V(\theta_{ijk}|I_1) = \frac{\mathrm{E}(\theta_{ijk}|I_1)\{1 - \mathrm{E}(\theta_{ijk}|I_1)\}}{\alpha_{ij} + n(\pi_{ij}) + 1}$$

with a local precision α_{ij} on $\boldsymbol{\theta}_{ij}$ which is increased by the frequency of parents observed in the configuration π_{ij}.

Example 4.10 (continued). Data in the contingency Table 4.2 are extracted from the British General Election Panel Survey (April 1992). The frequencies are displayed according to Sex (Y_2: 1=male and 2=female), Social Class (Y_3: 1=low, 2=middle and 3=high), and the class variable Voting Intention (Y_1: 1=Conservative, 2=Labour, 3=Liberal Democrat and 4=Other). The task is to classify the voters into four mutually exclusive classes of voting intentions on the basis of their attributes (Sex and Social Class). For this purpose, we can therefore define a Naive Bayesian Classifier similar to the one displayed in Figure 4.7 and estimate the conditional probability distributions associated to its dependency.

Let $\boldsymbol{\theta}_1 = (\theta_{11}, \ldots, \theta_{14})$ be the parameter vector associated to the marginal distribution of Y_1, and let $\boldsymbol{\theta}_{2j} = (\theta_{2j1}, \theta_{2j2})$ and $\boldsymbol{\theta}_{3j} = (\theta_{3j1}, \theta_{3j2}, \theta_{3j3})$ be the parameter vectors associated to the conditional distributions of $Y_2|y_{1j}$ and $Y_3|y_{1j}$, $j = 1, \ldots, 4$. A global prior precision $\alpha = 12$ and the assumption of uniform prior probabilities for y_{1k}, $y_{2k}|y_{1j}$ and $y_{3k}|y_{1j}$ induce a prior distribution $D(3,3,3,3)$ for the parameter $\boldsymbol{\theta}_1$, on letting $\alpha_{1k} = 12/4$, prior distributions $D(1.5, 1.5)$ for $\boldsymbol{\theta}_{2j}$, on letting $\alpha_{2jk} = 12/(4 \times 2)$ and $D(1,1,1)$ for the parameters $\boldsymbol{\theta}_{3j}$, on letting $\alpha_{3jk} = 12/(4 \times 3)$. The frequencies in Table 4.2 are used to update the hyper-parameters, so that after the updating, the posterior distributions of the parameters are:

$$\begin{array}{ll} & \boldsymbol{\theta}_1|I_1 \sim D(400, 300, 152, 29) \\ \boldsymbol{\theta}_{21}|I_1 \sim D(202.5, 197.5) & \boldsymbol{\theta}_{32}|I_1 \sim D(30, 319, 51) \\ \boldsymbol{\theta}_{22}|I_1 \sim D(154.5, 145.5) & \boldsymbol{\theta}_{32}|I_1 \sim D(10, 201, 89) \\ \boldsymbol{\theta}_{23}|I_1 \sim D(78.5, 73.5) & \boldsymbol{\theta}_{33}|I_1 \sim D(8, 108, 36) \\ \boldsymbol{\theta}_{24}|I_1 \sim D(16.5, 12.5) & \boldsymbol{\theta}_{34}|I_1 \sim D(2, 21, 6) \end{array}$$

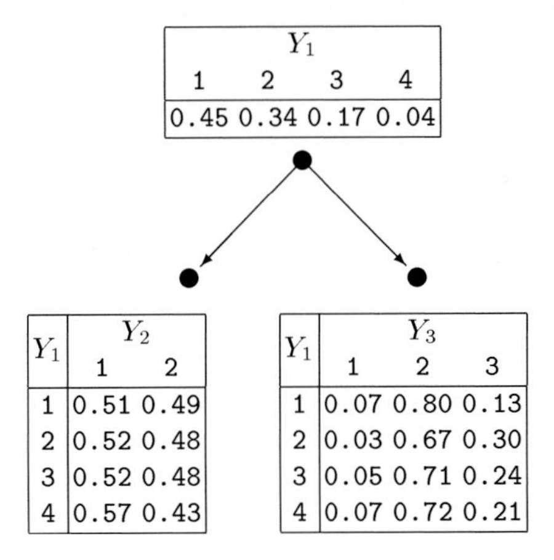

Fig. 4.9. The BBN induced by the data in Table 4.2.

from which the probabilities of $y_{1k}|I_1$, $y_{2k}|y_{1j}, I_1$ and $y_{3k}|y_{1j}, I_1$ are computed as posterior expectations, and are reported in Figure 4.9. It is worth noting that the so defined BBN, when coupled with the classification procedure defined in Example 4.10, turns out to be a complete classification system: we can train the network with the data using the procedure just described and then classify future cases using the algorithm described in Example 4.10. Using the same procedure, we can calculate the distributions of $Y_1|y_{2j}, y_{3k}, I_1$ as shown in Table 4.3, given a set of attribute values. We then discover that the fundamental attribute for classification turns out to be the Social Class (Y_3): high and middle class intend to vote for the Conservative party, while the lower social class has a clear preference for the Labour party.

Model Selection Suppose now that the graphical model M has to be induced from the data. As in Section 4.4.1, let $\mathcal{M} = \{M_0, M_1, \ldots, M_m\}$ be the set of models that are believed, a priori, to contain the true model of dependence among

Table 4.3. Classification of data in Table 4.2 using the BBN in Figure 4.9.

Y_2	Y_3	$P(Y_1\|y_{2j}, y_{3k}, I_1)$ 1	2	3	4	Classify as
1	1	0.59	0.20	0.16	0.05	1
1	2	0.49	0.31	0.17	0.03	1
1	3	0.28	0.49	0.20	0.03	2
2	1	0.61	0.20	0.16	0.03	1
2	2	0.50	0.31	0.16	0.03	1
2	3	0.28	0.49	0.20	0.03	2

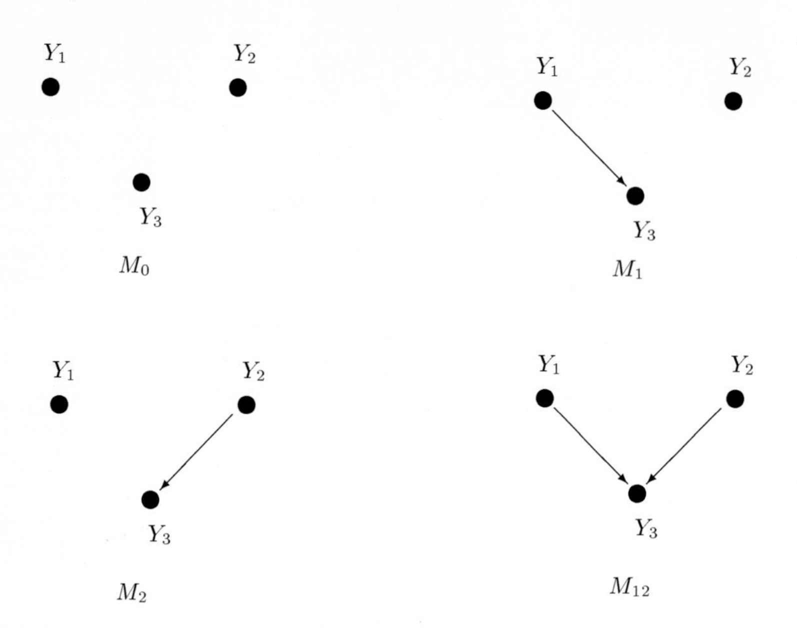

Fig. 4.10. A set of possible models.

the variables in $\mathcal{Y}$. For instance, let $\mathcal{Y} = \{Y_1, Y_2, Y_3\}$ and suppose that Y_1, Y_2 are known to be marginally independent, and that they can be both parents of Y_3, but Y_3 cannot be parent of Y_1, Y_2. These assumptions limit the set of possible models to be explored to $\mathcal{M} = \{M_0, M_1, M_2, M_{12}\}$ which are given in Figure 4.10.

Each model in $\mathcal{M}$ is assigned a prior probability $p(M_j|I_0)$. Let $\boldsymbol{\theta}^{(j)}$ be the parameter vector associated to the conditional dependencies specified by M_j. The sample information is used to compute the posterior probabilities $p(M_j|I_1)$ from which the most probable model in $\mathcal{M}$ can be selected. Recall from Section 4.4.1 that, by Bayes' Theorem, we have

$$p(M_j|I_1) \propto p(M_j|I_0)p(\mathbf{y}|M_j)$$

where $p(\mathbf{y}|M_j)$ is the marginal likelihood of M_j. In order to select the most probable model, it is therefore sufficient to compute the marginal likelihood $p(\mathbf{y}|M_j)$ which is

$$p(\mathbf{y}|M_j) = \int p(\boldsymbol{\theta}^{(j)}|M_j)p(\mathbf{y}|\boldsymbol{\theta}^{(j)})d\boldsymbol{\theta}^{(j)} \tag{4.18}$$

where $p(\boldsymbol{\theta}^{(j)}|M_j)$ is the prior density of $\boldsymbol{\theta}^{(j)}$ and $p(\mathbf{y}|\boldsymbol{\theta}^{(j)})$ is the likelihood function, when the model of dependence assumed is M_j.

It is shown in [126] that (4.18) has a closed form solution when:

1. *The sample is complete, i.e. there are not cases reported as unknown;*
2. *The cases are independent, given the parameter vector* $\boldsymbol{\theta}^{(j)}$ *associated to* M_j*;*
3. *The prior distribution of the parameters is conjugate to the sampling model* $p(\mathbf{y}|\theta^{(j)})$, *that is* $\boldsymbol{\theta}_{ij}^{(j)} \sim D(\alpha_{ij1}, \ldots, \alpha_{ijc_i})$ *and the parameters are locally and globally independent.*

Under these assumptions, the marginal likelihood of M_j is:

$$p(\mathbf{y}|M_j) = \prod_{i=1}^{I} \prod_{j=1}^{q_i} \frac{\Gamma(\alpha_{ij})}{\Gamma(\alpha_{ij} + n(\pi_{ij}))} \prod_{k=1}^{c_i} \frac{\Gamma(\alpha_{ijk} + n(y_{ik}|\pi_{ij}))}{\Gamma(\alpha_{ijk})} \tag{4.19}$$

where $\Gamma(\cdot)$ is the Gamma function [543]. When the database is complete, (4.19) can be efficiently computed using the hyper-parameters $\alpha_{ijk} + n(y_{ik}|\pi_{ij})$ and the precision $\alpha_{ij} + n(\pi_{ij})$ of the posterior distributions of $\boldsymbol{\theta}_{ij}$.

Example 4.11 (Model Discrimination).
Suppose we have two categorical variables Y_1 and Y_2, and a random sample of n cases. Both Y_1 and Y_2 are binary variables, and it is known that Y_1 cannot be parent of Y_2. This assumption leaves us with two possible models to be explored:

Model M_0**:** specifies that the two variables are independent and, conditional on M_0, we can parameterize $p(y_{11}|\boldsymbol{\theta}^{(0)}) = \theta_{11}$ and $p(y_{21}|\boldsymbol{\theta}^{(0)}) = \theta_{21}$ where $\boldsymbol{\theta}^{(0)}$ is the parameter vector associated to M_0.
Model M_1**:** specifies that Y_2 is a parent of Y_1, so that we can define $p(y_{21}|\boldsymbol{\theta}^{(1)}) = \theta_{21}$ and $p(y_1|y_{2j}, \boldsymbol{\theta}^{(1)}) = \theta_{1j1}$.

We assume that, given M_0, $\boldsymbol{\theta}_2 = (\theta_{21}, \theta_{22}) \sim D(2,2)$ and $\boldsymbol{\theta}_1 = (\theta_{11}, \theta_{12}) \sim D(2,2)$, and they are independent. Given M_1, we assume that $\boldsymbol{\theta}_{1j} = (\theta_{1j1}, \theta_{1j2}) \sim D(1,1)$, and they are independent. Thus, a priori, the marginal probabilities of y_{2j}, y_{1k} and $y_{1k}|y_{2j}$ are all uniform and are based on a global prior precision α=4. Suppose we collect a random sample, and we report the summary statistics in Table 4.4.

Table 4.4. Contingency table.

Y_2	Y_1		
	1	2	Total
1	$n(y_{11}|y_{21})$	$n(y_{12}|y_{21})$	$n(y_{21})$
2	$n(y_{11}|y_{22})$	$n(y_{12}|y_{22})$	$n(y_{22})$
Total	$n(y_{11})$	$n(y_{12})$	n

With complete data, the marginal likelihood under models M_0 and M_1 are found by applying Equation (4.19) and are:

$$p(\mathbf{y}|M_0) = \prod_{j=1}^{2} \frac{\Gamma(4)\Gamma(2+n(y_{2j}))}{\Gamma(4+n)\Gamma(2)} \prod_{k=1}^{2} \frac{\Gamma(4)\Gamma(2+n(y_{1k}))}{\Gamma(4+n)\Gamma(2)};$$
$$p(\mathbf{y}|M_1) = \prod_{j=1}^{2} \frac{\Gamma(4)\Gamma(2+n(y_{2j}))}{\Gamma(4+n)\Gamma(2)} \prod_{k=1}^{2} \frac{\Gamma(2)\Gamma(1+n(y_{1k}|y_{2j}))}{\Gamma(2+n(y_{2j}))\Gamma(1)}$$

and the model choice is based on the value of the ratio

$$r = \frac{p(M_0|I_0)p(\mathbf{y}|M_0)}{p(M_1|I_0)p(\mathbf{y}|M_1)},$$

from which the following decision rule is derived: if $r < 1$, model M_1 is chosen; if $r > 1$ model M_0 is chosen; and if $r = 1$ then the two models are equivalent.

Unfortunately, as the number of variables increases, the evaluation of all possible models becomes infeasible, and heuristic methods have to be used to limit the search process to a subset of models. The most common heuristic search limits its attention to the subset of models that are consistent with a partial ordering among the variables: $Y_i \prec Y_j$ if Y_i cannot be parent of Y_j. Furthermore, the fact that (4.19) is a product of terms that measure the evidence of each parents-child dependence can be exploited to develop search algorithms that work locally. This heuristic method was originally proposed by Cooper and Herskovitz [126]: they describe an algorithm — the K2 algorithm — which uses this heuristic to fully exploit the decomposability of (4.19). Denote the local contribution of a node Y_i and its parents Π_i to the overall joint probability $p(\mathbf{y}|M_j)$ by

$$g(Y_i, \Pi_i) = \prod_{j=1}^{q_i} \frac{\Gamma(\alpha_{ij})}{\Gamma(\alpha_{ij} + n(\pi_{ij}))} \prod_{k=1}^{c_i} \frac{\Gamma(\alpha_{ijk} + n(x_{ik}|\pi_{ij}))}{\Gamma(\alpha_{ijk})}. \tag{4.20}$$

For each node Y_i, the algorithm proceeds by adding a parent at a time and by computing $g(Y_i, \Pi_i)$. The set Π_i is expanded to include the parent node that gives the largest contribution to $g(Y_i, \Pi_i)$, and stops if the probability does not increase any longer. In the next example we use synthetic data to show the algorithm at work in a simple application.

Example 4.12 (Model Search). Let $\mathcal{Y} = \{Y_1, Y_2, Y_3\}$, where Y_i are binary variables, and suppose that $Y_3 \prec Y_2 \prec Y_1$. The order assumed implies that Y_3 cannot be parent of Y_2, Y_1 and Y_2 cannot be parent of Y_1. Thus, the node Y_3 can have Y_1, Y_2 as parents, and Y_2 can have Y_1 as parents. These are the dependencies that we are exploring. Suppose further that all models consistent with this ordering have the same prior probability, and that the parameterization induced by each of these models is based on a global precision $\alpha = 6$ which is then distributed uniformly across the parameters. Data collected are reported in Table 4.5.

Table 4.5. An artificial sample.

Y_1	Y_2	Y_3 = 1	Y_3 = 2
1	1	2	10
	2	5	3
2	1	6	3
	2	10	2

The algorithm starts by exploring the dependencies of Y_3 as child node, and results are

$$\begin{array}{ll} \Pi_3 = \emptyset & \log g(Y_3, \Pi_3) = -29.212 \\ \Pi_3 = Y_1 & \log g(Y_3, \Pi_3) = -27.055 \\ \Pi_3 = Y_2 & \log g(Y_3, \Pi_3) = -27.716 \end{array}$$

so that the node Y_1 is selected as parent of Y_3. Next, both nodes Y_1, Y_2 are linked to Y_3 and $\log g(Y_3, (Y_1, Y_2)) = -26.814$. Since this value is larger than -27.055, this model of local dependence is selected. Then, the node Y_2 is chosen as child node, and the dependence from Y_1 is explored:

$$\begin{array}{ll} \Pi_2 = \emptyset & \log g(Y_2, \Pi_2) = -29.474 \\ \Pi_2 = Y_1 & \log g(Y_2, \Pi_2) = -30.058 \end{array}$$

and since the model with Y_2 independent of Y_1 gives the largest contribution, the model selected is Y_1, Y_2 independent, both parents of Y_3.

4.6. Conclusion

In the description of learning BBNs from data, we have made the assumption that the sample is complete, that is there are no cases reported as unknown. When the sample is incomplete, each missing datum can be replaced by any value of the associated variable. Therefore, an incomplete sample induces a set of possible databases given by the combination of all possible values for each missing datum. Exact Bayesian analysis requires the computation of the posterior distribution of the parameters $\boldsymbol{\theta}$ as a mixture of the posterior distributions that can be obtained from all these possible databases. Clearly, this approach becomes infeasible as the proportion of missing data increases and we must resort to approximate methods. Popular approaches use either asymptotic approximations based on the use of the ML estimates or GML estimates, that can be computed using iterative methods as the EM algorithm [149], or stochastic approximations as Gs (see [258] for a review.) The Bound and Collapse method by [439] provides an efficient deterministic algorithm to approximate the posterior distribution of incomplete samples which can be further used for model selection [437,

470]. This method is implemented in the computer program Bayesian Knowledge Discoverer (BKD).

In this chapter, we have limited our attention to BBNs with discrete variables. Methods for learning and reasoning with BBNs with continuous, or mixed variables can be found in [336] and a recent review is reported in [101].

Chapter 5
Support Vector and Kernel Methods

Nello Cristianini* and John Shawe-Taylor**
*University of California at Davis, USA
**Royal Holloway, University of London, UK

Kernel Methods (KM) are a relatively new family of algorithms that presents a series of useful features for pattern analysis in datasets. In recent years, their simplicity, versatility and efficiency have made them a standard tool for practitioners, and a fundamental topic in many data analysis courses. We will outline some of their important features in this Chapter, referring the interested reader to more detailed articles and books for a deeper discussion (see for example [135] and references therein).

KMs combine the simplicity and computational efficiency of linear algorithms, such as the perceptron algorithm or ridge regression, with the flexibility of non-linear systems, such as for example neural networks, and the rigour of statistical approaches such as regularization methods in multivariate statistics. As a result of the special way they represent functions, these algorithms typically reduce the learning step to a convex optimization problem, that can always be solved in polynomial time, avoiding the problem of local minima typical of neural networks, decision trees and other non-linear approaches.

Their foundation in the principles of Statistical Learning Theory make them remarkably resistant to overfitting especially in regimes where other methods are affected by the 'curse of dimensionality'. It is for this reason that they have become popular in bioinformatics and text analysis. Another important feature for applications is that they can naturally accept input data that are not in the form of vectors, such as for example strings, trees and images.

Their characteristically modular design makes them amenable to theoretical analysis but also well suited to a software engineering approach: a general purpose learning module is combined with a data specific 'kernel function' that provides the interface with the data and incorporates domain knowledge. Many learning modules can be used depending on whether the task is one of classification, regression, clustering, novelty detection, ranking, etc. At the same time many kernel functions have been designed: for protein sequences, for text and hypertext documents, for images, time series, etc. The result is that this method can be used for dealing with rather exotic tasks, such as ranking strings, or clustering graphs, in addition to such classical tasks as classifying vectors. We will delay the definition of a kernel till the next section even though kernels form the core of this contribution.

In this Chapter, we will introduce the main concepts behind this approach to data analysis by discussing some simple examples. We will start with the simplest algorithm and the simplest kernel function, so as to illustrate the basic concepts. Then we will discuss the issue of overfitting, the role of generalization bounds and how they suggest more effective strategies, leading to the Support Vector Machine (SVM) algorithm. In the conclusion, we will briefly discuss other pattern recognition algorithms that exploit the same ideas, for example Principal Components Analysis (PCA), Canonical Correlation Analysis (CCA), and extensions of the SVM algorithm to regression and novelty detection. In this short chapter we err in favour of giving a detailed description of a few standard methods, rather than a superficial survey of the majority.

The problem we will use as an example throughout the chapter is the one of learning a binary classification function using a real-valued function $f : X \subseteq \mathbb{R}^n \to \mathbb{R}$ in the following way: the input $\mathbf{x} = (x_1, \ldots, x_n)'$ is assigned to the positive class, if $f(\mathbf{x}) \geq 0$, and otherwise to the negative class. We are interested in the case where $f(\mathbf{x})$ is a non-linear function of $\mathbf{x} \in X$, though we will solve the non-linear problem by using linear $f(\mathbf{x})$ in a space that is the image of a non-linear mapping.

We will use X to denote the input space and Y to denote the output domain. Usually we will have $X \subseteq \mathbb{R}^n$, while for binary classification $Y = \{-1, 1\}$ and for regression $Y \subseteq \mathbb{R}$. The *training set* is a collection of *training examples*, which are also called *training data*. It is denoted by

$$S = ((\mathbf{x}_1, y_1), \ldots, (\mathbf{x}_m, y_m)) \subseteq (X \times Y)^m ,$$

where m is the number of examples. We refer to the $\mathbf{x}_i$ as *examples* or *instances* and the y_i as their *labels*. Note that if X is a vector space, the input vectors are column vectors as are the weight vectors. If we wish to form a row vector from $\mathbf{x}_i$ we can take the transpose $\mathbf{x}_i'$. We denote by $\langle \mathbf{x}, \mathbf{w} \rangle = \mathbf{x}'\mathbf{w} = \sum_i \mathbf{x}_i \mathbf{w}_i$ the inner product between the vectors $\mathbf{x}$ and $\mathbf{w}$.

5.1. Example: Kernel Perceptron

The main idea of Kernel Methods is to first embed the data into a suitable vector space, and then use simple linear methods to detect relevant patterns in the resulting set of points. If the embedding map is non-linear, this enables us to discover non-linear relations using linear algorithms. Hence, we consider a map from the input space X to a feature space F,

$$\phi : x \in X \longmapsto \phi(x) \in F.$$

Such a mapping of itself will not solve the problem, but it can become very effective if coupled with the following two observations: 1) the information about the relative positions of the data points in the embedding space encoded in the inner products between them is all that is needed by many pattern analysis algorithms; 2) the inner products between the projections of data inputs into high dimensional embedding spaces can often be efficiently computed directly from the inputs using a so-called kernel function. We will illustrate these two points by means of the example of the kernel perceptron.

5.1.1 Primal and Dual Representation

A simple rewriting of the perceptron rule yields an alternative representation for functions and learning algorithms, known as the *dual representation.* In the dual representation, all that is needed are the inner products between data points. There are many linear learning algorithms that can be represented in this way.

Primal Perceptron. As already seen in Chapter 8, the perceptron algorithm learns a binary classification function using a real-valued linear function $f : X \subseteq \mathbb{R}^n \rightarrow \mathbb{R}$ that can be written as

$$\begin{aligned} f(\mathbf{x}) &= \langle \mathbf{w}, \mathbf{x} \rangle + b \\ &= \sum_{i=1}^{n} w_i x_i + b \end{aligned}$$

where $(\mathbf{w}, b) \in \mathbb{R}^n \times \mathbb{R}$ are the parameters that control the function and the decision rule is given by $\text{sgn}\,(f(\mathbf{x}))$. These parameters must be learned from the data, and are the output of the perceptron algorithm.

A geometric interpretation of this kind of hypothesis is that the input space X is split into two parts by the hyperplane defined by the equation $\langle \mathbf{w}, \mathbf{x} \rangle + b = 0$. A hyperplane is an affine subspace of dimension $n - 1$ which divides the space into two half spaces corresponding to the inputs from the two classes. The vector $\mathbf{w}$ defines a direction perpendicular to the hyperplane, while the value of b determines the distance of the hyperplane from the origin of the coordinate system. It is therefore clear that a representation involving $n+1$ free parameters is natural, if one wants to represent all possible hyperplanes in $\mathbb{R}^n$.

Both statisticians and neural network researchers have frequently used this simple kind of classifier, calling them respectively *linear discriminants* and *perceptrons.* The theory of linear discriminants was developed by Fisher in 1936, while neural network researchers studied perceptrons in the early 1960s, mainly due to the work of Rosenblatt [451]. We will refer to the quantities $\mathbf{w}$ and b as the *weight vector* and *bias*, terms borrowed from the neural network's literature. Sometimes $-b$ is replaced by θ, a quantity known as the *threshold.*

The simplest iterative algorithm for learning linear classifications is the procedure proposed by Frank Rosenblatt in 1959 for the perceptron [451]. The algorithm created a great deal of interest when it was first introduced. It is an 'on-line' and 'mistake-driven' procedure, which starts with an initial weight vector $\mathbf{w}_0$ (usually $\mathbf{w}_0 = \mathbf{0}$ the all zero vector) and adapts it each time a training point is misclassified by the current weights. The algorithm is shown in Table 5.1.

Table 5.1. The Perceptron Algorithm (primal form)

```
Given a linearly separable training set S
w_0 ← 0; b_0 ← 0; k ← 0
R ← max_{1≤j≤m} ||x_j||
repeat
        for j = 1 to m
                if y_j(⟨w_k, x_j⟩ + b_k) ≤ 0 then
                        w_{k+1} ← w_k + y_j x_j
                        b_{k+1} ← b_k + y_j R^2
                        k ← k + 1
                end if
        end for
until no mistakes made within the for loop
return (w_k, b_k)
```

The algorithm updates the weight vector and bias directly, something that we will refer to as the primal form in contrast to an alternative dual representation which we will introduce below.

This procedure is guaranteed to converge provided there exists a hyperplane that correctly classifies the training data. In this case we say that the data are *linearly separable*. If no such hyperplane exists the data are said to be non-separable.

Dual Representation. It is important to note that the perceptron algorithm works by adding misclassified positive training examples or subtracting misclassified negative examples to an initial arbitrary weight vector. If we take the initial weight vector to be the zero vector, this implies that the final weight vector will be a linear combination of the training points:

$$\mathbf{w} = \sum_{j=1}^{m} \alpha_j y_j \mathbf{x}_j,$$

where, since the sign of the coefficient of $\mathbf{x}_j$ is given by the classification y_j, the α_j are positive integral values equal to the number of times misclassification of $\mathbf{x}_j$ has caused the weight to be updated. Points that have caused fewer mistakes will have smaller α_j, whereas difficult points will have large values. Once a sample S has been fixed, the vector α is a representation of the weight vector in different coordinates, known as the dual representation. This expansion is however not unique: different α can correspond to the same weight vector $\mathbf{w}$. Intuitively, one can also regard α_j as an indication of the information content of the example $\mathbf{x}_j$. In the case of non-separable data, the coefficients of misclassified points grow indefinitely. In the dual representation the decision function can be expressed as follows:

$$\begin{aligned} h(\mathbf{x}) &= \operatorname{sgn}\left(\langle \mathbf{w}, \mathbf{x} \rangle + b\right) \\ &= \operatorname{sgn}\left(\left\langle \sum_{j=1}^{m} \alpha_j y_j \mathbf{x}_j, \mathbf{x} \right\rangle + b\right) \\ &= \operatorname{sgn}\left(\sum_{j=1}^{m} \alpha_j y_j \langle \mathbf{x}_j, \mathbf{x} \rangle + b\right). \end{aligned} \tag{5.1}$$

Furthermore, the perceptron algorithm can also be implemented entirely in this dual form as shown in Table 5.2.

Table 5.2. The Perceptron Algorithm (dual form)

Given training set S
$\alpha \leftarrow \mathbf{0}$; $b \leftarrow 0$
$R \leftarrow \max_{1 \le i \le m} \|\mathbf{x}_i\|$
repeat
    for $i = 1$ to m
        if $y_i \left(\sum_{j=1}^{m} \alpha_j y_j \langle \mathbf{x}_j, \mathbf{x}_i \rangle + b\right) \le 0$ then
            $\alpha_i \leftarrow \alpha_i + 1$
            $b \leftarrow b + y_i R^2$
        end if
    end for
until no mistakes made within the *for* loop
return (α, b) to define function $h(\mathbf{x})$ of equation (5.1)

This alternative formulation of the perceptron algorithm and its decision function has many interesting properties. For example, the fact that the points that are harder to learn have larger α_i can be used to rank the data according to their information content.

It is important to note that the information from the training data only enters the algorithm through inner products of the type $\langle \mathbf{x}_i, \mathbf{x}_j \rangle$: in other words we do not need the coordinates of the points, just the inner products between all pairs. We will see in the next section that, in order to run this algorithm in a feature space, it will be sufficient to compute the value of the inner products between the data in that space, and that these can often be efficiently computed using a special function known as a kernel.

5.1.2 Implicit Embedding via Kernel Functions

Despite the deep theoretical understanding of their statistical and computational properties, the power of linear pattern recognition systems like the Perceptron in real applications is very limited. With some important exceptions, usually real data require non-linear decision boundaries. In practice, more results have been obtained by using less efficient and less well understood non-linear heuristics (such as neural networks or decision trees). In this section we will see how to exploit the dual representation of the perceptron algorithm derived in the previous section in order to obtain a non-linear algorithm. The technique presented here is very general and can be applied to many other algorithms.

Implicit mapping by kernels. In order to transform a linear algorithm into a non-linear one, one can first embed the data into a more powerful feature space F, where a linear machine is sufficiently expressive to capture the relevant relations in the data, by means of a map ϕ:

$$\phi : x \in X \mapsto \phi(x) \in F$$

This task presents both statistical and computational challenges. In this section we address the computational ones, leaving the statistical issues to later in the chapter. We have remarked that one important property of the dual representation of the perceptron obtained above is that the weight vector can be expressed as a linear combination of the training points, so that the decision rule can be evaluated using just inner products between the test point and the training points:

$$f(\mathbf{x}) = \sum_{i=1}^{m} \alpha_i y_i \langle \phi(\mathbf{x}_i), \phi(\mathbf{x}) \rangle + b.$$

Similarly, we have seen that the training (learning the coefficients α_i) can be implemented using just inner products between training points. This makes it possible to use the technique of implicit mapping by replacing every inner product between the feature mapped data points with a kernel function that directly computes the inner product $\langle \phi(\mathbf{x}_i), \phi(\mathbf{x}) \rangle$ as a function of the original input points. In this way it becomes possible to merge the two steps needed to build a non-linear learning machine. We call such a direct computation method a *kernel* function [76, 11].

Definition 5.1. *A* kernel *is a function K, such that for all $\mathbf{x}, \mathbf{z} \in X$*

$$K(\mathbf{x}, \mathbf{z}) = \langle \phi(\mathbf{x}), \phi(\mathbf{z}) \rangle.$$

where ϕ is a mapping from X to an (inner product) feature space F.

The name 'kernel' is derived from integral operator theory, which underpins much of the theory of the relation between kernels and their corresponding feature spaces. An important consequence of the dual representation is that the dimension of the feature space need not affect the computation. As the feature vectors are not represented explicitly, the number of operations required to compute the inner product by evaluating the kernel function is not necessarily proportional to the number of features. The use of kernels makes it possible to map the data implicitly into a feature space, train a linear machine in that space and subsequently evaluate it on new examples, potentially side-stepping the computational problems inherent in computing the feature map.

Example 5.1. As a first example of a kernel function, consider two points $\mathbf{x} = (x_1, x_2)$ and $\mathbf{z} = (z_1, z_2)$ in a 2-dimensional space, and the function $K(\mathbf{x}, \mathbf{z}) = \langle \mathbf{x}, \mathbf{z} \rangle^2$:

$$\begin{aligned}
\langle \mathbf{x}, \mathbf{z} \rangle^2 &= \langle (x_1, x_2), (z_1, z_2) \rangle^2 \\
&= (x_1 z_1 + x_2 z_2)^2 \\
&= x_1^2 z_1^2 + x_2^2 z_2^2 + 2 x_1 x_2 z_1 z_2 \\
&= \langle (x_1^2, x_2^2, \sqrt{2} x_1 x_2), (z_1^2, z_2^2, \sqrt{2} z_1 z_2) \rangle
\end{aligned}$$

Hence, this can be regarded as the inner product between two vectors in a feature space with corresponding feature map

$$(x_1, x_2) \longmapsto \phi(x_1, x_2) = (x_1^2, x_2^2, \sqrt{2} x_1 x_2),$$

so that

$$K(\mathbf{x}, \mathbf{z}) = \langle \mathbf{x}, \mathbf{z} \rangle^2 = \langle \phi(\mathbf{x}), \phi(\mathbf{z}) \rangle$$

In the same way using the easy to compute kernel $K(\mathbf{x}, \mathbf{z}) = \langle \mathbf{x}, \mathbf{z} \rangle^d$ gives a feature space of $\binom{n+d-1}{d}$ dimensions, calculating $\phi(x)$ would soon become computationally infeasible for reasonable n and d.

Figure 5.1.2 shows an example of a feature mapping from a two dimensional input space to a two dimensional feature space, where the data cannot be separated by a linear function in the input space, but can be in the feature space. The aim of this section is to show how such mappings can be generated into very high dimensional spaces where linear separation becomes much easier.

One can use kernels even without knowing the feature mapping ϕ associated with them, as long as it is possible to prove that such a mapping exists (i.e. if it is possible to prove that the function being used is really a kernel). An important mathematical characterisation of kernels is given by Mercer's theorem [11], which provides sufficient conditions for a function $K(\mathbf{x}, \mathbf{z})$ to be a valid kernel.

Theorem 5.1. *(Mercer) Let X be a compact subset of $\mathbb{R}^n$. Suppose K is a continuous symmetric function such that the integral operator $T_K : L_2(X) \to L_2(X)$,*

$$(T_K f)(\cdot) = \int_X K(\cdot, \mathbf{x}) f(\mathbf{x}) d\mathbf{x},$$

is positive, that is

$$\int_{X \times X} K(\mathbf{x}, \mathbf{z}) f(\mathbf{x}) f(\mathbf{z}) d\mathbf{x} d\mathbf{z} \geq 0,$$

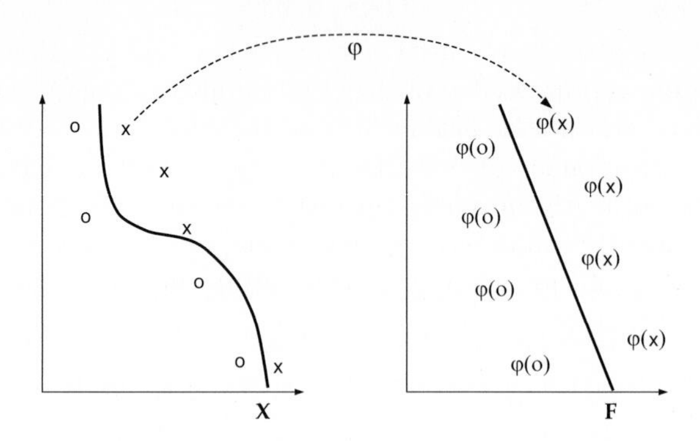

Fig. 5.1. An example of a feature mapping.

for all $f \in L_2(X)$. Then we can expand $K(\mathbf{x}, \mathbf{z})$ in a uniformly convergent series (on $X \times X$) in terms of T_Ks eigen-functions $\phi_j \in L_2(X)$, normalized in such a way that $||\phi_j||_{L_2} = 1$, and positive associated eigenvalues $\lambda_j \geq 0$,

$$K(\mathbf{x}, \mathbf{z}) = \sum_{j=1}^{\infty} \lambda_j \phi_j(\mathbf{x}) \phi_j(\mathbf{z}).$$

The image of a vector $\mathbf{x}$ via the implicit mapping defined by a Mercer kernel is hence $\sum_{j=1}^{\infty} \sqrt{\lambda_j} \phi_j(\mathbf{x})$. Standard kernel functions are the gaussian

$$K(\mathbf{x}, \mathbf{z}) = \exp\left(-\frac{||\mathbf{x} - \mathbf{z}||^2}{2\sigma^2}\right)$$

and the polynomial kernel of degree d:

$$K(\mathbf{x}, \mathbf{z}) = (\langle \mathbf{x}, \mathbf{z} \rangle + 1)^d$$

More information on the rich theory of kernel functions can be found in [135, Chapter 3].

The efficient access to high-dimensional spaces by means of kernels solves the computational problem of dealing with non-linear relations, but leaves open the statistical problem of overfitting. When the number of dimensions exceeds the number of training points, it is always possible to find a linear separation for any partitioning of a non-degenerate dataset. The resulting function is unlikely to generalize well on unseen data. In order to safely use such a flexible class of functions, it is necessary to introduce tools from Statistical Learning Theory.

5.2. Overfitting and Generalization Bounds

When mining for relations in a dataset, we need to pay particular attention to the risk of detecting 'spurious' relations, that are the effect of chance and do not reflect any underlying property of the data. Such relations can also be defined as 'unsignificant' or unstable, and their characterization is a rather subtle matter. In general, one would like relations that are going to be found also in future datasets generated by the same source, so that they can be used to make predictions.

Statistical Learning Theory is a framework in which it is possible to study the predictive power of relations found in the data under the assumption that the dataset has been generated probabilistically in an independent, identically distributed way. Such model can then be used to design algorithms that are less prone to fitting irrelevant aspects of the data, or 'overfitting'.

We will use Learning Theory to obtain insight into which aspects of a learning algorithm need to be controlled in order to improve its performance on test points. Such indications take the form of upper bounds on the risk of mislabeling a test point (based on some assumptions) that depend on some observable features of the learning system or of the learned rule itself. This section presents a number of results in this sense, that will be used to motivate the algorithms introduced later.

The central assumption is that all the data points (training and test set) are drawn independently from the same (fixed but unknown) distribution. The fundamental quantities that control the generalization power of the system are the training set size and the 'effective capacity' of the hypothesis class used by the system. This quantity is roughly speaking a measure of the flexibility of the algorithm, counting how many different labellings of a dataset the algorithm could successfully separate using a function that is 'equivalent' to the solution obtained.

In order to maintain the focus on the main algorithmic issues, we will simply cite the fundamental bounds that apply to linear classification, referring the interested reader to Chapter 4 of [135] for more details.

We will now define a fundamental quantity in Statistical Learning Theory, the margin of a separating hyperplane with respect to a given labeled set of points. Although technical, the essential idea is that the (geometric) margin is the distance between the separating hyperplane and the nearest point or, equivalently, proportional to the distance between the convex hull of the positive points and the one of the negative points. The functional margin is the smallest value assumed by the linear function on the separated data set. Note that functional and geometric margin are directly proportional.

Definition 5.2. *We define the* (functional) margin of an example $(\mathbf{x}_i, y_i)$ with respect to a hyperplane $\mathbf{w}$ *to be the quantity*

$$\gamma_i = y_i(\langle \mathbf{w}, \mathbf{x}_i \rangle + b).$$

Note that $\gamma_i > 0$ *implies correct classification of* $(\mathbf{x}_i, y_i)$. *The* (functional) margin distribution of a hyperplane $\mathbf{w}$ with respect to a training set S *is the distribution of the margins of the examples in* S. *We sometimes refer to the minimum of the margin distribution as the* (functional) margin of a hyperplane $\mathbf{w}$ with respect to

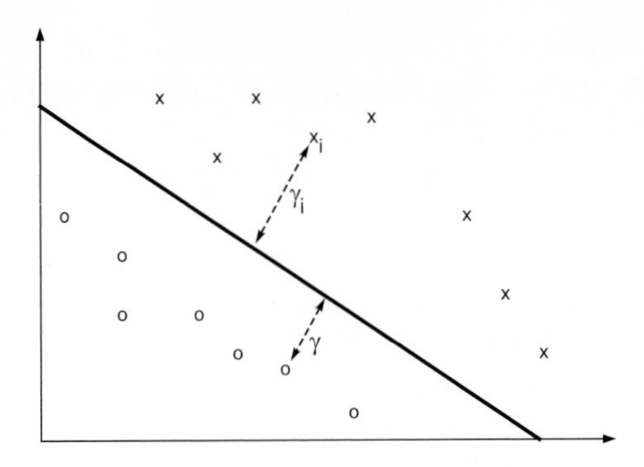

Fig. 5.2. The geometric margin of two points.

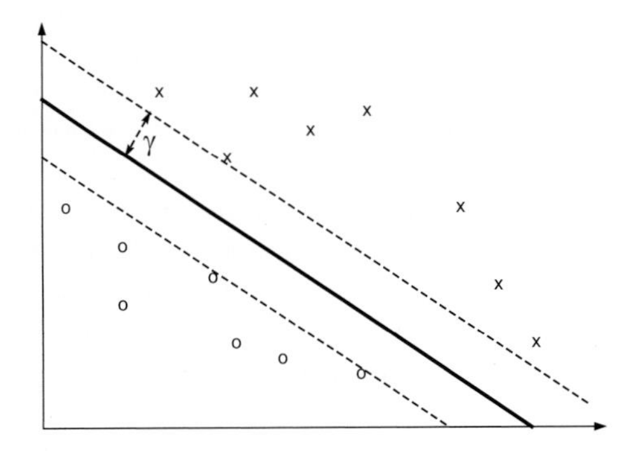

Fig. 5.3. The margin of a training set.

a training set S. *In both definitions if we replace functional margin by* geometric margin *we obtain the equivalent quantity for the normalised linear function* $\left(\frac{1}{\|\mathbf{w}\|}\mathbf{w}, \frac{1}{\|\mathbf{w}\|}b\right)$, *which therefore measures the Euclidean distances of the points from the decision boundary in the input space (see Figure 5.2 for an example). Finally, the* margin of a training set S *is the maximum geometric margin over all hyperplanes. A hyperplane realising this maximum is known as a* maximal margin hyperplane. *The size of its margin will be positive for a linearly separable training set (Figure 5.3).*

The importance of this definition is that the margin (and the related quantities) can be used to obtain information on the predictive power of the solution. The following theorem states that the risk of misclassification of a new point for a separating hyperplane can be bounded by $\varepsilon = O(\frac{R^2}{m\gamma^2})$, where R is the radius of the sphere containing the data, m is the sample size and γ is the (geometric) margin of the separation. In this case we assume the data to be linearly separable.

Theorem 5.2. *([476]) Consider thresholding real-valued linear functions $\mathcal{L}$ with unit weight vectors on an inner product space X and fix $\gamma \in \mathrm{I\!R}^+$. For any* *probability distribution $\mathcal{D}$ on $X \times \{-1, 1\}$ with support in a ball of radius R around the origin, with probability $1-\delta$ over m random examples S, any hypothesis $f \in \mathcal{L}$ that has margin $m_S(f) \geq \gamma$ on S has error no more than*

$$err_{\mathcal{D}}(f) \leq \varepsilon(m, \mathcal{L}, \delta, \gamma) = \frac{2}{m}\left(\frac{64R^2}{\gamma^2} \log \frac{em\gamma}{8R^2} \log \frac{32m}{\gamma^2} + \log \frac{4}{\delta}\right),$$

provided $m > 2/\varepsilon$ and $64R^2/\gamma^2 < m$.

All the log factors can be neglected at a first reading, being partly the effect of the proof techniques employed. The important *qualitative* aspect of this result is that the dimension of the input space does not appear, indeed the result also applies to infinite dimensional spaces. This type of result is sometimes said to be *dimension free*, as it suggests that the bound may overcome the curse of dimensionality. It is important to note that avoidance of the curse of dimensionality will only be possible if the distribution generating the examples is sufficiently benign and renders the task of identifying the particular target function correspondingly easier. In other words, only if the margin happens to be sufficiently large is this bound useful, while the worst case bound will always depend on the dimensionality as well.

However, we consider here the 'lucky' case in which the margin is large, or -equivalently- the case where the kernel was chosen to match the problem at hand. In such cases the bound gives an assurance that with high probability we will make few errors on randomly chosen test examples. It is in this sense that we can view γ as providing a measure of how benign the distribution is and therefore how much better we can expect to generalise than the worst case.

Theorem 5.2 becomes trivial and hence gives no information for the case where the data are non-separable or noise in the data causes the margin to be very small. The next result discusses a method which can handle these situations by taking a different measure of the margin distribution.

Definition 5.3. *Consider using a class $\mathcal{F}$ of real-valued functions on an input space X for classification by thresholding at 0. We define the* margin slack variable *of an example $(\mathbf{x}_i, y_i) \in X \times \{-1, 1\}$ with respect to a function $f \in \mathcal{F}$ and target margin γ to be the quantity*

$$\xi\left(\left(\mathbf{x}_i, y_i\right), f, \gamma\right) = \xi_i = \max\left(0, \gamma - y_i f(\mathbf{x}_i)\right).$$

Note that $\xi_i > \gamma$ implies incorrect classification of $(\mathbf{x}_i, y_i)$. The margin slack vector *$\xi(S, f, \gamma)$ of a training set*

$$S = \left(\left(\mathbf{x}_1, y_1\right), \ldots, \left(\mathbf{x}_m, y_m\right)\right)$$

with respect to a function f and target margin γ contains the margin slack variables

$$\xi = \xi(S, f, \gamma) = \left(\xi_1, \ldots, \xi_m\right),$$

where the dependence on S, f, and γ is dropped when it is clear from the context.

We can think of the slack variables as measures of noise in the data which has caused individual training points to have smaller or even negative margins. The approach derived from taking account of the slack variables is suitable for handling noisy data.

With this definition, we can give two extensions of the previous result to the noisy case. They will directly motivate the algorithms presented in the following section.

Theorem 5.3. *([477]) Consider thresholding real-valued linear functions $\mathcal{L}$ with unit weight vectors on an inner product space X and fix $\gamma \in \mathbb{R}^+$. There is a constant c, such that for any probability distribution $\mathcal{D}$ on $X \times \{-1, 1\}$ with support in a ball of radius R around the origin, with probability $1 - \delta$ over m random examples S, any hypothesis $f \in \mathcal{L}$ has error no more than*

$$err_{\mathcal{D}}(f) \leq \frac{c}{m}\left(\frac{R^2 + \|\xi\|_2^2}{\gamma^2}\log^2 m + \log\frac{1}{\delta}\right),$$

where $\xi = \xi(f, S, \gamma)$ is the margin slack vector with respect to f and γ.

An analogous result can be proven using the 1-norm of the slack variables.

Theorem 5.4. *Consider thresholding real-valued linear functions $\mathcal{L}$ with unit weight vectors on an inner product space X and fix $\gamma \in \mathbb{R}^+$. There is a constant c, such that for any probability distribution $\mathcal{D}$ on $X \times \{-1, 1\}$ with support in a ball of radius R around the origin, with probability $1-\delta$ over m random examples S, any hypothesis $f \in \mathcal{L}$ has error no more than*

$$err_{\mathcal{D}}(f) \leq \frac{c}{m}\left(\frac{R^2}{\gamma^2}\log^2 m + \frac{\|\xi\|_1}{\gamma}\log m + \log\frac{1}{\delta}\right),$$

where $\xi = \xi(f, S, \gamma)$ is the margin slack vector with respect to f and γ.

The conclusion to be drawn from the previous theorems is that the generalisation error bound takes into account the amount by which points fail to meet a target margin γ. The bound is in terms of a norm of the slack variable vector suggesting that this quantity should be minimised in order to improve performance. The bound does not rely on the training points being linearly separable and hence can also handle the case when the data are corrupted by noise or the function class cannot capture the full complexity of the decision rule. Optimising the norm of the margin slack vector does not necessarily mean minimising the number of misclassifications. This fact will be important, as we shall see that minimising the number of misclassifications is computationally more demanding than optimising the norms of the margin slack vector.

Optimising the norms of the margin slack vector has a diffuse effect on the margin. For this reason it is referred to as a *soft margin* in contrast to the maximal margin, which depends critically on a small subset of points and is therefore often called a *hard margin*. We will refer to the bound in terms of the 2-norm of the margin slack vector as the 2-norm soft margin bound, and similarly for the 1-norm soft margin.

The next section will give a number of algorithms that directly minimise the (upper bound on the) risk of making wrong predictions. Importantly, such algorithms reduce to optimizing a convex function.

5.3. Support Vector Machines

We are now ready to derive the main algorithm of this chapter: the celebrated Support Vector Machine, introduced by Vapnik and coworkers in 1992 [76, 525]. The aim of Support Vector classification is to devise a computationally efficient way of learning 'good' separating hyperplanes in a high dimensional feature space, where by 'good' hyperplanes we will understand ones that optimise the generalisation bounds described in the previous section, and by 'computationally efficient' we will mean algorithms able to deal with sample sizes of hundreds of thousands of examples. The generalisation theory gives clear guidance about how to control capacity and hence prevent overfitting by controlling the hyperplane margin measures. Optimisation theory will provide the mathematical techniques necessary to find hyperplanes optimising these measures, and to study their properties.

Different generalisation bounds exist, motivating different algorithms: one can for example optimise the maximal margin, the margin distribution, and other quantities defined later in this section. We will first consider the most common and well-established approach, which reduces the hard margin problem to minimising the norm of the weight vector.

The aim of this algorithm is to find a (generally nonlinear) separation between the data by using kernels, and to ensure that such separation has strong statistical properties aimed at preventing overfitting. This is done efficiently, by finding the global minimum of a convex cost function.

5.3.1 The Maximal Margin Classifier

The simplest type of Support Vector Machine, which was also the first to be introduced, is the so-called maximal margin classifier that optimises the hard margin. It works only for data which are linearly separable in the feature space, and hence cannot be used in many real-world situations. Nonetheless it is the easiest algorithm to understand, and it forms the main building block for the more complex Support Vector Machines. It exhibits the key features that characterise this kind of learning machine, and its description is therefore useful for understanding the more realistic systems introduced later.

Theorem 5.2 bounds the generalisation error of linear machines in terms of the margin $M_S(f)$ of the hypothesis f with respect to the training set S. The dimensionality of the space in which the data are separated does not appear in this theorem. The maximal margin classifier optimises this bound by separating the data with the maximal margin hyperplane, and given that the bound does not depend on the dimensionality of the space, this separation can be sought in any kernel-induced feature space. The maximal margin classifier forms the strategy of the first Support Vector Machine, namely to separate the data by using the maximal margin hyperplane in an appropriately chosen kernel-induced feature space.

This strategy is implemented by reducing it to a convex optimisation problem: minimising a quadratic function under linear inequality constraints. The cost function depends on the norm of the weight vector, although this may not be immediately obvious.

In order to see it, first we note that in the definition of linear classifiers there is an inherent degree of freedom, as the function associated with the hyperplane $(\mathbf{w}, b)$ does not change if we rescale the hyperplane to $(\lambda\mathbf{w}, \lambda b)$, for $\lambda \in \mathbb{R}^+$. There will, however, be a change in the functional margin as opposed to the geometric margin. Theorem 5.2 involves the *geometric margin*, that is the functional margin of a normalised weight vector. Hence, we can equally well optimise the geometric margin by fixing the functional margin to be equal to 1 (hyperplanes with functional margin 1 are sometimes known as *canonical hyperplanes*) and minimising the norm of the weight vector. The resulting geometric margin will be equal to $1/\|\mathbf{w}\|_2$, since this will be the functional margin if we scale by this amount to create the required unit weight vector. The following proposition summarises the situation.

Proposition 5.1. *Given a linearly separable training sample*

$$S = ((\mathbf{x}_1, y_1), \ldots, (\mathbf{x}_m, y_m))$$

the hyperplane $(\mathbf{w}, b)$ *that solves the optimisation problem*

$$\begin{array}{ll} \textit{minimise}_{\mathbf{w},b} & \langle \mathbf{w}, \mathbf{w} \rangle, \\ \textit{subject to} & y_i\left(\langle \mathbf{w}_i, \mathbf{x}_i \rangle + b\right) \geq 1, \\ & i = 1, \ldots, m, \end{array}$$

realises the maximal margin hyperplane with geometric margin $\gamma = 1/\|\mathbf{w}\|_2$.

The solution of this problem requires the use of quadratic optimization theory. Surprisingly, it naturally produces a dual representation analogous to the one encountered for the dual perceptron. The primal Lagrangian is

$$L(\mathbf{w}, b, \alpha) = \frac{1}{2}\langle \mathbf{w}, \mathbf{w} \rangle - \sum_{i=1}^{m} \alpha_i \left[y_i \left(\langle \mathbf{w}_i, \mathbf{x}_i \rangle + b \right) - 1 \right]$$

where $\alpha_i \geq 0$ are Lagrange multipliers, and the factor $\frac{1}{2}$ has been inserted for algebraic convenience and does not affect the generality of the result. The corresponding dual is found by differentiating with respect to $\mathbf{w}$ and b, imposing stationarity,

$$\frac{\partial L(\mathbf{w}, b, \alpha)}{\partial \mathbf{w}} = \mathbf{w} - \sum_{i=1}^{m} y_i \alpha_i \mathbf{x}_i = \mathbf{0},$$

$$\frac{\partial L(\mathbf{w}, b, \alpha)}{\partial b} = \sum_{i=1}^{m} y_i \alpha_i = 0,$$

and resubstituting the relations obtained,

$$\mathbf{w} = \sum_{i=1}^{m} y_i \alpha_i \mathbf{x}_i,$$

$$0 = \sum_{i=1}^{m} y_i \alpha_i,$$

into the primal we obtain

$$\begin{aligned} L(\mathbf{w}, b, \alpha) &= \frac{1}{2} \langle \mathbf{w}, \mathbf{w} \rangle - \sum_{i=1}^{m} \alpha_i \left[y_i \left(\langle \mathbf{w}_i, \mathbf{x}_i \rangle + b \right) - 1 \right] \\ &= \frac{1}{2} \sum_{i,j=1}^{m} y_i y_j \alpha_i \alpha_j \langle \mathbf{x}_i, \mathbf{x}_j \rangle - \sum_{i,j=1}^{m} y_i y_j \alpha_i \alpha_j \langle \mathbf{x}_i, \mathbf{x}_j \rangle + \sum_{i=1}^{m} \alpha_i \\ &= \sum_{i=1}^{m} \alpha_i - \frac{1}{2} \sum_{i,j=1}^{m} y_i y_j \alpha_i \alpha_j \langle \mathbf{x}_i, \mathbf{x}_j \rangle . \end{aligned}$$

The first of the substitution relations shows that the hypothesis can be described as a linear combination of the training points: the application of optimisation theory naturally leads to the dual representation already encountered for the perceptron. This means we can use maximal margin hyperplanes in combination with with kernels.

It is rather remarkable that one can achieve a nonlinear separation of this type by solving a convex problem, since neural networks often used for this purpose are affected by local minima in the training. The situation is summarized in the following proposition (following from 5.1).

Proposition 5.2. *Consider a linearly separable training sample*

$$S = ((\mathbf{x}_1, y_1), \ldots, (\mathbf{x}_m, y_m)),$$

and suppose the parameters α^* *solve the following quadratic optimisation problem:*

$$\left.\begin{aligned} &\text{maximise } W(\alpha) = \textstyle\sum_{i=1}^{m} \alpha_i - \frac{1}{2} \sum_{i,j=1}^{m} y_i y_j \alpha_i \alpha_j \langle \mathbf{x}_i, \mathbf{x}_j \rangle, \\ &\text{subject to } \textstyle\sum_{i=1}^{m} y_i \alpha_i = 0, \\ &\qquad\qquad \alpha_i \geq 0, \ i = 1, \ldots, m. \end{aligned}\right\} \quad (5.2)$$

Then the weight vector $\mathbf{w}^* = \sum_{i=1}^{m} y_i \alpha_i^* \mathbf{x}_i$ *realises the maximal margin hyperplane with geometric margin*

$$\gamma = 1 / \|\mathbf{w}^*\|_2 .$$

The value of b does not appear in the dual problem and so b^* must be found making use of the primal constraints:

$$b^* = -\frac{\max_{y_i=-1}\left(\langle \mathbf{w}^*, \mathbf{x}_i \rangle\right) + \min_{y_i=1}\left(\langle \mathbf{w}^*, \mathbf{x}_i \rangle\right)}{2}$$

The Kuhn-Tucker Theorem from optimization theory provides further information about the structure of the solution. The Karush-Kuhn-Tucker (KKT) conditions state that the optimal solutions α^*, $(\mathbf{w}^*, b^*)$ must satisfy

$$\alpha_i^* \left[y_i \left(\langle \mathbf{w}_i^*, \mathbf{x}_i \rangle + b^* \right) - 1 \right] = 0, \qquad i = 1, \ldots, m.$$

This implies that only for inputs $\mathbf{x}_i$ for which the functional margin is one and that therefore lie closest to the hyperplane are the corresponding α_i^* non-zero. All the other parameters α_i^* are zero. Hence, in the expression for the weight vector only these points are involved. It is for this reason that they are called *support vectors*. We will denote the set of indices of the support vectors with sv.

In other words, the optimal hyperplane can be expressed in the dual representation in terms of this subset of the parameters:

$$\begin{aligned} f(\mathbf{x}, \alpha^*, b^*) &= \sum_{i=1}^{m} y_i \alpha_i^* \langle \mathbf{x}_i, \mathbf{x} \rangle + b^* \\ &= \sum_{i \in \mathrm{sv}} y_i \alpha_i^* \langle \mathbf{x}_i, \mathbf{x} \rangle + b^*. \end{aligned}$$

The Lagrange multipliers associated with each point become the dual variables, giving them an intuitive interpretation quantifying how important a given training point is in forming the final solution. Points that are not support vectors have no influence, so that in non-degenerate cases slight perturbations of such points will not affect the solution. A similar meaning was found in the case of the dual representations for the perceptron learning algorithm, where the dual variable was proportional to the number of mistakes made by the algorithm on a given point during the training.

Another important consequence of the Karush Kuhn Tucker complementarity conditions is that for $j \in \mathrm{sv}$,

$$y_j f(\mathbf{x}_j, \alpha^*, b^*) = y_j \left(\sum_{i \in \mathrm{sv}} y_i \alpha_i^* \langle \mathbf{x}_i, \mathbf{x}_j \rangle + b^* \right) = 1,$$

and therefore we can express the norm of the weight vector and the margin separating the data as a function of the multipliers α:

$$\begin{aligned}\langle \mathbf{w}^*, \mathbf{w}^* \rangle &= \sum_{i,j=1}^{m} y_i y_j \alpha_i^* \alpha_j^* \langle \mathbf{x}_i, \mathbf{x}_j \rangle \\ &= \sum_{j \in \mathrm{SV}} \alpha_j^* y_j \sum_{i \in \mathrm{SV}} y_i \alpha_i^* \langle \mathbf{x}_i, \mathbf{x}_j \rangle \\ &= \sum_{j \in \mathrm{SV}} \alpha_j^* (1 - y_j b^*) \\ &= \sum_{i \in \mathrm{SV}} \alpha_i^* .\end{aligned}$$

This gives us the following remarkable proposition, connecting all the main quantities of the algorithm:

Proposition 5.3. *Consider a linearly separable training sample*

$$S = ((\mathbf{x}_1, y_1), \ldots, (\mathbf{x}_m, y_m)),$$

and suppose the parameters α^* and b^* solve the dual optimisation problem (5.2). Then the weight vector $\mathbf{w} = \sum_{i=1}^{m} y_i \alpha_i^* \mathbf{x}_i$ realises the maximal margin hyperplane with geometric margin

$$\gamma = 1/\left\|\mathbf{w}\right\|_2 = \left(\sum_{i \in sv} \alpha_i^* \right)^{-1/2} .$$

For convenience of the Reader, we summarize all the results obtained above in a single proposition, and we express them by explicitly using kernel functions.

Proposition 5.4. *(**Maximal Margin Algorithm**) Consider a training sample*

$$S = ((\mathbf{x}_1, y_1), \ldots, (\mathbf{x}_m, y_m))$$

that is linearly separable in the feature space implicitly defined by the kernel $K(\mathbf{x}, \mathbf{z})$ and suppose the parameters α^* and b^* solve the following quadratic optimisation problem:

$$\left.\begin{array}{l} \text{maximise } W(\alpha) = \sum_{i=1}^{m} \alpha_i - \frac{1}{2} \sum_{i,j=1}^{m} y_i y_j \alpha_i \alpha_j K(\mathbf{x}_i, \mathbf{x}_j), \\ \text{subject to } \sum_{i=1}^{m} y_i \alpha_i = 0, \\ \qquad \alpha_i \geq 0,\ i = 1, \ldots, m. \end{array}\right\} \tag{5.3}$$

Then the decision rule given by sgn($f(\mathbf{x})$), where $f(\mathbf{x}) = \sum_{i=1}^{m} y_i \alpha_i^* K(\mathbf{x}_i, \mathbf{x}) + b^*$, is equivalent to the maximal margin hyperplane in the feature space implicitly defined by the kernel $K(\mathbf{x}, \mathbf{z})$ and that hyperplane has geometric margin

$$\gamma = \left(\sum_{i \in sv} \alpha_i^* \right)^{-1/2} .$$

Note that the requirement that the kernel satisfy Mercer's conditions is equivalent to the requirement that the matrix with entries $(K(\mathbf{x}_i, \mathbf{x}_j))_{i,j=1}^m$ be positive definite for all training sets. This in turn means that the optimisation problem (5.3) is convex since the matrix $(y_i y_j K(\mathbf{x}_i, \mathbf{x}_j))_{i,j=1}^m$ is also positive definite. Hence, the property required for a kernel function to define a feature space also ensures that the maximal margin optimisation problem has a unique solution that can be found efficiently. This rules out the problem of local minima encountered in training neural networks.

The fact that only a subset of the Lagrange multipliers is non-zero is often referred to as *sparseness*, and means that the support vectors contain all the information necessary to reconstruct the hyperplane. Even if all of the other points were removed the same maximal separating hyperplane would be found for the remaining subset of the support vectors. This can also be seen from the dual problem, since removing rows and columns corresponding to non-support vectors leaves the same optimisation problem for the remaining submatrix. Hence, the optimal solution remains unchanged. This shows that the maximal margin hyperplane is a compression scheme in the sense that we can reconstruct the maximal margin hyperplane from just the support vectors. This fact makes it possible to prove the following theorem, connecting generalization power of the function with its sparseness.

Theorem 5.5. *Consider thresholding real-valued linear functions $\mathcal{L}$ with unit weight vectors on an inner product space X. For any probability distribution $\mathcal{D}$ on $X \times \{-1, 1\}$, with probability $1 - \delta$ over m random examples S, the maximal margin hyperplane has error no more than*

$$err_{\mathcal{D}}(f) \leq \frac{1}{m-d}\left(d \log \frac{em}{d} + \log \frac{m}{\delta}\right),$$

where $d = \#sv$ is the number of support vectors.

The only degree of freedom left in the maximal margin algorithm is the choice of kernel, which amounts to model selection. Any prior knowledge we have of the problem can help in choosing a parametrised kernel family, and then model selection is reduced to adjusting the parameters. For most classes of kernels, for example polynomial or Gaussian, it is always possible to find a kernel parameter for which the data become separable. In general, however, forcing separation of the data can easily lead to overfitting, particularly when noise is present in the data.

In this case, outliers would typically be characterised by a large Lagrange multiplier, and the procedure could be used for data cleaning, since it can rank the training data according to how difficult they are to classify correctly.

This algorithm provides the starting point for the many variations on this theme proposed in the last few years and attempting to address some of its weaknesses: that it is sensitive to the presence of noise; that it only considers two classes; that it is not expressly designed to achieve sparse solutions.

5.3.2 Soft Margin Optimisation

The maximal margin classifier described in the previous subsection is an important concept, as a starting point for the analysis and construction of more sophisticated Support Vector Machines, but it cannot be used in many real-world problems: if the data are noisy, there will in general be no linear separation in the feature space (unless we are ready to use very powerful kernels, and hence overfit the data). The main problem with the maximal margin classifier is that it always produces a consistent classification function, that is a classification function with no training error. This is of course a result of its motivation in terms of a bound that depends on the margin, a quantity that is negative unless the data are perfectly separated.

The dependence on a quantity like the margin opens the system up to the danger of falling hostage to the idiosyncracies of a few points. In real data, where noise can always be present, this can result in a brittle (non robust) estimator. Furthermore, in the cases where the data are not linearly separable in the feature space, the optimisation problem cannot be solved as the primal has an empty feasible region and the dual an unbounded objective function. These problems motivate using the more robust measures of the *margin distribution* introduced above. Such measures can tolerate noise and outliers, and take into consideration the positions of more training points than just those closest to the boundary.

Algorithms motivated by such robust bounds are often called Soft Margin Classifiers and will be presented in this subsection. Recall that the primal optimisation problem for the maximal margin case is the following:

$$\begin{aligned}&\text{minimise}_{\mathbf{w},b}\ \langle \mathbf{w}, \mathbf{w} \rangle,\\ &\text{subject to}\quad y_i\left(\langle \mathbf{w}_i, \mathbf{x}_i \rangle + b\right) \geq 1,\ i = 1, \ldots, m.\end{aligned}$$

In order to optimise the margin slack vector we need to introduce slack variables to allow the margin constraints to be violated, so that the constraints become:

$$\begin{aligned}\text{subject to } & y_i\left(\langle \mathbf{w}_i, \mathbf{x}_i \rangle + b\right) \geq 1 - \xi_i,\ i = 1, \ldots, m,\\ & \xi_i \geq 0,\ i = 1, \ldots, m.\end{aligned}$$

and at the same time we need to somehow limit the overall size of the violations ξ_i.

Notice that optimizing a combination of the violations and the margin is directly equivalent to optimizing the statistical bounds given in Section 5.2. This can be easily seen, for example in the case of Theorem 5.3, that bounds the generalisation error in terms of the 2-norm of the margin slack vector, the so-called 2-norm soft margin, which contains the ξ_i scaled by the norm of the weight vector $\mathbf{w}$. The equivalent expression on which the generalisation depends is $\frac{R^2 + \frac{\|\xi\|_2^2}{\|\mathbf{w}\|_2^2}}{\gamma^2} = \|\mathbf{w}\|_2^2 \left(R^2 + \frac{\|\xi\|_2^2}{\|\mathbf{w}\|_2^2}\right) = \|\mathbf{w}\|_2^2 R^2 + \|\xi\|_2^2$, motivating the following optimization problem:

$$\left.\begin{aligned}&\text{minimise}_{\xi,\mathbf{w},b}\ \langle \mathbf{w}, \mathbf{w} \rangle + C \textstyle\sum_{i=1}^{m} \xi_i^2,\\ &\text{subject to}\quad y_i\left(\langle \mathbf{w}_i, \mathbf{x}_i \rangle + b\right) \geq 1 - \xi_i,\ i = 1, \ldots, m,\end{aligned}\right\} \tag{5.4}$$

The parameter C controls the trade-off between classification errors and margin. In practice it is varied through a wide range of values and the optimal performance assessed using a separate validation set or techniques of cross-validation. As the parameter C runs through a range of values, the norm of the solution $\|\mathbf{w}\|_2$ varies smoothly through a corresponding range. Hence, for a particular problem, choosing a particular value for C corresponds to choosing a value for $\|\mathbf{w}\|_2$, and then minimising $\|\xi\|_2$ for that size of $\mathbf{w}$. This approach is also adopted in the 1-norm case where the optimisation problem minimises a combination of the norm of the weights and the 1-norm of the slack variables that does not exactly match that found in Theorem 5.4:

$$\left.\begin{array}{ll}\text{minimise}_{\xi,\mathbf{w},b} & \langle \mathbf{w}, \mathbf{w}\rangle + C\sum_{i=1}^{m} \xi_i,\\ \text{subject to} & y_i\left(\langle \mathbf{w}_i, \mathbf{x}_i\rangle + b\right) \geq 1 - \xi_i,\ i = 1, \ldots, m,\\ & \xi_i \geq 0,\ i = 1, \ldots, m.\end{array}\right\} \tag{5.5}$$

Since there is a value of C corresponding to the optimal choice of $\|\mathbf{w}\|_2$, that value of C will give the optimal bound as it will correspond to finding the minimum of $\|\xi\|_1$ with the given value for $\|\mathbf{w}\|_2$. We will devote the next two subsections to investigating the duals of the two margin slack vector problems creating the so-called soft margin algorithms. This will give considerable insight into the structure and the properties of the solutions, as well as being practically important in the implementation.

2-Norm Soft Margin. The primal Lagrangian for the problem of equation (5.4) is

$$L(\mathbf{w}, b, \xi, \alpha) = \frac{1}{2}\langle \mathbf{w}, \mathbf{w}\rangle + \frac{C}{2}\sum_{i=1}^{m} \xi_i^2 - \sum_{i=1}^{m} \alpha_i\left[y_i\left(\langle \mathbf{w}_i, \mathbf{x}_i\rangle + b\right) - 1 + \xi_i\right]$$

where $\alpha_i \geq 0$ are the Lagrange multipliers. The corresponding dual is found by differentiating with respect to $\mathbf{w}$, ξ and b, imposing stationarity,

$$\frac{\partial L(\mathbf{w}, b, \xi, \alpha)}{\partial \mathbf{w}} = \mathbf{w} - \sum_{i=1}^{m} y_i \alpha_i \mathbf{x}_i = \mathbf{0},$$

$$\frac{\partial L(\mathbf{w}, b, \xi, \alpha)}{\partial \xi} = C\xi - \alpha = \mathbf{0},$$

$$\frac{\partial L(\mathbf{w}, b, \xi, \alpha)}{\partial b} = \sum_{i=1}^{m} y_i \alpha_i = 0,$$

and resubstituting the relations obtained into the primal to obtain the following objective function:

$$\begin{aligned} L(\mathbf{w}, b, \xi, \alpha) &= \sum_{i=1}^{m} \alpha_i - \frac{1}{2}\sum_{i,j=1}^{m} y_i y_j \alpha_i \alpha_j \langle \mathbf{x}_i, \mathbf{x}_j\rangle + \frac{1}{2C}\langle \alpha, \alpha\rangle - \frac{1}{C}\langle \alpha, \alpha\rangle \\ &= \sum_{i=1}^{m} \alpha_i - \frac{1}{2}\sum_{i,j=1}^{m} y_i y_j \alpha_i \alpha_j \langle \mathbf{x}_i, \mathbf{x}_j\rangle - \frac{1}{2C}\langle \alpha, \alpha\rangle . \end{aligned}$$

Hence, maximising the above objective over α is equivalent to maximising

$$W(\alpha) = \sum_{i=1}^{m} \alpha_i - \frac{1}{2} \sum_{i,j=1}^{m} y_i y_j \alpha_i \alpha_j \left(\langle \mathbf{x}_i, \mathbf{x}_j \rangle + \frac{1}{C} \delta_{ij} \right),$$

where δ_{ij} is the Kronecker δ defined to be 1 if $i = j$ and 0 otherwise. The corresponding Karush Kuhn Tucker complementarity conditions are

$$\alpha_i \left[y_i(\langle \mathbf{x}_i, \mathbf{w} \rangle + b) - 1 + \xi_i \right] = 0, \; i = 1, \ldots, m.$$

Therefore, we have the following result (where we use kernels instead of inner products):

Proposition 5.5. *(2-Norm Soft Margin Classifier) Consider classifying a training sample*

$$S = ((\mathbf{x}_1, y_1), \ldots, (\mathbf{x}_m, y_m)),$$

using the feature space implicitly defined by the kernel $K(\mathbf{x}, \mathbf{z})$, and suppose the parameters α^* solve the following quadratic optimisation problem:

$$\begin{aligned} \text{maximise } & W(\alpha) = \textstyle\sum_{i=1}^{m} \alpha_i - \frac{1}{2} \sum_{i,j=1}^{m} y_i y_j \alpha_i \alpha_j \left(K(\mathbf{x}_i, \mathbf{x}_j) + \frac{1}{C} \delta_{ij} \right), \\ \text{subject to } & \textstyle\sum_{i=1}^{m} y_i \alpha_i = 0, \\ & \alpha_i \geq 0, \; i = 1, \ldots, m. \end{aligned}$$

Let $f(\mathbf{x}) = \sum_{i=1}^{m} y_i \alpha_i^* K(\mathbf{x}_i, \mathbf{x}) + b^*$, where b^* is chosen so that $y_i f(\mathbf{x}_i) = 1 - \alpha_i^*/C$ for any i with $\alpha_i^* \neq 0$. Then the decision rule given by sgn($f(\mathbf{x})$) is equivalent to the hyperplane in the feature space implicitly defined by the kernel $K(\mathbf{x}, \mathbf{z})$ which solves the optimisation problem (5.4), where the slack variables are defined relative to the geometric margin

$$\gamma = \left(\sum_{i \in sv} \alpha_i^* - \frac{1}{C} \langle \alpha^*, \alpha^* \rangle \right)^{-1/2}.$$

Proof. The value of b^* is chosen using the relation $\alpha_i = C\xi_i$ and by reference to the primal constraints which by the Karush Kuhn Tucker complementarity conditions

$$\alpha_i \left[y_i \left(\langle \mathbf{w}_i, \mathbf{x}_i \rangle + b \right) - 1 + \xi_i \right] = 0, \; i = 1, \ldots, m,$$

must be equalities for non-zero α_i. It remains to compute the norm of $\mathbf{w}^*$ which defines the size of the geometric margin.

$$\begin{aligned}
\langle \mathbf{w}^*, \mathbf{w}^* \rangle &= \sum_{i,j=1}^{m} y_i y_j \alpha_i^* \alpha_j^* K(\mathbf{x}_i, \mathbf{x}_j) \\
&= \sum_{j \in \mathrm{SV}} \alpha_j^* y_j \sum_{i \in \mathrm{SV}} y_i \alpha_i^* K(\mathbf{x}_i, \mathbf{x}_j) \\
&= \sum_{j \in \mathrm{SV}} \alpha_j^* \left(1 - \xi_j^* - y_j b^*\right) \\
&= \sum_{i \in \mathrm{SV}} \alpha_i^* - \sum_{i \in \mathrm{SV}} \alpha_i^* \xi_i^* \\
&= \sum_{i \in \mathrm{SV}} \alpha_i^* - \frac{1}{C} \langle \alpha^*, \alpha^* \rangle .
\end{aligned}$$

This is still a quadratic programming problem, and can be solved with the same methods used for the maximal margin hyperplane. The only change is the addition of $1/C$ to the diagonal of the kernel matrix. This has the effect of adding $1/C$ to the eigenvalues of the matrix, rendering the problem better conditioned. We can therefore view the 2-norm soft margin as simply a change of kernel

$$\hat{K}(\mathbf{x}, \mathbf{z}) = K(\mathbf{x}, \mathbf{z}) + \frac{1}{C} \delta_{\mathbf{x}}(\mathbf{z}).$$

1-Norm Soft Margin. The Lagrangian for the 1-norm soft margin optimisation problem presented above is

$$\begin{aligned}
L(\mathbf{w}, b, \xi, \alpha, \mathbf{r}) = {}& \frac{1}{2} \langle \mathbf{w}, \mathbf{w} \rangle + C \sum_{i=1}^{m} \xi_i \\
& - \sum_{i=1}^{m} \alpha_i \left[y_i (\langle \mathbf{x}_i, \mathbf{w} \rangle + b) - 1 + \xi_i \right] - \sum_{i=1}^{m} r_i \xi_i
\end{aligned}$$

with $\alpha_i \geq 0$ and $r_i \geq 0$. The corresponding dual is found by differentiating with respect to $\mathbf{w}$, ξ and b, imposing stationarity,

$$\begin{aligned}
\frac{\partial L(\mathbf{w}, b, \xi, \alpha, \mathbf{r})}{\partial \mathbf{w}} &= \mathbf{w} - \sum_{i=1}^{m} y_i \alpha_i \mathbf{x}_i = \mathbf{0}, \\
\frac{\partial L(\mathbf{w}, b, \xi, \alpha, \mathbf{r})}{\partial \xi_i} &= C - \alpha_i - r_i = 0, \\
\frac{\partial L(\mathbf{w}, b, \xi, \alpha, \mathbf{r})}{\partial b} &= \sum_{i=1}^{m} y_i \alpha_i = 0,
\end{aligned}$$

and resubstituting the relations obtained into the primal. We obtain the following objective function:

$$L(\mathbf{w}, b, \xi, \alpha, \mathbf{r}) = \sum_{i=1}^{m} \alpha_i - \frac{1}{2} \sum_{i,j=1}^{m} y_i y_j \alpha_i \alpha_j \langle \mathbf{x}_i, \mathbf{x}_j \rangle,$$

which curiously is identical to that for the maximal margin. The only difference is that the constraint $C-\alpha_i-r_i=0$, together with $r_i \geq 0$, enforces $\alpha_i \leq C$, while $\xi_i \neq 0$ only if $r_i = 0$ and therefore $\alpha_i = C$. The Karush Kuhn Tucker complementarity conditions are therefore

$$\begin{aligned} &\alpha_i \left[y_i(\langle \mathbf{x}_i, \mathbf{w} \rangle + b) - 1 + \xi_i \right] = 0, \; i = 1, \ldots, m, \\ &\xi_i \left(\alpha_i - C \right) = 0, \qquad\qquad\qquad\qquad i = 1, \ldots, m. \end{aligned}$$

Notice that the KKT conditions imply that non-zero slack variables can only occur when $\alpha_i = C$. The points with non-zero slack variables are $1/\left\|\mathbf{w}\right\|$-margin errors, as their geometric margin is less than $1/\left\|\mathbf{w}\right\|$. Points for which $0 < \alpha_i < C$ lie at the target distance of $1/\left\|\mathbf{w}\right\|$ from the hyperplane. We therefore have the following proposition.

Proposition 5.6. *(**1-Norm Soft Margin Classifier**) Consider classifying a training sample*

$$S = ((\mathbf{x}_1, y_1), \ldots, (\mathbf{x}_m, y_m)),$$

using the feature space implicitly defined by the kernel $K(\mathbf{x}, \mathbf{z})$, and suppose the parameters α^* solve the following quadratic optimisation problem:

$$\left. \begin{aligned} &\textit{maximise } W(\alpha) = \textstyle\sum_{i=1}^{m} \alpha_i - \frac{1}{2} \sum_{i,j=1}^{m} y_i y_j \alpha_i \alpha_j K(\mathbf{x}_i, \mathbf{x}_j), \\ &\textit{subject to } \textstyle\sum_{i=1}^{m} y_i \alpha_i = 0, \\ &\qquad\qquad C \geq \alpha_i \geq 0, \; i = 1, \ldots, m. \end{aligned} \right\} \tag{5.6}$$

Let $f(\mathbf{x}) = \sum_{i=1}^{m} y_i \alpha_i^* K(\mathbf{x}_i, \mathbf{x}) + b^*$, where b^* is chosen so that $y_i f(\mathbf{x}_i) = 1$ for any i with $C > \alpha_i^* > 0$. Then the decision rule given by sgn($f(\mathbf{x})$) is equivalent to the hyperplane in the feature space implicitly defined by the kernel $K(\mathbf{x}, \mathbf{z})$ that solves the optimisation problem (5.5), where the slack variables are defined relative to the geometric margin

$$\gamma = \left(\sum_{i,j \in sv} y_i y_j \alpha_i^* \alpha_j^* K(\mathbf{x}_i, \mathbf{x}_j) \right)^{-1/2}.$$

The value of b^* is chosen using the Karush Kuhn Tucker complementarity conditions which imply that if $C > \alpha_i^* > 0$ both $\xi_i^* = 0$ and

$$y_i(\langle \mathbf{x}_i, \mathbf{w}^* \rangle + b^*) - 1 + \xi_i^* = 0.$$

The norm of $\mathbf{w}^*$ is clearly given by the expression

$$\begin{aligned} \langle \mathbf{w}^*, \mathbf{w}^* \rangle &= \sum_{i,j=1}^{m} y_i y_j \alpha_i^* \alpha_j^* K(\mathbf{x}_i, \mathbf{x}_j) \\ &= \sum_{j \in \mathrm{SV}} \sum_{i \in \mathrm{SV}} y_i y_j \alpha_i^* \alpha_j^* K(\mathbf{x}_i, \mathbf{x}_j). \end{aligned}$$

So surprisingly this problem is equivalent to the maximal margin hyperplane, with the additional constraint that all the α_i are upper bounded by C. This gives rise to the name *box constraint* that is frequently used to refer to this formulation, since the vector α is constrained to lie inside the box with side length C in the positive orthant. The trade-off parameter between accuracy and regularisation directly controls the size of the α_i. This makes sense intuitively as the box constraints limit the influence of outliers, which would otherwise have large Lagrange multipliers. The constraint also ensures that the feasible region is bounded and hence that the primal always has a non-empty feasible region.

It is now easy to see why the maximal (or hard) margin case is an important concept in the solution of more sophisticated versions of the machine: both the 1- and the 2-norm soft margin machines lead to optimisation problems that are solved by relating them to the maximal margin case.

One problem with the soft margin approach suggested is the choice of parameter C. Typically a range of values must be tried before the best choice for a particular training set can be selected. Furthermore the scale of the parameter is affected by the choice of feature space. It has been shown, however, that the solutions obtained for different values of C in the optimisation problem (5.6) are the same as those obtained as ν is varied between 0 and 1 in the optimisation problem

$$\begin{aligned} &\text{maximise } W(\alpha) = -\tfrac{1}{2}\textstyle\sum_{i,j=1}^{m} y_i y_j \alpha_i \alpha_j K(\mathbf{x}_i, \mathbf{x}_j) \\ &\text{subj. to } \textstyle\sum_{i=1}^{m} y_i \alpha_i = 0, \\ &\qquad\quad \textstyle\sum_{i=1}^{m} \alpha_i \geq \nu \\ &\qquad\quad 1/m \geq \alpha_i \geq 0,\ i = 1, \ldots, m. \end{aligned}$$

In this parametrisation ν places a lower bound on the sum of the α_i, which causes the linear term to be dropped from the objective function. It can be shown that the proportion of the training set that are margin errors is upper bounded by ν, while ν provides a lower bound on the total number of support vectors. Therefore ν gives a more transparent parametrisation of the problem which does not depend on the scaling of the feature space, but only on the noise level in the data.

5.3.3 Kernel Ridge Regression

We now discuss a classical approach to regression known as Ridge Regression, that generalizes Least Squares regression, and we will show how this simple method can be used in combination with kernels to obtain an algorithm that is equivalent to a statistical method known as a Gaussian Process. We will give it an independent derivation, which highlights the connections with the systems from the Support Vector family [460].

The problem is the one of finding a linear regression function $f(\mathbf{x}) = \langle \mathbf{w}, \mathbf{x} \rangle$ that fits a dataset $S = ((\mathbf{x}_1, y_1), \ldots, (\mathbf{x}_m, y_m))$ where the labels are real numbers: $y_i \in \mathbb{R}$. The quality of the fit is measured by the squares of the deviations $y_i - \langle \mathbf{w}, \mathbf{x}_i \rangle$ between the predicted and the given labels, and at the same time we attempt to keep the norm of the function as small as possible.

The resulting trade-off can be stated as the following optimization problem:

$$\left.\begin{array}{l}\text{minimise } \lambda \|\mathbf{w}\|^2 + \sum_{i=1}^m \xi_i^2, \\ \text{subject to } y_i - \langle \mathbf{w}, \mathbf{x}_i \rangle = \xi_i,\ i = 1, \ldots, m,\end{array}\right\} \tag{5.7}$$

from which we derive the following Lagrangian

$$\text{minimise } L(\mathbf{w}, \xi, \alpha) = \lambda \|\mathbf{w}\|^2 + \sum_{i=1}^m \xi_i^2 + \sum_{i=1}^m \alpha_i (y_i - \langle \mathbf{w}, \mathbf{x}_i \rangle - \xi_i).$$

Differentiating and imposing stationarity, we obtain that

$$\mathbf{w} = \frac{1}{2\lambda} \sum_{i=1}^m \alpha_i \mathbf{x}_i \text{ and } \xi_i = \frac{\alpha_i}{2}.$$

Resubstituting these relations gives the following (unconstrained) dual problem:

$$\text{maximise } W(\alpha) = \sum_{i=1}^m y_i \alpha_i - \tfrac{1}{4\lambda} \sum_{i,j=1}^m \alpha_i \alpha_j \langle \mathbf{x}_i, \mathbf{x}_j \rangle - \tfrac{1}{4} \sum \alpha_i^2,$$

that for convenience we rewrite in vector form:

$$W(\alpha) = \mathbf{y}'\alpha - \frac{1}{4\lambda} \alpha' \mathbf{K} \alpha - \frac{1}{4} \alpha' \alpha,$$

where $\mathbf{K}$ denotes the Gram matrix $\mathbf{K}_{ij} = \langle \mathbf{x}_i, \mathbf{x}_j \rangle$, or the kernel matrix $\mathbf{K}_{ij} = K(\mathbf{x}_i, \mathbf{x}_j)$, if we are working in a kernel-induced feature space. Differentiating with respect to α and imposing stationarity we obtain the condition

$$-\frac{1}{2\lambda} \mathbf{K}\alpha - \frac{1}{2}\alpha + \mathbf{y} = 0,$$

giving the solution

$$\alpha = 2\lambda (\mathbf{K} + \lambda \mathbf{I})^{-1} \mathbf{y}$$

and the corresponding regression function

$$f(\mathbf{x}) = \langle \mathbf{w}, \mathbf{x} \rangle = \mathbf{y}' (\mathbf{K} + \lambda \mathbf{I})^{-1} \mathbf{k}$$

where $\mathbf{k}$ is the vector with entries $k_i = \langle \mathbf{x}_i, \mathbf{x} \rangle$, $i = 1, \ldots, m$. Hence, we have the following result.

Proposition 5.7. ***(Kernel Ridge-Regression)*** *Suppose that we wish to perform regression on a training sample*

$$S = ((\mathbf{x}_1, y_1), \ldots, (\mathbf{x}_m, y_m)),$$

using the feature space implicitly defined by the kernel $K(\mathbf{x}, \mathbf{z})$*, and let* $f(\mathbf{x}) = \mathbf{y}'(\mathbf{K} + \lambda \mathbf{I})^{-1}\mathbf{k}$*, where* $\mathbf{K}$ *is the* $m \times m$ *matrix with entries* $\mathbf{K}_{ij} = K(\mathbf{x}_i, \mathbf{x}_j)$ *and* $\mathbf{k}$ *is the vector with entries* $\mathbf{k}_i = K(\mathbf{x}_i, \mathbf{x})$*. Then the function* $f(\mathbf{x})$ *is equivalent to the hyperplane in the feature space implicitly defined by the kernel* $K(\mathbf{x}, \mathbf{z})$ *that solves the ridge regression optimisation problem (5.7).*

This algorithm has appeared independently under a number of different names. Apart from being the Gaussian Process solution, it is also known as Krieging and the solutions are known as regularisation networks, where the regulariser has been implicitly selected by the choice of kernel. It is very simple to implement, essentially requiring only a matrix inversion, and is a very effective way to solve non-linear regression problems. Also in this case, it is important to note how the training is not affected by local minima.

5.4. Kernel PCA and CCA

In this section we briefly describe two techniques for discovering hidden structure respectively in a set of unlabeled data and in a 'paired' dataset, that is formed by two datasets in bijection. In their linear version they were both introduced by Hotelling, and both can be adapted for use with kernel functions. The first one is the classical Principal Components Analysis (PCA), aimed at finding a low dimensional representation of the data, the second one is Canonical Correlation Analysis (CCA), aimed at finding correlations between a dataset formed by pairs of vectors or, equivalently, between two 'matched' datasets, whose elements are in bijection [329, 36, 465].

Principal Components Analysis is a classical technique for analysing high dimensional data, and extracting hidden structure by finding a (small) set of coordinates that carry most of the information in the data. It can be proven that the most informative directions are given by the k principal eigenvectors of the data, in the sense that this choice minimizes - for any fixed k - the mean square distance between the original and the reconstructed data (see discussion in Chapter 3). The eigenvectors are called the *principal axes* of the data, and the new coordinates of each point are obtained by projecting it onto the first k principal axes.

As before, we will represent such vectors in dual form, as linear combinations of data vectors $\mathbf{v} = \sum_i \alpha_i \mathbf{x}_i$ and we will need to find the parameters α_i. Given a set of (unlabeled) observations $S = \{\mathbf{x}_1, \ldots, \mathbf{x}_m\}$ that are centered, $\sum_i \mathbf{x}_i = 0$, the (empirical) covariance matrix is defined as:

$$C = \frac{1}{m-1} \sum_i \mathbf{x}_i \mathbf{x}_i^T$$

and is a positive semi-definite matrix. Its eigenvectors and eigenvalues can be written as:

$$\lambda \mathbf{v} = C\mathbf{v} = \sum_i \langle \mathbf{x}_i, \mathbf{v} \rangle \mathbf{x}_i$$

that is each eigenvector can be written as linear combination of the training points,

$$\mathbf{v} = \sum_i \alpha_i \mathbf{x}_i$$

for some α, hence allowing a dual representation and the use of kernels. The new coordinates of a point are then given by projecting it onto the eigenvectors, and it is possible to prove that for any k using the first k eigenvectors gives the best approximation in that it minimises the sum of the 2-norms of the residuals of the training points.

By performing the same operation in the feature space, that is using the images of the points $\phi(\mathbf{x}_i)$, with simple manipulations we can find that the coefficients α^n of the n-th eigenvector can be obtained by solving the eigenvalue problem

$$m\lambda\alpha = K\alpha$$

and subsequently imposing the normalization $1 = \lambda_n \langle \alpha^n, \alpha^n \rangle$, $n = 1, \ldots, n$. Although we do not have the explicit coordinates of the eigenvectors, we can always use them for calculating the projections of the data points onto the n-th eigenvector $\mathbf{v}^n$, as follows

$$\langle \phi(\mathbf{x}), \mathbf{v}^n \rangle = \sum_{i=1}^{m} \alpha_i^n K(\mathbf{x}_i, \mathbf{x})$$

and this information is all we need for recoding our data and extracting hidden regularities.

Canonical Correlation Analysis (CCA) is a technique (also introduced by Hotelling) that can be used when we have two datasets that might have some underlying correlation. Assume there is a bijection between the elements of the two sets, possibly corresponding to two alternative descriptions of the same object (e.g., two views of the same 3D object; or two versions of the same document in 2 languages).

Given a set of pairs $S = \{(\mathbf{x}^1, \mathbf{x}^2)_i\}$, the task of CCA is to find linear combinations of variables in each of the two sets that have the maximum mutual correlation. Given two real valued variables a and b with zero mean, we define their correlation as

$$r = \frac{\sum_i a_i b_i}{\sqrt{\sum_i a_i^2 \sum_i b_i^2}}.$$

where $\{a_i\}$ and $\{b_i\}$ are realizations of the random variable. The problem of CCA can be formalized as follows: given two sets of paired vectors, $\mathbf{x}_i^1 \in X_1$ and $\mathbf{x}_i^2 \in X_2$, $i = 1, \ldots, m$, find vectors $\mathbf{w}^1 \in X_1$ and $\mathbf{w}^2 \in X_2$ such that the projections of the data onto these vectors $a_i = \langle \mathbf{x}_i^1, \mathbf{w}^1 \rangle$ and $b_i = \langle \mathbf{x}_i^2, \mathbf{w}^2 \rangle$ have maximal correlation.

Solving this task can be transformed into the following generalized eigenvector problem

$$\begin{bmatrix} 0 & C_{12} \\ C_{21} & 0 \end{bmatrix} \begin{bmatrix} \mathbf{w}^1 \\ \mathbf{w}^2 \end{bmatrix} = \lambda \begin{bmatrix} C_{11} & 0 \\ 0 & C_{22} \end{bmatrix} \begin{bmatrix} \mathbf{w}^1 \\ \mathbf{w}^2 \end{bmatrix},$$

where

$$C_{jk} = \sum_{i=1}^{m} \mathbf{x}_i^j \left(\mathbf{x}_i^k\right)^T, \quad j, k = 1, 2.$$

The approach can be kernelized following the same procedures discussed above. We obtain that $\mathbf{w}^1 = \sum_i \alpha_i^1 \phi(\mathbf{x}_i^1)$ and $\mathbf{w}^2 = \sum_i \alpha_i^2 \phi(\mathbf{x}_i^2)$ and this leads to another generalized eigenvalue problem to find the dual variables α:

$$\begin{bmatrix} 0 & K_1 K_2 \\ K_2 K_1 & 0 \end{bmatrix} \begin{bmatrix} \alpha^1 \\ \alpha^2 \end{bmatrix} = \lambda \begin{bmatrix} K_1^2 & 0 \\ 0 & K_2^2 \end{bmatrix} \begin{bmatrix} \alpha^1 \\ \alpha^2 \end{bmatrix},$$

where K_1 and K_2 are the kernel matrices for the vectors $\mathbf{x}_i^1 \in X_1$ and $\mathbf{x}_i^2 \in X_2$, that by assumption are in bijection, that is the ij-th entry in each matrix corresponds to the same pair of points.

By solving this problem, one can find nonlinear transformations of the data both in the first and in the second set that maximise the correlation between them. One use of this approach can be to analyze two different representations of the same object, possibly translations in different languages of the same documents, or different views of the same object in machine vision.

5.5. Conclusion

We have seen how the problem of finding nonlinear relations between points in a space can be reduced to finding linear relations in a kernel-induced feature space, and how this can first be reduced to solving a convex optimization problem. This approach includes methods for classification, regression and unsupervised learning. Furthermore, we have shown how the use of concepts from Statistical Learning Theory can give insight into which parameters of the algorithm should be controlled in order to avoid overfitting.

The result is a very versatile family of algorithms, known as Kernel Methods, that combine the statistical robustness provided by rigorous analysis with the computational efficiency given by convex optimization. The absence of local minima, however, is not the only reason for their fast uptake among data analysts. Their modular design, for example, by which the algorithm and the kernel function can be implemented and studied separately, means that any algorithm can work with any kernel, and hence previous work can be easily reused and prior knowledge naturally incorporated.

We have not discussed another important feature of such approach, that is worth mentioning at this point. Kernels can be defined between pairs of data items that are not vectors: we can consider embedding two symbol sequences of different length into some vector space where their inner product is defined, even if the sequences themselves are not vectors. This can be done efficiently by means of special string-matching kernels, and has proven useful in the analysis of biological data. Similarly, kernels exist for text, images, graphs and other data structures other than vectors. The design of a good kernel function requires some prior knowledge of the domain and the task, and once a good kernel for a type of data has been found, it can be used for a number of tasks, from classification to regression to PCA and many more.

Data Analysis practitioners can now rely on a rich toolbox containing several kernel-based algorithms together with a choice of kernels, each of which can be combined in a modular way. The development of such systems in the last few years, within a common theoretical framework, opens unprecedented opportunities for intelligent data analysis. In this Chapter we have presented just Support Vector Machines, kernel Ridge Regression and, briefly, kernel Principal Components and Canonical Correlation Analysis. Methods for anomaly detection, ranking, Time Series analysis, even Reinforcement Learning have been designed for kernels. The interested Reader is referred to the papers and websites in the bibliography, but in particular to **www.kernel-machines.org** and **www.support-vector.net** for further information, papers, books and online software related to this fast growing literature [134, 488, 135].

Chapter 6
Analysis of Time Series

Elizabeth Bradley
University of Colorado, USA

6.1. Introduction

Intelligent data analysis often requires one to extract meaningful conclusions about a complicated system using time-series data from a single sensor. If the system is linear, a wealth of well-established, powerful techniques is available to the analyst. If it is not, the problem is much harder and one must resort to nonlinear dynamics theory in order to infer useful information from the data. Either way, the problem is often complicated by a simultaneous overabundance and lack of data: megabytes of time-series data about the voltage output of a power substation, for instance, but no information about other important quantities, such as the temperatures inside the transformers. Data-mining techniques [177] provide some useful ways to deal successfully with the sheer volume of information that constitutes one part of this problem. The second part of the problem is much harder. If the target system is highly complex—say, an electromechanical device whose *dynamics* is governed by three metal blocks, two springs, a pulley, several magnets, and a battery—but only one of its important properties (e.g., the position of one of the masses) is sensor-accessible, the data analysis procedure would appear to be fundamentally limited.

Fig. 6.1 shows a simple example of the kind of problem that this chapter addresses: a mechanical spring/mass system and two time-series data sets gathered by sensors that measure the position and velocity of the mass. This system is linear: it responds *in proportion to* changes. Pulling the mass twice as far down, for instance, will elicit an oscillation that is twice as large, not one that is $2^{1.5}$ as large or $\log 2$ times as large. A pendulum, in contrast, reacts *nonlinearly*: if it is hanging straight down, a small change in its angle will have little effect, but if it is balanced at the inverted point, small changes have large effects. This

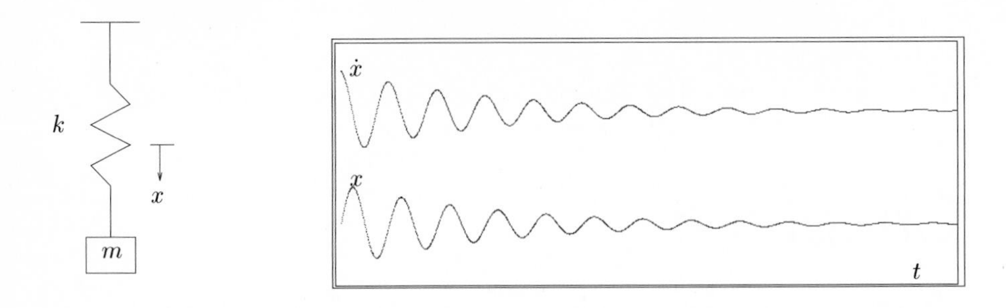

Fig. 6.1. A simple example: A spring/mass system and a time series of the vertical position and velocity of the mass, measured by two sensors

distinction is extremely important to science in general and data analysis in particular. If the system under examination is linear, data analysis is comparatively straightforward and the tools—the topic of section 6.2 of this chapter—are well developed. One can characterize the data using statistics (mean, standard deviation, etc.), fit curves to them (functional approximation), and plot various kinds of graphs to aid one's understanding of the behavior. If a more-detailed analysis is required, one typically represents the system in an "input + transfer function → output" manner using any of a wide variety of time- or frequency-domain models. This kind of formalism admits a large collection of powerful reasoning techniques, such as superposition and the notion of transforming back and forth between the time and frequency domains. The latter is particularly powerful, as many signal processing operations are much easier in one domain than the other.

Nonlinear systems pose an important challenge to intelligent data analysis. Not only are they ubiquitous in science and engineering, but their mathematics is also vastly harder, and many standard time-series analysis techniques simply do not apply to nonlinear problems. Chaotic systems, for instance, exhibit broad-band behavior, which makes many traditional signal processing operations useless. One cannot decompose chaotic problems in the standard "input + transfer function → output" manner, nor can one simply low-pass filter the data to remove noise, as the high-frequency components are essential elements of the signal. The concept of a discrete set of spectral components does not make sense in many nonlinear problems, so using transforms to move between time and frequency domains—a standard technique that lets one transform differential equations into algebraic ones and vice versa, making the former much easier to work with—does not work. For these and related reasons, nonlinear dynamicists eschew most forms of spectral analysis. Because they are soundly based in nonlinear dynamics theory and rest firmly on the formal definition of invariants, however, the analysis methods described in section 6.3 of this chapter do not suffer from the kinds of limitations that apply to traditional linear analysis methods.

Another common complication in data analysis is *observability*: whether or not one has access to enough information to fully describe the system. The spring/mass system in Fig. 6.1, for instance, has two *state variables*—the position and velocity of the mass—and one must measure both of them in order to know the state of the system. (One can, to be sure, reconstruct velocity data from the position time series in the Figure using divided differences[1], but that kind of operation magnifies noise and numerical error, and thus is impractical.) Delay-coordinate embedding is one way to get around this problem; it lets one reconstruct the internal dynamics of a complicated nonlinear system from a *single* time series—e.g. inferring useful information about internal (and unmeasurable) transformer temperatures from their output voltages. The reconstruction produced by delay-coordinate embedding is not, of course, completely equivalent to the internal dynamics in all situations, or embedding would amount to a general solution to control theory's *observer problem*: how to identify all of the internal state variables of a system and infer their values from the signals that *can* be observed. However, a single-sensor reconstruction, if done right, can still be extremely useful because its results are guaranteed to be *topologically* (i.e., qualitatively) identical to the internal dynamics. This means that conclusions drawn about the reconstructed dynamics are also true of the internal dynamics of the system inside the black box. All of this is important for intelligent data analysis because fully *observable* systems are rare in science and engineering practice; as a rule, many—often, most—of a system's state variables either are physically inaccessible or cannot be measured with available sensors. Worse yet, the true state variables may not be known to the user; temperature, for instance, can play an important and often unanticipated role in the behavior of an electronic circuit. The delay-coordinate embedding methods covered in section 6.4 of this chapter not only yield useful information about the behavior of the unmeasured variables, but also give some indication of how many independent state variables actually exist inside the black box.

Although the vast majority of natural and man-made systems is nonlinear, almost all textbook time-series analysis techniques are limited to linear systems. The objective of this chapter is to present a more broadly useful arsenal of time-series analysis techniques—tools that can be applied to *any* system, linear or nonlinear. The techniques that have been developed by the nonlinear dynamics community over the past decade play a leading role in this presentation, but many other communities have developed different approaches to nonlinear time-series analysis. One of the more famous is Tukey's "exploratory data analysis," a sleuthing approach that emphasizes (and supports) visual examination over blind, brute-force digestion of data into statistics and regression curves [518]. Some of the more-recent developments in this field attempt to aid—or even augment—the analyst's abilities in unconventional ways, ranging from 3D virtual-reality displays to haptics (representing the data as a touch pattern, which has been proposed for reading mammograms [344]) or data sonification.

[1] e.g., dividing the difference between successive positions by the time interval between the measurements

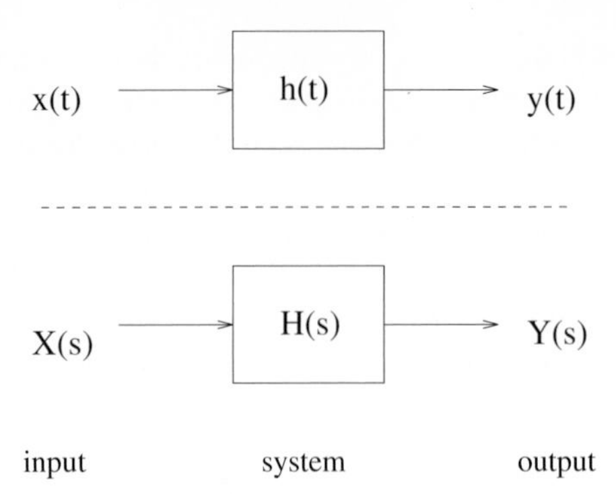

Fig. 6.2. The "input + transfer function → output" framework of traditional signal processing. Top: time domain. Bottom: frequency domain.

The sections that follow are organized as follows. Section 6.2 quickly reviews some of the traditional methods that apply to *linear* systems. Section 6.3 covers the bare essentials of dynamical systems theory and practice, with a specific emphasis on how those techniques are useful in IDA applications. This material forms the basis of the general theory of dynamics that applies to any system, linear or nonlinear. If all of the important properties of the target system can be identified and measured and the data are basically noise-free, these techniques, alone, can provide a very good solution to many nonlinear data-analysis problems. If there are fewer sensors than state variables, however, one must call upon the methods described in section 6.4 in order to reconstruct the dynamics before one can apply the section 6.3 methods. Noise is a much more difficult problem. There exist techniques that "filter" nonlinear time-series data, turning the nonlinearity to advantage and reducing the noise by a exponentially large factor [173], but the mathematics of this is well beyond the scope of this discussion. This chapter continues with two extended examples that demonstrate both the analysis methods of section 6.3 and the delay-coordinate reconstruction techniques of section 6.4, and concludes with some discussion of the utility of these methods in intelligent data analysis.

6.2. Linear Systems Analysis

The basic framework of traditional signal analysis [480] is schematized in Fig. 6.2; in it, an input signal is applied to a system to yield an output. One can describe this process in the time domain, using the *impulse response* $h(t)$ to model the system, or in the frequency domain, using the *frequency response* transfer function $H(s)$. The impulse response of a system is its transient response to a quick kick ($x(t_0) = 1$; $x(t) = 0 \, \forall \, t \neq t_0$); the frequency response $H(s)$ describes, for all s, what the system does to a sinusoidal input of frequency s. $H(s)$ is a complex function; it is most frequently written (and plotted) in magnitude ($|H(s)|$) and angle ("$H(s)$) form, but sometimes appears as $Re\{H(s)\}$ and $Im\{H(s)\}$.

Decomposing a problem in this "input + transfer function → output" manner is very useful; among other things, it allows one to apply powerful reasoning techniques like superposition[2]. The problem with Fig. 6.2 is that systems can react very differently to different inputs at different times—that is, $h(t)$ and $H(s)$ may depend on the magnitude of x, or they may have time-dependent coefficients. Either situation negates almost all of the advantages of both parts of the framework shown in the Figure. Nonlinearity (the former case) and nonstationarity (the latter) are treated later in this chapter; in the remainder of this section, we assume linearity and time invariance.

The top paradigm in Fig. 6.2 is easier to think about, but the bottom is mathematically much easier to work with. In particular, deriving $y(t)$ from $x(t)$ and $h(t)$ involves a convolution:

$$\begin{aligned} y(t) &= x(t) * h(t) \\ &= \int_{-\infty}^{+\infty} x(\tau)h(t-\tau)d\tau \end{aligned}$$

whereas the frequency-domain calculation only requires multiplication:

$$Y(s) = X(s)H(s)$$

The frequency domain has a variety of other powerful features. The spectrum is easy to interpret; the peaks of $|H(s)|$ correspond to the natural frequencies ("modes") of the system and hence, loosely speaking, to the number of degrees of freedom. Differential equations become algebraic equations when transformed into the frequency domain, and signal separation is a trivial operation. Because of these advantages, engineers are trained to transform problems into the frequency domain, perform any required manipulations (e.g., filtering) in that domain, and then reverse-transform the results back into the time domain.

Traditional analysis methods characterize a linear system by describing $h(t)$ or $H(s)$. Depending on the demands of the application, this description—the "model"—can range from the highly abstract to the very detailed:

1. descriptive models: e.g., the sentence "as water flows out of a bathtub, the level in the tub decreases"
2. numerical models: a table of the water level in the tub versus time
3. graphical models: the same information, but in pictorial form
4. statistical models: the mean, standard deviation, and/or trend of the water level
5. functional models: a least-squares fit of a line to the water level data
6. analytic models: an equation, algebraic or differential, that relates outflow and water level

The simplicity of the first item on the list is deceptive. Qualitative models like this are quite powerful—indeed, they are the basis for most human reasoning about the physical world. A circuit designer, for instance, reasons about the

[2] If the inputs x_1 and x_2 produce the outputs y_1 and y_2, respectively, then the input $x_1 + x_2$ will produce the output $y_1 + y_2$.

gain-bandwidth tradeoff of a circuit, and understands the system in terms of a balance between these two quantities: "if the gain goes up, the bandwidth, and hence the speed, goes down...". Many traditional analysis methods are also based on qualitative models. One can, for instance, compute the location of the natural frequencies of a system from the ring frequency and decay time of its impulse response $h(t)$ or the shape of its frequency response $H(s)$; the latter also lets one compute the speed (rise time) and stability (gain or phase margin) of the system. *Step* and *ramp* response—how the system reacts to inputs of the form

$$\begin{aligned} x(t) &= 0 \quad t < 0 \\ x(t) &= 1 \quad t \geq 0 \end{aligned}$$

and

$$\begin{aligned} x(t) &= 0 \quad t < 0 \\ x(t) &= t \quad t \geq 0 \end{aligned}$$

respectively—also yield useful data analysis results; see [449] for details. Though qualitative models are very powerful, they are also very difficult to represent and work with explicitly; doing so effectively is the focus of the qualitative reasoning/qualitative physics community [538].

As noted and discussed by many authors (e.g., [517]), tables of numbers are much more useful to humans when they are presented in graphical form. For this reason, numerical models—item 2 in the list above—are rarely used, and many IDA researchers, among others, have devoted much effort to finding and codifying systematic methods for portraying a data set graphically and highlighting its important features. Another way to make numbers more useful is to digest them into statistical values [554] like means, medians, and standard deviations, or to use the methods of functional approximation (e.g., chapter 10 of [209]) and regression to fit some kind of curve to the data. Statisticians sometimes apply transformations to data sets for the purpose of stabilizing the variance or forcing the distribution into a normal form. These methods—which can be found in any basic text on statistical methods, such as [387]—can make data analysis easier, but one has to remember how the transformed data have been manipulated and be careful not to draw unwarranted conclusions from it. It can also be hard to know what transformation to apply in a given situation; Box and Cox developed a formal solution to this, based on a parametric family of power transforms [79].

Sometimes, none of these abstractions and approximations is adequate for the task at hand and one must use an analytic model. Again, these come in many flavors, ranging from algebraic expressions to partial differential equations. One of the simplest ways to use an algebraic equation to describe a system's behavior is to model its output as a weighted sum of its current and previous inputs. That is, if one has a series of values $\{x_i(t)\}$ of some system input x_i—e.g., the position of a car's throttle, measured once per second—one predicts its output y (the car's speed) using the equation:

$$y(t) = \sum_{l=0}^{L} b_l x_i(t-l) \tag{6.1}$$

The technical task in fitting such an Lth-*order moving average* (MA) model to a data set involves choosing the window size L and finding appropriate values for the b_l. A weighted average of the last L values is a simple smoothing operation, so this equation represents a low-pass filter. The impulse response of such a filter—again, how it responds to a quick kick—is described by the coefficients b_l: as l goes from 0 to L, the impulse first "hits" b_0, then b_1, and so on. Because this response dies out after L timesteps, equation (6.1) is a member of the class of so-called *finite impulse response* (FIR) filters.

Autoregressive (AR) models are similar to MA models, but they are designed to account for feedback, where the output depends not only on the inputs, but also on the previous output of the system:

$$y(t) = \sum_{m=0}^{M} a_m y(t-m) + x_i(t) \tag{6.2}$$

Feedback loops are common in both natural and engineered systems; consider, for instance, a cruise control whose task is to stabilize the speed of a car at 100 kph by manipulating the throttle control. Traditional control strategies for this problem measure the difference between the current output and the desired set point, then use that difference to compute the input—e.g., opening the car's throttle x in proportion to the difference between the output y and the desired speed. Feedback also has many important implications for stability, in part because the loop from output to input means that the output y can continue to oscillate indefinitely even if the input is currently zero. (Consider, for example, the AR model $y(t) = -y(t-1) + x(t)$ if $x = 0$.) For this reason, AR models are sometimes called *infinite impulse response* (IIR) filters. The dependence of $y(t)$ on previous values of y also complicates the process of finding coefficients a_m that fit the model to a data set; see, e.g., [83] for more details.

The obvious next step is to combine MA and AR models:

$$y(t) = \sum_{l=0}^{L} b_l x_i(t-l) + \sum_{m=0}^{M} a_m y(t-m) \tag{6.3}$$

This "ARMA" model is both more general and more difficult to work with than its predecessors; one must choose L and M intelligently and use frequency-transform methods to find the coefficients; see [83] for this methodology. Despite these difficulties, ARMA models and their close relatives have "dominated all areas of time-series analysis and discrete-time signal processing for more than half a century" [536].

Models like those in the ARMA family capture the input/output behavior of a system. For some tasks, such as controller design, input/output models are inadequate and one really needs a model of the internal dynamics: a differential equation that accounts for the system's dependence on present and previous states. As an example, consider the spring/mass system of Fig. 6.1. If x is the deformation of the spring from its natural length, one can write a force balance at the mass as follows:

$$\Sigma F = ma$$
$$mg - kx = ma$$

Acceleration a is the second derivative of position ($a = x''$) and both are functions of time, so the force-balance equation can be rewritten as:

$$mx(t)'' = mg - kx(t) \tag{6.4}$$

This linear[3] differential equation expresses a set of constraints among the derivatives of an unknown function $x(t)$ and a set of constants. The mg term is gravity; the kx term is Hooke's law for the force exerted by a simple spring. The signs of mg and kx are opposite because gravity pulls in the direction of positive x and the spring pulls in the direction of negative x. Differential equations capture a system's physics in a general way: not only does their form mirror the physical laws, but their solutions also account for every possible behavior of the system. For any initial conditions for the position and velocity of the mass, for instance, the equation above completely describes where it will be at all times in the future. However, differential equations are much more difficult to work with than the algebraic models described in the previous paragraphs. They are also much more difficult to construct. Using observations of a black-box system's outputs to reverse-engineer its governing equations—i.e., figuring out a differential equation from partial knowledge about its *solutions*—is an extremely difficult task if one does not know what is inside the box. This procedure, which is known as *system identification* in the control-theory literature, is fairly straightforward if the system involved is linear; the textbook approach [297] is to choose a generic ordinary differential equation (ODE) system $\dot{\boldsymbol{x}}(t) = B\boldsymbol{x}(t)$—with $\boldsymbol{x}(t) = x_1(t), x_2(t), \ldots x_n(t)$—fast-Fourier-transform the sensor data, and use the characteristics of the resulting impulse response to determine the coefficients of the matrix B. The natural frequencies, which appear as spikes on the impulse response, yield the system's eigenvalues; the off-diagonal elements can be determined via an analysis of the shape of the impulse response curve between those spikes. See [297] or [356] for a full description of this procedure.

A linear, time-invariant system can be described quite nicely by the kinds of models that are described in this section, but nonstationarity or nonlinearity can throw a large wrench in the works. The standard textbook approach [94] to *nonstationary* data analysis involves special techniques that recognize the exact form of the nonstationarity (e.g., linear trend) and various machinations that transform the time series into stationary form, at which point one can use ARMA methods. *Nonlinearity* is not so easy to get around. It can be shown,

[3] The right-hand side of a linear differential equations is of the form $ax + b$

for instance, that ARMA coefficients and the power spectrum (i.e., Fourier coefficients) contain the same information. Two very different nonlinear systems, however, may have almost indistinguishable spectra, so methods in the ARMA family break down in these cases[4]. Spectral similarity of dissimilar systems also has important implications for signal separation. In linear systems, it is often safe to assume, and easy to recognize, that the "important" parts of the signal are lower down on the frequency scale and easily separable from the noise (which is assumed to be high frequency), and it is easy to implement digital filters that remove components of a signal above a specified cutoff frequency [410]. In nonlinear systems, as described in more detail in the following section, the important parts of the signal often cover the entire spectrum, making signal separation a difficult proposition. Nonlinearity is even more of a hurdle in system identification: constructing dynamic models of linear systems is relatively tractable, but human practitioners consider nonlinear system identification to be a "black art," and automating the process [85] is quite difficult.

6.3. Nonlinear Dynamics Basics

A dynamical system is something whose behavior evolves with time: binary stars, transistor radios, predator-prey populations, differential equations, the air stream past the cowl of a jet engine, and myriad other examples of interest to scientists and engineers in general and intelligent data analysts in particular. The bulk of an engineering or science education and the vast majority of the data analysis methods in current use, some of which are outlined in the previous section, are focused on *linear* systems, like a mass on a spring: systems whose governing equations do not include products, powers, transcendental functions, etc. Very few systems fit this mold, however, and the behavior of nonlinear systems is far richer than that of linear systems. This richness and generality makes nonlinear systems both much more difficult and much more interesting to analyze.

The *state variables* of a dynamical system are the fundamental quantities needed to describe it fully—angular position θ and velocity $\omega = \dot{\theta}$ for a pendulum, for instance, or capacitor voltages and inductor currents in an electronic circuit. The number n of state variables is known as the *dimension* of the system; a pendulum or a mass on a spring is a two-dimensional system, while a three-capacitor circuit has three dimensions. Simple systems like this that have a finite number of state variables can be described by *ordinary differential equation* (ODE) models like Equation (6.4) for the spring-mass system or

$$\ddot{\theta}(t) = -g \sin \theta(t) \tag{6.5}$$

for a pendulum moving under the influence of gravity g. Equation (6.4) is linear and equation (6.5), because of the sin term, is not; in both systems, $n = 2$.

[4] One can construct a patchwork of local-linear ARMA models [515] in situations like this, but such tactics contribute little to *global* system analysis and understanding.

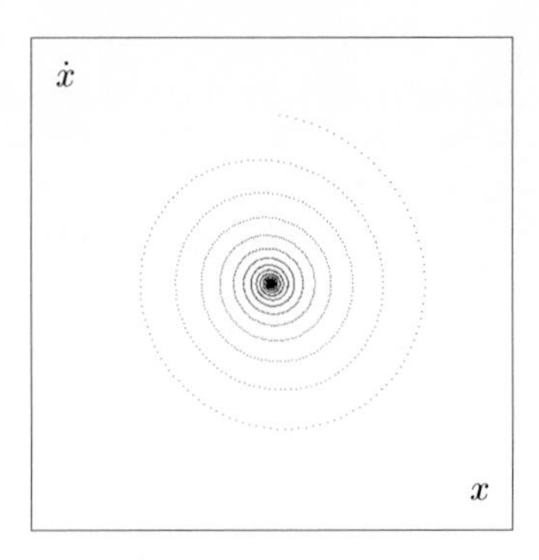

Fig. 6.3. A state-space trajectory representing the oscillation of the spring-mass system of Figure 6.1.

If the number of state variables in the system is infinite—e.g., a moving fluid, whose physics is influenced by the pressure, temperature and velocity at *every point*—the system is called *spatiotemporally extended*, and one must use *partial differential equation* (PDE) models [169] to describe it properly. In this chapter, we will confine our attention to finite-dimensional dynamical systems that admit ODE models. Because so many real-world problems are nonlinear, we will concentrate on methods that are general and powerful enough to handle *all* dynamical systems—not just linear ones. Finally, since most natural and man-made systems are not only nonlinear but also *dissipative*—that is, they lose some energy to processes like friction—we will not cover the methods of conservative or Hamiltonian dynamics [32, 360].

Much of traditional systems analysis, as described in the previous section, focuses on time-series or frequency-domain data. The nonlinear dynamics community, in contrast, relies primarily upon the *state-space* representation, plotting the behavior on the n-dimensional space (R^n) whose axes are the state variables. In this representation, the damped oscillation of a mass bouncing on a spring manifests not as a pair of decaying sinusoidal time-domain signals, as in Fig. 6.1, but rather as a spiral, as shown in Fig. 6.3. *State-space trajectories* like this—system behavior (i.e., ODE solutions) for particular initial conditions—only implicitly contain time information; as a result, they make the geometry of the equilibrium behavior easy to recognize and analyze.

Dissipative dynamical systems have *attractors*: invariant state-space structures that remain after transients have died out. A useful way to think about this is to envision the "flow" of the dynamics causing the state to evolve towards a "low point" in the state-space landscape (cf., a raindrop running downhill into an ocean). There are four different kinds of attractors:

- *fixed* or *equilibrium points*
- *periodic orbits* (a.k.a. *limit cycles*)

– *quasiperiodic* attractors
– *chaotic* or *"strange"* attractors

A variety of pictures of these different attractors appear in the later pages of this chapter. Fixed points—states from which the system does not move—can be stable or unstable. In the former case (cf., Fig. 6.3) perturbations will die out; in the latter, they will grow. A commonplace example of a stable fixed point is a marble at rest in the bottom of a bowl; the same marble balanced precariously on the rim of that bowl is at an unstable fixed point. Limit cycles are signals that are periodic in the time domain and closed curves in state space; an everyday example is the behavior of a healthy human heart. (One of the heart's pathological behaviors, termed *ventricular fibrillation*, is actually chaotic.) Quasiperiodic orbits and chaotic attractors are less familiar and harder to analyze, but no less common or interesting. The latter, in particular, are fascinating. They have a fixed, complicated, and highly characteristic geometry, much like an eddy in a stream, and yet nearby trajectories on a chaotic attractor move apart exponentially fast with time, much as two nearby wood chips will take very different paths through the same eddy. Trajectories cover chaotic attractors *densely*, visiting every point to within arbitrary ϵ, and yet they never quite repeat exactly. These properties translate to the very complex, almost-random, and yet highly structured behavior that has intrigued scientists and engineers for the last twenty years or so. Further discussion of chaotic systems, including a variety of examples, appears in section 6.5. Parameter changes can cause a nonlinear system's attractor to change drastically. A change in blood chemistry, for instance, can cause the heart's behavior to change from a normal sinus rythym to ventricular fibrillation; a change in temperature from 99.9 to 100.1 degrees Celsius radically alters the dynamical properties of a pot of water. These kinds of *topological* changes in its attractor are termed *bifurcations*.

Attractor type is an important nonlinear data analysis feature, and there are a variety of ways for computer algorithms to recognize it automatically from state-space data. One standard geometric classification approach is *cell dynamics* [272], wherein one divides the state space into uniform boxes. In Fig. 6.4, for example, the limit cycle trajectory—a sequence of two-vectors of floating-point numbers measured by a finite-precision sensor—can be represented as the *cell sequence*

$$[...(1,0)(2,0)(3,0)(4,0)(4,1)(5,1)(5,2)(4,2)(3,2)(3,3)(4,3)(4,4)...]$$

Because multiple trajectory points are mapped into each cell, this discretized representation of the dynamics is significantly more compact than the original series of floating-point numbers and therefore much easier to work with. This is particularly important when complex systems are involved, as the number of cells in the grid grows exponentially with the number of dimensions[5]. Though the approximate nature of this representation does abstract away much detailed

[5] The example of Fig. 6.4 is two-dimensional, but the cell dynamics formalism generalizes easily to arbitrary dimension.

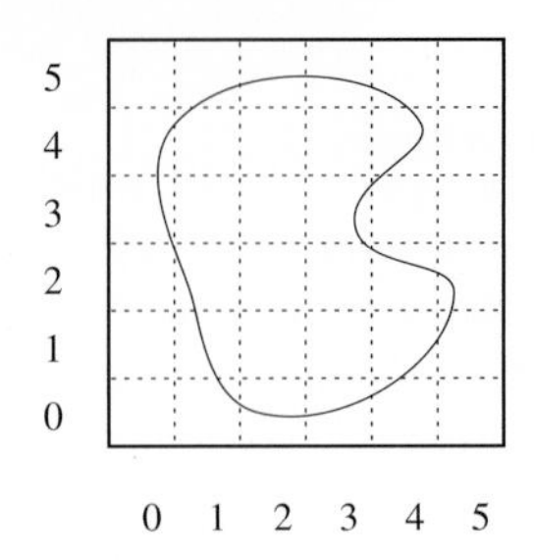

Fig. 6.4. Identifying a limit cycle using simple cell mapping

information about the dynamics, it preserves many of its important invariant properties; see [246] or [351] for more details. This point is critical to the utility of the method; it means that conclusions drawn from the discretized trajectory are also true of the real trajectory—for example, a repeating sequence of cells in the former, as in Fig. 6.4, implies that the full R^n dynamics is on a limit cycle.

Much as a bowl can have several low spots or a mountain range can include many drainages, nonlinear systems can have multiple attractors of different types. Each attractor lies in a unique *basin of attraction* (all the points in the bowl or mountain range from which a marble or raindrop will end up at that attractor), and those basins partition[6] the state space. A linear system, on the other hand, can have only one fixed point, and its basin—if it is stable—is all of R^n. Dissipation, the notion of transient behavior that dies out, and the requirement that attractors are *proper* subsets of their basins are linked. Dynamicists think about basin/attractor dynamics using the *state-space contraction* metaphor: initial conditions anywhere inside the boundary of a basin of attraction will converge to the associated attractor, so one envisions a volume of initial conditions spread out across the basin, all eventually converging to the attractor. (*Conservative* systems—those in which energy is conserved—preserve state-space volumes and do not have attractors.) Basins are very important for nonlinear data analysis. Attractors in neighboring basins can be quite different, and so small differences in initial conditions matter; a raindrop a millimeter away from a sharp mountain ridge will take a radically different path if a light breeze comes up. This can be a useful way to approach the analysis of a system that appears to have several behavior modes. Basin boundaries can be computed using the grid-based techniques described in the previous paragraph, as well as a variety of other approaches; see [223] or section 10.3.3 of [414] for more details.

The fixed nature of an attractor of a dynamical system is critically important to the approach to intelligent data analysis that is outlined in this chapter; it implies that the *dynamical invariants* of such attractors—their immutable mathematical properties—do not depend on how these attractors are viewed[7], and

[6] This is a slight abuse of the technical term "partition;" nonattracting sets—which have no basins of attraction—can exist in dynamical systems, and basins technically do not include their boundaries.

[7] within some limits, of course

therefore that analysis techniques that measure those invariants should yield the same results in the face of transformations like coordinate changes, for instance. Stability is such an invariant: a stable fixed point should not become unstable if one recalibrates a sensor. Topological dimension is another: a fixed point should not appear as a limit cycle when viewed from another angle. The nonlinear dynamics literature defines dozens of other dynamical invariants and proposes hundreds of algorithms for computing them; see [2] for a readable and comprehensive introduction. The two most common invariants in this list are the *Lyapunov exponent* λ, which measures how fast neighboring trajectories diverge, and the family of *fractal dimensions*, so named because they can take on non-integer (fractional → "fractal") values, which measure how much of R^n a trajectory actually occupies.

The Lyapunov exponent is defined as:

$$\lambda = \lim_{t \to \infty} \frac{1}{t} \ln |s_i(t)| \tag{6.6}$$

where the $s_i(t)$ are the eigenvalues of the variational system (the matrix-valued linear differential equation that governs the growth of a small variation in the initial condition; see appendix B of [414] for details). A n-dimensional system has n λs, each measuring the expansion rate, in one "direction," of the distance between two neighboring trajectories. λ is the nonlinear generalization of the real part of an eigenvalue; a positive λ implies exponential growth of a perturbation along the *unstable manifold*, the nonlinear generalization of the eigenvector associated with a positive-real-part eigenvalue. A negative λ implies exponential shrinkage of the perturbation along the *stable manifold* that is the nonlinear analog of the stable eigenvector. A system that has all negative λs in some region is said to be "stable in the sense of Lyapunov," and its trajectories relax to some proper subset of that region (the attractor). A system with all *positive* λs is unstable in all directions. A zero λ implies less-than-exponential growth, which generally takes place along the attractor. State-space contraction, part of the formal definition of dissipation, requires that $\Sigma\lambda_i < 0$ for any dissipative system.

The point of retooling the definition of dimension to allow for non-integer values is to be able to accurately characterize objects that are "between" two topological dimensions. A Cantor set, for example—constructed by removing the middle portion of a line segment *ad infinitum*, as shown in Fig. 6.5—contains an infinite number of zero-dimensional objects (points) but its topological dimension is still zero. Fractal dimensions capture this property; one standard measure of the fractal dimension of the middle-third removed Cantor set, for example, is 0.63. This invariant is common in the nonlinear dynamics community because many (not all) chaotic attractors have fractal state-space structure—that is, their attractors have non-integer values of the fractal dimension. The most-common algorithm for computing any fractal dimension of a set A, loosely described, is to discretize state space into ϵ-boxes, count the number of boxes[8] occupied by

[8] Hence the term "box-counting dimension."

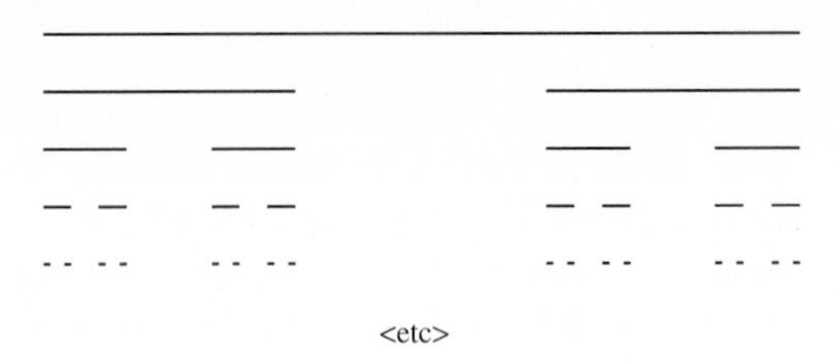

Fig. 6.5. A middle-third-removed Cantor set

A, and let $\epsilon \to 0$:

$$d_c = \lim_{\epsilon \to 0} \left\{ \frac{\log(N(A,\epsilon))}{\log(1/\epsilon)} \right\} \tag{6.7}$$

where $N(A, \epsilon)$ is the number of closed balls of radius $\epsilon > 0$ needed to cover A. (Strictly speaking, one doesn't just *count* the boxes, but rather accumulates the value of some measure on each box; see the discussion of equation (6.8) in section 6.4.2.) In reality, floating-point arithmetic and computational complexity place obvious limits on the $\epsilon \to 0$ part of equation (6.7); in practice, one repeats the dimension calculation for a range of ϵs and finds the power-law asymptote in the middle of the log-log plot of dimension versus ϵ.

Dynamical invariants like λ and d_c can be used to classify attractors. In a n-dimensional system, there are n Lyapunov exponents λ_i and:

- A stable fixed point has n negative λs (since perturbations in any direction will die out) and a fractal dimension of zero.
- An attracting limit cycle has one zero λ and $n - 1$ negative λs (since perturbations off the attractor will die out, and a perturbation along the orbit will remain constant) and a fractal dimension of one.
- A chaotic attractor has one zero λ (along the attractor), at least one positive λ and—generally but not always—a non-integer fractal dimension. The positive λ reflects chaos's hallmark "sensitive dependence on initial conditions:" the system's tendency to force neighboring trajectories apart.

Intelligent data analysis tools that target attractor type, basin geometry, dynamical invariants, etc. are harder to implement than the kinds of techniques that one can apply to a linear system, and their implications are generally less wide-ranging. If the system under consideration is linear, as mentioned previously, data analysis is relatively easy and one can make more (and more-powerful) inferences from the results. Where nonlinear systems are concerned, however, traditional methods often do not apply; in these problems, time-series analysis is much harder and the conclusions one can draw from the results are fundamentally limited in range. This stems from the inherent mathematical difficulties of the domain, and it is essentially unavoidable. If one is faced with a fundamentally nonlinear problem, one has no choice but to use the more difficult (and perhaps unfamiliar) methods covered in this chapter. The reader who is interested in delving deeper into this field should consult any of the dozens of good nonlinear dynamics books that are currently in print. An excellent overall starting point is [504], the basic mathematics is covered particularly well in [264], a comprehensive collection of algorithms appears in [414], and an entertaining popular overview may be found in [500].

6.4. Delay-Coordinate Embedding

Given a time series from a sensor on a single state variable $x_i(t)$ in a n-dimensional dynamical system, delay-coordinate embedding lets one reconstruct a useful version of the internal dynamics[9] of that system. If the embedding is performed correctly, the theorems involved guarantee that the reconstructed dynamics is topologically (i.e., qualitatively) identical to the true dynamics of the system, and therefore that the dynamical invariants are also identical. This is an extremely powerful correspondence; it implies that conclusions drawn from the *embedded* or *reconstruction-space* dynamics are also true of the real—unmeasured—dynamics. This implies, for example, that one can reconstruct the dynamics of the earth's weather simply by setting a thermometer on a windowsill.

There are, of course, some important caveats. Among other things, a correct embedding requires at least twice as many dimensions as the internal dynamics—a requirement that makes reconstruction of the weather thoroughly impractical, as it is a spatially extended system and thus of infinite dimension. Moreover, even if the dynamics of the system under examination is simple, its precise dimension is often very hard to measure and rarely known *a priori*. This is the main source of the hard problems of delay-coordinate embedding, which are discussed in more detail—together with some solutions—in the following sections.

6.4.1 Embedding: the basic ideas

Consider a data set comprised of samples $x_i(t)$ of a single state variable x_i in a n-dimensional system, measured once every Δt seconds, such as the example sensor time series shown in Table 6.1. To embed such a data set, one constructs d_E-dimensional *reconstruction-space* vectors $\boldsymbol{r}(t)$ from d_E time-delayed samples of the $x_i(t)$, such that

$$\boldsymbol{r}(t) = [x_i(t),\ x_i(t-\tau),\ x_i(t-2\tau),\ \ldots\ ,\ x_i(t-(m-1)\tau)]$$

or

$$\boldsymbol{r}(t) = [x_i(t),\ x_i(t+\tau),\ x_i(t+2\tau),\ \ldots\ ,\ x_i(t+(m-1)\tau)]$$

For example, if the time series in Table 6.1 is embedded in two dimensions ($d_E = 2$) with a delay $\tau = 0.005$, the first few points in the reconstruction-space trajectory are:

```
(1.6352 1.6260)
(1.6337 1.6230)
```

[9] That is, the state-space trajectory $\{\boldsymbol{x}(t)\}$, where $\boldsymbol{x} = \{x_1, x_2, \ldots x_n\}$ is the vector of state variables

```
(1.6322 1.6214)
(1.6306 1.6214)
(1.6276 1.6183)
(1.6260 1.6183)
...
```

If $d_E = 5$ and $\tau = 0.003$, the first few points of the trajectory are:

```
(1.6352 1.6306 1.6230 1.6183 1.6137)
(1.6337 1.6276 1.6214 1.6183 1.6107)
(1.6322 1.6260 1.6214 1.6168 1.6076)
(1.6306 1.6230 1.6183 1.6137 1.6045)
...
```

The act of sampling a single system state variable $x_i(t)$ is equivalent to projecting an n-dimensional state-space dynamics down onto a single axis; the embedding process demonstrated above is akin to "unfolding" or "reinflating" such a projection, albeit on different axes: the d_E delay coordinates $x_i(t)$, $x_i(t-\tau)$, $x_i(t-2\tau)$, etc. instead of the n true state variables $x_1(t)$, $x_2(t)$, ... , $x_n(t)$. The central theorem [508] relating such embeddings to the true internal dynamics, which is generally attributed to Takens, was proved in [413] and made practical in [459]; informally, it states that given enough dimensions (d_E) and the right delay (τ), the reconstruction-space dynamics and the true, unobserved state-space dynamics are topologically identical. More formally, the reconstruction-space and state-space trajectories are guaranteed to be diffeomorphic if $d_E = 2n+1$, where n is the true dimension of the system[10].

Diffeomorphisms—transformations that are invertible, differentiable, and that possess differentiable inverses—preserve topology but not necessarily geometry. This means that an attractor reconstructed using delay-coordinate embedding may look very different from the true attractor, but the former can be stretched

[10] τ is missing from these requirements because the theoretical conditions upon it are far less stringent and limiting, as described in the second paragraph of the next section.

Table 6.1. An example data set: samples of one state variable x_i, measured every $\Delta t = 0.001$ seconds.

$x_i(t)$	t	$x_i(t)$	t
1.6352	0.000	1.6214	0.008
1.6337	0.001	1.6183	0.009
1.6322	0.002	1.6183	0.010
1.6306	0.003	1.6168	0.011
1.6276	0.004	1.6137	0.012
1.6260	0.005	1.6107	0.013
1.6230	0.006	1.6076	0.014
1.6214	0.007	1.6045	0.015

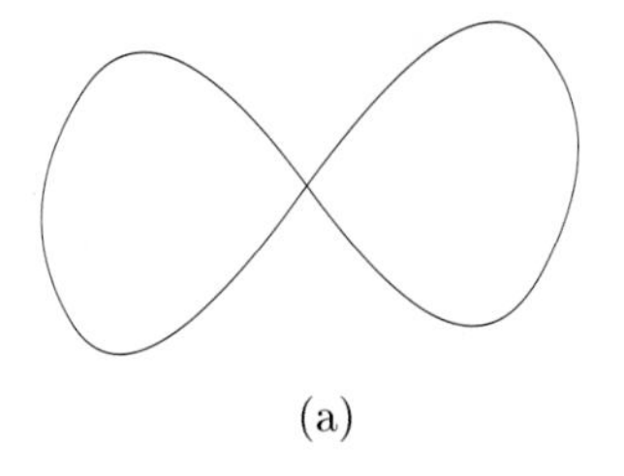

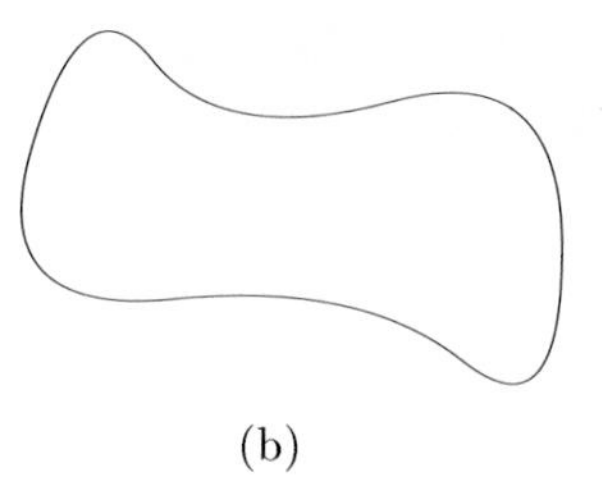

Fig. 6.6. A closed curve in 3D, viewed from (a) the top and (b) the side. The latter projection is is topologically conjugate to a circle; because of the self-intersection, the projection in (a) is not.

and bent into the shape of the latter without "crossing over" itself. The $2n + 1$ requirement of the theorem is really a brute-force worst-case limit for eliminating projection-induced crossings. The self-intersection point in Fig. 6.6(a), for example, makes the 2D projection of that curve not diffeomorphic to a circle; viewed from another angle, however, as in part (b), the curve is indeed smoothly deformable into a circle. $2n + 1$ is simply the minimum number of dimensions required to eliminate all such crossings, so lower-dimension embeddings may well be correct. This can, in fact, be exploited in deriving a tighter and easy-to-compute bound on d_E that is valid in "almost every" situation [459].

The topological equivalence guaranteed by the Takens theorem is a powerful concept: it lets one draw sensible, justifiable conclusions about the full dynamics of an n-dimensional system using only the output of a single sensor. In particular, many properties of the dynamics are preserved by diffeomorphisms; if one computes them from a correct embedding, the answer will hold for the true internal dynamics as well. There are, of course, some important conditions on the theorem, and the difficulties that they pose are the source of most of the effort and subtlety in these types of methods. Specifically, in order to embed a data set, one needs d_E and τ, and neither of these parameters can be measured or derived from the data set, either directly or indirectly, so algorithms like those described in the following section rely on numeric and geometric heuristics to estimate them.

From a qualitative standpoint, embedding is not as outlandish as it may initially appear. The state variables in a nonlinear system are generally coupled to one another temporally by the dynamics, so using quantities that resemble forward differences as the axes of a reconstruction space makes some sense. (As mentioned before, techniques like divided differences can, in theory, be used to derive velocities from position data; in practice, however, these methods often fail because the associated arithmetic magnifies sensor error.) One can think of the $x_i(t)$, $x_i(t - \tau)$, etc., as independent coordinates that are nonlinearly related to the true state variables. The specifics of that relationship may not—and *need* not—be obvious; the important point is that the form of that relationship ensures that the reconstructed dynamics $\boldsymbol{r}(t) \in R^{d_E}$ is diffeomorphic to the true dynamics $\boldsymbol{x}(t) \in R^n$.

6.4.2 Finding appropriate embedding parameters

The time-series analysis literature contains scores of methods that use a variety of heuristics to solve the central problem of delay-coordinate reconstruction: given a scalar time series from a dynamical system of unknown dimension, estimate values for the dimension d_E and delay τ that will guarantee a correct embedding. Many of these algorithms are somewhat *ad hoc*; almost all are computationally expensive and highly sensitive to sensor and algorithm parameters, and different ones produce surprisingly different results, even on the same data set. See [2] for a recent summary and the FAQ for the newsgroup `sci.nonlinear` [563] for a list of public-domain software implementations of many of these algorithms. This chapter covers only a few of the most widely accepted and/or interesting representatives of this body of work.

The delay τ governs whether or not the coordinates $x(t-j\tau)$ are indeed independent. If τ is small, the reconstruction-space trajectory will lie very near the main diagonal. As long as the structure is not infinitely thin, this type of embedding is *theoretically* correct; in practice, however, finite-precision arithmetic on fixed-length (and possibly noisy) trajectories can easily generate apparent crossings in situations like this. If τ is too large, on the other hand, successive points $\boldsymbol{r}(t)$ and $\boldsymbol{r}(t + \Delta t)$, where Δt is the sampling interval, will be uncorrelated and the larger spacing of the points in $\boldsymbol{r}(t)$ again interferes numerically with topological equivalence. Ideally, then, one wants a time window for τ that is long enough for the system state to evolve to a visible (with respect to floating-point arithmetic) but not excessive extent.

One way to compute such an estimate is to perform some sort of averaged autocorrelation of successive points in the time series $x_i(t)$ or in the embedded trajectory $\boldsymbol{r}(t)$—e.g., average mutual information [194]—as a function of τ. For very small τ, these statistics will be close to 1.0, since successive reconstruction-space trajectory points are very close to one another[11]. For larger τ, successive points become increasingly uncorrelated. The first minimum in the distribution is a sensible choice for τ: qualitatively, it corresponds to the smallest τ for which the dynamics has caused nearby trajectory points to become *somewhat* uncorrelated (i.e., new information has been introduced between samples). This choice was originally proposed [194] by Fraser; other authors suggest using other features of the autocorrelation curve to choose good values for τ—e.g., the first *maximum*, with the rationale that these "close returns" correspond to natural periods of the system. Note that since one can compute average mutual information (AMI) from one- and two-embeddings (that is, $d_E = 1$ and $d_E = 2$), this kind of procedure does *not* require one to first find a correct value for d_E.

The Pineda-Sommerer (P-S) algorithm [423], which solves both halves of the embedding parameter problem at once, is more esoteric and complicated.

[11] Note that $\boldsymbol{r}(t) = x_i(t)$ if $d_E = 1$.

Its input is a time series; its outputs are a delay τ and a variety of different estimates of the dimension d_E. The procedure has three major steps: it estimates τ using the mutual information function, uses that estimated value τ_0 to compute a temporary estimate E of the embedding dimension, and uses E and τ_0 to compute the *generalized dimensions* D_q, members of a parametrized family of fractal dimensions. Generalized dimensions are defined as

$$D_q = \frac{1}{q-1} \limsup_{\epsilon \to 0} \frac{\log \sum_i p_i^q}{\log \epsilon} \tag{6.8}$$

where p_i is some measure of the trajectory on box i. D_0, D_1, and D_2 are known, respectively, as the capacity, information, and correlation dimensions. The actual details of the P-S algorithm are quite involved; we will only give a qualitative description:

- Construct one- and two-embeddings of the data for a range of τs and compute the saturation dimension D_1^* of each; the first minimum in this function is τ_0. The D_1^* computation entails:
  - Computing the information dimension D_1 for a range of embedding dimensions E and identifying the saturation point of this curve, which occurs at embedding dimension D_1^*. The D_1 computation entails:
    - Embedding the data in E-dimensional space, dividing that space into E-cubes that are ϵ on a side, and computing D_1 using equation (6.8) with $q = 1$.

P-S incorporates an ingenious complexity-reduction technique in the fractal dimension calculation: the ϵs (see equation (6.7)) are chosen to be of the form 2^{-k} for integers k and the data are integerized, allowing most of the mathematical operations to proceed at the bit level and vastly accelerating the algorithm.

The false near neighbor (FNN) algorithm [305], which takes a τ and a time series and produces a lower bound on d_E, is far simpler than P-S. (As mentioned above, *upper* bounds for d_E are often chosen to be the smallest integer greater than twice the capacity dimension, D_0, of the data, in accordance with [459].) FNN is based on the observation that neighboring points may in reality be projections of points that are very far apart, as shown in Fig. 6.7. [t] The algorithm starts with $d_E = 1$, finds each point's nearest neighbor, and then embeds the data with $d_E = 2$. If the point separations change abruptly between the one- and two-embeddings, then the points were *false* neighbors (like A and C in the x-projection of Fig. 6.7). The FNN algorithm continues adding dimensions and re-embedding until an acceptably small[12] number of false near neighbors remains, and returns the last d_E-value as the estimated dimension. This algorithm is computationally quite complex; finding the nearest neighbors of m points requires $O(m^2)$ distance calculations and comparisons. This can be reduced to $O(m \log m)$ using a K-D tree implementation [200].

As should be obvious from the content and tone of this introduction, estimating τ and d_E is algorithmically *ad hoc*, computationally complex, and numerically sensitive. For this reason, among others, nonlinear time-series analysis

[12] An algorithm that removes *all* false near neighbors can be unduly sensitive to noise.

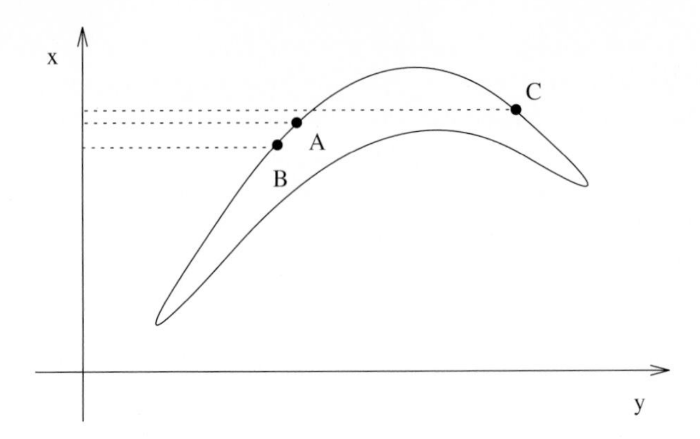

Fig. 6.7. The geometric basis of the FNN algorithm. If this curve is projected onto the x axis, the points A, B, and C appear to be near neighbors, even though C is quite distant in the 2D view. Differences between one- and two-embeddings of these data will expose *false near neighbors* like the [A,C] pair.

techniques that do not require embedding are extremely attractive. Recent evidence [283] suggests that the *recurrence plot*—a two-dimensional representation of a single trajectory wherein the time series spans both ordinate and abscissa and each point (i, j) on the plane is shaded according to the distance between the two corresponding trajectory points y_i and y_j—may be such a technique. Among their other advantages, recurrence plots also work well on nonstationary data; see the following section for an example (Fig. 6.11) and more discussion.

6.5. Examples

In this section, we demonstrate some of the concepts and algorithms described in the previous two sections using two examples, one simulated and one real.

6.5.1 The Lorenz system

In the early 1960s [357], Edward Lorenz derived a simple model of the physics of a fluid that is being heated from below:

$$\dot{\boldsymbol{x}}(t) = \frac{d}{dt}\boldsymbol{x}(t) = \begin{bmatrix} \dot{x}(t) \\ \dot{y}(t) \\ \dot{z}(t) \end{bmatrix} = \begin{bmatrix} a(y(t) - x(t)) \\ rx(t) - y(t) - x(t)z(t) \\ x(t)y(t) - bz(t) \end{bmatrix} \tag{6.9}$$

This 3rd-order ($n = 3$) ODE system is a rough approximation of a much more complex model: the Navier-Stokes PDEs for fluid flow. The state variables x, y, and z are convective intensity, temperature variation, and the amount of deviation from linearity in the vertical convection profile, respectively; the coefficients a and r are physical parameters of the fluid—the Prandtl and Rayleigh

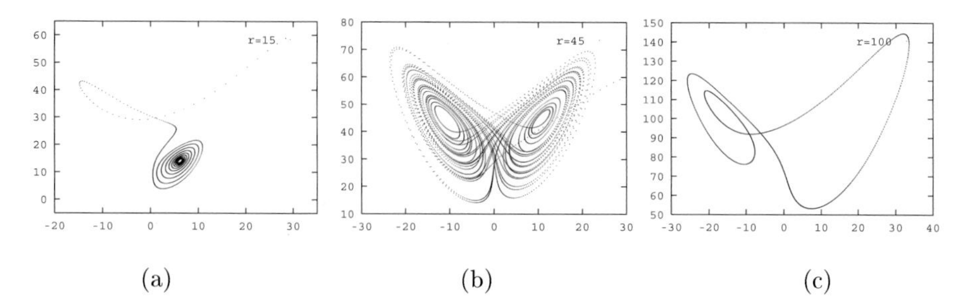

(a) (b) (c)

Fig. 6.8. State-space plots of Lorenz system behavior with $a = 10$ and $b = 8/3$: (a) a stable fixed point for $r = 15$ (b) a chaotic attractor for $r = 45$ (c) a periodic orbit for $r = 100$. All three plots are two-dimensional $(x - z)$ projections of three-dimensional attractors.

numbers—and b is the aspect ratio. This set of equations is one of the most common examples in the nonlinear dynamics literature. At low r values, its solutions exhibit damped oscillations to simple fixed-point equilibria, the first category on the list of attractor types on page 208, as shown in Fig. 6.8(a). For higher r—which translates to a higher heat input—the convection rolls in the modeled fluid persist, in a complicated, highly structured, and nonperiodic way; see part (b) of Fig. 6.8 for an example. This behavior, reported in a 1963 paper entitled "Deterministic Nonperiodic Flow," led Lorenz to recognize the classic "sensitive dependence on initial conditions" in the context of a fixed attractor geometry that is now a well-known hallmark of chaos. (The term "chaos" was coined twelve years later [350].) If r is raised further, the convection rolls become periodic—the second category in the list on page 208. See part (c) of the Figure for an example.

The trajectories plotted in Fig. 6.8 include complete information about all three of the state variables. In the analysis of a real system, this may be an overly optimistic scenario; while temperature is not hard to measure, the other state variables are not so easy, so a full state-space picture of the dynamics—information that is amenable to the techniques of section 6.3—may well be unavailable. Using the theory and techniques described in section 6.4, however, one can reconstruct the internal dynamics of this system from a time-series sampling of *one* of its state variables—say, the x coordinate of the chaotic attractor in part (b) of Fig. 6.8, which is plotted in time-domain form in Fig. 6.9(a). After embedding those data in delay coordinates, one can apply the nonlinear state-space analysis methods of section 6.3 to the results. The first step in the embedding process is to decide upon a delay, τ. The first minimum in the AMI results shown in Fig. 6.9 falls at roughly $\tau = 0.09$ seconds[13]. Using this τ, the false-near neighbor results (part (b) of Fig. 6.9) suggest an embedding dimension of two or three, depending on one's interpretation of the heuristic "acceptably small percentage" threshold in the algorithm. The box-counting dimension of

[13] The x-axis of the plot is measured in multiples of the sample interval of 0.002 second.

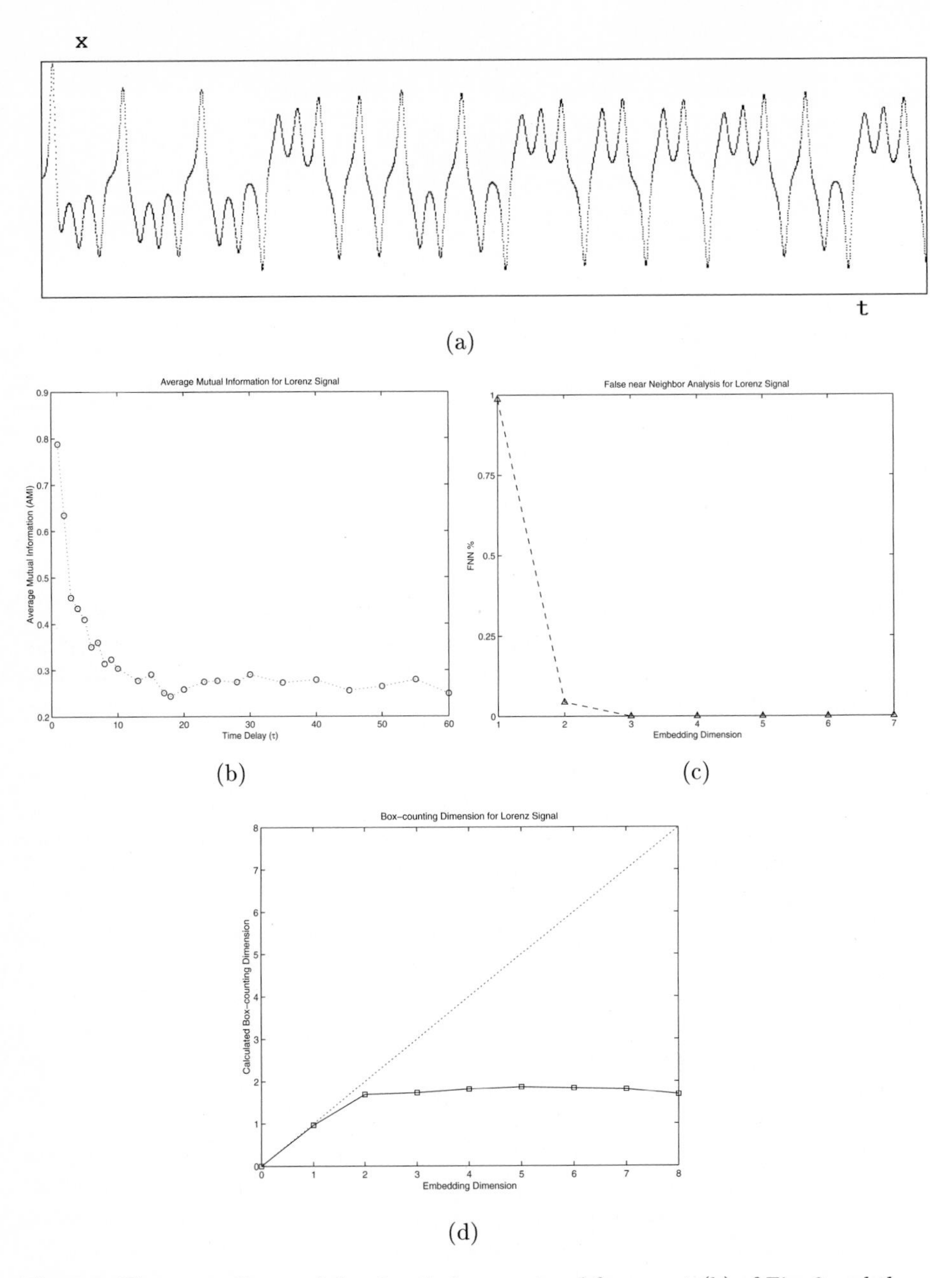

Fig. 6.9. The x coordinate of the chaotic Lorenz signal from part (b) of Fig. 8 and the corresponding embedding parameter analysis: (a) time series (b) average mutual information (AMI) as a function of the delay τ (c) false-near neighbor (FNN) percentage as a function of embedding dimension d_E (d) box-counting dimension (D_0) as a function of d_E. AMI, FNN and D_0 results courtesy of Joe Iwanski.

this data set levels off at roughly 1.8 for $d_E = 2$ and above, as can be seen in part (c) of the Figure. Following [459], this would imply an upper-bound embedding dimension of four.

It can be difficult to keep this menagerie of dimensions straight. In this example, the true dimension is known: $n = 3$. The time series $x(t)$ in Fig. 6.9(a) is a one-dimensional projection of the R^3 trajectory in Fig. 6.8(b) onto the x axis. In the worst case, the Takens theorem tells us that an accurate reconstruction may require as many as $d_E = 2n + 1 = 7$ embedding dimensions in order to assure topological conjugacy to the true dynamics. Recall that this is a very pessimistic upper bound; in practice, slightly more opportunistic algorithms like the one proposed in [459] are able to make better bounds estimates—values for d_E that are lower than $2n + 1$ *and*, at the same time, that avoid projection-induced topological inequivalencies between the true and reconstructed dynamics. In making such estimates, many of these algorithms make use of the fact that attractors do not occupy *all* of R^n. The fractal dimension of the $a = 10$, $r = 45$, $b = 8/3$ Lorenz attractor, for instance, is somewhere between 1 and 2, depending upon which algorithm one uses; the calculated capacity dimension D_0 of the trajectory in Fig. 6.8(b), in particular, is 1.8, implying an upper bound of $d_E = 4$. Even this estimate is somewhat pessimistic. Fractal dimension is a highly digested piece of information: a lumped parameter that compresses all the geometric information of an attractor into a single number. Because the FNN algorithm is based upon a more-detailed examination of the geometry, its results ($d_E = 3$, in this case) are a better *lower* bound.

Fig. 6.10 shows embeddings of the Lorenz time series of Fig. 6.9 with $d_E = 3$ and various τs. Note how this reconstructed attractor starts out as a thin band near the main diagonal and "inflates" with increasing τ. The sample interval in this data set was not much smaller than the τ returned by the AMI algorithm, so the thinnest reconstruction is fairly wide. Note, too, the resemblance of these reconstructed attractors to the true state-space trajectory in Fig. 6.8(b) and how that resemblance changes with τ. The whole point of doing an embedding is that the former can be deformed smoothly into the latter—even the $\tau = 0.5$ reconstruction, where the similarity (let alone the diffeomorphism!) is hard to visualize—and that the dynamical invariants of true (Fig. 6.8(b)) and reconstructed (Fig. 6.10) attractors are identical. That is, a fixed point in the reconstructed dynamics implies that there is a fixed point in the true dynamics, and so on. As noted before, this is the power of delay-coordinate embedding: one can use nonlinear dynamics analysis techniques on its results and safely extend those conclusions to the hidden internal dynamics of the system under examination.

It would, of course, be ideal if one could avoid all of these embedding machinations and analyze the scalar time series directly. As mentioned at the end of section 6.4, recurrence plots (RPs) are relatively new and potentially quite powerful nonlinear time-series analysis tools whose results appear to be independent of embedding dimension in some cases [283]. An RP is a two-dimensional representation of a single trajectory; the time series is spread out along both x

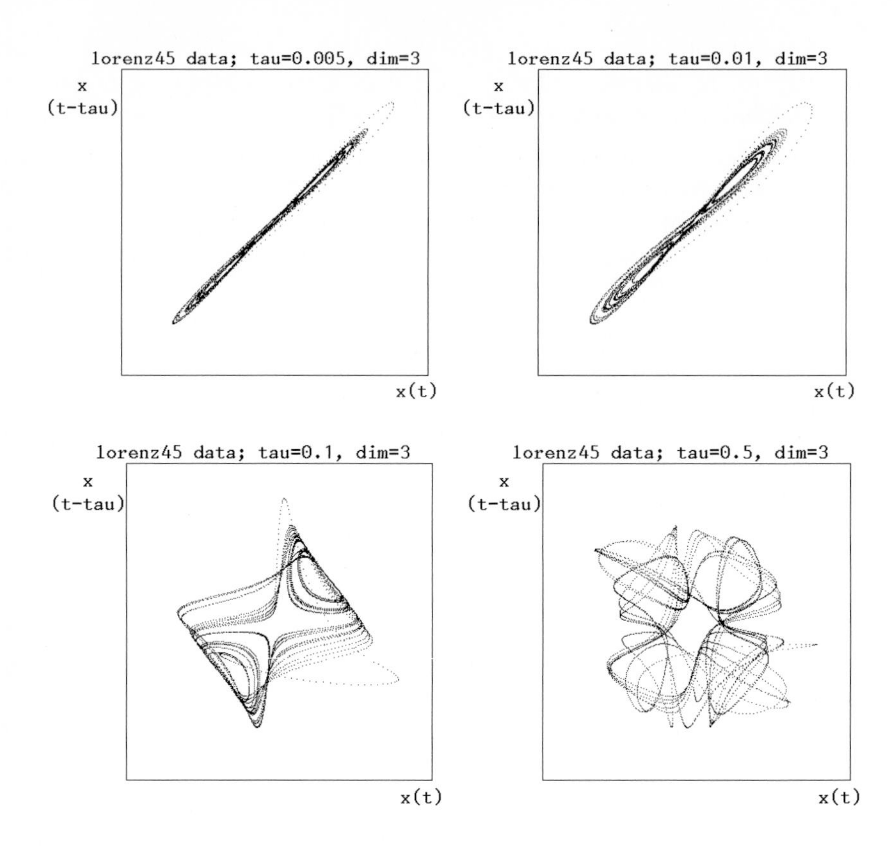

Fig. 6.10. Embeddings of the chaotic Lorenz signal from Fig. 9(a) with $d_E = 3$ and various delays, plotted in 2D projection. The formal requirements of the embedding process—which these attractors meet—guarantees that they are topologically identical to the true attractor in Fig. 8(b).

and y axes of the plot, and each pixel is shaded according to the distance between the corresponding points—that is, if the 117th point on the trajectory is 14 distance units away from the 9435th point and the distance range 13–15 corresponds to the color red, the pixel lying at (117, 9435) on the RP will be shaded red. Fig. 6.11 shows a recurrence plot (RP) of a short segment of the the Lorenz signal in part (a) of Fig. 6.9. Different types of attractors leave clear and suggestive signatures in RPs; it is easy to recognize a periodic signal, for instance, and chaotic attractors exhibit the type of intricate patterns that are visible in Fig. 6.11. Formalized classification of these signatures, however, is a difficult problem—and a current research topic. There are well-developed statistical approaches [283, 516], but structural/metric analysis (e.g., via pattern recognition) is still an open problem, although some recent progress has been made [86, 206].

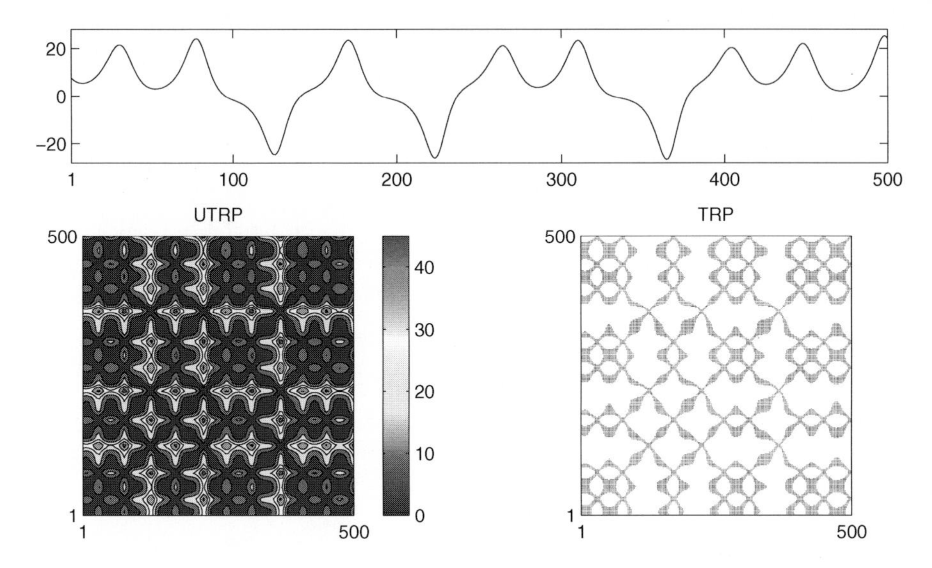

Fig. 6.11. Recurrence plots (RPs) of a short segment (top) of the Lorenz data from part (a) of Fig. 9. The pixel at i, j is shaded to reflect the distance between the ith and jth point in the time series. On the unthresholded recurrence plot (UTRP) on the bottom left, each pixel is coded according to the color bar shown to the right of the UTRP; in the thresholded RP to the bottom right, pixels are black if the distance falls within some prescribed *threshold corridor* and white otherwise. Results courtesy of Joe Iwanski.

6.5.2 The driven pendulum

A time-series plot of a data set from an angle sensor on a parametrically forced pendulum—a solid aluminum arm that rotates freely on a standard bearing, driven vertically by a motor through a simple linkage—is shown in part (a) of Fig. 6.12.

An actuator controls the drive frequency and a sensor (an optical encoder) measures its angular position. The behavior of this apparently simple device is really quite complicated and interesting: for low drive frequencies, it has a single stable fixed point, but as the drive frequency is raised, the attractor undergoes a series of bifurcations. In the sensor data, this manifests as interleaved chaotic and periodic regimes [151]. The driven pendulum is also interesting from a modeling standpoint; at high resolutions, the backlash in the bearings invalidates the standard textbook model. Modeling these effects is critical, for instance, to the accurate control of robot arms.

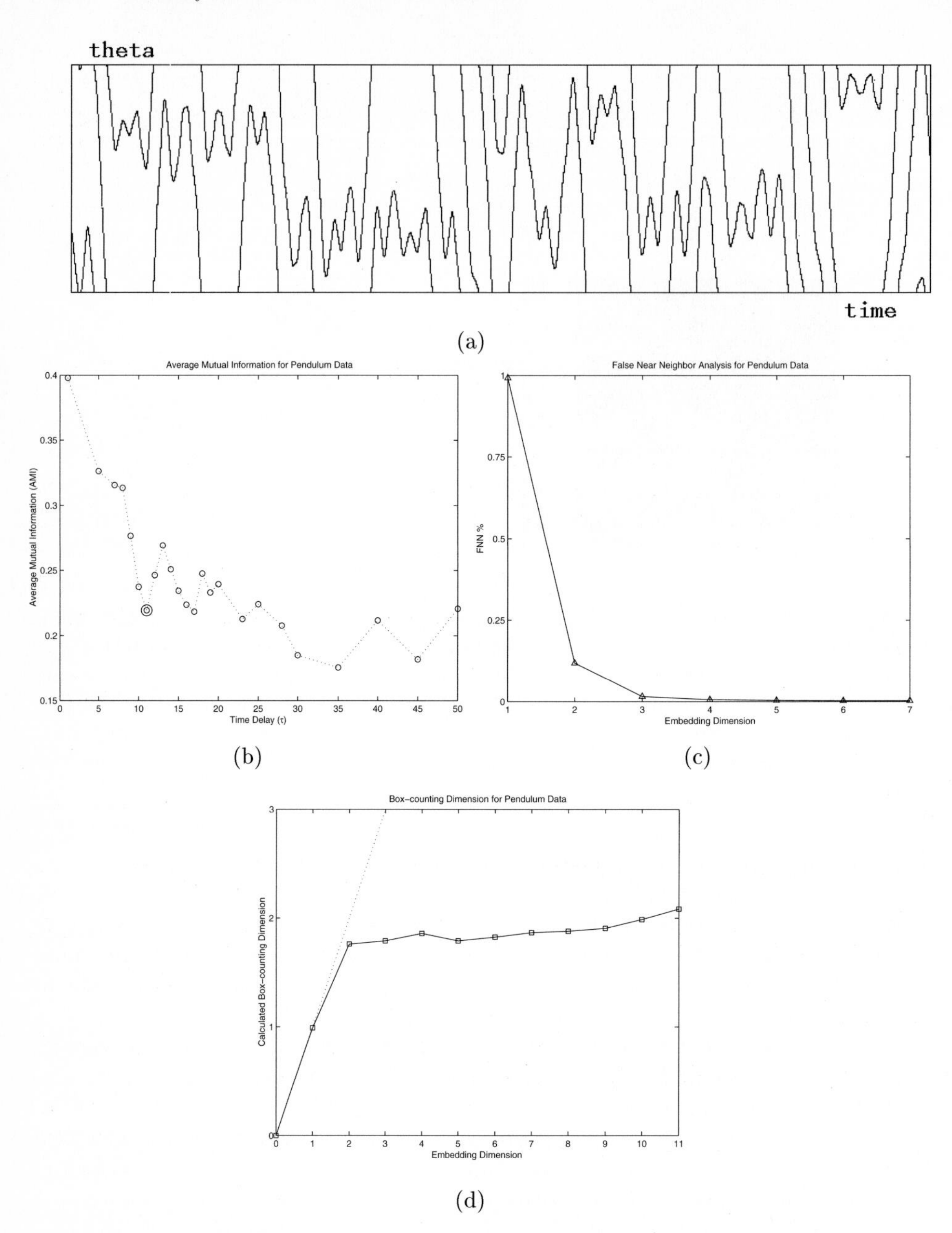

Fig. 6.12. A chaotic sensor data set from a parametrically forced pendulum: (a) time-domain plot of the bob angle, measured modulo 2π (b) AMI (c) FNN and (d) D_0 results, all courtesy of Joe Iwanski.

The test run plotted in Fig. 6.12 was chosen for this example because the pendulum is oscillating in a chaotic manner, which rules out many traditional time-series analysis methods. The chaos manifests as seemingly structured, almost-periodic patterns in the time-series signal: oscillations that are quite similar but not identical and that almost (but not quite) repeat. Though these patterns are highly suggestive, they are very difficult to describe or classify in the time domain; in a state-space view, however, the characteristic structure of the pendulum's chaotic attractor becomes patently obvious. Unfortunately, direct state-space analysis of this system is impossible. Only angle data are available; there is no angular velocity sensor and attempts to compute angular velocity via divided differences from the angle data yield numerically obscured results because the associated arithmetic magnifies the discretization error in angle (from the sensor resolution) and time (from timebase variation in the data channel).

Delay-coordinate embedding, however, produces a clean, easily analyzable picture of the dynamics that is guaranteed to be diffeomorphic to the system's true dynamics. As in the Lorenz example, the embedding procedure begins with an estimation of τ. AMI results on the chaotic pendulum data set, shown in part (b) of Fig. 6.12, suggest a delay of 0.022 seconds (roughly 11 clicks at a sample interval of 0.002 seconds). FNN results constructed using this τ, shown in Fig. 6.11(c), suggest an embedding dimension of $d_E = 3$. The capacity dimension D_0—part (d)—varies between 1.7 and 2.1, implying an upper bound of $d_E = 5$, following [459].

In the Lorenz example of the previous section, the true dimension n was known. In the experimental pendulum setup, this is not the case. Presumably, three of the state variables are the bob angle θ, the angular velocity ω, and the time[14] t; if, however, the device is shaking the lab bench or contracting and expanding with ambient temperature, other forces may come into play and other state variables may have important roles in the dynamics. The results described in the previous paragraph, which suggest that the dynamical behavior of the pendulum is low-dimensional ($d_E = 3 - 5$, specifically), imply that the system is probably not influenced by variables like lab bench position or temperature. Higher d_E values from the estimation algorithms would suggest otherwise. This kind of high-level information, a natural result of delay-coordinate reconstruction and nonlinear dynamics analysis, is extremely useful for intelligent data analysis.

Fig. 6.13 shows embeddings for various τs; note how a small τ, as in the Lorenz example, creates a reconstruction that hugs the main diagonal, and how that reconstructed attractor unfolds as τ grows. The pendulum data were greatly oversampled, so it is possible to create a thinner embedding than in the Lorenz example, as shown in part (a) of this Figure. This is the type of reconstruction whose topologically conjugacy to the true dynamics is effectively destroyed by noise and numerical problems; note the apparent overlap of trajectories and sprinkling of noisy points just outside the true attractor in the $\tau = 0.01$ and $\tau = 0.02$ embeddings.

[14] In a driven or *nonautonomous* system, time is an exogenous variable.

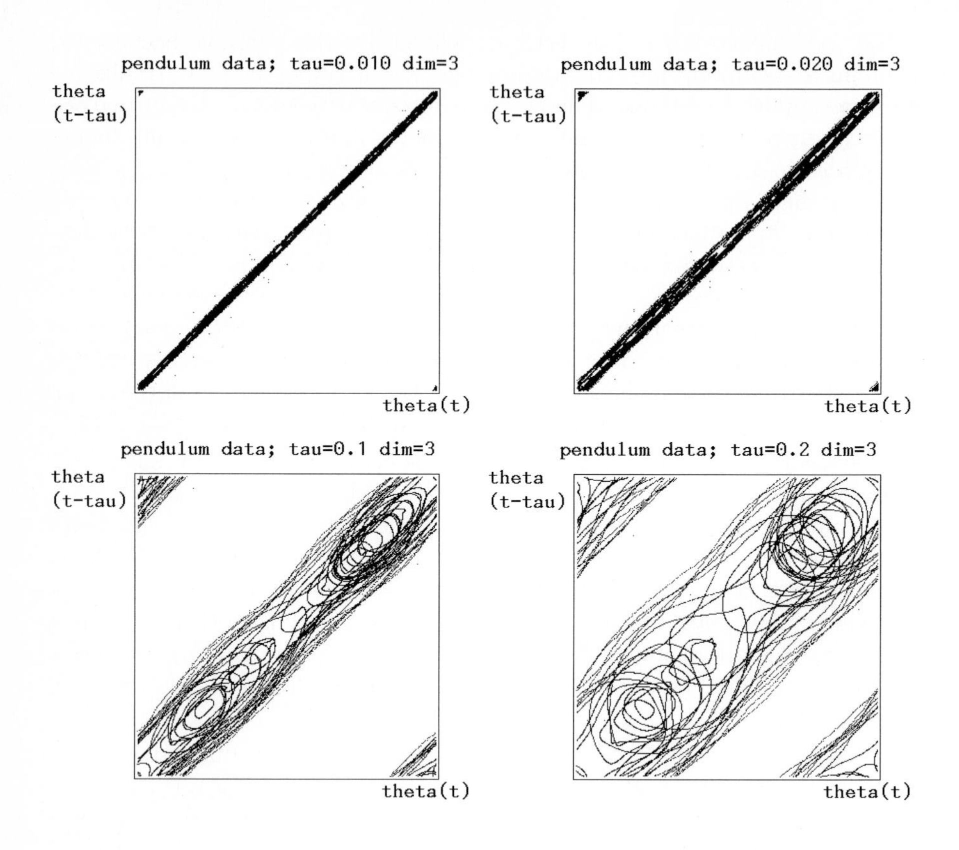

Fig. 6.13. Embeddings of the pendulum data set from part (a) of Fig. 12.

As before, once one has a successful reconstruction of the dynamics, all of the analysis tools described in section 6.3 can be brought to bear upon it, and their conclusions can be assumed to hold for the system's full underlying behavior.

6.6. Conclusion

One of the more common—and more difficult—problems faced by an engineer or scientist is to analyze the dynamics of a complicated nonlinear system, given only measurements of one state variable. The techniques described in section 6.4 of this chapter, coupled with the theory covered in section 6.3, make significant inroads on this problem, allowing one to draw useful, justifiable, and sensible conclusions about a nonlinear system from the output of a single sensor. Specifically, a correct embedding of a data set from a single sensor on a black-box system is guaranteed to have the same dynamical invariants as the n-dimensional dynamics of the system inside the black box, and those invariants are useful tools for intelligent data analysis. Time-series analysis tools for *linear* systems are much easier to understand, implement, and use, but the universe is by and large nonlinear, so the application range of those kinds of tools is severely limited. Filtering out noise, for example, is fairly straightforward when one is working with data from a linear system: one simply transforms the data into the frequency domain and uses a low-pass filter. In nonlinear systems, separating signal from noise is problematic, as the former is often broad band and thus the two are intermingled. (Noise, incidentally, is infinite-dimensional, so its implications for embedding dimension calculations are dire; recall the $2n + 1$ requirement in the embedding theorems.) There has been some recent work on nonlinear "filtering" algorithms [232], including *filtered delay-coordinate embedding* [459] and an intriguing technique that exploits the stable and unstable manifold structure of a chaotic attractor to compress the noise ball. The latter method requires complete knowledge of the dynamics—the ODEs that govern the system. Since reverse-engineering ODEs from time-series samples of their solutions is an open problem for nonlinear systems, this filtering approach is hard to put into practice. One can, however, *approximate* the ODEs with local-linear models and get some reasonable results; see [173] for more details. In some cases, noise can actually be turned to advantage; its presence in a time series can allow the modeler to "explore" more of the state space [87].

One popular technique that may be conspicuous by its absence from this chapter is the neural net. Neural nets [257], which are discussed in Chapter 8 of this volume, are essentially nonlinear regression networks that model the input/output behavior of a system. They are very good at learning the patterns in a data set, and hence are very effective at predicting what a system will do next. However, they do *not* model the underlying physics in a human-comprehensible form. It is very difficult to learn anything useful about a system by examining a neural net that has been "trained" on that system, so this technique has been omitted from this discussion. Their ability to predict, however, makes neural nets potentially useful to intelligent data analysis in a somewhat counterintuitive fashion: if one needs more data, one can train a neural net on the time series and then use it to augment that data set, generating new points that are consistent with the dynamics [96].

Nonlinear dynamics techniques like the ones described in this chapter may be more difficult to understand and use than the more-familiar linear ones, but they are more broadly applicable—indeed, the latter can be viewed as a subset of the former. This family of theory and technique is valuable not only for time-series analysis, but also for many other tasks, such as modeling and prediction [106]. The kinds of models mentioned in the first paragraph of this section, for instance, have been successfully used to predict the behavior of systems ranging from roulette wheels [46] to physiological disease patterns, currencies markets, and Bach fugues [536].

Chapter 7
Rule Induction

Peter Flach* and Nada Lavrač**
*University of Bristol, UK and **Jožef Stefan Institute, Slovenia

7.1. Introduction

Machine learning is an important form of intelligent data analysis. It is common in machine learning to distinguish symbolic and sub-symbolic approaches. Symbolic approaches employ some kind of description language in which the learned knowledge is expressed. The machine learning methods described in this chapter construct explicit symbolic classification rules that generalise the training cases, and are thus instances of symbolic machine learning. This area of machine learning is called *classification rule induction* [376, 378, 381]. Non-symbolic learning approaches are covered in other chapters. One of the main attractions of rule induction is that the rules are much more transparent and easier to interpret than, say, a regression model or a trained neural network.

The classification rule learning task can be defined as follows: Given a set of training examples (instances for which the classification is known), find a set of classification rules that can be used for prediction or classification of new instances, i.e., cases that haven't been presented to the learner before. A more formal definition of the classification rule learning task has to take into account the restrictions imposed by the language used to describe the data (data description language) and the language used to describe the induced set of rules (hypothesis description language). The *language bias* refers to the restrictions imposed by the languages defining the format and scope of data and knowledge representation.

A generic classification rule learning task can now be defined, given a binary classification problem of classifying instances into classes named *positive* and *negative*. The task of learning a set of rules defining the class *positive* is defined as follows.

Given:

- a data description language, imposing a bias on the form of data
- training examples, i.e., a set of classified instances described in the data description language
- a hypothesis language, imposing a bias on the form of induced rules
- a coverage function, defining when an instance is covered by a rule

Find:

- a hypothesis as a set of rules described in the hypothesis language, that is
  - consistent, i.e., does not cover any negative example
  - complete, i.e., covers all positive examples

This definition distinguishes between the terms *examples* and *instances*. Examples usually refer to instances labelled by a class label, where instances themselves bear no class label. This is how we will use the two terms in this chapter. Another term appearing in the definition is *hypothesis*, used to denote the output of learning. Due to the hypothetical nature of induction, the output of inductive learning can be falsified by new evidence presented to the learner. In this chapter, the induced hypotheses will always be represented as sets of rules.

Generally speaking, a rule is an expression of the form $Head \leftarrow Body$, where $Body$ describes the conditions under which the rule 'fires', and $Head$ is typically a class label. (In the simplified learning setting above we learn rules for only one class, so there the head of the rule is strictly speaking redundant.) An instance is *covered* by such a rule if it satisfies the conditions in $Body$. An example can be either correctly covered (if it is covered and the class label in $Head$ corresponds with the class label of the example), incorrectly covered ($Head$ assigns a different class), or not covered.

A simple extension of this definition allows us to deal with multi-class learning problems. Suppose that training instances are labelled with three class labels: c_1, c_2, and c_3. The above definition of the learning task can be applied if we form three different learning tasks. In the first task, instances labelled with class c_1 are treated as the positive examples, and instances labelled c_2 and c_3 are the negative examples. In the next run, class c_2 will be considered as the positive class, and finally, in the third run, rules for class c_3 will be learned. Due to this simple transformation of a multi-class learning problem into several binary classification problems, the rest of this chapter will mostly deal with binary classification tasks, unless explicitly stated otherwise. Notice that this procedure could also be applied in the case of binary classification problems to learn an explicit definition of both the positive and the negative class.

Generally speaking, consistency and completeness – as required in the above task definition – are very strict conditions. They are unrealistic in learning from large datasets which may contain errors. *Noise* refers to random errors in the data, either due to incorrect class labels or errors in instance descriptions. It is also possible that the hypothesis language is not expressive enough to allow a complete and consistent hypothesis, in which case the target class needs to be approximated. Another complication is caused by target classes that are not strictly disjoint. To deal with these cases, the consistency and completeness requirements need to be relaxed and replaced with some other evaluation criteria, such as sufficient *coverage* of positive examples, high *predictive accuracy* of the hypothesis or its *significance* above the requested, predefined threshold. These measures can be used both as heuristics to guide rule construction and as measures to evaluate the quality of induced hypotheses. Some of these measures and related issues will be discussed in more detail in Section 7.4.

The above definition of the learning task assumes that the learner has no prior knowledge about the problem and that it learns exclusively from examples. However, difficult learning problems typically require a substantial body of prior knowledge. We refer to declarative prior knowledge as *background knowledge*. Using background knowledge, the learner may express the induced hypotheses in a more natural and concise manner. In this chapter we mostly disregard background knowledge, except in the process of constructing features used as ingredients in forming hypotheses. This issue is discussed in Sections 7.3.1 and 7.6.3.

This chapter provides an overview of two main approaches to classification rule induction: propositional rule learning (Section 7.2) and relational rule learning (Section 7.6). One of our main aims in this chapter is to provide a logical connection between the two, showing how the more advanced setting of relational rule learning can be obtained from the propositional framework.

Section 7.2 presents the classical form of rule induction, where the training examples can be represented in a single table, and the output are if-then rules. This simplest setting of classification rule induction is usually called *propositional rule induction* or *attribute-value rule learning*. Good representatives of this class of learners are CN2 [118, 117, 162] and RIPPER [121]. An example of rule learning from the statistics literature is PRIM [197]. We will describe CN2 in more detail in Secion 7.5.

Sections 7.6 and 7.7 present a more elaborate setting, where data is represented in several tables, and the output are relational rules, possibly in the form of Prolog clauses. This approach is usually called *induction of relational rules*, *multi-relational data mining*, or *inductive logic programming* (ILP). We describe several representative examples, including the top-down rule learner FOIL [434] and the bottom-up rule learner GOLEM [394]. We also demonstrate that some relational learning problems can be transformed to a propositional rule induction problem. These approaches are usually referred to as *propositionalisation approaches*. The representative described in this chapter is LINUS [340] which transforms the input from a multiple table representation to a single table training set representation.

Section 7.3 and 7.4 discuss issues that are important for both frameworks, presenting rule learning as a search problem. Section 7.3 introduces the important notions of generalisation and specialisation, and describes algorithms for rule and hypothesis construction. Section 7.4 discusses measures for evaluating the quality of rules.

7.2. Propositional rule learning

7.2.1 Data representation

The input to a classification rule learner consists of a set of instances with known classifications. An *instance* is described by values of a fixed collection of *attributes*: A_i, $i \in \{1, \ldots, n\}$. An attribute can either have a finite set of values (*discrete*) or take real numbers as values (*continuous*). An *example* e_j is a vector of attribute values labelled by a class label $e_j = (v_{1,j}, \ldots, v_{n,j}, c_j)$, where each $v_{i,j}$ is a value of attribute A_i, and $c_j \in \{c_1, \ldots, c_k\}$ is one of the k possible values of class attribute *Class*. A *dataset* is a set of examples. We will normally organise a dataset in tabular form, with columns for the attributes and rows or tuples for the examples.[1]

Suppose we are given a dataset of patient records with corresponding diagnoses, from which we wish to induce a set of diagnostic or prognostic rules. In medical problems, patient records are typically described by some attributes which are continuous (for instance, age or duration of illness) and others that have discrete values (e.g., sex of a patient that is either male or female). Table 7.1 contains patient records for a simplified problem of contact lens prescriptions [547]. In this table, patients are described by four attributes: *Age*, *Spectacle prescription*, *Astigmatism* and *Tear production rate*. Patient records are labelled with three possible class labels, denoting the type of lens prescribed to an individual patient: *none*, *hard*, and *soft*.

The reader will have noticed that this dataset is *complete*, in the sense that all possible combinations of attribute values are present. In such a case, the learning algorithm cannot generalise beyond the classifications already in the training set. Rather, the goal of learning is to transform the data into a more compressed form without losing classification power.

Notice furthermore that all four attributes in Table 7.1 have discrete values. For three attributes, *Spectacle prescription*, *Astigmatism* and *Tear production rate*, this is the most natural representation, whereas one would expect *Age* to be an attribute with numeric values. For simplicity, this attribute has been discretised into three values *young*, *pre-presbyopic* and *presbyopic*. This discretisation was done by an expert, to indicate the age categories of patients with specific properties. Normally, attribute *Age* would have numeric values. Its values could be discretised into value intervals either in data preprocessing or in the process of rule learning itself. Discretisation of attributes is an important knowledge representation issue, but is out of the main scope of this chapter.

[1] In this chapter we use the terminology and notation from the machine learning literature. In most other chapters in this book, attributes A_i are called input variables x_i, and the class attribute *Class* is called the output variable y.

Table 7.1. The contact lens dataset.

Age	Spectacle prescription	Astigmatism	Tear production rate	Recommended lenses
young	myope	no	reduced	none
young	myope	no	normal	soft
young	myope	yes	reduced	none
young	myope	yes	normal	hard
young	hypermetrope	no	reduced	none
young	hypermetrope	no	normal	soft
young	hypermetrope	yes	reduced	none
young	hypermetrope	yes	normal	hard
pre-presbyopic	myope	no	reduced	none
pre-presbyopic	myope	no	normal	soft
pre-presbyopic	myope	yes	reduced	none
pre-presbyopic	myope	yes	normal	hard
pre-presbyopic	hypermetrope	no	reduced	none
pre-presbyopic	hypermetrope	no	normal	soft
pre-presbyopic	hypermetrope	yes	reduced	none
pre-presbyopic	hypermetrope	yes	normal	none
presbyopic	myope	no	reduced	none
presbyopic	myope	no	normal	none
presbyopic	myope	yes	reduced	none
presbyopic	myope	yes	normal	hard
presbyopic	hypermetrope	no	reduced	none
presbyopic	hypermetrope	no	normal	soft
presbyopic	hypermetrope	yes	reduced	none
presbyopic	hypermetrope	yes	normal	none

7.2.2 Attribute-value language of if-then rules

Given a set of classified examples, a rule learning system constructs a set of if-then rules. An if-then rule has the form:

$$\text{IF } Conditions \text{ THEN } Class.$$

Conditions contains one or more attribute tests, i.e., *features* of the form $A_i = v_{ij}$ for discrete attributes, and $A_i < v$ or $A_i \geq v$ for continuous attributes (here, v is a threshold value that does not need to correspond to a value of the attribute observed in examples). The conclusion part has the form $Class = c_i$, assigning a particular value c_i to $Class$.

An alternative rule syntax that we will sometimes use is $Class \leftarrow Conditions$, or, in its most general form: $Head \leftarrow Body$. The latter form is most frequently used in predicate logic as a general form of rules where *Body* (also called the antecedent) is a conjunction of conditions or literals, and *Head* is the consequent which in this chapter is a single literal (in the general case it may be a disjunction).

Table 7.2. Classification rules induced from the contact lens dataset.

```
IF     TearProduction = reduced
THEN   ContactLenses = none               [#soft=0 #hard=0 #none=12]

IF     TearProduction = normal
  AND  Astigmatism = no
THEN   ContactLenses = soft               [#soft=5 #hard=0 #none=1]

IF     TearProduction = normal
  AND  Astigmatism = yes
  AND  SpectaclePrescription = myope
THEN   ContactLenses = hard               [#soft=0 #hard=3 #none=0]

IF     TearProduction = normal
  AND  Astigmatism = yes
  AND  SpectaclePrescription = hypermetrope
THEN   ContactLenses = none               [#soft=0 #hard=1 #none=2]
```

An example set of rules induced in the domain of contact lens prescriptions is given in Table 7.2. The numbers between brackets indicate the number of covered examples from each class. The first and the third rule are consistent, while the other two rules misclassify one example each. The second rule is complete with regard to the class `soft`.

It is interesting to note that in the second rule, the condition `TearProduction = normal` is the negation of the condition in the first rule, and that this condition is included in all subsequent rules. Similarly, the condition `Astigmatism = yes` in the third rule is the negation of the second condition in the second rule. As it turns out, the set of rules can equivalently be represented by a so-called *decision list*, as in Table 7.3. Alternatively, the whole hypothesis can be seen as a binary tree where at each node the left branch leads to a leaf. Rule sets, rule lists and trees are all closely related representations and each of them is being used in symbolic learning. We will not cover learning classification trees (or *decision trees* as they are commonly called) in this chapter (see appendix B for more details), but we will give algorithms for inducing unordered sets of rules as well as ordered decision lists.

Consider now a more realistic medical diagnostic problem concerning early diagnosis of rheumatic diseases with several diagnostic classes, and several hundreds of patient records as input to a learning algorithm. In this problem patients are classified into eight diagnostic categories, including degenerative spine disease (referred to as class c_1), degenerative joint disease (c_2), inflammatory spine disease (c_3), and so on. One of the rules induced by the CN2 learning algorithm (that will be described later in this chapter) is shown in Table 7.4.

Conditions of the rule in Table 7.4 consist of a conjunction of three features. Two features assign a discrete value to an attribute, `Sex = male` and `SpinalPain = spondylitic`, whereas one conjunct is of the form $A_i < v$ for

Table 7.3. Decision list induced from the contact lens dataset.

```
IF TearProduction = reduced THEN ContactLenses = none
   ELSE /* TearProduction = normal */
       IF Astigmatism = no THEN ContactLenses = soft
       ELSE /* Astigmatism = yes */
           IF SpectaclePrescription = myope THEN ContactLenses = hard
           ELSE /*SpectaclePrescription = hypermetrope */
               ContactLenses = none
```

Table 7.4. A sample classification rule induced by CN2 for the problem of early diagnosis of rheumatic diseases [340]. The number of covered examples of each of the 8 classes is also given (here, we have left out the class names, as CN2 would).

```
 IF    Sex = male
   AND NumberOfPainfulJoints < 3
   AND SpinalPain = spondylitic
 THEN  Diagnosis = inflammatory-spine-disease   [9 0 12 1 0 0 0 1]
```

the continuous attribute `NumberOfPainfulJoints`. As before, the rule has the following form:

IF *Conditions* THEN *Class* [*ClassDistribution*]

Numbers in the *ClassDistribution* list denote, for each individual class, how many training examples of this class are covered by the rule. This form of if-then rules is induced by the CN2 rule learner. The rule in Table 7.4 should be read as follows: "The rule covers 9 cases of class c_1, no cases of class c_2, 12 cases of class c_3, ..., and one case of class c_8. Out of eight diagnostic classes, the most probable diagnosis for a patient description satisfying the conditions of the rule is inflammatory spine disease (class c_3). Class c_3 is assigned, since this is the majority class among the covered examples."

If we want we can simply ignore the numbers in the class distribution, and interpret the rule categorically. We can also use the numbers to assess the reliability or significance of the rule: for instance, we may consider the above rule unreliable since the second-largest class is not much smaller than the majority class. Finally, we can use the numbers to output a probability distribution over all classes rather than making a single categorical prediction. For instance, if the patient description satisfies the conditions of the above rule, we may predict the diagnosis 'inflammatory spine disease' with probability 12/23=0.52, 'degenerative spine disease' with probability 9/23=0.39, and so on. In general, hypotheses induced by CN2 have the form

$$\{R1, R2, R3, \ldots, DefaultRule\}$$

These hypotheses can be either unordered if-then rules or ordered if-then-else rules. In the first case, several rules may cover an instance to be classified, and their predictions need to be combined, for instance by using a voting scheme. In the second case, the first rule that covers the instance is selected. In both cases, if no induced rule fires, a default rule is invoked which predicts the majority class of uncovered training instances. It should be noted that the learning algorithm for unordered and ordered rule sets is different, as will be detailed in the next section.

7.3. Rule learning as search

As introduced in Section 7.2.2, rules have the following general form:

`IF` *Conditions* `THEN` *Class*,

where *Conditions* is a conjunction of features, and conclusion has the form $Class = c_i$, assigning a particular value c_i to $Class$. Alternative representations for rules are $Class \leftarrow Conditions$, or $Head \leftarrow Body$. The output of learning is a hypothesis represented as a set rules. Constructing such a hypothesis takes place in several stages.

Hypothesis construction. To construct a hypothesis, the learner has to find a set of rules. In propositional rule learning, this can be simplified by learning the rules sequentially and independently, for instance by employing the covering algorithm (Section 7.3.4). In first-order rule learning the situation is more complex if recursion is employed, in which case rules cannot be learned independently.

Rule construction. An individual rule of the form $Head \leftarrow Body$ is usually constructed by fixing the $Head$ to a given class value $Class = c_i$, and heuristically searching for the best rule body, as outlined in the **LearnOneRule** algorithm in Section 7.3.3.

Body construction. Typically, a rule body is a conjunction of features. In the **LearnOneRule** algorithm, body construction is performed by adding features to the initially empty rule body.

Feature construction. In the simplest case, features have the form of simple literals $A_i = v_{ij}$ where A_i is an attribute and v_{ij} is one of its predefined values. With ordered or continuous attributes, one can also consider inequalities of the form $A_i < v$ or $A_i \geq v$, where v is a threshold value to be constructed (i.e., it may be a value not directly observed in the training data). More sophisticated forms of feature construction exist as well.

Perhaps surprisingly, the crucial role of the feature construction process is not often acknowledged in the literature. We consider it a bit more closely in the next subsection.

7.3.1 Feature construction

Ingredients for learning are either simple features defined by the hypothesis language, or combined features defined by the user as background knowledge available to the learner. Features can also be constructed by a separate feature construction algorithm. Sophisticated feature construction is known as *constructive induction* in the machine learning literature [371].

Let us illustrate an approach to simple binary feature construction, based on the analysis of values of examples in the training set. For each attribute A_i, let v_{ix} stand for an arbitrary value of the attribute found in a positive example, and let w_{iy} stand for any value found in a negative example. A simple feature construction results in the following set of binary features, for all appropriate values of x and y.

- For discrete attributes A_i, features of the form $A_i = v_{ix}$ and $A_i \neq w_{iy}$ are generated.
- For continuous attributes A_i, features of the form $A_i < (v_{ix} + w_{iy})/2$ are created for all neighbouring value pairs (v_{ix}, w_{iy}), and features $A_i \geq (v_{ix} + w_{iy})/2$ for all neighbouring pairs (w_{iy}, v_{ix}). A similar approach is suggested by [176].
- For integer-valued attributes A_i, features are generated as if A_i were both discrete and continuous, resulting in features of four different forms: $A_i \leq (v_{ix} + w_{iy})/2$, $A_i > (v_{ix} + w_{iy})/2$, $A_i = v_{ix}$, and $A_i \neq w_{iy}$.

More complex approaches to propositional feature construction could construct the following: features that test the relations $A_i = A_j$, $A_i \neq A_j$ for attributes of the same type (same sets of values, or continuous attributes); features introducing internal disjunctions $A_i = [v_{ik} \vee v_{il}]$, or intervals $A_i \in [v_{ik}, v_{il}]$; conjunctions of features $(A_i = v_{ik}) \wedge (A_j = v_{jl})$; features that test values of functions defined in the background knowledge, e.g., $f(A_i, A_j, ...) \leq v$; features using the relations defined in the background knowledge $r(A_i, A_j)$, and so on. An obvious way to construct new terms is also by generating negations of features: for every $r(A_i, A_j)$ construct a feature $\neg r(A_i, A_j)$.

7.3.2 Structuring the rule space by generality

Learning classification rules can be regarded as a search problem, where the space of possible solutions is determined by the language bias. In if-then rule learning, this is the space of all rules of the form $Class \leftarrow Conditions$, with *Class* being one of the classes, and *Conditions* being a conjunction of features as described above. For simplicity, in this section we limit attention to features of the form $A_i = v_{ij}$ (the equality may be replaced by an inequality for continuous attributes).

In viewing rule learning as a search problem, an evaluation criterion (e.g. rule accuracy or significance) needs to be defined in order to decide whether a candidate rule (or a rule set) is a solution to a given learning problem. Enumerating the whole space of possible rules is clearly inefficient, therefore a structure which enables pruning is required. Nearly all symbolic inductive learning techniques structure the search by means of the dual notions of generalisation and specialisation [381].

Generality is most easily defined in terms of coverage. Let $covers(R)$ stand for the set of instances covered by rule R. We define rule R to be *more general than* rule R' if (i) both have the same consequent, and (ii) $covers(R) \supseteq covers(R')$. This is sometimes called a semantic notion of generality, as it requires us to evaluate rules against a given dataset. For attribute-value if-then rule learning, a simple syntactic criterion can be used instead: given the same rule consequent, rule R is more general than rule R' if the antecedent of R' imposes at least the same constraints as the antecedent of R. We also say that R' is *more specific than* R. To illustrate this notion take two rules from the contact lens example.

```
IF     TearProduction = normal
  AND Astigmatism = yes
THEN  ContactLenses = hard          [#soft=0 #hard=4 #none=2]

IF     TearProduction = normal
  AND Astigmatism = yes
  AND SpectaclePrescription = myope
THEN  ContactLenses = hard          [#soft=0 #hard=3 #none=0]
```

Clearly, the second rule imposes at least the same constraints as the first, and is therefore more specific. In terms of coverage, the first rule covers 6 instances in Table 7.1, whereas the second rule covers only 3 of these. In case of continuous attributes, conditions involving inequalities are compared in the obvious way: e.g., the condition `NumberOfPainfulJoints < 3` is more general than `NumberOfPainfulJoints < 2`, which is more general than `NumberOfPainful-Joints = 1`.

Notice that generality ignores the class labels of the examples. On the other hand, since a more specific rule will cover the same or a subset of the examples of each class covered by a more general one, making a rule more specific (or *specialising it*) is a way to obtain consistent (pure) rules which cover only instances of one class. For instance, in comparison with the first rule above, the second rule managed to get rid of the two incorrectly covered examples, at the expense of losing one correctly covered example as well. Rule construction can be seen as a balancing act between accuracy (the proportion of examples correctly classified) and coverage. We will address rule evaluation in Section 7.4.

It is easily seen that the relation of generality between rules is reflexive, antisymmetric, and transitive, hence a partial ordering. The generality relation is useful for induction because:

- when generalising a rule R' to R, all training examples covered by R' will also be covered by R,
- when specialising a rule R to R', all training examples not covered by R will also not be covered by R'.

Here, generalisation is just the opposite of specialisation. The two properties can be used to prune large parts of the search space. The second property is used in conjunction with positive examples. If a rule does not cover a positive example, all specialisations of that rule can be pruned, as they cannot cover the example. Similarly, the first property is used with negative examples: if a rule covers a negative example, all its generalisations can be pruned since they will cover that negative example as well.

7.3.3 Learning individual rules: Search strategies

In learning a single rule, most learners use one of the following search strategies.

- *General-to-specific* or top-down learners start from the most general rules and repeatedly specialise them as long as they still cover negative examples. Specialisation stops when a rule is no longer inconsistent with the negative examples. During the search, general-to-specific learners ensure that the rules considered cover at least one positive example. When building a rule, learners employ a *refinement operator* that computes a set of specialisations of a rule.
- *Specific-to-general* or bottom-up learners start from the most specific rule that covers a given example; they then generalise the rule until it cannot further be generalised without covering negative examples.

The first approach generates rules from the top of the generality ordering downwards, and the learned rules are usually more general than those generated by the second approach. Therefore the first approach is less cautious and makes larger inductive leaps (i.e., generalisations on unseen data) than the second. General-to-specific search is very well suited for learning in the presence of noise because it can easily be guided by heuristics. Specific to general search strategies seem better suited for situations where fewer examples are available and for interactive and incremental processing.

As mentioned above, in general-to-specific approaches to learning of individual rules the notion of a refinement operator (first introduced by Shapiro [474, 475]) is of paramount importance. Formally, a *refinement operator* ρ for a language bias L is a mapping from L to 2^L such that $\forall R \in L : \rho(R)$ is a set of proper maximally general specialisations of R (i.e., each element in $\rho(R)$ is strictly more specific than R, and there is no rule strictly more general than an element of $\rho(R)$ and strictly more specific than R). Rules $R' \in \rho(R)$ are called *refinements* of rule R.

In attribute-value rule learning, for rule R being $Class = c_i \leftarrow Conditions$, refinements R' of R are obtained by adding a feature $Cond$ to the conjunction of conditions in rule body, so that R' becomes $c_i \leftarrow Conditions$ AND $Cond$. Using a refinement operator, it is easy to define a simple general-to-specific search algorithm for finding individual rules. A possible implementation of the **LearnOneRule** algorithm is sketched in Figure 7.1. The algorithm uses a heuristic to select the best refinement of the current rule at each step (e.g., the one with maximal coverage of positive examples, or alternatively, the one that maximises some other criterion, e.g., reduction in impurity of the covered examples). This amounts to a greedy *hill-climbing* search strategy. If appropriate, the algorithm can be modified to return n best rules (beam search), which is applied in CN2.

```
procedure LearnOneRule(E_i, Rule)
Input: E_i = P_i ∪ N_i: a set of positive and negative examples for class c_i
Procedure:
  Rule := Class = c_i ← Conditions, where Conditions := ∅
  repeat
    build refinements
        ρ(Rule) = {R', where R' := Class = c_i ← Conditions AND Cond}
    evaluate all R' ∈ ρ(Rule) according to a quality criterion
    Rule := the best refinement in ρ(Rule)
  until Rule satisfies a quality threshold or covers no examples from N_i
Output: Rule
```

Fig. 7.1. A general-to-specific algorithm for single rule learning.

7.3.4 Learning a set of rules: The covering algorithm

Both general-to-specific learners and specific-to-general learners repeat the procedure of single rule learning on a reduced example set (or on an example set with different weights assigned to covered and non-covered examples), if the constructed rule by itself does not cover all positive examples. They use thus an iterative process to compute disjunctive hypotheses consisting of more than one rule. One of the most commonly used approaches to rule set construction is the so-called *covering algorithm* outlined in Figure 7.2. The covering algorithm repeatedly generates different rules until all positive examples P_i for class c_i are covered by the hypothesis *RuleSet*, or until some other predefined quality threshold is satisfied. Once a rule is added to the hypothesis, all positive examples covered by that rule are deleted from the current set of positive examples P_i^{cur}. To find one rule, the **LearnOneRule** algorithm is applied.

```
procedure LearnSetOfRules(E_i, RuleSet)
Input: E_i = P_i ∪ N_i: a set of positive and negative examples for class c_i
Procedure:
  E_i^cur := E_i, P_i^cur := P_i
  RuleSet := ∅
  repeat
    call LearnOneRule(E_i^cur, Rule)
    RuleSet := RuleSet ∪ Rule
    P_i^cur := P_i^cur \ P_i^covered, delete from P_i^cur the positive examples covered by Rule
    E_i^cur := P_i^cur ∪ N_i
  until RuleSet satisfies a quality threshold or P_i^cur is empty
Output: RuleSet
```

Fig. 7.2. The covering algorithm. P_i^{cur} denotes the current training set of all uncovered positive examples for class c_i, and E_i^{cur} denotes all uncovered positives together with all negatives (for class c_i).

Some discussion of the covering algorithm is in order. Each rule in a hypothesis covers part of the instance space and assigns a class to it. The role of a positive example (for a certain class) is to draw the attention of the rule learner to a particular area of the instance space, while the role of negative examples is to prevent overgeneralisation, i.e. to make sure that certain other areas of instance space are not covered by the rule. Positive examples are to be covered by at least one rule, while negative examples are not to be covered by any of the rules. This explains while covered positives are removed from the instance set after a new rule has been constructed, while covered negatives are kept because they are required for preventing overgeneralisation when constructing the next rule.

It should be noted that removing examples during training distorts the training set statistics and introduces order dependencies between rules. For instance, the last rule learned is heavily dependent on the previous rules and the positives they cover, and it may not be meaningful (or statistically significant) when interpreted individually. A less drastic form of the covering algorithm can be obtained by using example weights, which decrease with increasing number of covering rules [343].

In order to learn a *RuleBase*, i.e., a set of rules for all classes c_i, the **LearnSetOfRules** procedure is simply iterated for each of the k classes c_i. earlier. In each iteration the current positive class will be learned against the negatives provided by all other classes. This procedure is shown in Figure 7.3.

This concludes our high-level description of a top-down rule learner. What is still missing is a specification of the heuristics used in rule and rule set construction. In particular, we need heuristics for

- evaluating the quality of a single rule;
- deciding when to stop refining a rule;
- deciding when to stop adding rules to a rule set for one class.

The latter two criteria can be replaced by the simple criteria of consistency and completeness, but only if the data is noise-free. Also, a common phenomenon

procedure **LearnRuleBase**(E, *RuleBase*)
Input: E set of training examples
Procedure:
  RuleBase := $\emptyset$
  for each class c_i, $i = 1$ **to** k
    P_i := {subset of examples in E with class label c_i}
    N_i := {subset of examples in E with other class labels}
    call **LearnSetOfRules**($P_i \cup N_i$, *RuleSet*)
    RuleBase := *RuleBase* $\cup$ *RuleSet*
  endfor
Output: *RuleBase*

Fig. 7.3. Constructing a set of rules for all classes.

in data analysis is *overfitting*, which occurs when the learning algorithm tries so hard to fit the training data that the learned rules do not generalise well to yet unseen data. A learning algorithm is particularly prone to overfitting when the induced models have many degrees of freedom or, in the case of rule learning, when the hypothesis language is detailed enough to represent arbitrary classifiers. In such cases, heuristics are needed to trade off the quality of a rule or rule set with other factors, such as their complexity. We continue by discussing possible heuristics to evaluate the quality of single rules.

7.4. Evaluating the quality of rules

Numerous measures are used for rule evaluation in machine learning and knowledge discovery. In classification rule induction, the most frequently used measure is *classification accuracy*. Other standard measures include *precision* and *recall* in information retrieval, and *sensitivity* and *specificity* in medical data analysis. In this section we explore the quality measures that are most frequently used as heuristics to evaluate individual rules and guide rule induction. To do so, the $Head \leftarrow Body$ form of if-then rules will be used. Rule evaluation measures are intended to give an indication of the strength of the (hypothetical) association between *Body* and *Head* expressed by a rule.

Some of the terminology we use has been borrowed from the *confusion matrices* used to evaluate classifiers (rather than single rules). In a two-class problem, examples are either positive or negative, and each example is predicted by the classifier to be positive or negative. Positive examples correctly predicted to be positive are called *true positives*, correctly predicted negative examples are called *true negatives*, positives incorrectly predicted as negative are called *false negatives*, and negatives predicted as positive are called *false positives*.

This terminology is not directly applicable when we evaluate single rules, since a rule does not make negative predictions (an example not covered by a rule may be predicted to be positive by another rule). Furthermore, it is sometimes necessary to generalise the notions of positive and negative examples as some of the measures under study are used outside a classification framework. This is the reason for using the general rule format $Head \leftarrow Body$. Predicted positives correspond to those instances for which the body is true, and actual positives correspond to instances for which the head is true. Similarly, predicted negatives correspond to instances for which the body is false, and actual negatives correspond to instances for which the head is false. We call the resulting table a *contingency table*.

Table 7.5 demonstrates the analogy between confusion matrices and contingency tables. B denotes the set of instances for which the body of the rule is true, and $\overline{B}$ denotes its complement (the set of instances for which the body is false); similarly for H and $\overline{H}$. HB then denotes $H \cap B$, $\overline{H}B$ denotes $\overline{H} \cap B$, and so on. We use $n(X)$ to denote the cardinality of set X, e.g., $n(\overline{H}B)$ is the number of instances for which H is false and B is true (i.e., the number of *counter-instances* of the rule). N denotes the total number of training examples. The relative fre-

Table 7.5. A confusion matrix (left) and a contingency table (right).

	predicted positive	predicted negative	
actual positive	TP	FN	Pos
actual negative	FP	TN	Neg
	$PredPos$	$PredNeg$	N

	B	$\overline{B}$	
H	$n(HB)$	$n(H\overline{B})$	$n(H)$
$\overline{H}$	$n(\overline{H}B)$	$n(\overline{H}\overline{B})$	$n(\overline{H})$
	$n(B)$	$n(\overline{B})$	N

quency $\frac{n(X)}{N}$ associated with X is denoted by $p(X)$; these sample frequencies can be interpreted as estimates of population probabilities (see below for alternative probability estimates). All rule evaluation measures considered in this section are defined in terms of frequencies from the contingency table only.

Let us first formally define *coverage* as the fraction of instances covered by the body of a rule. As such it is a measure of *generality* of a rule.

$$Cov(H \leftarrow B) = p(B)$$

Support of a rule is a related measure known from association rule learning [6], also called *frequency*.

$$Sup(H \leftarrow B) = p(HB)$$

7.4.1 Rule accuracy and related measures

Rule accuracy is the most important measure of rule quality.

$$Acc(H \leftarrow B) = p(H|B)$$

Rule accuracy is defined as the conditional probability that the head is true given that the body is true, and thus measures the fraction of predicted positives that are true positives in the case of binary classification problems:

$$\frac{TP}{TP + FP} = \frac{n(HB)}{n(HB) + n(\overline{H}B)} = \frac{n(HB)}{n(B)} = p(H|B)$$

Rule accuracy is called *precision* in information retrieval [523], and *confidence* in association rule learning [6].

Related measures are prediction error and informativity. The prediction error of a rule is defined as $Err(H \leftarrow B) = 1 - Acc(H \leftarrow B) = p(\overline{H}|B)$. A measure known from information theory, called *informativity*, is defined as follows:

$$Inf(R) = -\log_2 Acc(R)$$

Informally, informativity is related to the amount of information (measured in bits) that needs to be transmitted to a receiver who knows the rule and wants to know the correct classifications of the instances covered by the rule. Clearly, only the misclassifications need to be transmitted: a completely accurate rule requires no additional information.

The *accuracy gain* can be defined as the difference in rule accuracy between two rules $R = H \leftarrow B$ and $R' = H \leftarrow B'$.

$$AccG(R', R) = Acc(R') - Acc(R) = p(H|B') - p(H|B)$$

This is a well-known measure that is typically used to assess the gain achieved when refining rule R into R' (obtained by adding a condition to B). Similarly, *information gain* can be defined as the increase of informativity obtained by going from R to R'.

Relative accuracy is a special case of accuracy gain:

$$RAcc(H \leftarrow B) = p(H|B) - p(H)$$

Relative accuracy measures the increase in accuracy of rule $H \leftarrow B$ relative to the *default* rule $H \leftarrow true$. The latter rule predicts all instances to satisfy H; a rule is only interesting if it improves upon this default accuracy. Another way of viewing relative accuracy is that it measures the utility of connecting body B with a given head H.

7.4.2 Trading off accuracy and generality

All of these accuracy-related measures aim at minimising the number of counter-instances of a rule. Clearly, given two rules with the same accuracy, the more general rule is preferred. However, none of the above measures is able to make this distinction. Therefore, weighted variants are often used, which trade off accuracy and coverage. One such measure is *weighted accuracy gain*, defined as

$$WAccG(R', R) = \frac{p(B')}{p(B)}(p(H|B') - p(H|B))$$

The relative version is weighted relative accuracy, defined as

$$WRAcc(H \leftarrow B) = p(B)(p(H|B) - p(H))$$

These measures are able to prefer a slightly inaccurate but very general rule over a fully accurate but very specific rule.

Another way to achieve such a trade-off is by incorporating profits for true positives and costs for false positives. For instance, by setting the profit per true positive at twice the cost per false positive, each additionally covered positive example is worth two covered negative examples. From this perspective, the above unweighted measures assume that the cost per false positive is infinitely higher than the profit per true positive. The weighted versions assume a profit/cost ratio that is derived from the class distribution in the training set.[2]

[2] The profit per true positive is $\frac{p(\overline{H})}{p(H)}$ times the cost per false positive; these measures thus favour more accurate rules on majority classes and more general rules on minority classes.

It should be noted that the covering algorithm (Figure 7.2) itself is biased towards accurate rules, and does not combine well with the weighted or cost-sensitive measures discussed above. The reason is that after each newly constructed rule all its true positives are removed, thus changing the statistics of the training set. Other algorithms exist which yield sets of general and overlapping rules for each class [343]. The covering algorithm does achieve some bias towards more general rules, not through the rule evaluation heuristic, but by virtue of the refinement operator which constructs rules in a general-to-specific fashion, as well as the use of a stopping criterion which allows some negatives to be covered.

Yet another way to achieve a trade-off between accuracy and generality is to employ a second measure which is in some sense orthogonal to rule accuracy. In information retrieval, this second measure is *recall*, defined as the fraction of positives that are covered by the rule:

$$Recall(H \leftarrow B) = p(B|H)$$

A good rule achieves both high accuracy (called *precision* in information retrieval) and high recall. Precision and recall can be traded off using the so-called *F-measure* [523], which has a parameter indicating the relative importance of the two measures. Alternatively, rules can be plotted and compared in a *precision-recall* diagram, which allows to determine the conditions under which a particular rule is optimal (rather than selecting a single rule for all possible class and cost distributions).

In medical data analysis, the two measures used are sensitivity and specificity. *Sensitivity* is the same as recall, and *Specificity* is the fraction of negatives not covered by the rule (recall of negatives):

$$Spec(H \leftarrow B) = p(\overline{B}|\overline{H})$$

Again, a sensitivity vs. specificity diagram can be used to determine optimality conditions for each rule. A variant of this is the *ROC diagram* (for Receiver Operating Characteristic [429]), which uses the *true positive rate* (the same as recall and sensitivity) and the *false positive rate*, which is equal to $1-Specificity$.

7.4.3 Estimating probabilities

Rules are evaluated on a training sample, but we are interested in estimates of their performance on the whole population. Above we assumed for simplicity that population probabilities were estimated by sample relative frequencies. However, these are likely to lead to overfitting. We now briefly describe two very simple approaches that are routinely used in rule learning to make the probability estimates more robust: the *Laplace estimate* and the *m-estimate* [109, 162]. [3] Both of these apply a correction to the relative frequency, which can be interpreted from a Bayesian perspective as starting from a certain prior probability over the

[3] Much more sophisticated methods have been developed in the statistical literature.

random variable being estimated. The Laplace correction, which is used in CN2, employs a uniform prior distribution, while the m-estimate is a generalisation which is able to take non-uniform prior distributions into account.

For simplicity we restrict the discussion to estimating rule accuracy, but the procedure is applicable to arbitrary cases where a population probability is to be estimated from sample frequencies. Suppose a rule covers p positive and n negative examples. The relative frequency estimate of its accuracy is computed as $\frac{p}{p+n}$. If both p and n are low, this estimate is not very robust in that 1 extra true positive or false positive will make a big change. Moreover, some learning algorithms can easily be dominated by extreme probability estimates of 1 or 0 (e.g., naive Bayes).

The Laplace estimate is obtained by adapting the relative frequency to $\frac{p+1}{p+n+k}$, where k is the number of classes. This is equivalent to adding k 'virtual' covered examples, one for each class. For example, in binary classification ($k = 2$) a rule with one true positive ($p = 1$) and no false positives ($n = 0$) will have an estimated population accuracy of $2/3$, rather than 1. This estimate may asymptotically go to 1 if the number of true positives increases, but with finite amounts of data the probability estimate will never be exactly 1 (or 0). From a Bayesian perspective, a uniform prior distribution over the classes is assumed, and the sample frequencies are used to obtain posterior probabilities.

The m-estimate generalises this further by assuming a pre-specified prior probability p_i for each class c_i. The m-estimate is equal to $\frac{p+mp_i}{p+n+m}$, where m is a parameter. This corresponds to adding m virtual examples distributed according to the given prior. The Laplace estimate is a special case obtained by setting m to k and p_i to $1/k$ for all classes (uniform distribution). The parameter m controls the role of the prior probabilities and the evidence provided by the examples: higher m gives more weight to the prior probabilities and less to the examples. Higher values of m are thus appropriate for datasets that contain more noise. The prior p_i can be estimated from the training set using relative frequency.

7.5. Propositional rule induction at work

Putting everything together, we will now show an example of best-first rule induction at work, and briefly discuss the CN2 rule induction system.

7.5.1 Example

We take the lens prescription dataset in Table 7.1, and rule accuracy as a measure of rule quality. Consider calling **LearnOneRule** to learn the first rule for the class `ContactLenses = hard`. We initialise the rule to

```
IF
THEN  ContactLenses = hard        [#soft=5 #hard=4 #none=15]
```

i.e., the rule with empty body which classifies all examples into class `hard`. This rule covers all 5 examples of class `soft`, all 4 examples of class `hard`, and all

15 examples of class `none`. Given 4 true positives and 20 false positives it has accuracy $4/24 = 0.17$ (the Laplace estimate $5/27 = 0.19$ is slightly higher – we use relative frequency in this example for simplicity).

In the next run of the **repeat** loop the algorithm will need to select the most promising refinement by conjunctively adding the best feature to the current empty rule body. In this case there are as many refinements as there are values for all attributes. Here are the refinements concerning the attribute `TearProduction`:

```
IF     TearProduction = reduced
THEN   ContactLenses = hard        [#soft=0 #hard=0 #none=12]

IF     TearProduction = normal
THEN   ContactLenses = hard        [#soft=5 #hard=4 #none=3]
```

Clearly the second refinement is better than the first (which, incidentally, could be turned into a good rule for class `none`). Its accuracy is estimated at $4/12 = 0.33$, and its accuracy gain is 0.17. As it turns out, this rule is the best one overall and we proceed to refine it further. The best refinement turns out to be the following rule:

```
IF     TearProduction = normal
  AND  Astigmatism = yes
THEN   ContactLenses = hard        [#soft=0 #hard=4 #none=2]
```

This rule has accuracy $4/6 = 0.67$ and accuracy gain 0.33. For completeness, we give all its refinements.

```
IF     TearProduction = normal
  AND  Astigmatism = yes
  AND  Age = young
THEN   ContactLenses = hard        [#soft=0 #hard=2 #none=0]

IF     TearProduction = normal
  AND  Astigmatism = yes
  AND  Age = pre-presbyopic
THEN   ContactLenses = hard        [#soft=0 #hard=1 #none=1]

IF     TearProduction = normal
  AND  Astigmatism = yes
  AND  Age = presbyopic
THEN   ContactLenses = hard        [#soft=0 #hard=1 #none=1]

IF     TearProduction = normal
  AND  Astigmatism = yes
  AND  SpectaclePrescription = myope
THEN   ContactLenses = hard        [#soft=0 #hard=3 #none=0]
```

Table 7.6. Refinement steps of **LearnOneRule** The three right-most numbers indicate rule accuracy, relative accuracy, and weighted relative accuracy, respectively.

```
IF
THEN  ContactLenses = hard  [#soft=5 #hard=4 #none=15]  (0.17, 0.0, 0.0)

IF    TearProduction = normal
THEN  ContactLenses = hard  [#soft=5 #hard=4 #none=3]   (0.33, 0.17, 0.08)

IF    TearProduction = normal
  AND Astigmatism = yes
THEN  ContactLenses = hard  [#soft=0 #hard=4 #none=2]   (0.67, 0.5, 0.13)

IF    TearProduction = normal
  AND Astigmatism = yes
  AND SpectaclePrescription = myope
THEN  ContactLenses = hard  [#soft=0 #hard=3 #none=0]   (1.0, 0.83, 0.13)
```

```
IF    TearProduction = normal
  AND Astigmatism = yes
  AND SpectaclePrescription = hypermetrope
THEN  ContactLenses = hard      [#soft=0 #hard=1 #none=2]
```

Clearly, the fourth rule is the best, with accuracy 1 (accuracy gain 0.33). Also notice that its recall is 0.75 (3 out of 4 positives covered). Since no negatives are covered **LearnOneRule** terminates and returns the indicated rule.

This leaves one example of class `hard` uncovered. Depending on the stopping criterion for rule set construction, the algorithm may decide to find another rule for this class, or proceed with the next class. (In the case of the rule base in Table 7.2, the latter option was chosen.)

To summarise, we give the complete sequence of refinements selected in each iteration of **LearnOneRule** (Table 7.6). Also indicated are (from left to right) rule accuracy, relative accuracy, and weighted relative accuracy. Notice that the third measure does not distinguish between the last two rules.[4]

7.5.2 Rule induction with CN2

CN2 is an algorithm for inducing propositional classification rules. It is more or less an instantiation of the above generic algorithms **LearnOneRule** (Figure 7.1), **LearnSetOfRules** (Figure 7.2), and **LearnRuleBase** (Figure 7.3). We first describe a more recent version of CN2 [117] which learns unordered rule sets. After that, we describe an earlier version [118] which learns ordered decision lists.

[4] Weighted relative accuracy is a number between -0.25 and 0.25.

Learning unordered rule sets. The main difference with **LearnRuleBase** is that, after learning rule sets for all classes, a default rule is added to the learned rule base, assigning the majority class of the entire example set E to instances which are not covered by any of the induced rules. When classifying a new instance, all the rules are tried and predictions of those that cover the example are collected. A voting mechanism is used to obtain the final prediction. Conflicting decisions are resolved by taking into account the number of examples of each class (from the training set) covered by each rule. As an illustration, suppose we have a two-class problem and two rules with coverage [10,2] and [4,40] apply, i.e., the first rule covers 10 examples of class c_1 and 2 examples of class c_2, while the second covers 4 examples of class c_1 and 40 examples of class c_2. The 'summed' coverage would be [14,42] and the example is assigned class c_2. A recent version of CN2 [162] can give probabilistic classifications: in the example above, we divide the coverage [14,42] with the total number of examples covered (56) and obtain as answer the probability distribution [0.25,0.75]. This means that the probability of the example belonging to class c_1 is estimated as $1/4$, while for c_2 that probability is $3/4$.

CN2 can use a significance measure to enforce the induction of reliable rules. If CN2 is used to induce sets of unordered rules, the rules are usually required to be highly *significant* (at the 99% level), and thus reliable, representing a regularity unlikely to have occurred by chance. To test significance, CN2 uses the likelihood ratio statistic [118] that compares the class probability distributions in the set of covered examples with the distribution over the training set. A rule is deemed reliable if these two distributions are significantly different. Suppose the rule covers r_i examples of class c_i, $i \in \{1, \ldots, k\}$, and let $q_i = r_i/R$ where $R = \sum_{i=1}^{k} r_i$. Furthermore, let p_i be the relative frequency of training examples of class c_i. The value of the likelihood ratio statistic is then

$$2R \sum_{i=1}^{k} q_i \log_2(q_i/p_i)$$

This statistic is distributed as χ^2 with $k-1$ degrees of freedom. The rule is only considered significant if its likelihood ratio is above a specified significance threshold.

Learning ordered decision lists. In the decision list version of CN2, **LearnOneRule** is modified to learn rule bodies rather than complete rules. The best rule body minimises the entropy of the covered examples. *Entropy* is calculated as $\sum_{i=1}^{k} q_i \log_2 q_i$, where $q_i = r_i/R$ is the proportion of covered examples of class c_i as before. The entropy is minimal if all covered examples belong to the same class. Once the best rule body has been found, a rule is built which assigns the majority class among the covered examples.

Another difference with the previous algorithm is that before starting another iteration of **LearnSetOfRules**, *all* examples covered by the induced rule are removed – not just the true positives. The reason is that with an ordered decision list the first rule which covers the example fires, hence there is no need to try to cover false negatives by another rule assigning the correct class. Like for the unordered case, a default rule is added which applies if none of the induced rules fire; however, in the ordered case the default class assigned by the default rule is the majority class among all non-covered training examples (and not the majority class in the entire training set, as with the unordered algorithm).

7.6. Learning first-order rules

So far, we have only considered the attribute-value format for rules. *First-order logic* provides a much richer representation formalism, which allows classification of objects whose structure is relevant to the classification task. *Inductive logic programming* (ILP) is a form of rule learning employing clausal logic as the representation language [393, 339, 147, 163]. Other declarative languages can be used as well [185, 260]. This section illustrates some basic techniques employed in first-order rule learning, while the next section will briefly describe some ILP systems by example.

7.6.1 When is first-order rule learning needed?

To explain why inductive logic programming and related approaches should be of interest to those involved in intelligent data analysis, we start by pointing out their main distinguishing features compared to propositional machine learning.

If the available data has a standard tabular form, with rows being individual records (training examples) and columns being properties (attributes) used to describe the data, a data analyst has no reason to become interested in ILP since standard machine learning and statistical data mining tools will do the job. However, consider other scenarios. Suppose that your data is stored in several tables, e.g., it has a *relational database* form. In this case the data has to be transformed into a single table in order to use standard data mining techniques. The most common data transformation approach is to select one table as the main table to be used for learning, and try to incorporate the contents of other tables by summarising the information contained in the table into some summary attributes, added to the main table. The problem with such single-table transformations is that some information may be lost while the summarisation may also introduce artefacts, possibly leading to inappropriate data mining results. What one would like to do is to leave data conceptually unchanged and rather use data mining tools that can deal with multi-relational data. As will be shown in this chapter, ILP is intended at solving multi-relational data mining tasks.

Consider another data mining scenario where there is an abundance of expert knowledge available, in addition to the experimental data. Incorporating expert knowledge in propositional machine learning is usually done by introducing a new attribute, whose values are computed from existing attribute values. Existing data mining tools provide simple means of defining new columns as functions of other data columns (e.g., Clementine, a data mining tool of SPSS). This may not always be sufficient since the available expert knowledge may be structural or relational in nature. Consider chemical molecules and their properties, involving single or double bonds between atoms constituting individual molecules, benzene rings, and a host of other structural features. It is hard to imagine how to express this type of knowledge in a functional form, given that it is structural, and can be most naturally expressed in a relational form.

To summarise, inductive logic programming is to be used for data mining in multi-relational data mining tasks with data stored in relational databases and tasks with abundant expert knowledge (in machine learning called *background knowledge*) of a relational or structural nature. Compared to traditional inductive learning techniques, inductive logic programming is thus more powerful in several respects. First, ILP uses an expressive first-order rule formalism enabling the representation of concepts and hypotheses that cannot be represented in an attribute-value framework of traditional inductive machine learning. Next, it facilitates the representation and use of background knowledge which broadens the class of problems for which inductive learning techniques are applicable. Furthermore, many techniques and theoretical results from computational logic can be used and adapted for the needs of inductively generating theories from specific observations and background knowledge.

Where does the term inductive logic programming (ILP) come from? The term was coined in the early 90s to represent a new research area [393, 339], at that time usually described as being at the 'intersection of inductive machine learning and logic programming'. In ILP, *inductive* denotes that algorithms follow the inductive machine learning paradigm of learning from examples, and *logic programming* denotes that first-order representational formalisms developed within the research area of logic programming – Prolog in particular – are used for describing data and induced hypotheses. The term was intended to emphasise that ILP has its roots in two research areas: inheriting an experimental approach and orientation towards practical applications from inductive machine learning, and its sound formal and theoretical basis from computational logic and logic programming.

Having lived through the newborn and infant era of inductive logic programming, which resulted in many implemented systems and successful applications, nowadays many people like to use terms such as *relational data mining* or *multi-relational data mining* instead of inductive logic programming, mainly because research directions and application interests have changed. Early ILP systems concentrated almost exclusively on logic program synthesis,[5] while recent ILP systems are capable of solving many other types of data mining problems, such as learning association rules [148], first-order clustering [308], database restructuring [183], subgroup discovery [187, 551], and other forms of so-called *descriptive* ILP. Descriptive ILP differs from the more classical *predictive* ILP because its

[5] The early work of Ehud Shapiro on the MIS system [474, 475], is rightfully considered as one of the most influential ancestors of contemporary ILP.

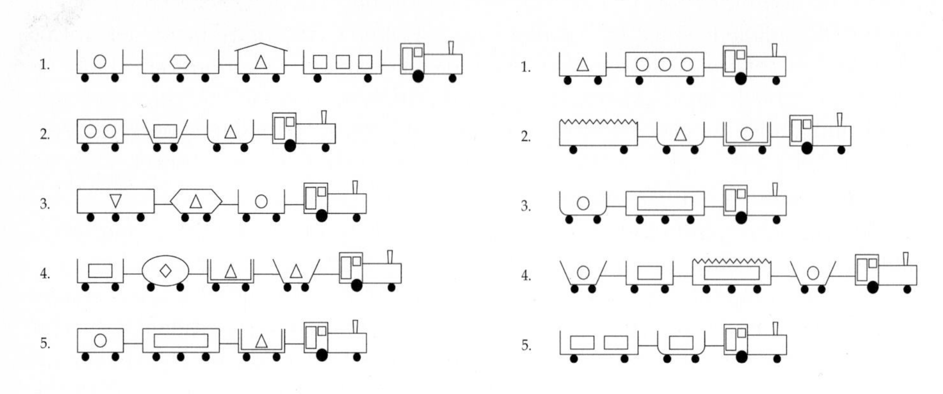

Fig. 7.4. The ten train East-West challenge.

goal is not learning of classification rules but rather learning of 'clausal theories', i.e., sets of logical 'clauses' representing various properties of the data.

That being said, the scope of this chapter allows us only to present some selected approaches to predictive ILP in which first-order classification or prediction rules are induced from examples and background knowledge. The next subsection introduces some basic terminology and representational formalisms, and shows how propositional learning can be upgraded to solve ILP tasks through first-order feature construction.

7.6.2 Knowledge representation for first-order rule learning

Logic is a powerful and versatile knowledge representation formalism. However, its versatility also means that there are usually many different ways of representing the same knowledge. What is the best representation depends on the task at hand. In this section we discuss several ways of representing a particular first-order[6] rule learning task in logic, pointing out the strengths and weaknesses of each.

As a running example we use a learning problem from [377]. The learning task is to induce a Prolog program classifying trains as eastbound or westbound. The problem is illustrated in Figure 7.4. Each train consists of 2–4 cars; the cars have attributes like shape (rectangular, oval, u-shaped, ...), length (long, short), number of wheels (2, 3), type of roof (no roof, peaked, jagged, ...), shape of load (circle, triangle, rectangle, ...), and number of loads (1-3). For this small

[6] The term 'first-order logic' is an abbreviation of *first-order predicate logic.* The main ingredients of predicate logic are predicates, referring to relations in the domain, and terms, referring to objects in the domain. The main difference with propositional logic is the use of variables and quantification to refer to arbitrary objects. The term 'first-order' indicates that quantification is restricted to variables ranging over domain objects, while higher-order logic allows variables to range over relations.

10-instance dataset, a possible rule distinguishing between eastbound and westbound trains is 'a train is eastbound if and only if it contains a short closed car'.

Prolog. A suitable language for expressing first-order rules is the declarative programming language Prolog [184]. The above classification rule can be expressed in Prolog as follows:

```
eastbound(T):-hasCar(T,C),clength(C,short),not croof(C,no_roof).
```

A Prolog rule is also called a *clause*. The symbol ':-' is the Prolog notation for the backward implication $\leftarrow$, and the comma in the body of the rule denotes conjunction. Head and body are made up of *literals*, which are applications of predicates to terms. For instance, in the literal `clength(C,short)` the predicate `clength` is applied to the terms `C` (a variable) and `short` (a constant). Each predicate has an associated *arity* which specifies the number of terms the predicate is applied to: e.g., the predicate `clength` has arity 2 (we also say that it is a binary predicate). For the moment, variables and constants are the only kinds of terms; later on, we will see how to construct aggregate terms (tuples and lists) by means of functors. The last literal in the above rule is a *negative* literal, in which the predicate is preceded by the negation symbol `not`. Note the Prolog convention of starting variables with an uppercase letter and everything else with a lowercase letter.

The predicate `eastbound/1`, where `/1` denotes the predicate's arity, is a unary predicate describing the property 'a train heading east'. This is the property of interest in our learning problem: we want to learn the definition of predicate `eastbound(T)` in terms of other predicates. Predicate `hasCar/2` is a binary predicate, describing a relation between trains and cars: `hasCar(T,C)` is true if and only if train `T` contains car `C`, for arbitrary trains and cars (all variables in a clause are implicitly universally quantified). If variables `T` and `C` are replaced by constants denoting a particular train and a particular car, we obtain the *fact* `hasCar(t1,c11)` which tells us that train `t1` contains car `c11`. In logic programming terminology, `hasCar(t1,c11)` is called a *ground fact* since it doesn't contain any variables. A fact can be seen as a rule with an empty body.

In order to make learning of Prolog classification rules feasible, we need some further information that is not formally part of the Prolog language. For instance, even though `eastbound(T):-hasCar(T,C),eastbound(C)` is a syntactically well-formed Prolog rule, we would normally have different predicates applying to trains and cars. ILP systems therefore often use some form of typing, assigning a type to each argument of a predicate. Furthermore, in classification we want to learn clauses with a distinct predicate in the head (here `eastbound/1`), which has exactly one variable indicating the object or *individual* we are generalising over (here the individuals are trains).[7]

[7] In multi-class problems the head predicate could be binary (i.e., having two arguments), with the second argument a constant indicating one of the classes.

It is worth pointing out that under these restrictions, all but one variables in a clause occur only in the body: these variables refer to parts of individuals (cars, loads), and are often called *existential* variables because a rule like 'for all trains T and cars C, if C occurs in T and C is short and closed then T is eastbound' can be reformulated as 'for all trains T, if there exists a car C in T which is short and closed, then T is eastbound'. That is, for a variable only occurring in the body of a rule, quantifying it universally over the whole rule is equivalent to quantifying it existentially over the body of the rule.

A set of rules with the same predicate in the head is called a *predicate definition*. Occasionally a predicate definition consists entirely of ground facts, enumerating all the instances of the predicate. This is referred to as an *extensional* predicate definition. Predicate definitions involving variables are called *intensional*. We can now define learning of first-order classification rules as learning an intensional definition of a target predicate from

- a (usually incomplete) extensional definition of the predicate, e.g., `eastbound(t1)`, `eastbound(t2)`, ..., `eastbound(t5)`;
- additional extensional definitions of other predicates referring to individuals in the domain, e.g., `hasCar(t1,c11)`, `croof(c11,no_roof)`, ..., `hasCar(t6, c61)`, ...;
- optionally, additional knowledge about the domain in the form of intensional clauses; for instance, if long cars always have three wheels and short cars two, this might be expressed by means of the clauses `cwheels(C,3):-clength(C, long)` and `cwheels(C,2):-clength(C,short)`.

The first kind of input to the learning algorithm are usually called examples, while the second and third kinds are commonly referred to as *background knowledge*. We will see below that in some representations the descriptions of individuals (second case above) can be incorporated into the examples, leaving the third case as true background knowledge.

We already indicated that, in addition to this purely logical knowledge, an ILP system needs further knowledge about predicates such as their types in order to generate meaningful hypotheses. We will now show that such knowledge can be derived from the structure of the domain in the form of a data model.

Data models. A data model describes the structure of the data, and can for instance be expressed as an entity-relationship diagram (Figure 7.5). The boxes in this diagram indicate *entities*, which are individuals or parts of individuals. Here, the Train entity is the individual, each Car is part of a train, and each Load is part of a car. The ovals denote attributes of entities. The diamonds indicate *relationships* between entities. There is a one-to-many relationship from Train to Car, indicating that each train can have an arbitrary number of cars but each car is contained in exactly one train; and a one-to-one relationship between Car and Load, indicating that each car has exactly one load and each load is part of exactly one car.

The main point about entity-relationship diagrams is that they can be used to choose a proper logical representation for our data. For instance, if we store

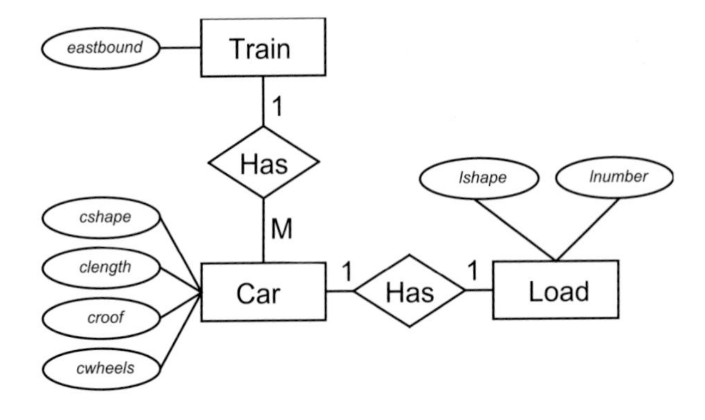

Fig. 7.5. Entity-relationship diagram for the East-West challenge.

the data in a relational database the most obvious representation is to have a separate table for each entity in the domain, with relationships being expressed by foreign keys. This is not the only possibility: for instance, since the relationship between Car and Load is one-to-one, both entities can be combined in a single table. Entities linked by a one-to-many relationship cannot be combined without either introducing significant redundancy (the size of the combined table is determined by the longest train) or loss of information. One-to-many relationships are what distinguishes ILP from propositional learning. We will take a closer look at constructing a single aggregated table (often called *propositionalisation*) in Section 7.7.1.

If we use Prolog as representation language, we can use the entity-relationship diagram to define types of objects in our domain. Basically, each entity will correspond to a distinct type. In both the database and Prolog representations, the data model constitutes a language bias that can be used to restrict the hypothesis space and guide the search. We will elaborate these representation issues below. Before we move on, however, we would like to point out that ILP classification problems rarely use the full power of entity-relationship modelling. In most problems only individuals and their parts exist as entities, which means that the entity-relationship model has a tree-structure with the individual entity at the root and only one-to-one or one-to-many relations in the downward direction. Representations with this restriction are called *individual-centred representations*. We will use this restriction in Section 7.6.3 to define an appropriate language bias.

Database representation. A relational database representation of the train dataset is given in Figure 7.6. The train attribute in the CAR relation is a foreign key referring to *trainID* in TRAIN, and the car attribute in the LOAD relation is a foreign key to *carID* in CAR. These foreign keys correspond to the relationships in the entity-relationship diagram. As expected, the first foreign key is many-to-one (from CAR to TRAIN), and the second one is one-to-one.

TRAIN

trainID	eastbound
t1	true
...	...

CAR

carID	cshape	clength	croof	cwheels	train
c11	rect	long	no_roof	2	t1
c12	rect	short	peak	2	t1
c13	rect	long	no_roof	3	t1
c14	rect	short	no_roof	2	t1
...	...	...	...	...	...

LOAD

loadID	lshape	lnumber	car
l11	rect	3	c14
l12	tria	1	c13
l13	hexa	1	c12
l14	circ	1	c11
...	...	...	...

Fig. 7.6. A relational database representation of the East-West challenge.

If our data is stored in a relational database, we can employ the database query language SQL for multi-relational data mining. For instance, the hypothesis 'trains with a short closed car are eastbound' can be expressed in SQL as

```
SELECT DISTINCT TRAIN.trainID FROM TRAIN, CAR WHERE
    TRAIN.trainID = CAR.train AND
    CAR.clength = 'short' AND
    CAR.croof != 'no_roof'
```

This query performs a join of the TRAIN and CAR tables over *trainID*, selecting only short closed cars.[8] To prevent trains that have more than one such car to be included several times, the DISTINCT construct is used. The query represents only the body of a classification rule, and so the head (the assigned class) must be fixed in advance – one could say that the goal of multi-relational data mining is to construct the correct query for each class. An instance is *covered* by a query if it is included in the computed table returned by the query.

Datalog representation. Mapping a database representation to Prolog is fairly straightforward. Since database attributes are atomic values, we only need

[8] Put differently, SQL takes the Cartesian product of the tables in the FROM clause, selects the tuples that meet the conditions in the WHERE clause, and projects on the attributes in the SELECT clause.

Table 7.7. Datalog representation of the East-West challenge.

`eastbound(t1).`	
`hasCar(t1,c11).`	`hasCar(t1,c12).`
`cshape(c11,rect).`	`cshape(c12,rect).`
`clength(c11,long).`	`clength(c12,short).`
`croof(c11,no_roof).`	`croof(c12,peak).`
`cwheels(c11,2).`	`cwheels(c12,2).`
`hasLoad(c11,l11).`	`hasLoad(c12,l12).`
`lshape(l11,rect).`	`lshape(l12,tria).`
`lnumber(l11,3).`	`lnumber(l12,1).`
`hasCar(t1,c13).`	`hasCar(t1,c14).`
`cshape(c13,rect).`	`cshape(c14,rect).`
`clength(c13,long).`	`clength(c14,short).`
`croof(c13,no_roof).`	`croof(c14,no_roof).`
`cwheels(c13,3).`	`cwheels(c14,2).`
`hasLoad(c13,l13).`	`hasLoad(c14,l14).`
`lshape(l13,hexa).`	`lshape(l14,circ).`
`lnumber(l13,1).`	`lnumber(l14,1).`

terms in the form of constants and variables; under this restriction, Prolog is often referred to as Datalog. One possible mapping is as follows: each table is mapped to a predicate, with as many arguments as the table has columns or attributes, i.e., `train(TrainID,Eastbound)`, `car(CarID,Cshape,Clength,Croof,Cwheels,Train)`, and `load(LoadID,Lshape,Lnumber,Car)`. Then, each row or tuple in each table corresponds to a ground fact, e.g., `train(t1,true)`, `car(c11,rect,long,no_roof,2,t1)`, and `load(l11,rect,3,c11)`.

A variation of this representation, in which each predicate essentially corresponds to an attribute, is usually preferred in ILP, although the two representations are equivalent. The mapping from the database to this alternative Datalog representation is as follows: each non-key attribute in each table is mapped to a binary predicate involving that attribute and the key, i.e., `eastbound(TrainID,Eastbound)`, `cshape(CarID,Cshape)`, `clength(CarID,Clength)`, ..., `lshape(LoadID,Lshape)`, `lnumber(LoadID,Lnumber)`. For the foreign keys it is slightly more natural to use the direction from the entity-relationship diagram, i.e., `hasCar(TrainID,CarID)` and `hasLoad(CarID,LoadID)`, although this is not essential. A small simplification can be furthermore obtained by representing Boolean attributes as unary rather than binary predicates, since the predicate itself can be seen as a function mapping into Booleans. For instance, instead of `eastbound(t1,true)` we can simply write `eastbound(t1)`. The resulting predicates are then extensionally populated as in Table 7.7. Using this representation, the induced hypothesis would be written as

```
eastbound(T):-hasCar(T,C),clength(C,short),not croof(C,no_roof).
```

Table 7.8. Prolog term representation of the East-West challenge.

```
eastbound([car(rect,long, no_roof,2,load(rect,3)),
           car(rect,short,peak, 2,load(tria,1)),
           car(rect,long, no_roof,3,load(hexa,1)),
           car(rect,short,no_roof,2,load(circ,1))]).
```

Testing whether instance `t1` is *covered* by this hypothesis amounts to deducing `eastbound(t1)` from the hypothesis and the description of the example (i.e., all ground facts without the fact providing its classification) and possibly other background knowledge. In order to deduce this we can make use of standard Prolog SLD resolution. One of the great advantages of using Prolog for rule learning is that the rule interpreter comes for free.

Term representation. In full Prolog we can use terms to represent individuals. The representation in Table 7.8 uses functors `car` and `load` to represent cars and loads as tuples, and lists (i.e., the binary head-tail functor) to represent a train as a sequence of cars. In this representation, the hypothesis given before is expressed as follows:

```
eastbound(T):-member(C,T),arg(2,C,short),not arg(3,C,no_roof).
```

Here we use the standard Prolog predicates `member(E,L)` which is true if `E` occurs in list `L`, and `arg(N,T,A)` which is true if `A` is the `N`-th argument of complex term `T`. These predicates are assumed to be part of the background knowledge – without them, lists and tuples cannot be decomposed and would have to be considered atomic values. Again, testing whether a particular instance `Train` (instantiated to a ground term representing a complete description of the instance) is *covered* by a hypothesis amounts to deducing `eastbound(Train)` from the hypothesis and the background knowledge.

The term representation enables us to keep all the information pertaining to a single example together in one term. Naming of individuals and their parts, as in the database and Datalog representations, can (usually) be avoided. Moreover, the structure of the terms can be used to guide hypothesis construction, as there is an immediate connection between the type of an individual (e.g., lists of tuples) and the predicate(s) used to refer to parts of the individuals (i.e., `member/2` and `arg/3`). In this respect, the term representation is superior to both the Datalog and the database representation. On the other hand, while the term representation works very nicely on tree-structured data, naming cannot be avoided when the individuals are graphs (e.g., molecules consisting of atoms, and atoms connected by bonds). Moreover, the term representation can be inflexible if we want to learn on a different level of individuals, e.g., if we want to learn on the level of cars rather than trains. Generally speaking, there is no single representation which is best in all cases, and representation engineering is an important part of every successful ILP application.

Notice that the Datalog and term representations appear very different regarding examples, and very similar concerning hypotheses: `hasCar(T,C)` corresponds to `member(C,T)`, `clength(C,short)` corresponds to `arg(2,C,short)`, and `croof(C,no_roof)` corresponds to `arg(3,C,no_roof)`. The Datalog representation is often called a *flattening* of the term representation [453]. Notice also that `hasCar/2` and `member/2` introduce a new existential or local variable, while the other predicates consume this variable. We will make use of this distinction below.

7.6.3 First-order features

Consider the following two clauses.

```
eastbound(T):-hasCar(T,C),clength(C,short),not croof(C,no_roof).

eastbound(T):-hasCar(T,C1),clength(C1,short),
              hasCar(T,C2),not croof(C2,no_roof).
```

The first clause, which we have seen several times already, expresses that trains which have a short closed car are going east. The second clause states that trains which have a short car and a closed car are going east. The second clause is clearly more general than the first, covering all instances that the first one covers, and in addition instances where the short car is different from the closed car.

We say that the body of the first clause consists of a single *first-order feature*, while the body of the second clause contains two distinct features. Formally, a feature is defined as a minimal set of literals such that no local (i.e., existential) variable occurs both inside and outside that set. The main point of first-order features is that they localise variable sharing: the only variable which is shared among features is the global variable occurring in the head. This can be made explicit by naming the features:

```
hasShortCar(T):-hasCar(T,C),clength(C,short).
hasClosedCar(T):-hasCar(T,C),not croof(C,no_roof).
```

Using these named features as background predicates, the second clause above can now be translated into a clause without local variables:

```
eastbound(T):-hasShortCar(T),hasClosedCar(T).
```

This clause only refers to properties of trains, and hence could be captured extensionally by a single table describing trains in terms of those properties. It is therefore called a *semi-propositional* rule (it is not strictly a propositional rule, because propositional logic doesn't have variables). We described the process of propositionalisation above as transforming a multi-table dataset into a single-table one – equivalently, we can see it as pre-defining the set of first-order features to be considered, and making them available in the background knowledge. If an

ILP system does not perform propositionalisation, it must construct the features during learning – a hard task because of the variable sharing between literals.

We now recall the four steps of rule construction, and indicate how the case of first-order rule learning differs from propositional rule learning.

Hypothesis construction. To construct a set of first-order rules the covering algorithm, outlined in the algorithm `LearnSetOfRules` in Figure 7.2, can be used, although the situation becomes more complex if recursion is allowed, since recursion introduces dependencies between rules.

Rule construction. An individual first-order rule of the form $Head \leftarrow Body$ is constructed by fixing the $Head$ to a given predicate to be learned, e.g., `eastbound(T)`, and heuristically searching for the best rule body, as outlined in the `LearnOneRule` algorithm in Figure 7.1.

Body construction. The rule body is a conjunction of features. As in the `LearnOneRule` algorithm body construction is performed by body refinement, i.e., by adding features to the initially empty rule body.

Feature construction. While propositional features are usually single literals, first-order features consist of several literals. Individual first-order features are constructed using a feature refinement operator, as will be described below.

We should add that, in practice, body construction is often done in a single step by adding literals one-by-one. However, this can be problematic because a single literal such as `hasCar(T,C)` is not really meaningful unless it is followed by a literal consuming the local variable `C`. Furthermore, it is unlikely to improve the quality of the rule (unless it is explicitly meant to distinguish between trains with and trains without cars). In practice, therefore, some sort of lookahead is applied in order to add several literals at a time. The notion of first-order feature formalises the need for lookahead in a systematic way.

The notion of a refinement operator was defined in Section 7.3.3. If we distinguish between body construction and feature construction, we need a refinement operator for both. Because on the level of features each rule is semi-propositional, the body refinement operator is fairly simple as it considers all possible ways to add a single feature to the body of the rule. A feature refinement operator, on the other hand, constructs features consisting of several literals sharing local variables and hence needs to assume some kind of language bias. Here, we will describe a feature bias inspired by the entity-relationship data model, that has been used by several ILP systems employing propositionalisation [186, 342]. Using refinement operators, it is easy to define a simple general-to-specific search algorithm for constructing new features.[9]

---

[9] It should be noted that in first-order logic the semantic and syntactic notions of generality do not necessarily coincide. The most widely used syntactic notion of generality is θ-subsumption, which states that a clause C_1 is more general than a clause C_2 if, after applying a substitution to C_1, all its literals occur in C_2. This is known to be incomplete in the case of recursive clauses. Furthermore, there are various ways to incorporate background knowledge into the generalisation hierarchy. The interested reader is referred to [277, 278].

In the definition of the feature bias, we will distinguish between two types of predicates:

1. *structural* predicates, which introduce variables, and
2. *properties*, which are predicates which consume variables.

From the perspective of the entity-relationship data model, predicates defining properties correspond to attributes of entities, while structural predicates correspond to relationships between entities. A *first-order feature* of an individual is a conjunction of structural predicates and properties as follows:

1. there is exactly one free variable which will play the role of the global variable in rules;
2. each structural predicate introduces a new existentially quantified local variable, and uses either the global variable or one of the local variables introduced by other structural predicates;
3. properties do not introduce new variables (this typically means that one of their arguments is required to be instantiated);
4. all variables are used either by a structural predicate or a property;
5. the feature is minimal in the sense that it cannot be partitioned into subsets without violating any of the previous rules.

The actual first-order feature construction can be restricted by parameters that define the maximum number of literals constituting a feature, maximum number of variables, and the number of occurrences of individual predicates. The following first-order feature can be constructed in the trains example, allowing for 4 literals and 3 variables:

```
trainFeature42(T):-hasCar(T,C),hasLoad(C,L),lshape(L,tria).
```

As can be seen here, a typical feature contains a chain of structural predicates, closed off by one or more properties. Properties can also establish relations between parts of the individual: e.g., the following feature expresses the property of 'having a car whose shape is the same as the shape of its load'.

```
trainFeature978(T):-hasCar(T,C),cshape(C,CShape),
                    hasLoad(C,L),lshape(L,LShape),
                    CShape = LShape.
```

In this section we have seen various ways in which ILP problems can be represented. We have concentrated on individual-centred domains, and used the notion of first-order features to present ILP in such a way that the relations and differences with propositional rule learning is most easily seen. As a consequence, some aspects of first-order rule learning may seem simpler than they actually are. First of all, the set of first-order features is potentially infinite. For instance, as soon as an individual can have an arbitrary number of parts of a particular kind (e.g., trains can have an arbitrary number of cars) there is an existential feature of the kind we discussed above ('train having a short car') for every *value* of every attribute applicable to that part **and** to its sub-parts. Secondly,

Table 7.9. Propositional form of the `eastbound` learning problem.

T	f1(T)	f2(T)	f3(T)	f4(T)	...	f190(T)	eastbound(T)
t1	true	false	false	true	...	false	true
t2	true	true	false	true	...	true	true
...	...	...	...	...	...		
t6	false	false	true	true	...	true	false
t7	true	false	false	false	...	false	false
...	...	...	...	...	...		

we can choose other aggregation mechanisms instead of existential quantification, such as universal quantification or counting. Thirdly, things become much more complicated when recursion is involved. We have deliberately simplified these issues in order to concentrate on what propositional and first-order rule learning have in common.

7.7. Some ILP systems at work

In this section we illustrate the ILP concepts introduced previously through example runs with three (simplified) ILP systems: LINUS, FOIL, and GOLEM.

7.7.1 Propositionalisation with LINUS

LINUS [341, 342] is a first-order rule learner, performing *propositionalisation* by transforming a multi-relational learning problem into a problem of learning from a single table. By taking into account the types of the arguments of the target predicate, applications of background predicates and functions are considered as attributes for attribute-value learning. Existing attribute-value learners can then be used to induce if-then rules. The example below shows how by using first-order features, a relational learning problem is transformed into a propositional one. First-order features bound the scope of local variables, and hence constructing bodies from features is essentially a propositional process that can be solved by a propositional rule learner such as CN2.

We demonstrate LINUS by applying it to the East-West trains example. Propositionalisation is performed by providing LINUS with first-order features defined as background predicates, such as 'has a car with a triangular load'. In the experiment below we simply provided LINUS with all features using maximally two structural predicates introducing local variables, and maximally two predicates describing properties (there are 190 of such features). This means that, for each train, the truthvalue of these 190 first-order features is recorded in a single propositional table (Table 7.9).

Taking Table 7.9 as input (without the first column), an attribute-value rule learner such as CN2 can be used to induce if-then rules like the following:

```
IF    f1(T) = true
  AND f4(T) = true
THEN  eastbound(T) = true
```

or, translated back into Prolog:

```
eastbound(T):-f1(T),f4(T).
```

The actual result of learning eastbound trains applying CN2 on a table with 10 rows and 190 columns describing the dataset from Figure 7.4 are the rules shown below (with feature names edited for clarity).

```
eastbound(T,true):-
  hasCarHasLoadSingleTriangle(T),
  not hasCarLongJagged(T),
  not hasCarLongHasLoadCircle(T).
eastbound(T,false):-
  not hasCarEllipse(T),
  not hasCarShortFlat(T),
  not hasCarPeakedTwo(T).
```

Expanding the first-order features, this translates into the following clauses:

```
eastbound(T,true):-
  hasCar(T,C1),hasLoad(C1,L1),lshape(L1,tria),lnumber(L1,1),
  not (hasCar(T,C2),clength(C2,long),croof(C2,jagged)),
  not (hasCar(T,C3),hasLoad(C3,L3),clength(C3,long),
       lshape(L3,circ)).
eastbound(T,false):-
  not (hasCar(T,C1),cshape(C1,ellipse)),
  not (hasCar(T,C2),clength(C2,short),croof(C2,flat)),
  not (hasCar(T,C3),croof(C3,peak),cwheels(C3,2)).
```

In the first rule, the first feature is a property shared by all eastbound trains but also by two westbound trains, and the two negated features are needed to exclude the latter two. Apparently, it is harder to characterise the westbound class: the first condition in the second rule excludes one eastbound train, the second excludes two, and the third excludes one, so the rule covers all westbound trains but also one eastbound train.

The reason that we didn't find the simple hypothesis 'a train is eastbound if and only if it contains a short closed car' is that in the above feature bias negation was not allowed within the features. If we relax the feature bias accordingly, we obtain the expected hypothesis:

```
eastbound(T,true):-
  hasCar(T,C),clength(C,short),not croof(C,no_roof).
eastbound(T,false):-
  not (hasCar(T,C),clength(C,short),not croof(C,no_roof)).
```

It should be added that the number of first-order features grows exponentially with the number of literals and variables allowed in the features, and thus a simple approach like the above is likely to be (i) computationally expensive, and (ii) prone to overfitting. More practical approaches would filter the features before learning or during propositionalisation.

7.7.2 Top-down first-order rule learning with FOIL

Using refinement operators, it is easy to define a simple general-to-specific search algorithm for finding hypotheses. FOIL [434] is such a top-down ILP learner, employing the Datalog representation explained in the previous section, and using the covering algorithm explained in Section 7.3.4. In a repeat loop, FOIL generates different clauses until all positive examples of the selected target predicate are covered by the hypothesis or the encoding length restriction is violated.[10] Once a clause is added to the hypothesis, all positive examples covered by that clause are deleted from the current set of positive examples. FOIL is heuristically guided, ordering the refinements of an overly general clause according to the weighted information gain criterion [434]. FOIL applies greedy hill-climbing search, i.e., it always chooses the best refinement and never backtracks, but this could be easily changed into, for instance, a beam search strategy as employed by CN2.

A clause is constructed by repeatedly applying a refinement operator, until the clause is consistent with all negative examples for the predicate (no negative example is covered) or the encoding length restriction is violated. In FOIL, the search for the predicate definition of target predicate p starts with the maximally general clause `p(X1,...,Xn):-true`, where `true` stands for the empty body. In each refinement step a literal is added to the body of the clause, having one of the following four forms: (i) `q(Y1,...,Yk)` or (ii) `not q(Y1,...,Yk)`, where `q` is a background predicate or possibly the target predicate; (iii) `Yi=Yj` or (iv) `Yi` $\neq$ `Yj`. It is required that at least one of the variables in the new literal has appeared before in the clause. This ensures that the clause is 'linked', i.e., there is a connection between each body literal and the head of the clause. The remaining variables in the new literal are new local variables. Some further restrictions are applied if the new literal contains the target predicate, in order to avoid possible non-termination problems.

The chief difference between FOIL and CN2 is the way the current sets of positive and negative examples are extended if new local variables are added to the clause. For instance, in the East-West trains problem the initial hypothesis is

---

[10] The encoding length restriction stops clause construction when encoding of a new clause would require more bits that the encoding of yet uncovered examples themselves.

eastbound(T):-true. This clause covers all positive examples t1,...,t5 as well as all negative examples t6,...,t10. Suppose that we add the literal hasCar(T,C) to the clause, then the covered instances are extended to pairs of constants, one for each of the variables in the current clause. So the covered examples are now

```
+  <t1,c11>   <t1,c12>   <t1,c13>   <t1,c14>
   <t2,c21>   <t2,c22>   <t2,c23>
   <t3,c31>   <t3,c32>   <t3,c33>
   <t4,c41>   <t4,c42>   <t4,c43>   <t4,c44>
   <t5,c51>   <t5,c52>   <t5,c53>
-  <t6,c61>   <t6,c62>
   <t7,c71>   <t7,c72>   <t7,c73>
   <t8,c81>   <t8,c82>
   <t9,c91>   <t9,c92>   <t9,c93>   <t9,c94>
   <t10,c101> <t10,c102>
```

Extending the clause to eastbound(T):-hasCar(T,C),clength(C,short) removes 4 positive examples and 5 negative examples, resulting in the following covered examples:

```
+  <t1,c12>   <t1,c14>
   <t2,c21>   <t2,c22>   <t2,c23>
   <t3,c31>   <t3,c32>
   <t4,c41>   <t4,c42>   <t4,c43>   <t4,c44>
   <t5,c51>   <t5,c53>
-  <t6,c62>
   <t7,c71>   <t7,c72>
   <t8,c82>
   <t9,c91>   <t9,c93>   <t9,c94>
   <t10,c101>
```

Finally, adding the literal not croof(C,no_roof) removes all the remaining negative examples, and the clause is added to the hypothesis.

A few points are worth noting here. First, structural predicates like hasCar/2 can be easily overlooked by a greedy system such as FOIL, because they do not help to distinguish between eastbound and westbound trains, but facilitate the inclusion of properties which do discriminate. Secondly, extending the notion of example in the above way (i.e., as a tuple of constants for *all* variables in the clause) can lead to counter-intuitive results: e.g., if there are many more cars in westbound trains than in eastbound trains, the clause

eastbound(T):-hasCar(T,C)

will get a much worse evaluation than

eastbound(T):-true,

whereas if eastbound trains tend to be longer the situation is opposite. Adapting FOIL to work with first-order features would solve both these problems.

7.7.3 Bottom-up first-order rule learning with GOLEM

Until now we have concentrated on top-down approaches to rule learning. The final example we give in this chapter concerns the bottom-up ILP system GOLEM [394]. Bottom-up systems start directly from the examples and construct specific rules which are subsequently generalised.

We illustrate GOLEM on a simplified version of the trains example, in which cars are only described by their length and roof-type. Suppose the first eastbound train `t1` has two cars, `c11` which is long with a flat roof, and `c12` short and open. A rule covering exactly this positive example is the following:

```
eastbound(t1):-
     hasCar(t1,c11),clength(c11,long),croof(c11,flat),
     hasCar(t1,c12),clength(c12,short),croof(c12,no_roof).
```

Similarly, let the second eastbound train `t2` have two cars, `c21` which is short and open, and `c22` short with a flat roof. A rule covering only `t2` is the following:

```
eastbound(t2):-
     hasCar(t2,c21),clength(c21,short),croof(c21,no_roof),
     hasCar(t2,c22),clength(c22,short),croof(c22,flat).
```

These rules are obviously too specific to be useful in learning. However, we can construct a single clause covering both positive examples, and potentially others as well, by an algorithm called *least general generalisation* or LGG. Essentially, this involves comparing the two clauses, and in particular each pair of literals, one from each clause, with the same predicate: for each argument of the predicate, if they are the same constant then we copy that constant to the LGG, else the LGG will have a variable for that argument, such that the same variable will be used whenever that pair of constants is encountered. The LGG of the two ground clauses above is the following.

```
eastbound(T):-hasCar(T,CA),clength(CA,L1),croof(CA,R1),
              hasCar(T,CB),clength(CB,L1),croof(CB,flat),
              hasCar(T,CC),clength(CC,short),croof(CC,no_roof),
              hasCar(T,CD),clength(CD,short),croof(CD,R2).
```

The variable `CA` results from comparing `c11` and `c21`, `CB` stands for `c11` and `c22`, and so on. The variable `L1` results from comparing a short car in the first train with a long car in the second, and thus occurs twice. However, these are variables in the Prolog sense, and thus match any constant, not just the ones from which they were constructed.

It may appear that the generalised rule requires four cars in an eastbound train, but since some of the features in the rule may coincide upon instantiation, this is not really the case. For instance, note that if `L1` is instantiated to `short` and `R1` to `no_roof`, the first and the third feature become equivalent (and essentially describe car `c21`). By construction, the LGG covers `t1` and `t2` which are trains with only two cars. Also, note that a clause can be simplified if it has a feature which is a more specific version of another feature: in that case, the latter is redundant. This results in the following simplified clause:

```
eastbound(T):-hasCar(T,CB),clength(CB,L1),croof(CB,flat),
              hasCar(T,CC),clength(CC,short),croof(CC,no_roof).
```

That is, any train with a short open car and a car with a flat roof is eastbound, which is a very sensible generalisation.

Notice that the LGG grows quadratically with the size of the examples. In more realistic cases the LGG will still contain a lot of redundant information. In this case, GOLEM uses the negative examples to search for literals that can be removed without covering negatives. The overall GOLEM algorithm randomly selects a few pairs of positive examples, constructs their LGG and selects the best one of those. It then reduces the LGG in the way just described, removes all the positives covered by the clause as in the covering algorithm, and iterates. GOLEM can take background knowledge into account, but only in the form of ground facts. Before constructing the LGG, it adds all background facts to the bodies of both ground clauses: the resulting LGG is called a *relative* least general generalisation or RLGG.

7.8. Conclusion

In this chapter we have given an introduction to propositional rule learning and inductive logic programming. While the two subjects are usually treated as rather distinct, we have tried to present them in such a way that the connections become apparent. We believe that this emphasis leads to a better understanding of the nature of ILP, and how work in propositional rule learning, for instance on heuristics, can be carried over to the ILP domain. On the other hand, our perspective has been biased towards approaches that are more easily connected with propositional learning, and many exciting topics such as learning recursive rules and logic program synthesis had to be left out.

One subject that, while close to rule learning, hasn't been covered in this chapter is decision tree learning. In a decision tree, interior nodes stand for attributes or conditions on attributes, arcs stand for values of attributes or outcomes of those conditions, and leaves assign classes. A decision list like the one in Table 7.3 can be seen as a right-branching decision tree. Conversely, each branch from root to a leaf in a decision tree can be seen as a rule. The standard algorithm for learning decision trees is much like the CN2 algorithm for learning decision lists explained in Section 7.5.2, in that the aim of extending a decision tree with another split is to reduce the class impurity in the leaves (usually measured by entropy or the Gini-index). Appendix B explains such an algorithm in more detail. Decision tree learning has also been upgraded to the first-order case [72, 78].

Acknowledgements: Writing of this chapter was partially supported by the British Council through Partnership for Science PSP18, the Slovenian Ministry of Science and Technology, and the EU Framework V project Data Mining and Decision Support for Business Competitiveness: A European Virtual Enterprise (IST-1999-11495).

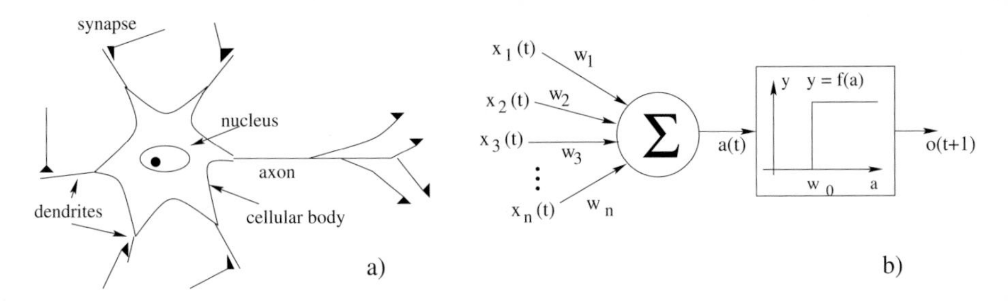

Fig. 8.1. a) The biological neuron and b) the artificial neuron of McCulloch and Pitts.

The final equation modeling the artificial neuron of McCulloch and Pitts is as follows (eq. 8.2):

$$o(t+1) = f\left(\sum_{i=1}^{n} w_i \; x_i(t)\right) \tag{8.2}$$

$f(a)$ in eq. 8.1 is centered around w_0, but the same output can be obtained considering $f(a)$ centered around 0 and a new weight $-w_0$ permanently clamped to an input $x_0 = +1$ to be included in the sum in eq. 8.2.

The model is highly simplified with respect to biological neurons [261]. For example, biological neurons do not respond in a binary fashion, but in a continuous range and for continuous time. Many biological cells use a nonlinear weighting of their inputs. The reaction time of each neuron can be different from $t \to t+1$, so that neurons in a structure can act asynchronously. The precise inhibitory/excitatory behavior of a synapse can be unpredictable. Biological neurons do not respond with only one spike, but if excited produce sequences of pulses, the firing rate and pattern of which carry information about the input signals.

Nevertheless, the artificial neuron of McCulloch and Pitts is computationally a powerful device. If its weights and threshold are appropriately selected, it can perform the basic boolean operations NOT, OR, and AND. This gives a large computational power to any synchronous assembly of artificial neurons with a suitably chosen connection of weights. Another important feature of the McCulloch and Pitts' neuron consists of the nonlinear relationship between inputs and output, which is a common feature of biological neurons. For all these reasons the described artificial neuron, with many improvements and variations, has been adopted as the elementary unit for building Artificial Neural Networks.

A simple generalization of the McCulloch and Pitts' neuron can be obtained by using continuous input and output values and different threshold functions $f(a) \in [0,1]$, such as the linear function (eq. 8.3), the logistic sigmoid function (eq. 8.4, fig. 8.2), or the hyperbolic tangent (eq. 8.5), all exhibiting the very useful property of differentiability unlike the step activation function in eq. 8.1. The sigmoid function with $h = 1$ is usually adopted as an activation function. The corresponding bipolar activation functions with $f(a) \in [-1,+1]$ can also be

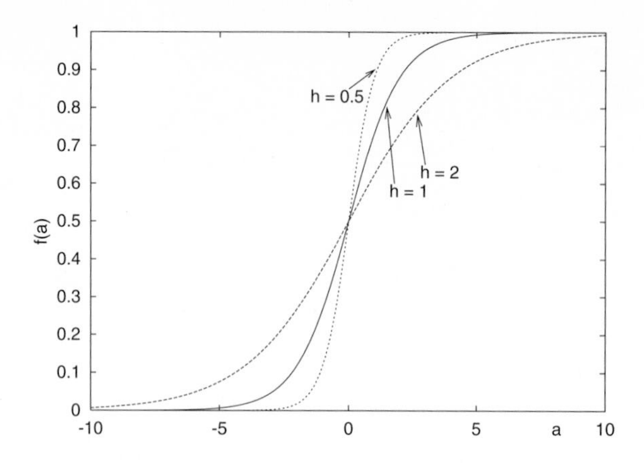

Fig. 8.2. The logistic sigmoid function for different values of h.

used instead of $f(a) \in [0, 1]$.

$$f(a) = a \tag{8.3}$$

$$f(a) = \frac{1}{1 + e^{-ha}} \tag{8.4}$$

$$f(a) = \tanh(a) = \frac{e^{a} - e^{-a}}{e^{a} + e^{-a}} \tag{8.5}$$

To take into account the unpredictable character of the synapses for a more faithful simulation of biological models, some ANN paradigms employ *stochastic units*, where the output of the artificial neuron is probabilistically defined as in eq. 8.6.

$$P\left(o(t+1) = \pm 1\right) = f(a) = \frac{1}{1 + e^{\mp 2\beta a(t)}} \tag{8.6}$$

8.2.2 The Learning Problem

As defined in the previous section, an artificial neuron has to be designed – that is its weights have to be carefully chosen – before the analysis is performed. A learning process however allows biological neural structures to adaptively modify their weights and consequently their input/output functions, to automatically learn how to implement at their best a desired task.

In [256] a physiologically motivated theory of learning was presented, based on two basic notions about the learning process of a neural cell:

- memory is stored in synapses and learning takes place by synaptic modifications;
- neurons become organized into larger configurations, which can perform more complex information processing.

As a result, several learning algorithms have been developed, in order to teach a large assembly of artificial neurons (an *Artificial Neural Network*) to perform a given task. All these learning algorithms modify the connection weights of the network (the synapses) to force the network to perform a desired task. Let us suppose that a sufficiently large set of examples (the *training set*) is available. Two main learning strategies can be adopted.

- If the target output values – the desired answers – of the network are known for all the input patterns of the training set, a *supervised learning* strategy can be applied. In supervised learning the network's answer to each input pattern is directly compared with the known desired answer, and a feedback is given to the network to correct possible errors.
- In other cases, the target answer of the network is unknown. Thus the *unsupervised learning* strategy teaches the network to discover by itself correlations and similarities among the input patterns of the training set and, based on that, to group them in different clusters. There is no feedback from the environment to say what the answer should be or whether it is correct.

In biological systems both learning processes, supervised and unsupervised, are observed. The supervised learning simulates the learning process where animals learn based on a reward that comes from a teacher. In this case, a teacher from the environment rewards/punishes the animal every time the task is correctly/wrongly performed; or, in case of a bad performance, the teacher explains again how the task should be correctly performed. The unsupervised learning approach represents all those unconscious learning processes, during which the brain neurons learn how to answer to a recurrent set of stimuli. For example different regions of the somatosensory map in the brain learn by themselves how to recognize different environment stimuli.

8.2.3 Perceptron

A simple neural network architecture allows only unidirectional forward connections among neurons and, because of that, it is called *feed-forward* neural network. The simplest type of feed-forward neural network, the *Perceptron* [452, 541], consists of only one layer of p neural units connected with a set of n input terminals as in figure 8.3. It is a common convention not to include the set of input terminals in the count of the network layers, because they do not play any active role in the information processing. The number p of outputs is the same as the number of neural units.

In a Perceptron each output o_i is an explicit function of the input vector $\boldsymbol{x} = [x_1, \ldots, x_n]^T$, that can be straightforward calculated after propagating the input values through the network as in eq. 8.7.

$$o_i = f(a_i) = f\left(\sum_{k=0}^{n} w_{ik}\, x_k\right) \qquad i = 1, \ldots, p \tag{8.7}$$

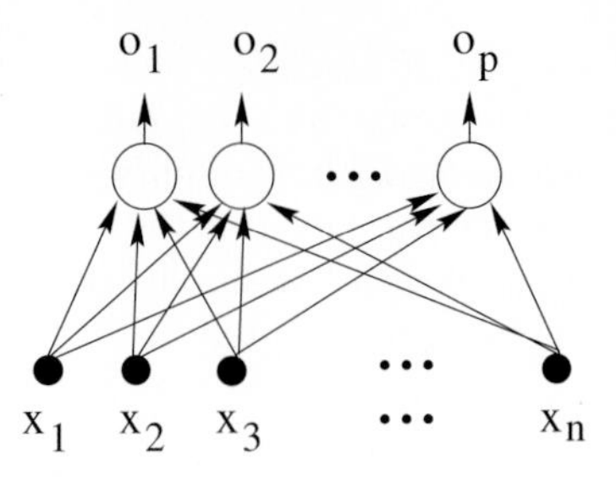

Fig. 8.3. A Perceptron.

a_i indicates the net input to neuron i and w_{ik} the weight connecting the i-th output unit o_i with the k-th input value x_k, being $x_k \in [0,1]$, $k = 1, \ldots, n$, and $o_i \in [0,1]$, $i = 1, \ldots, p$. Considering a virtual input parameter x_0 permanently set to $+1$, the threshold value w_{i0} of the activation function of neuron i has been inserted into the sum in eq. 8.7.

8.2.4 The Delta Rule

Let us suppose that a set T of m training examples $q = 1, \ldots, m$, associating an input vector $\boldsymbol{x}^q$ to a target pattern $\boldsymbol{d}^q$, $T = \{(\boldsymbol{x}^q, \boldsymbol{d}^q)|_{q=1,\ldots,m}\}$, is available. A learning procedure is supposed to adjust the Perceptron weights so that the network's output $\boldsymbol{o}^q$ to the input vector $\boldsymbol{x}^q$ becomes more and more similar to the corresponding target pattern $\boldsymbol{d}^q$ (supervised learning). The set of weights $\boldsymbol{W} = \{w_{ik}\}$, that produces in average the closest answer $\boldsymbol{o}^q$ to $\boldsymbol{d}^q$ for every input pattern $\boldsymbol{x}^q$ for $q = 1, \ldots, m$, represents the optimal weight matrix $\boldsymbol{W}^*$. Starting from initial random weight values, the learning rule should reach the solution $\boldsymbol{W}^*$, if it exists, in a finite number of steps. In this section a learning rule is reported for differentiable neuron activation functions, $f(a)$. The corresponding learning strategy in case of step activation functions, as in eq. 8.1, can be found in [261, 69, 402, 253].

First of all an *error measure* $E(\boldsymbol{W})$ needs to be defined as a function of the weight matrix $\boldsymbol{W}$ of the network. If the error function $E(\boldsymbol{W})$ is differentiable with respect to the elements w_{ik} of the weight matrix $\boldsymbol{W}$, then a variety of gradient-based optimization algorithms can be applied to find the optimal weight matrix $\boldsymbol{W}^*$ at the minimum of the error function. The best known of these numerical optimization methods is the *gradient descent* algorithm and the most commonly used error function is the sum-of-squares error, defined as:

$$E(\boldsymbol{W}) = \frac{1}{2} \sum_{q=1}^{m} \sum_{i=1}^{p} (o_i^q - d_i^q)^2 = \sum_{q=1}^{m} E^q(\boldsymbol{W}) \tag{8.8}$$

$$E^q(\boldsymbol{W}) = \frac{1}{2} \sum_{i=1}^{p} (o_i^q - d_i^q)^2 \tag{8.9}$$

where m is the number of examples in the training set T and p is the number of output units of the network.

Starting with an initial guess of $\boldsymbol{W}$ (usually random values of the weights w_{ik}), the gradient descent algorithm suggests to move in the opposite direction of the gradient, i.e. to change each w_{ik} proportionally to the negative gradient of $E(\boldsymbol{W})$ at the present location $\boldsymbol{W}$, as:

$$w_{ik}(u+1) = w_{ik}(u) + \Delta w_{ik}(u) \tag{8.10}$$

$$\Delta w_{ik}(u) = -\eta \frac{\partial E(u)}{\partial w_{ik}} = -\eta \sum_{q=1}^{m} \frac{\partial E^q(u)}{\partial w_{ik}} = \sum_{q=1}^{m} \Delta w_{ik}^q(u) \tag{8.11}$$

$$\Delta w_{ik}^q(u) = -\eta \frac{\partial E^q(u)}{\partial w_{ik}}. \tag{8.12}$$

η is a small positive number called *learning rate* parameter, and u indicates the current cycle of the procedure over the whole training set T. A cycle u is also referred to as an *epoch*. In the following, the procedure step u will be omitted for sake of clarity.

There are two main strategies for the network's weights update. In the *batch approach* the weights w_{ik} are changed according to eq. 8.11, after all the m training patterns of the training set T are presented to the network. A second strategy, the *incremental approach*, simply changes the weights w_{ik} after every training example q is presented to the network. In this case the *partial sum-of-squares error* in eq. 8.9 is adopted and the corresponding weight update is performed as in eq. 8.12. The incremental learning procedure then cycles through all the training examples until a satisfactory error value is reached. The batch weight update (eq. 8.11) consists just of the sum of the incremental updates Δw_{ik}^q (eq. 8.12) over all the m examples of the training set T.

If the neuron activation function $f(a)$ is differentiable, the error partial derivative for training example q becomes:

$$\frac{\partial E^q}{\partial w_{ik}} = \frac{\partial \frac{1}{2} \sum_{i=1}^{p} (o_i^q - d_i^q)^2}{\partial w_{ik}} = (o_i^q - d_i^q) \frac{\partial f(\sum_{k=0}^{n} w_{ik}\, x_k^q)}{\partial w_{ik}} =$$

$$= (o_i^q - d_i^q)\, f'(a_i^q)\, x_k^q = \delta_i^q\, x_k^q \tag{8.13}$$

$$\tag{8.14}$$

with:

$$\delta_i^q = (o_i^q - d_i^q)\, f'(a_i^q) \tag{8.15}$$

$$a_i^q = \sum_{k=0}^{n} w_{ik}\, x_k^q \tag{8.16}$$

$$\tag{8.17}$$

then:

$$\Delta w_{ik}^q = -\eta\, \delta_i^q\, x_k^q \tag{8.18}$$

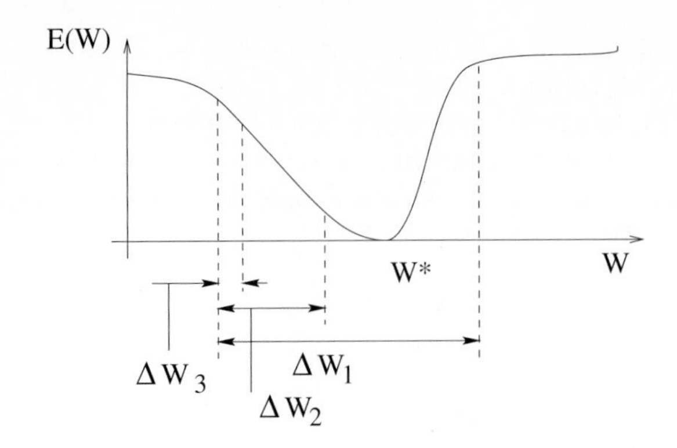

Fig. 8.4. The effect of different learning rates on the learning procedure.

Equation 8.18 defines an update rule for the Perceptron's weights, that minimizes the error function $E(\boldsymbol{W})$ with respect to the implementation of the mapping $\{\boldsymbol{x}^q, \boldsymbol{d}^q\}$. The described learning strategy is referred to as *delta rule* [369], Adaline rule or Least Mean Square rule [541].

The quantity δ_i^q, used for the calculation of Δw_{ik}^q, is a function only of the parameters of neuron i. This *locality* property simplifies the implementation of the learning algorithm. If the activation function is the logistic sigmoid in eq. 8.4 with $h = 1$, then $f'(a) = f(a)\,(1 - f(a))$ and the calculation of δ_i^q is particularly simple.

For the batch approach, it can be proven [261] that if the solution $\boldsymbol{W}^*$, $\{\boldsymbol{W}^* : \; E(\boldsymbol{W}^*) = \min_{\boldsymbol{W}} \; E(\boldsymbol{W})\}$, exists and if the learning rate η is small enough, then the delta rule converges to the solution $\boldsymbol{W}^*$. However only small differences are generally observed if the incremental approach is adopted.

The parameter η has to be chosen small, in order to guarantee to move $E(\boldsymbol{W})$ along the descent gradient direction towards a $\boldsymbol{W}$ configuration with smaller error. If η is too large, the $\Delta\boldsymbol{W}$ update, though moving along the decreasing gradient direction, can lead to a new $\boldsymbol{W}$ configuration with greater error. On the other hand, if the learning rate is too small, the learning procedure takes too long to converge. This phenomenon is illustrated in figure 8.4, where the whole $\boldsymbol{W}$ configuration is reported on the x-axis and $\Delta\boldsymbol{W}_1$, $\Delta\boldsymbol{W}_2$, and $\Delta\boldsymbol{W}_3$ are derived from a learning rate η_1 too large, η_2 reasonable, and η_3 too small respectively.

8.2.5 The Linear Separability Problem

With the introduction of the delta rule, Perceptrons showed to be able to learn several complex tasks only on the basis of the examples in the training set. But the initial enthusiastic reaction suddenly died away at the end of the 60's after the publication of [379]. In this book the authors made clear from a formal

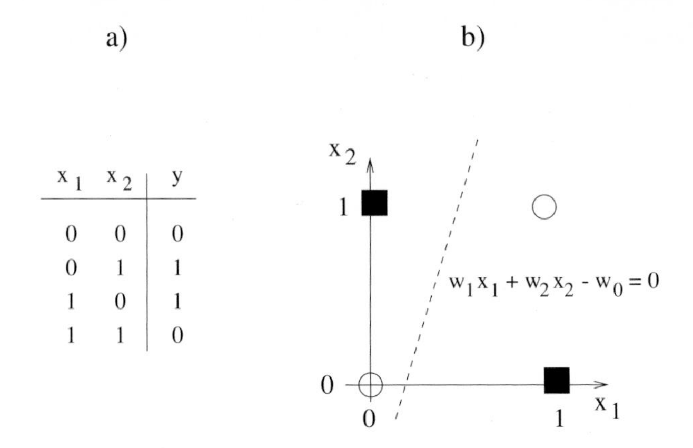

Fig. 8.5. The XOR function's a) truth table b) plot in $\{x_1, x_2\}$ space.

mathematical viewpoint that Perceptrons can not solve a large class of problems, even though they can easily solve apparently similar problems.

The classical example is given by the XOR function. As we have seen in section 8.2.1 the artificial neuron of McCulloch and Pitts, though very simple, is able to reproduce the OR, NOT, and AND boolean functions, considered the basics of the whole boolean arithmetic. The artificial neuron, however, is not able to model the exclusive-OR (XOR) function. The truth table of the XOR function is reported in figure 8.5.a and represented in a two-dimensional space in figure 8.5.b, where the 0 target answers are indicated with a circle and the 1s with a filled square.

The McCulloch and Pitts artificial neuron with two inputs x_1 and x_2 and one output o is adopted to implement the XOR function. The discrimination surface built by the McCulloch and Pitt neuron lies on the line $w_1\ x_1 + w_2\ x_2 \ = \ w_0$, represented by the dotted line in figure 8.5.b. The output o is high if $w_1\ x_1 + w_2\ x_2 \ \geq \ w_0$, low otherwise.

The solution of this problem consists of building an appropriate separation line between the two output classes. That can be generalized to a p-dimensional output produced by an array of p artificial neurons, i.e. a Perceptron. In case of p output units, the solution of the problem requires the definition of an appropriate separation line for every output dimension, that is the definition of a discriminative p-dimensional hyperplane in the input space. Problems where a hyperplane can be found in the input space, correctly separating the patterns into the desired output classes, are said to be *linearly separable.*

It is easy to see from figure 8.5.b that the XOR problem is not linearly separable, because there is no line that can separate the circles from the squares. On the other hand, if no discriminative line can be placed between the two output classes, an artificial neuron alone can not solve the XOR problem. If the neuron activation function $f(a)$ is a continuous nonlinear function, as in eq. 8.4 or in eq. 8.5, instead of a step function as in eq. 8.1, it is still impossible to separate the two classes with only one curve of the type $x_2 = f(x_1)$. If the problem is not linearly separable in a p-dimensional output space, at least one of the p neurons of the Perceptron will fail in defining the discriminative line in the input space. Perceptrons can not solve non linearly separable problems.

Because in the real world and in theory the amount of not linearly separable problems is considerable [69, 541], this is a significant drawback in terms of the computational power of the artificial neural networks. For example, in case of binary n-dimensional input patterns there are 2^n possible input patterns and in case of only 2 output classes there are $p = 2^{(2^n-1)}$ possible labels of these input patterns. Only an extremely small subset ($\frac{2^{n^2}}{n!}$) can be implemented by a Perceptron [69, 261, 253].

The problem now shifts from how to separate the output classes in the input space to how to transform the input space into a linearly separable problem. In [452] some possible transformations of the input patterns $\boldsymbol{\Phi}(\boldsymbol{x})$ are considered before feeding the network, but such transformations $\boldsymbol{\Phi}(\boldsymbol{x})$ have to be appropriately designed in advance for each particular problem. If the input dimensionality increases, the number of Perceptron units or the complexity of the transformations $\boldsymbol{\Phi}(\boldsymbol{x})$ have to increase too [379]. That is only simple problems with low dimension can be practically implemented by a Perceptron, unless an automatic definition of the input transform is somehow included in the learning process.

8.3. Multilayer Feedforward Neural Networks

8.3.1 Multilayer Perceptrons

The easiest way of transforming the input vector probably consists of introducing one or more layers of artificial neurons in the Perceptron architecture, so that the first layer of neurons pre-processes the input space and the second layer builds up the discrimination surfaces necessary to solve the problem.

Networks with more than one layer of artificial neurons, where only forward connections from the input towards the output are allowed, are called *Multi-Layer Perceptrons* (*MLP*) or *Multilayer Feedforward Neural Networks*. Each MLP consists of a set of input terminals, an output neural layer, and a number of layers of *hidden* neural units between the input terminals and the output layer.

The great computational power of multilayer feedforward neural networks was already known many years ago [379]. A two-layer feedforward neural network, like the one in figure 8.6.a, with step activation functions, can implement any Boolean function provided that the number of hidden neurons H is sufficiently large [369]. For example, several solutions for the XOR problem (fig. 8.5.a) have been proposed, after introducing one hidden layer between the input terminals and the output layer in the Perceptron architecture. One of those is shown in figure 8.6.b [261].

If the input variables are not binary but take continuous values in $[0, 1]$, one layer of artificial neurons builds a discriminative hyperplane in the input space (section 8.2.5). In a MLP the second layer organizes these discrimination

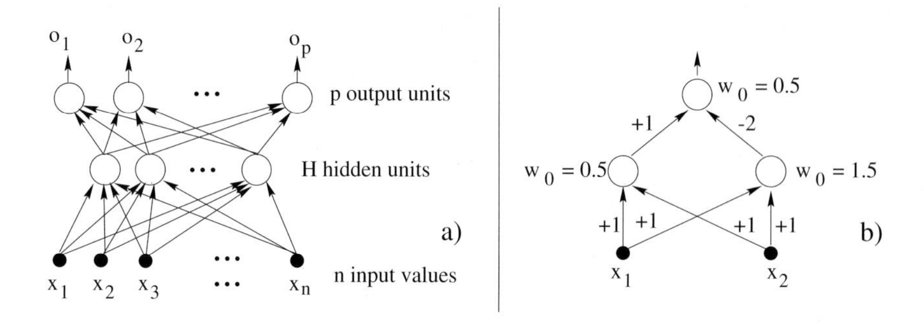

Fig. 8.6. a) A two-layer feedforward neural network can implement any Boolean function b) A two-layer network that can solve the XOR problem (reproduced from [261]).

hyperplanes, so that arbitrarily complex convex separation surfaces can be approximated. A third neural layer would allow the definition of also non convex and disjoint separation surfaces in the input space [69]. Moreover, if the activation function is the logistic sigmoid, it can be proven [137] that any continuous decision boundary – and therefore any generally smooth mapping – can be approximated arbitrarily close by a two-layer Perceptron with a sufficient number of hidden neurons. Thus MLPs provide universal nonlinear discriminant functions. This feature is called the *universality property.*

Let us suppose that we are dealing with a fully connected multilayer Perceptron with L layers, $l = 0, 1, \ldots, L$, with $l = 0$ denoting the set of input terminals, and $l = L$ the output layer. Each layer l has $n(l)$ neurons. The output value o_i of each unit i in layer l can still be calculated by means of eq. 8.7, where inputs x_k to unit i correspond to outputs o_k in layer $l - 1$. Thus eq. 8.7 becomes:

$$o_i = f_i(a_i) = f_i\left(\sum_{k=0}^{n(l-1)} w_{ik}\ o_k\right) \qquad i = 1, \ldots, n(l) \tag{8.19}$$

where a_i represents the net input to unit i, $f_i()$ its activation function and $n(l-1)$ the number of units in layer $l-1$. If layers $l-1$ and l are only partially connected, the sum in eq. 8.19 has to be changed to cover only the units k in layer $l - 1$ that feed unit i in layer l.

In particular for a two-layer feedforward neural network ($L = 2$) with only one output unit o_i ($p = 1$) with linear activation function (eq. 8.3), the mapping function from the input to the output can be written as:

$$o_i = \sum_{k=0}^{n(1)} w_{ik}\ o_k \qquad \begin{array}{l} o_i \in \text{ output layer} \\ o_k \in \text{ hidden layer} \end{array} \tag{8.20}$$

$$o_k = f_k \left(\sum_{j=0}^{n(0)} v_{kj}\ x_j \right) \qquad x_j \in \text{input vector} \tag{8.21}$$

$$o_i = \sum_{k=0}^{n(1)} w_{ik}\ f_k \left(\sum_{j=0}^{n(0)} v_{kj}\ x_j \right) = \sum_{k=1}^{n(1)} w_{ik}\ f_k \left(\boldsymbol{v}_k^T\ \boldsymbol{x} + v_{k0} \right) + w_{i0} \tag{8.22}$$

where $\boldsymbol{v}_k^T$ represents the vector of weight connections from the input vector $\boldsymbol{x}$ in layer $l = 0$ to unit k in the hidden layer $l = 1$, and w_{ik} the weight connection between unit k in the hidden layer $l = 1$ and output unit i in the output layer $l = 2$. Eq. 8.22 looks similar to a general nonlinear regression mapping (chapter 3). The main difference is that in neural networks usually hidden neurons refer to the same activation function ($f_k() = f()\ \forall k$), while in nonlinear regression techniques different functions $f_k()$ can be used for different k. Indeed multilayer Perceptrons can be interpreted as a special case of nonlinear regression techniques [276, 69], and in particular they appear to have the same universal approximation capability.

Even though multilayer feedforward neural networks offer great computational promises, they can not be trained by using the delta rule because the target patterns of the hidden layer is unknown. If an output unit produces an incorrect answer for a given input vector, it is impossible to detect which of the hidden units makes a mistake and consequently which weights should be changed.

8.3.2 The Back-Propagation Learning Algorithm

Indeed to extend the delta rule to MLPs, it is not necessary to know the target pattern of the hidden layer. If the error function and the activation functions of the network units are differentiable, the gradient descent strategy can still be applied in order to find the network's weight configuration $\boldsymbol{W}^*$ that minimizes the error.

The algorithm that implements the gradient descent learning strategy for multilayer feedforward neural networks, independently reinvented multiple times [455, 424], is known as *Back-Propagation*, because it consists of two steps:

1. a forward propagation of the input pattern from the input to the output layer of the network;
2. a back-propagation of the error vector from the output to the input layer of the network.

To apply the gradient descent strategy for the minimization of the error function, an arbitrary differentiable error function can be chosen. In this section a fully connected MLP is considered, the partial sum-of-squares error $E^q(\boldsymbol{W})$ (eq. 8.9) is adopted and the incremental approach (eq. 8.12) is applied on a training set T of m examples, $T = \{(\boldsymbol{x}^q, \boldsymbol{d}^q)\,|_{q=1,\ldots,m}\}$.

The first step of the algorithm consists of applying the input pattern $\boldsymbol{x}^q$ and of calculating through the network's layers the output values $o_1^q, \ldots, o_p^q$ (fig. 8.7.a). The evaluation of the error function $E^q(\boldsymbol{W})$ (eq. 8.9) on the output layer L concludes this forward propagation phase.

If the output error $E^q(\boldsymbol{W})$ is different from 0, an update $\Delta\boldsymbol{W}^q$ of the weight matrix is required. Like for the delta rule, following the gradient descent strategy, the weight update reported in eq. 8.23 is applied to every network weight w_{ik} connecting unit k in layer $l-1$ to unit i in layer l.

$$\Delta w_{ik}^q = -\eta \frac{\partial E^q}{\partial w_{ik}} \tag{8.23}$$

Using the chain rule, the partial derivative can be expressed as:

$$\frac{\partial E^q}{\partial w_{ik}} = \frac{\partial E^q}{\partial a_i^q} \frac{\partial a_i^q}{\partial w_{ik}} \tag{8.24}$$

From eq. 8.19, we obtain:

$$\frac{\partial a_i^q}{\partial w_{ik}} = o_k^q \tag{8.25}$$

and using

$$\delta_i^q = \frac{\partial E^q}{\partial a_i^q} \tag{8.26}$$

we derive

$$\frac{\partial E^q}{\partial w_{ik}} = \delta_i^q \, o_k^q \tag{8.27}$$

$$\Delta w_{ik}^q = -\eta \, \delta_i^q \, o_k^q \qquad \begin{array}{l} i \in \text{layer } l \\ k \in \text{layer } l-1 \end{array} \tag{8.28}$$

which has the same form as the delta rule in eq. 8.18 for the weight updating in a single-layer Perceptron.

For the output units, that is if unit $i \in$ layer L, the delta rule (eq. 8.18) is found again:

$$\delta_i^q = \frac{\partial E^q}{\partial a_i^q} = f'(a_i^q)\,(o_i^q - d_i^q) \qquad i \in \text{output layer } L \tag{8.29}$$

The problem now is how to evaluate $\frac{\partial E^q}{\partial a_i^q}$ for *hidden units*, that is for unit $i \in$ layer l with $l < L$. Using again the chain rule for partial derivatives, we can introduce a contribution from layer $l+1$ to the calculation of δ_i^q:

$$\delta_i^q = \frac{\partial E^q}{\partial a_i^q} = \sum_{j=1}^{n(l+1)} \frac{\partial E^q}{\partial a_j^q} \frac{\partial a_j^q}{\partial a_i^q} \tag{8.30}$$

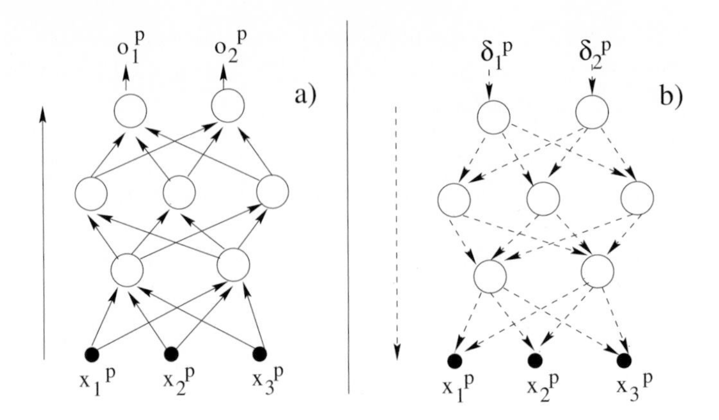

Fig. 8.7. a) Forward propagation of the input vector to calculate the error and b) back propagation of the error vector to update the weight matrix.

where $n(l+1)$ is the number of units j in layer $l+1$. The term $\frac{\partial E^q}{\partial a_j^q}$ is exactly the quantity δ_j^q defined for units j in layer $l+1$ and, considering eq. 8.19, we have $\frac{\partial a_j^q}{\partial a_i^q} = f'(a_i^q)\, w_{ji}$. Then for hidden units i:

$$\delta_i^q = f'(a_i^q) \sum_{j=1}^{n(l+1)} w_{ji}\, \delta_j^q \qquad \begin{array}{l} i \in \text{ layer } l < L \\ j \in \text{ layer } l+1 \end{array} \tag{8.31}$$

Starting with the calculation of the δ values in the output layer L (eq. 8.29), it is possible to calculate the δ values of the hidden units in an intermediate layer $l < L$ by using the δ values of layer $l+1$ (eq. 8.31), as shown in figure 8.7.b for a three layer feedforward fully connected neural network. This is equivalent to the back propagation of the error vector from the output layer towards the input terminals.

When all the weights are updated with Δw_{ik}^q in eq. 8.28, the next training pattern $\boldsymbol{x}^q$ is presented to the network and the procedure starts again. The stopping criterion can be either a threshold on the error function or a maximum on the number of performed epochs.

In both phases, forward and backward, only quantities already available from previous calculations are adopted. For this reason, the Back-Propagation algorithm is said to be *local*. This provides implementations with low computational complexity [366]. The overall computational cost has been estimated as $O(N_w)$, N_w being the number of weights of the network, against the $O(N_w^2)$ required to directly calculate all the N_w involved derivatives [69]. The locality property of the Back-Propagation algorithm also makes it appropriate for parallel computation. The Back-Propagation procedure can be summarized as follows.

1. Initialize the weights w_{ik} with random values.

2. Apply training pattern $\boldsymbol{x}^q$ to the input layer.
3. Propagate $\boldsymbol{x}^q$ forward from the input terminals to the output layer, using eq. 8.19.
4. Calculate error $E^q(\boldsymbol{W})$ on the output layer according to eq. 8.9.
5. Compute the δ's of the output layer as in eq. 8.29.
6. Compute the δ's of the preceding layers, by propagating the δ's backward (eq. 8.31).
7. Use $\Delta w_{ik}^q = -\eta\, \delta_i^q o_k^q$, where $i \in$ layer l and $k \in$ layer $l-1$, for all w_{ik} of the network.
8. $q \to q+1$ and go to 2.

Both batch and incremental approach can be applied, as described in section 8.2.4, for the weight matrix update. The relative effectiveness of the two approaches depends on the problem, but the incremental approach seems to be superior in most cases especially for very regular or redundant training sets [261].

Defining $n(l+1)$ in eq. 8.30 as the connected neurons in the next layers and $n(l-1)$ in eq. 8.19 as the connected neurons in the previous layers for each unit i, the algorithm is easily extensible to partially connected multilayer Perceptrons and to networks with connections between units not in consecutive layers. A similar algorithm can be implemented to train MLPs with stochastic units (eq. 8.6) by measuring the average error of the network [261].

The Back-Propagation algorithm has been used with satisfactory results for a very wide range of classification, prediction, and function approximation problems, and it represents perhaps the most famous ANN training procedure.

8.4. Learning and Generalization

8.4.1 Local Minima

The Back-Propagation learning algorithm (section 8.3.2) is supposed to converge to the optimal weight configuration $\boldsymbol{W}^*$, that represents the location of the absolute minimum on the error surface $E(\boldsymbol{W})$, $\{\boldsymbol{W}^* : E(\boldsymbol{W}^*) = \min_{\boldsymbol{W}} E(\boldsymbol{W})\}$. Indeed Back-Propagation, like every gradient descent procedure, converges to any configuration of the weight matrix $\boldsymbol{W}$ with zero error gradient, that is to any $\boldsymbol{W} : \nabla E(\boldsymbol{W}) = 0$. If the network has a small number of weights, usually with only one layer of neurons, there is only one point $\boldsymbol{W}^* : \nabla E(\boldsymbol{W}^*) = 0$, corresponding to the absolute minimum of the error surface $E(\boldsymbol{W})$. If the network has a higher number of layers, many other minima and stationary points, called *local minima*, arise on the error surface (fig. 8.8), all corresponding to a weight matrix $\boldsymbol{W}$ with $\nabla E(\boldsymbol{W}) = 0$. In the local minima the value of the error function might not be satisfactorily low and consequently the network performance might not be sufficiently close to the desired behavior.

The initial weight configuration determines which global or local minimum the learning algorithm will converge to. Thus the setting of the initial weight values plays a very important role. Usually small random initial values are chosen for the weight matrix. They have to be random to avoid early local minima due

LIVERPOOL
JOHN MOORES UNIVERSITY
AVRIL ROBARTS LRC
TEL 0151 231 4022

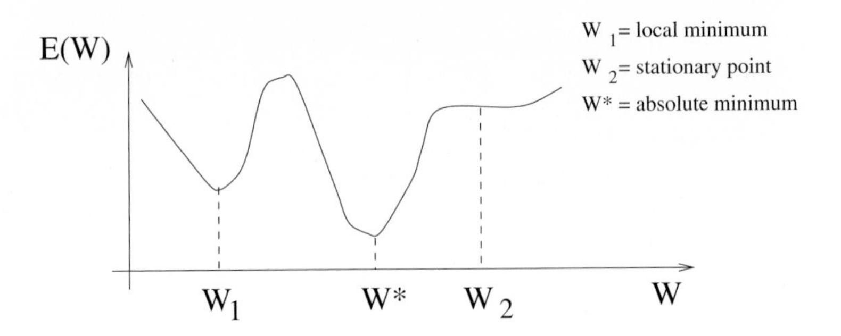

Fig. 8.8. The general error surface as a function of the weight matrix W.

to symmetries in the weight configuration. They have to be small, otherwise the sigmoidal activation functions risk saturating to ± 1 from the beginning and the learning procedure gets stuck in a local minimum close to the starting point, i.e. still with a very high value of the error function. In general a Gaussian distribution with zero mean and variance $\sigma = n(l-1)^{-\frac{1}{2}}$ leads to an appropriate initialization of weights w_{ik} ($i \in$ layer l, $k \in$ layer $l-1$) [69].

If the learning procedure ends in a local minimum located on a quite flat valley, small perturbations (*noise*) to the weight values can help to move out of the valley looking for a better minimum of the error function. In case of incremental update of the weight matrix, a simple and very efficient approach to simulate this kind of noise consists of presenting the training patterns to the network in a random order.

Because of this dependence of the final network's state on the initial weight configuration, it is common practice to train the same network many times using different weight initializations. This leads to a set of different networks. The one with the smallest error on the training set is chosen.

To avoid the local minima problem and to make the learning procedure more effective, a number of more powerful learning techniques have been introduced, like the *Newton's* and *quasi-Newton method*, the *Line search routine*, the *Polak-Ribiere rule*, the *Levenberg-Marquardt algorithm*, the *conjugate gradients methods*, and many others. More details are reported in [69]. While improving the learning performance, such algorithms often give up the locality property of the Back-Propagation algorithm.

Finally, new differentiable error functions have been introduced to implement different input/output mappings. However, all these alternative definitions of the error function are still more or less sensitive to the local minima problem.

8.4.2 Other Error Functions

The choice of a particular error function defines the kind of input/output mapping the network is going to learn, whether classification, prediction, or probability estimation [69].

Usually the sum-of-squares error (eq. 8.8), or its averaged variation the Root Mean Square error $RMS(\boldsymbol{W})$ in eq. 8.32, with p being the number of output units, is adopted. The $RMS(\boldsymbol{W})$ error function has the same features as the sum-of-squares error in eq. 8.8, but the division by the number m of training examples does not allow the error to grow with the size of the training set.

$$RMS(\boldsymbol{W}) = \frac{1}{m} \sum_{q=1}^{m} \sum_{i=1}^{p} (o_i^q - d_i^q)^2 \tag{8.32}$$

Given a multilayer feedforward neural network with a sufficient number of hidden units and a set of m training examples describing the problem with sufficient accuracy and uniformly distributed across the output classes, the network at the minimum of the error function, as defined in eq. 8.8 or in eq. 8.32, learns the conditional average of the target vectors $\boldsymbol{d}^q$ with respect to the input vectors $\boldsymbol{x}^q$ [69].

Other norms, as in eq. 8.33, can be used instead of the Euclidean distance. For high values of R though, the training examples with largest error are privileged during the learning procedure and the solution $\boldsymbol{W}^*$ can be dominated by a small number of outliers or incorrectly labeled data.

$$E(\boldsymbol{W}) = \sum_{q=1}^{m} \sum_{i=1}^{p} (o_i^q - d_i^q)^R \tag{8.33}$$

In [69] the log-likelihood error function in eq. 8.34 is applied, to model the conditional density distribution of the target pattern $\boldsymbol{d}^q$ given the input vector $\boldsymbol{x}^q$. Such conditional distribution $p(\boldsymbol{d}|\boldsymbol{x})$ is modeled by means of a linear combination of M probability models $\Phi_j(\boldsymbol{d}|\boldsymbol{x})$, usually taken as Gaussians, as in eq. 8.35. The mixing coefficients $\alpha_j(\boldsymbol{x})$ must satisfy the condition $\sum_{j=1}^{M} \alpha_j(\boldsymbol{x}) = 1$ to ensure that the distribution in eq. 8.35 is correctly normalized. A multilayer Perceptron can be used to estimate the parameters of the M probability models, $\Phi_j(\boldsymbol{d}|\boldsymbol{x})$, by using eq. 8.34 as error function, that is the negative sum of the logarithm of the conditional probability density $p(\boldsymbol{d}|\boldsymbol{x})$ over the m examples of the training set T. The minimization of such error function leads to a statistical estimation of the parameters of conditional density $p(\boldsymbol{d}|\boldsymbol{x})$ of target data $\boldsymbol{d}$ given input $\boldsymbol{x}$.

$$E(\boldsymbol{W}) = -\sum_{q=1}^{m} \ln \left(\sum_{j=1}^{M} \alpha_j(\boldsymbol{x}^q) \; \Phi_j(\boldsymbol{d}^q|\boldsymbol{x}^q) \right) \tag{8.34}$$

$$p(\boldsymbol{d}|\boldsymbol{x}) = \sum_{j=1}^{M} \alpha_j(\boldsymbol{x}) \; \Phi_j(\boldsymbol{d}|\boldsymbol{x}) \tag{8.35}$$

The *relative entropy* function in eq. 8.36 forces the network to learn the hypothesis represented by output unit o_i with a probability $\frac{1}{2}\,(1 + o_i^q)$ that hypothesis o_i is true [261]. The advantage of the relative entropy error function is that it rapidly diverges if the output of one neural unit saturates. The sum-of-squares error in the same case would just stay constant for a very long time. It is usually used for probabilistic training data.

$$E(\boldsymbol{W}) = \sum_{q=1}^{m} \sum_{i=1}^{p} \left[\frac{1}{2} \, (1 + d_i^q) \, \ln \frac{1 + d_i^q}{1 + o_i^q} \; + \; \frac{1}{2} \, (1 - d_i^q) \, \ln \frac{1 - d_i^q}{1 - o_i^q} \right] \tag{8.36}$$

Other error functions, as the one in eq. 8.37 with $0 < \gamma < 1$, are differently defined at the beginning and at the end of the learning procedure. To avoid early local minima, the error function is larger for those units o_i that are not yet close enough to the desired answer d_i [261].

$$E(\boldsymbol{W}) = \sum_{q=1}^{m} \sum_{i=1}^{p} \left\{ \begin{array}{ll} \gamma (o_i^q - d_i^q)^2 & \text{if } \operatorname{sgn}(d_i^q) = \operatorname{sgn}(o_i^q) \\ (o_i^q - d_i^q)^2 & \text{if } \operatorname{sgn}(d_i^q) = -\operatorname{sgn}(o_i^q) \end{array} \right\} \tag{8.37}$$

8.4.3 Learning Rate and Momentum Term

As introduced in section 8.2.4 (fig. 8.4), the learning rate value η has to be carefully chosen inside the open interval $(0, 1)$. When broad minima yield small gradient values, a large value of η results in a more rapid convergence. However, in presence of steep and narrow minima, a small value of η avoids overshooting the solution.

In general, a learning rule too strongly dependent on the learning rate parameter can produce oscillations around each minimum of the error function $E(\boldsymbol{W})$ and as a final consequence slow down or even prevent the convergence process. To avoid this, the formula in eq. 8.38 is generally adopted instead of eq. 8.11 for updating the weight matrix at each step u of the training procedure.

$$\Delta w_{ik}(u) \; = -\eta \frac{\partial E(u)}{\partial w_{ik}} \; + \; \alpha \; \Delta w_{ik}(u-1) \tag{8.38}$$

α is an arbitrary positive constant, usually chosen between 0.1 and 0.8, called *momentum term*. The idea is to give each connection w_{ik} some inertia or momentum so that each change $\Delta w_{ik}(u)$ does not move the weight w_{ik} too far from the overall update direction. The parameter α avoids drastic changes in the weight w_{ik} due to a high value of $\eta \, \frac{\partial E(u)}{\partial w_{ik}}$, just by keeping track of the last weight change $\Delta w_{ik}(u-1)$. The momentum term typically helps to speed up convergence and to achieve an efficient and more reliable learning profile [560].

Even with the help of the momentum term, the choice of an appropriate value of the learning rate is still important. It would be preferable to have a high value of η at the beginning of the training procedure and to make it smaller and smaller on the way towards the error function minimum (*adaptive learning rate*). In this case the learning rate is increased if the error $E(\boldsymbol{W})$ has been decreasing during all the last U steps; and it is decreased if the error $E(\boldsymbol{W})$ increased at the previous training step $u - 1$; finally it is kept constant if no changes in the error function are observed. Given $\Delta E(\boldsymbol{W}, u) = E(\boldsymbol{W}, u) - E(\boldsymbol{W}, u - 1)$ as the

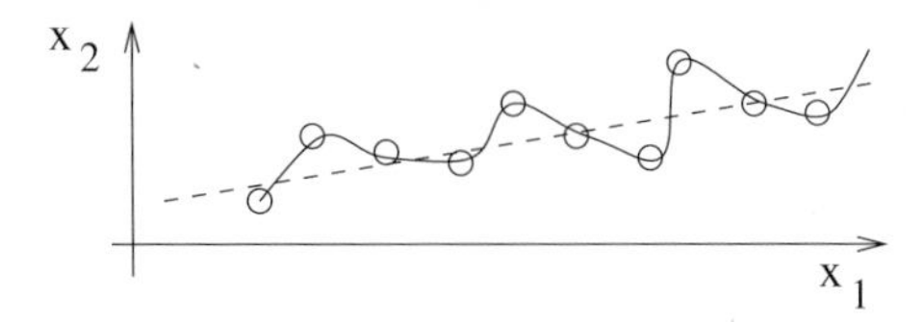

Fig. 8.9. An Example of Overfitting.

cost function change between training step u and $u-1$, and a and b appropriate positive constants, an adaptive learning rate can be defined as in eq. 8.39.

$$\Delta\eta(u) = \begin{cases} +a\eta(u-1) & \text{if} \quad \Delta E(\mathbf{W}, z) < 0 \text{ for } z = u-1, \ldots, u-U \\ -b\eta(u-1) & \text{if} \quad \Delta E(\mathbf{W}, z) > 0 \text{ for } z = u-1 \\ 0 & \text{otherwise} \end{cases} \tag{8.39}$$

Other theoretically more founded learning rules have been introduced, such as *delta-delta*, *delta-bar-delta* and *Quickprop* [69].

8.4.4 Generalization and Overfitting

Unlike the XOR problem in section 8.2.5, the main reason for using an ANN is most often not to memorize the examples of the training set, but to build a general model of the input/output relationships based on the training examples. A general model means that the set of input/output relationships, derived from the training set, apply equally well to new sets of data from the same problem not included in the training set. The main goal of a neural network is thus the *generalization* to new data of the relationships learned on the training set.

The composition of the training set represents a key point for the generalization property. The examples included in the training set have to fully and accurately describe those rules the network is supposed to learn. This means a sufficient number of clean and correctly labeled data equally distributed across the output classes. In practice, the data available for training are usually noisy, partly incorrectly labeled, and do not describe exhaustively all the particular relationships among the input patterns.

A too large number of parameters can memorize all the examples of the training set with the associated noise, errors, and inconsistencies, and therefore perform a poor generalization on new data. This phenomenon is known as *overfitting* and it is illustrated in figure 8.9. The dotted line represents the underlying function to learn, the circles the training set data, and the continuous line the estimated function by an oversized parametric model. Overfitting is thought to happen when the model has more degrees of freedom (roughly the number of weights in an ANN) than constraints (roughly the number of independent training examples). On the contrary, as in every approximation technique, a too small number of model parameters can prevent the error function from reaching satisfying values and the model from learning. How can a good trade-off between learning and generalization be reached?

LIVERPOOL JOHN MOORES UNIVERSITY
LEARNING & INFORMATION SERVICES

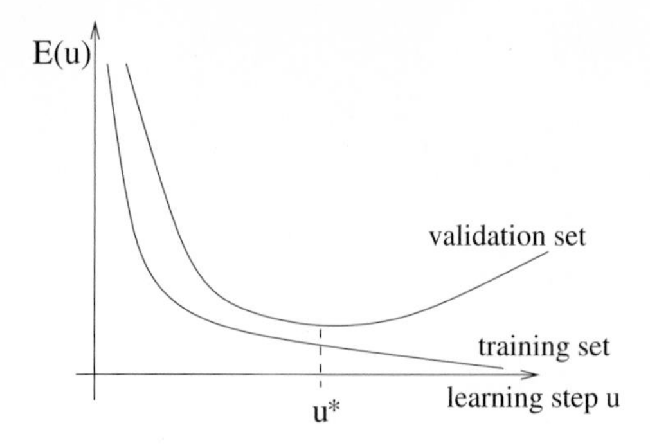

Fig. 8.10. The error plot on the training set and on the validation set as a function of the learning step u.

8.4.5 Validation Set

One of the most common techniques to avoid overfitting in neural networks employs a small set of independent data, called *validation set*, together with the training set. During the learning procedure, the error shows at first a decrease on the validation set as well as on the training set, as the network generalizes from the training data to the underlying input/output function. After some training steps u^*, usually in the later stages of learning, the network starts overfitting the spurious and misleading features of the training set. Consequently the error on the validation set increases, while the error on the training set keeps decreasing (fig. 8.10). The learning procedure can therefore be stopped at this step u^* with the smallest error on the validation set, because at this point the network is expected to yield the best generalization performance. This technique, known also as *early stopping*, however, may not be practical when only a small amount of data is available, since a requirement is that the validation set can not be used for training.

Another approach, called *training with noise*, adds noise to each training input pattern, to smooth out the training data so that overfitting does not occur. A new random vector is added to each input pattern before it feeds the network. For the network it becomes more and more difficult to fit individual data points precisely.

8.4.6 Network Dimension

The overfitting problem depends on the model size – the number of free parameters – with respect to the number of constraints – the number of independent training examples. A rule of thumb for obtaining good generalization is to use the smallest network that fits the training data. A small network, besides the better expected generalization, is also faster to train. Unfortunately, the smallest size of a MLP for a given task is not known a priori. What is the smallest number of hidden layers and units per layer necessary to approximate a particular set of functions with a given accuracy? Regarding the number of layers it was proven that with at most two hidden layers with a *sufficient number of units* it is possible to approximate any function with arbitrary accuracy. If the function is continuous, only one hidden layer is sufficient for this purpose [137].

Even though in theory a two-layer neural network is able to approximate any function, in practice it is always advisable to perform an adequate pre-processing of the data for the best information representation. A bad representation of the data features could lead to a very complicated error surface with many local minima and therefore to a very difficult learning process. Similarly the outputs of a neural network are often post-processed to produce more significant output values [114].

It remains to define what "a sufficient number of hidden units" is. No general conclusions have been stated so far. Some theorems about the minimal dimension of a multilayer Perceptron are proven only for binary input/output variables and for networks with step activation functions (eq. 8.1). Those theorems generally establish a connection between the number of independent training examples and the network size in the worst-case performance, as in [48]. In practice a good generalization can be achieved even with a smaller number of training data or with a smaller network.

Two kinds of iterative techniques have been proposed to optimize the network size.

Growing Algorithms. This group of algorithms begins with training a relatively small neural architecture and allows new units and connections to be added during the training process, when necessary. Three growing algorithms are commonly applied: the upstart algorithm, the tiling algorithm, and the cascade correlation [261]. The first two apply to binary input/output variables and networks with step activation function (eq. 8.1). The third one, which is applicable to problems with continuous input/output variables and with units with sigmoidal activation function, keeps adding units into the hidden layer until a satisfying error value is reached on the training set.

Pruning Algorithms. Unlike the growing algorithm the general pruning approach consists of training a relatively large network and gradually removing either weights or complete units that seem not to be necessary. The large initial size allows the network to learn quickly and with a lower sensitivity to initial conditions and local minima. The reduced final size helps to improve generalization [446]. There are basically two ways of reducing the size of the original network.

1. *Sensitivity methods.* After learning the sensitivity of the error function to the removal of every element (unit or weight) is estimated: the element with the least effect can be removed.
2. *Penalty-term methods.* Weight decay terms are added to the error function, to reward the network for choosing efficient solutions. That is networks with

small weight values are privileged. At the end of the learning process, the weights with smallest values can be removed, but, even in case they are not, a network with several weights close to 0 already acts as a smaller system.

In general, sensitivity methods modify a trained network: the network is trained, the sensitivity of each parameter is estimated, and the weights or nodes are removed consequently. Penalty-term methods modify the error function so that the Back-Propagation algorithm drives unnecessary weights to zero and, in effect, removes them during training. A more detailed review of pruning techniques can be found in [446].

8.4.7 Knowledge Extraction from Multilayer Perceptrons

MLP architectures, trained with Back-Propagation to solve classification, approximation, or prediction problems, very frequently outperformed more traditional methods in several areas [31], such as handwritten character recognition [347], speech recognition [389], robotics [400], and medical diagnostics [483]. One of the reasons of MLP's success is its ability to develop an internal representation of the knowledge necessary to solve a given problem.

However, such internal knowledge representation is very difficult to understand and to translate into symbolic knowledge, due to its distributed nature. At the end of the learning process, the network's knowledge is spread all over its weights and units. In addition, even if a translation into symbolic rules is possible it might not have physical meaning, because the network's computation does not take into account the physical ranges of the input variables. Finally the obtained knowledge representation is not unique. The final status of the network depends on the initial conditions, leading to different symbolic rules according to the initial random values of the weights. Generally the internal knowledge representation of MLPs presents a very low degree of human comprehensibility and, for this reason, it has often been described as *opaque* to the outside world. More explicitly neural networks have often been referred to as *black boxes*, describing their unexplainable capability of solving problems.

This lack of comprehension of how decisions are made inside a neural network definitely represents a strong limitation for the application of ANNs to intelligent data analysis. Several real world applications need an explanation of how a given decision is reached. Let us assume that in a bank a MLP analyzes credit requests and decides whether or not to assign a loan. Even if the network is always right in predicting who is worth it to receive a loan and who is not, a good explanation is necessary for the customer whose loan is refused. In the same way any physician would be reluctant to propose a diagnosis to a patient without understanding the underlying decision process. Many other examples could be cited.

A variety of approaches for extracting knowledge from MLP architectures have been proposed [513, 133, 202]. Mainly they concentrate on compiling symbolic rules from the final neural network status and, because of that, go under the name of *rule extraction* procedures. All the rule extraction procedures can be roughly grouped in two categories.

One approach, called *global*, extracts a set of rules characterizing the behavior of the whole network in terms of input/output mapping. A tree of candidate rules is defined. The node at the top of the tree represents the most general rule – for example: "all patterns belong to class $\mathcal{C}_1$" – and the nodes at the bottom of the tree represent the most specific rules, which cover one example each. Each candidate symbolic rule is tested against the network's behavior, to see whether such a rule can apply. The process of rule verification continues until most of the training set is covered. One of the problems connected with this approach is that the number of candidate rules can become huge when the rule space becomes more detailed. Some heuristics have been developed to limit the number of rules that can be generated from a parent node [133, 513].

The second approach, called *local*, decomposes the original multilayer network into a collection of smaller, usually single-layered, sub-networks, whose input/output mapping might be easier to model in terms of symbolic rules. Based on the assumption that hidden and output units, though sigmoidal, can be approximated by threshold functions, individual units inside each sub-network are modeled by interpreting the incoming weights as the antecedent of a symbolic rule. The resulting symbolic rules are gradually combined together to define a more general set of rules that describes the network as a whole. The monotonicity of the activation function is required, to limit the number of candidate symbolic rules for each unit [202]. Local rule-extraction methods usually employ a special error function and/or a modified learning algorithm, to encourage hidden and output units to stay in a range consistent with possible rules and to achieve networks with the smallest number of units and weights.

Local methods draw a clear one-to-one correspondence between the rule system and the neural network architecture. This should make the ANN's decision process more transparent. However, individual hidden units do not typically represent clear logical entities because of the distributed nature of the knowledge representation in a neural network. On the other hand, although global methods are more generally applicable, they provide a map of the input/output behavior of the network and fail to provide any deeper insight into the internal knowledge representation.

All cited methods for the extraction of symbolic rules from neural networks require quite intensive computational efforts. They also do not address the problem of the non-uniqueness of the final solution as well as of the practical applicability of the obtained symbolic rules. All of these reasons have pushed the research towards different neural network architectures, which offer a higher degree of interpretability.

8.4.8 Other Learning Algorithms

Other learning algorithms have been developed at the same time as Back Propagation, either to solve the learning problem of multilayer feedforward neural networks or to propose different neural structures from the multilayer Perceptron. A short overview is reported here. More details can be found in [261, 253].

A major theme is the *associative content addressable memory*, that associates input patterns with one another if they are sufficiently similar. In this field, Hopfield introduced some helpful physics concepts by emphasizing the notion of memories as stable attractors of dynamical systems and by associating an energy function to each network (*Hopfield's nets*).

Another large group of neural networks uses *reinforcement learning* as learning procedure. In the *associative reward-penalty* algorithm, an arbitrary architecture of stochastic units is used. Instead of known target patterns, the network is given only the reinforcement signal r. $r = +1$ means positive reinforcement (reward); $r = -1$ means negative reinforcement (penalty). Based on the r value, target outputs can be defined.

Boltzmann neural networks borrowed the concept of simulated annealing (chapter 10) from physics to train neural networks with hidden units. Stochastic units are adopted (eq. 8.6), where the parameter $\frac{1}{\beta} = T$ is the global temperature in the simulated annealing procedure. A set of desired probabilities can be defined as target. An energy value, instead of an error value, is associated with every particular configuration of the network. At high temperature the network will ignore small energy differences and will rapidly approach the equilibrium point, finding a good energy minimum. As the temperature is lowered, the network will begin to respond to smaller energy differences and will find a better energy minimum.

Adaptive Resonance Theory networks (*ART*) create prototypes of data clusters of the input space. If the input vector and a prototype are sufficiently similar, the network resonates. If an input pattern is not sufficiently similar to any existing prototype, a new category is formed having the input pattern as prototype and using a previously uncommitted output unit. If no free units are left, then any new input does not produce any response. The "sufficiently similar" is decided by a vigilance parameter $0 < \rho < 1$. When ρ is large, the input space is finely divided in many categories. If ρ is small, a more coarse categorization of the input space is obtained.

The ANN paradigms, described in this section, generally provide more faithful models to biological nervous systems. However, despite their theoretical importance, ANN architectures more specifically oriented to data analysis will be described in the next sections.

8.5. Radial Basis Function Networks

We have seen (section 8.2.5) that an adequate pre-processing of the input features would allow a single-layer Perceptron to approximate any boundary functions. A hidden layer of neural units performs this pre-processing of the input space in a MLP. This leads indeed to the universality property, but does not make the system's decisions transparent to the user (the black-box model). It would be desirable on one side to make the neural mapping easier to understand and to translate into symbolic knowledge and on the other side to keep valid the universality property. In this section a new ANN paradigm is described, where a hidden layer of a different kind of neural units performs the required pre-processing of the input space [385].

The underlying idea is to force each unit of the hidden layer to represent a given region of the input space. That is each hidden unit must contain a *prototype* of a cluster in the input space. When a new pattern is presented to the network, the hidden unit with the most similar prototype will activate a decisional path inside the network that will lead to the final result. Thus, the activation function of the new hidden units must include the concept of prototype of a region and the concept of similarity of an input pattern with this prototype. This can be translated into a measure of distance.

Several distance measures have been proposed and several training algorithms designed to define the input cluster and its prototype associated with each hidden unit of the network. In particular, a very successful approach introduced *Radial Basis Function* (*RBF*) units into the hidden layer. Besides the advantage of an easier interpretation of the system's results, RBF networks hold the universality property and are usually accompanied by much faster learning algorithms than Back Propagation [69].

8.5.1 The RBF Architecture

Radial Basis Function networks include one hidden layer of special units, that pre-process the input pattern and feed a single-layer Perceptron (fig. 8.11). Each unit k in the hidden layer contains the prototype $\boldsymbol{x}_k$ of a given region k of the input space. The corresponding nonlinear activation function $\Phi_k()$ expresses the similarity between any input pattern $\boldsymbol{x}$ and the prototype $\boldsymbol{x}_k$ by means of a distance measure (eq. 8.40).

With H hidden units and a p-dimensional output vector, as illustrated in figure 8.11, the input pattern $\boldsymbol{x}$ is transformed from the input space into the i-th dimension of the output space according to the mapping function in eq. 8.41. Weight w_{ik} represents the connection between the hidden unit k and the output unit i and w_{i0} the threshold value of output unit i. This can also be represented as a weight connected to an extra function Φ_0 constantly set to $+1$ and inserted into the sum in eq. 8.41. No more than one hidden layer of units $\Phi_k()$ is usually considered. Units in the output layer can have linear or sigmoidal activation functions.

$$\Phi_k(\boldsymbol{x}) = \Phi_k(\| \boldsymbol{x} - \boldsymbol{x}_k \|) \qquad k = 1, \ldots, H \qquad (8.40)$$

$$o_i(\boldsymbol{x}) = \sum_{k=1}^{H} w_{ik} \, \Phi_k(\boldsymbol{x}) + w_{i0} \qquad i = 1, \ldots, p \qquad (8.41)$$

Since the first work in RBF networks [426], it has been shown that many properties of the final input/output mapping $\boldsymbol{o}(\boldsymbol{x})$ are relatively insensitive to the form of the functions $\Phi_k()$. Any kind of nonlinear function, based on a distance measure as in eq. 8.40, can be adopted as $\Phi_k()$.

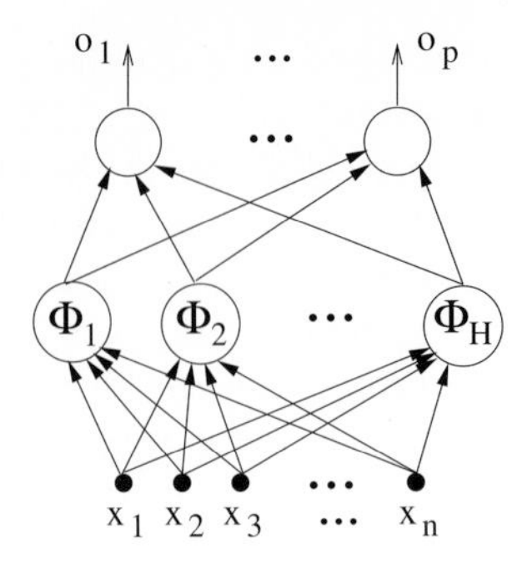

Fig. 8.11. The general architecture of a Radial Basis Function network.

Usually the hidden units are preferred to be *locally tuned*, i.e. to be active only for a delimited region of the input space with $\Phi \to 0$ as $\| \boldsymbol{x} \| \to \infty$. $\Phi_k()$ in eq. 8.41 are usually chosen as *radially-symmetric* functions – from which the name of Radial Basis Function units – with a single maximum at the origin, dropping off rapidly to zero for large distances. Locally tuned receptive fields are widely found in biology, even though they are not single-cell properties, but usually emerge from groups of cells.

The most commonly adopted Radial Basis Function is the unnormalized Gaussian (eq. 8.42). The center $\boldsymbol{\mu}_k$ defines the prototype of the input cluster k and the variance σ_k^2 its size. The usual Gaussian normalization can be included by means of weight w_{ik} or by using lateral connections. The set of Gaussian functions $\Phi_1, \ldots, \Phi_H$ represents a set of non orthonormal basis functions of the output space.

$$\Phi_k(\boldsymbol{x}) = e^{-\frac{\| \boldsymbol{x} - \boldsymbol{\mu}_k \|^2}{2\sigma_k^2}} \tag{8.42}$$

The Gaussian activation function can be generalized to allow an arbitrary covariance matrix Σ_k, as in eq. 8.43. In this case the number of independent adjustable parameters increases dramatically, and so does the training time.

$$\Phi_k(\boldsymbol{x}) = e^{-\frac{1}{2}\,(\boldsymbol{x} - \boldsymbol{\mu}_k)^T \Sigma_k^{-1} (\boldsymbol{x} - \boldsymbol{\mu}_k)} \tag{8.43}$$

The Gaussian function has the highly desirable property of *locality*, that is $\Phi \to 0$ as $\| \boldsymbol{x} \| \to \infty$. This makes the final decision process easy to interpret in contrast with the distributed nature of the knowledge representation in MLPs.

In addition, Back-Propagation training process updates all units independently from their contribution to the final outputs. The interferences among hidden units produce a highly nonlinear updating process with problems of local minima. This can lead to a very slow convergence of the training algorithm. On the other hand in a RBF neural structure for any given input vector $\boldsymbol{x}$ only a small fraction of the H hidden units – those with centers $\boldsymbol{\mu}_k$ very close to the input vector $\boldsymbol{x}$ – will respond with significantly large activation values. This limits

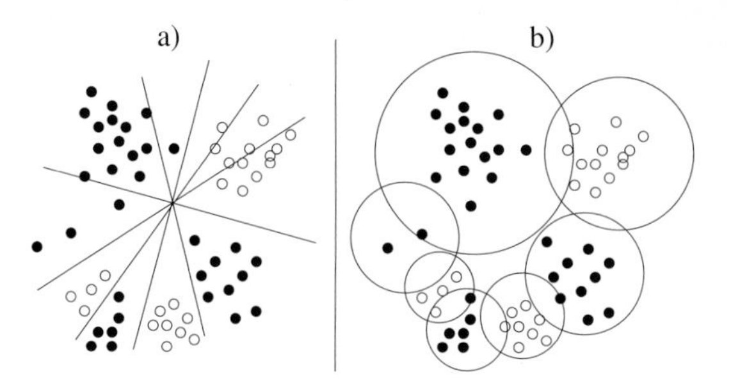

Fig. 8.12. An example of separation surfaces created by a) a multilayer Perceptron and b) an RBF network.

the number of hidden units that need to be evaluated for each training example $\boldsymbol{x}^q$ and leads to more efficient training algorithms than Back Propagation.

Another difference between MLP and RBF architectures consists of the separation surfaces implemented in classification problems. A MLP network separates classes by building hyperplanes in the input space (fig. 8.12.a). Usually several hyperplanes have to be combined together to form closed separation surfaces. RBF networks divide the input space into a number of sub-spaces and each subspace is represented by only a few hidden RBF units. The first layer activates the RBF units representing the input cluster, to which the input pattern $\boldsymbol{x}$ belongs. The second layer, the Perceptron, linearly combines the outputs of the RBF units to produce the appropriate output vector. This builds closed separation surfaces among groups of data in the input space (fig. 8.12.b). Because of the locality property, that leads to closed separation surfaces in classification problems, RBF networks are generally not well suited to learn logical mappings such as parity or boolean functions [385].

The universality property valid for a two-layer Perceptron (section 8.4.6) still holds for neural networks with a sufficient number of Gaussian basis functions, provided that the widths σ_k of the Gaussians are adjustable parameters [69]. However, as for the number of hidden units in MLPs, the minimum number of RBFs necessary to implement a given task is unknown and no practical procedures to construct an appropriate RBF architecture are given.

If the adopted RBFs are Gaussian, the interpretation of the network's decision becomes easy from a statistical point of view. A linear combination of Gaussian probabilities, each one modeling a given group of training data, produces the desired approximation or classification function. Moreover in the classification task, the basis function $\Phi_k()$ can be interpreted as the posterior probability of input vector $\boldsymbol{x}$ belonging to cluster k, $p(\text{cluster } k|\boldsymbol{x})$. The weights w_{ik} to the output layer can similarly be interpreted as the posterior probability of class $\mathcal{C}_i$ given the presence of cluster k, $p(\mathcal{C}_i|\text{cluster } k)$ [69].

8.5.2 The Training Algorithms

In [69] it is shown that the universality property still holds for RBF networks, if the parameters of the Gaussian activation functions are adjustable. Thus an updating rule for RBF parameters has to be included in the learning algorithm. Let us suppose that target outputs are available and that a supervised learning strategy has to be implemented, by using the sum-of-squares error defined in eq. 8.8.

Modified Back-Propagation One possible supervised learning strategy could be derived from extending the Back-Propagation algorithm (section 8.3.2) to update the RBF's parameters. The corresponding expressions of the partial derivatives of the error function with respect to the RBF's parameters have to be evaluated [69, 385] and included in the gradient descent procedure.

This supervised learning yields high precision results, but has some disadvantages. First of all, it is a nonlinear optimization problem, which is in general computationally expensive and can run into local minima of the error function. In addition, if the widths σ_k are not forced to be small, the locality property of the RBFs can be lost, with a high disadvantage for both computational speed and interpretation. More powerful hybrid training algorithms have been developed and have proven to be much faster than the modified Back-Propagation [385].

The Hybrid Learning Strategy The hybrid learning strategy for RBF networks [385] divides the training procedure in two stages.

1. In the first stage the input vectors $\boldsymbol{x}^q$ of the training set alone, without the corresponding target vectors $\boldsymbol{d}^q$, are used to determine the parameters of the RBFs $\Phi_k()$ in the hidden layer. This is $\boldsymbol{\mu}_k$ and σ_k, if the chosen basis functions are the Gaussians in eq. 8.42.
2. During the second stage the basis functions parameters are kept fixed and the weights between the hidden and the output layer are trained with a supervised algorithm as Back Propagation.

The first unsupervised stage serves to allocate network resources in a meaningful way, by placing the unit centers in only those regions of the input space where data are present. The second stage produces the association between regions of the input space and output classes.

The unsupervised part of the training consists of the determination of the receptive field centers $\boldsymbol{\mu}_k$ and widths σ_k. One simple way is to have as many RBF units as training examples [491], but this is not feasible for large data sets. The Gaussian centers $\boldsymbol{\mu}_k$ can also be chosen to represent random subsets of the training space. Another approach starts with one $\boldsymbol{\mu}_k$ on every data point, and selectively removes centers in such a way as to have the minimum disruption of the network performance. In all those cases, the widths σ_k have to be estimated separately and usually are set as the average distance of each cluster to its

Q nearest neighbors. All these algorithms, however, achieve only suboptimal solutions.

A more sophisticated approach [385] uses the *k-means clustering* algorithm, described in chapter 3, for the unsupervised part of the hybrid training procedure. The k-means procedure is used to divide the input space in a number H of clusters, H being decided in advance. The center $\boldsymbol{\mu}_k$ of the k-th RBF unit is set to represent the prototype of input cluster S_k and calculated as the average of the N_k training data points contained in S_k (eq. 8.44). This solution represents a local minimum of the total squared Euclidean distance J (eq. 8.45) between each of the m training patterns and the nearest $\boldsymbol{\mu}_k$ center.

$$\boldsymbol{\mu}_k = \frac{1}{N_k} \sum_{q \in S_k} \boldsymbol{x}^q \tag{8.44}$$

$$J = \sum_{k=1}^{H} \sum_{q \in S_k} \| \boldsymbol{x}^q - \boldsymbol{\mu}_k \|^2 \tag{8.45}$$

The algorithm begins by randomly assigning training points $\boldsymbol{x}^q$ to clusters S_k with $k = 1, \ldots, H$. The prototype $\boldsymbol{\mu}_k$ of each cluster S_k is calculated. Each point $\boldsymbol{x}^q$ is then reassigned to that cluster S_i that has the nearest prototype, $\boldsymbol{\mu}_i : \;\; \| \boldsymbol{x}^q - \boldsymbol{\mu}_i \| = \min_j \| \boldsymbol{x}^q - \boldsymbol{\mu}_j \|$. The prototype of every set S_k is again calculated, and so on until no more changes are observed.

The widths σ_k of the RBF units are determined using various nearest neighbor heuristics, to achieve a certain amount of response overlap between each RBF unit and its neighbors. In this way a smooth and continuous interpolation of the input space is produced [385]. A particularly simple example sets every σ_k to the global average σ of the Euclidean distance between each prototype $\boldsymbol{\mu}_i$ and its closest neighbor $\boldsymbol{\mu}_j$ (eq. 8.46).

$$\sigma = \frac{1}{H} \sum_{i=1}^{H} \| \boldsymbol{\mu}_i - \boldsymbol{\mu}_j \| \qquad \text{with } \boldsymbol{\mu}_j : \; \| \boldsymbol{\mu}_i - \boldsymbol{\mu}_j \| \; = \min_k \; \| \boldsymbol{\mu}_i - \boldsymbol{\mu}_k \| \tag{8.46}$$

There is also an adaptive version of the k-means algorithm. Again the initial clusters S_k are randomly built in the input space. Then for each training input pattern $\boldsymbol{x}^q$ the nearest cluster center $\boldsymbol{\mu}_k^q : \min_j \| \boldsymbol{x}^q - \boldsymbol{\mu}_j^q \|$ is moved by an amount as in eq. 8.47, until no more changes are observed.

$$\Delta \boldsymbol{\mu}_k^q = \eta \; (\boldsymbol{x}^q - \boldsymbol{\mu}_k^q) \tag{8.47}$$

The RBFs in the hidden layer can also be seen as the basis for a mixture of H Gaussians $o(\boldsymbol{x}) = \sum_{j=1}^{H} \alpha_j(\boldsymbol{x}) \; \Phi_j(\boldsymbol{x})$. The parameters $\alpha_j(\boldsymbol{x})$ and $\Phi_j(\boldsymbol{x})$ of the Gaussian mixture model can then be found by maximizing the likelihood function in eq. 8.48, $p(\boldsymbol{x}^q)$ being the probability of training example $\boldsymbol{x}^q$ to occur. Eq. 8.48 can be maximized using a standard maximization technique, like the expectation-maximization algorithm (chapter 2), both with respect to the mixing coefficients $\alpha_j(\boldsymbol{x})$ and to the parameters of the basis functions. At the end only the basis functions $\Phi_j(\boldsymbol{x})$ are kept and the coefficients $\alpha_j(\boldsymbol{x})$ are discarded [503].

$$\mathcal{L} = \prod_{q=1}^{m} p(\boldsymbol{x}^q) \tag{8.48}$$

To train the upcoming single-layer Perceptron, fed with the RBF output values, the delta rule (section 8.2.4) is applied, which converges already much faster than the Back Propagation algorithm applied to a MLP. Moreover, because of the unsupervised training of the hidden layer, an exhaustive representation of the pattern distribution in the input space is available and that makes even faster the convergence of the supervised learning in the second stage of the training procedure.

A similar approach (Counter-Propagation network) has been implemented by using a self-organizing feature map classifier (section 8.6.3) [311], instead of a RBF hidden layer, to represent the prototypes of the hidden layer spanning the input space.

Orthogonal Least Squares Algorithm Another suggested training algorithm for the RBF layer is based on the *orthogonal least squares* technique, where hidden RBF units are sequentially introduced. As first step, every RBF hidden unit is centered on one training input pattern $\boldsymbol{x}^q$ and the final classification is performed. The pattern $\boldsymbol{x}^j$ with the smallest residual error is retained as prototype of the first RBF unit. Then a second unit is added spanning all the $m-1$ left training patterns and again its center is chosen to be the training example $\boldsymbol{x}^i$ with smallest final residual error, and so on. In order to achieve good generalization, the algorithm must be stopped before having the same number of hidden units as training examples. This algorithm though requires very high computational expenses.

8.5.3 Probabilistic Neural Networks

The probabilistic interpretation of the RBF networks can be even more enhanced by using a limited number of connections between the RBF layer and each output unit. Such networks, applied only to classification problems, are called *Probabilistic Neural Networks* (*PNN*).

In a general PNN structure (fig. 8.13), the first hidden layer consists of RBF units, with generalized Gaussians (eq. 8.43) as activation functions. They estimate the posterior probability of the occurrence of cluster k in the input space given input vector $\boldsymbol{x}$: $\Phi_k(\boldsymbol{x}) = p(\text{cluster } k|\boldsymbol{x})$. Considering p possible output classes, h_i clusters in the input space are supposed to represent output class $\mathcal{C}_i$ for $i = 1, \ldots, p$. Thus a number $H = \sum_{i=1}^{p} h_i$ of RBF units is introduced into the hidden layer, each with activation function $\Phi(\boldsymbol{x}) = p_k^i(\boldsymbol{x}) = p(\text{cluster } k|_{k \in \mathcal{C}_i}|\boldsymbol{x})$ with $k = 1, \ldots, h_i$ and $i = 1, \ldots, p$.

The second hidden layer consists of linear units, called *summation units*. Each unit i represents an output class $\mathcal{C}_i$ and collects contributions only from

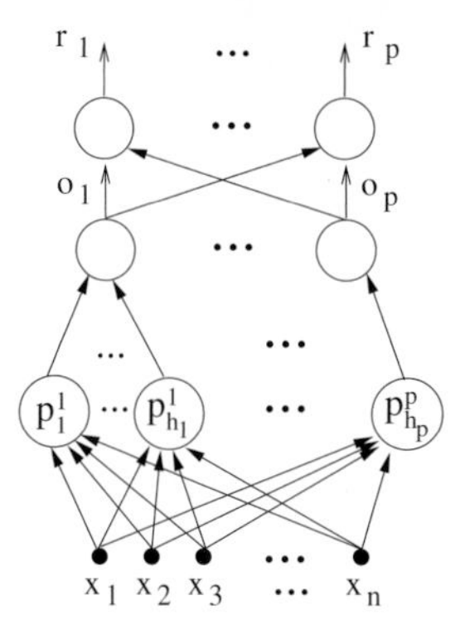

Fig. 8.13. A Probabilistic Neural Network.

the h_i RBF units associated with output class $\mathcal{C}_i$ (eq. 8.49). Summation unit i approximates the posterior probability function $p(\mathcal{C}_i|\boldsymbol{x})$ of class $\mathcal{C}_i$ given input vector $\boldsymbol{x}$, by means of weight w_{ik} connecting RBF unit k and summation unit i and representing the posterior probability $p(\mathcal{C}_i|\text{cluster } k)$ (eq. 8.49). The small number of connections between RBF and summation layer produces an even more localized response of the summation layer to a given input $\boldsymbol{x}$. Because of their probabilistic nature, PNN are generally used to implement statistical classifiers.

$$o_i = \sum_{k=1}^{h_i} w_{ik}\ p_k^i(\boldsymbol{x}) = \sum_{k=1}^{h_i} p(\mathcal{C}_i|\text{cluster } k)\ p(\text{cluster } k|\boldsymbol{x}) = p(\mathcal{C}_i|\boldsymbol{x}) \tag{8.49}$$

If known, it is also possible to assign a cost v_{jl} to the decision of assigning pattern $\boldsymbol{x}$ to class l instead of class j, to which $\boldsymbol{x}$ really belongs. The third layer of *decision units* (fig. 8.13) estimates the global decision risk r_j for class j, by using the decision cost v_{jl} as in eq. 8.50, where p represents the number of output classes, $p(\mathcal{C}_l)$ the a priori probability of class l, and $o_l(\boldsymbol{x})$ the output of summation unit l. $(p(\mathcal{C}_l)\ v_{jl})$ represents the connection weight between summation unit l and risk unit j. The goal is to choose the class C_j with minimum risk r_j [491, 61].

$$r_j = \sum_{l=1}^{p} \hat{w}_{jl}\ o_l(\boldsymbol{x}) \ = \ \sum_{l=1}^{p} v_{jl}\ p(\mathcal{C}_l)\ o_l(\boldsymbol{x}) \tag{8.50}$$

In [491] one RBF unit is introduced to represent each training example. Each RBF unit is then connected with the summation unit of the corresponding class. If the decision unit l in the third layer receives inputs from summation units of class $\mathcal{C}_i$ and $\mathcal{C}_j$, the weight $\hat{w}_{lj}$ to decision unit l from summation unit j is set from the beginning to:

$$\hat{w}_{lj} = -\frac{p(\mathcal{C}_j) r_{\mathcal{C}_j}}{p(\mathcal{C}_i) r_{\mathcal{C}_i}} \frac{n_{\mathcal{C}_i}}{n_{\mathcal{C}_j}} \tag{8.51}$$

where $p(\mathcal{C}_i)$ and $p(\mathcal{C}_j)$ represent the a priori probability respectively of class $\mathcal{C}_i$ and $\mathcal{C}_j$, $r_{\mathcal{C}_i}$ and $r_{\mathcal{C}_j}$ the corresponding decision risk, and $n_{\mathcal{C}_i}$ and $n_{\mathcal{C}_j}$ the number of training examples of the two classes. A negative value of the decision unit means class $\mathcal{C}_j$; a positive value class $\mathcal{C}_i$. The decision is then taken according to the probability ratio, the risk ratio and the representation ratio of the two classes in the training set.

Because of the obvious problems in dealing with large data sets, more recent algorithms [61] dynamically build PNNs with only one layer of RBF units, automatically optimizing the number of required RBF units during training. These approaches start with one Gaussian centered on the first training example. Other Gaussians are added and the existing ones are shrunk or extended, according to the occurrence of new example patterns of the same class or of different classes.

8.6. Competitive Learning

In the ANN paradigms described in the previous sections of this chapter a training set of labeled data is supposed to exist. This is not always the case. Sometimes, even though a large amount of data is available, there is no clue of how those data are organized, not even for a small subset that could work as training set. In this case the training set T consists only of examples of input patterns $\boldsymbol{x}^q$ without a corresponding target answer, that is $T = \{\boldsymbol{x}^q\}$. Without a target answer to learn, the neural network can only discover by itself typical patterns, regularities, clusters, or any other relationships of interest inside the training set, without using any feedback from the environment. This kind of learning is called *unsupervised.*

In order to learn without a teacher, neural networks need some *self-organization* property, that keeps track of previously seen patterns and answers on the basis of some *familiarity* criterion to the current input. The input patterns are grouped into clusters on the basis of each others' similarity and independently from the external environment. It is important to notice that unsupervised learning can happen only if there is *redundancy* in the data, because without an external feedback only redundancy can provide knowledge about the input space properties [158, 43].

Two main philosophies lead the unsupervised learning approaches: the *competitive learning* and the *Hebbian learning.* In the first case the resulting networks are mainly oriented to clustering or classification, while in the second case they are more oriented to measure familiarity or to project the input data onto their principal components.

This section describes some of the most popular ANN learning algorithms based on competitive learning. The best known of them originate from the standard competitive learning strategy, which is unsupervised. However, supervised extensions have also been implemented, such as the Linear Vector Quantization procedure. The most famous of the unsupervised competitive ANN paradigms most likely consists of the Kohonen's Self-Organizing Maps. This model owes its celebrity to its capability of presenting on the output layer a very intuitive description of the similarity among groups of data in the input space. This allows a faithful representation and an easy interpretation of the relationships among input patterns. A subsequent *calibration* phase can always produce labels for the clusters resulting from an unsupervised learning algorithm.

Generally, unsupervised ANN paradigms based on competitive learning have simpler architectures and faster learning algorithms than MLPs. In case of a high dimensional input space, it is often convenient to apply an unsupervised ANN first and only later a MLP, as in the case of the hybrid learning algorithm of Radial Basis Function networks (section 8.5.2). The organization of the input space and the dimensionality reduction provided by the unsupervised network would speed up the MLP's learning process. In addition, even if a labeled training set is available, an unsupervised analysis could still be helpful, to discover further unknown relationships among input patterns as well as to adapt to gradual changes or drifts in the input data.

Hebbian learning will be described in section 8.7.1 among the most frequently used ANN paradigms to perform the Principal Components Analysis (PCA) transform of the input space.

8.6.1 Standard Competitive Learning

The standard competitive learning process is the *winner-take-all* strategy. In a one-layer architecture, the neural units compete to be the winner for a given input pattern and only the winner unit is allowed to fire.

Let us consider a single layer of p units, fully connected to the n input terminals x_j by excitatory feedforward connections, $w_{ij} \geq 0$ (fig. 8.14.c). Each unit i will receive an input value a_i as:

$$a_i = \sum_{j=1}^{n} w_{ij} x_j = \boldsymbol{w}_i^T \boldsymbol{x} \tag{8.52}$$

The neural unit with highest input value a_i (the winner) will be the one to fire, that is with output $o_i = 1$.

$$\begin{cases} o_i = 1 & \text{if: } \boldsymbol{w}_i^T \boldsymbol{x} = \max_{k=1,\ldots,p} (\boldsymbol{w}_k^T \boldsymbol{x}) \\ o_i = 0 & \text{otherwise} \end{cases} \tag{8.53}$$

If all the weight vectors are normalized so that $\| \boldsymbol{w}_i \| = 1$, the winner-take-all condition in eq. 8.53 can also be expressed as:

$$\begin{cases} o_i = 1 & \text{if: } \| \boldsymbol{w}_i - \boldsymbol{x} \| = \min_{k=1,\ldots,p} \| \boldsymbol{w}_k - \boldsymbol{x} \| \\ o_i = 0 & \text{otherwise} \end{cases} \tag{8.54}$$

Conditions 8.53 and 8.54 reward the unit i with weight vector $\boldsymbol{w}_i$ closest to the input vector $\boldsymbol{x}$ as the winning unit. Thus in a winner-take-all strategy, each neuron represents a group of input patterns by means of its weight vector $\boldsymbol{w}_i$. The task of the learning algorithm is to choose such weight vectors as representative prototypes of data clusters in the input space.

Let us start with a random set of values for the weights w_{ij}. At each iteration u of the learning algorithm, when the input pattern $\boldsymbol{x}^q$ from the training set is presented to the network, the winner unit i^q is found by using eq. 8.53 or 8.54 by means of the associated weight vector $\boldsymbol{w}_i^q$. Unit i^q is then chosen to represent again this and similar input patterns in the future. That is the occurrence of training vector $\boldsymbol{x}^q$ during the next cycle $u+1$ of the training procedure on the training set must produce a higher input value a_i^q for unit i^q:

$$(\boldsymbol{w}_i^q(u))^T\ \boldsymbol{x}^q \leq (\boldsymbol{w}_i^q(u+1))^T\ \boldsymbol{x}^q = ((\boldsymbol{w}_i^q(u))^T + (\Delta\boldsymbol{w}_i^q(u))^T\ \boldsymbol{x}^q \qquad (8.55)$$

In order to achieve this goal, only the winner unit i^q is rewarded with a change in the weight vector $\Delta\boldsymbol{w}_i^q$ proportional to the current input $\boldsymbol{x}^q$. This update moves the weight vector $\boldsymbol{w}_i^q$ in the direction of the input vector $\boldsymbol{x}^q$, so that the current winner unit i^q will likely win again in the future for the same input pattern $\boldsymbol{x}^q$. This is represented in the first part of eq. 8.56. To avoid an uncontrolled growth of the weight vector $\boldsymbol{w}_i$ during the whole training process, a second inertial term, proportional to the current value of the weight vector $\boldsymbol{w}_i^q$, is subtracted from the standard competitive learning rule, leading to the updating rule in eq. 8.56. η is a small positive learning constant, heuristically selected usually between 0.1 and 0.7 [560].

$$\begin{cases} \Delta\boldsymbol{w}_i^q(u) = \eta(\boldsymbol{x}^q - \boldsymbol{w}_i^q(u)) & \text{if: } (\boldsymbol{w}_i^q(u))^T\ \boldsymbol{x}^q = \max_{k=1,\ldots,p}\ ((\boldsymbol{w}_k^q(u))^T\ \boldsymbol{x}^q) \\ \Delta\boldsymbol{w}_k^q(u) = 0 & \text{for } k^q \neq i^q \end{cases} \qquad (8.56)$$

By using eq. 8.56 and assuming normalized weight vectors, the proposition expressed in eq. 8.55 is fulfilled, because:

$$(\Delta\boldsymbol{w}_i^q(u))^T\ \boldsymbol{x}^q = \eta((\boldsymbol{x}^q)^T\boldsymbol{x}^q - (\boldsymbol{w}_i^q(u))^T\boldsymbol{x}^q) = \eta(\cos(0) - \cos(\boldsymbol{w}_i^q(u), \boldsymbol{x}^q)) \geq 0 \qquad (8.57)$$

The learning rule proposed in eq. 8.56 also corresponds to an update of the weight in the direction of the negative gradient of the function $f(\boldsymbol{x}) = \parallel \boldsymbol{x} - \boldsymbol{w}_i \parallel$ with respect to $\boldsymbol{w}_i$ [560].

The effect of the competitive learning process is traditionally [261, 560] geometrically described in a three-dimensional space, where normalized vectors lie on the surface of a sphere with radius 1 (fig. 8.14.a). The winner weight vector $\boldsymbol{w}_i$ is the closest one to the input vector $\boldsymbol{x}$. At each learning step, the current input pattern $\boldsymbol{x}^q$ attracts the winner weight vector $\boldsymbol{w}_i^q$ to move closer. If $\boldsymbol{x}^q$ is presented to the network more frequently than other vectors, its influence in the definition of the final positions of the weight vectors $\boldsymbol{w}_k$ is stronger. At the end of the learning process the weight vectors $\boldsymbol{w}_k$ are located inside the clusters they represent (fig. 8.14.b), in a position that depends on the probability of occurrence of the input vectors.

A more biologically inspired model includes lateral inhibitory (negative) and auto-excitatory (positive) connections in the single layer of the network. However oscillations in the network behavior can arise and the easy implementation of the winner-take-all algorithm is lost.

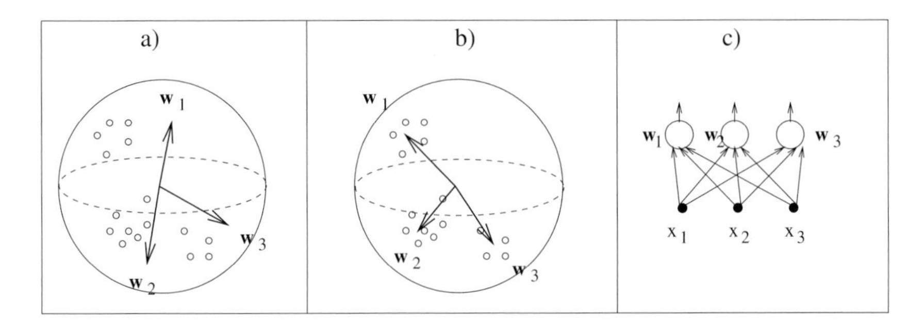

Fig. 8.14. a) The weight vectors at the beginning and b) at the end of the training process for c) a single layer neural structure trained with the standard competitive learning rule.

Some disadvantages are associated with the non distributed nature of the winner-take-all learning approach. Since only one unit is responsible for a given input cluster, the network is not robust against degradation or failure. If one unit fails, then a whole input cluster is lost. For the same reason, at least as many units are necessary as there are clusters in the input space, which can not be known a priori. On the other hand, if too many units are used it can happen that some of them never learn. Those are called *dead units*. As in the previous supervised architectures, criteria to build an a priori optimal architecture are not yet available.

If oversized networks are used, a modified learning strategy of the winner-take-all algorithm, the *leaky* or *soft competitive learning*, allows every unit to learn proportionally to its level of response. The problem of dead units is then partially contained and a more detailed clustering of the input space is obtained.

8.6.2 Learning Vector Quantization

The *Learning Vector Quantization* (*LVQ*) is a *supervised extension* of the winner-take-all learning algorithm. In this case a training set of labeled data is supposed to exist, where each input pattern $\boldsymbol{x}^q$ is associated with its target answer $\boldsymbol{d}^q$ and the number of output classes is known a priori. Generally, it is useful to consider several prototypes per class and consequently networks with a higher number of output units than output classes. A neural network with units with unit length weight vectors and activation function as in eq. 8.54 is considered.

Let us apply the input vector $\boldsymbol{x}^q$ to the network at the u-th iteration of the learning process and find the corresponding winner unit i^q with the associated weight vector $\boldsymbol{w}_i^q$. The weight update occurs as a reward in the direction of $\boldsymbol{x}^q$ if the output class of unit i^q is correct. An update in the opposite direction of the input vector $\boldsymbol{x}^q$ punishes the same unit i^q if its output class is incorrect

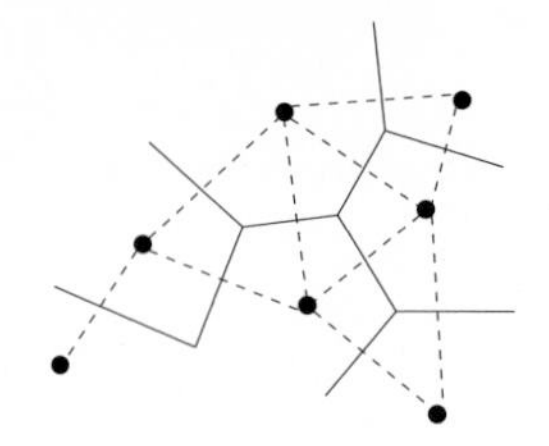

Fig. 8.15. An example for a Voronoi tessellation.

(eq. 8.58) [311]. The weight vectors $\boldsymbol{w}_k$ of non-winner units are not updated.

$$\begin{cases} \Delta \boldsymbol{w}_i^q = +\eta(u)(\boldsymbol{x}^q - \boldsymbol{w}_i^q) & \text{if class of unit } i^q \text{ is correct} \\ \Delta \boldsymbol{w}_i^q = -\eta(u)(\boldsymbol{x}^q - \boldsymbol{w}_i^q) & \text{if class of unit } i^q \text{ is incorrect} \\ \Delta \boldsymbol{w}_k^q = 0 & \text{if unit } k \text{ is not the winner} \end{cases} \tag{8.58}$$

$0 < \eta(u) < 1$ is the learning rate parameter and it is usually taken as a decreasing function of the algorithm iteration u. This allows the network to quickly reach the area of the solution and only subsequently to move around with smaller changes in the weights and find a more refined solution. It is also possible to define individual learning $\eta_i(u)$ for each unit i, recursively calculated from the previous value $\eta_i(u-1)$ [311], but precautions have to be taken to force the learning rate value into $[0, 1]$. If the learning rate value $\eta_i(u)$ stays in $[0, 1]$, the recursive definition of $\eta_i(u)$ provides the optimal learning rate for each unit and speeds up the learning process.

Because the Euclidean metric is used in eq. 8.54 to define the winner unit, at the end of the LVQ's learning process the input space will be divided into a *Voronoi tessellation*, where the class decision boundaries are perpendicular to the distance between two neighboring prototypes (the points in figure 8.15).

An improved version of the LVQ algorithm updates simultaneously the winner unit weight vector $\boldsymbol{w}_i^q$ and the closest prototype vector $\boldsymbol{w}_j^q$ for input pattern $\boldsymbol{x}^q$. That shifts the decision borders towards the Bayes limits (chapter 2) and ensures that each prototype $\boldsymbol{w}_k$ approximates, at least roughly, the distribution of input cluster k [311]. Two cases are distinguished.

1. The class of input vector $\boldsymbol{x}^q$ is different from the class represented by the winner unit i^q associated with prototype $\boldsymbol{w}_i^q$, and is the same as the class of the second highest unit j^q associated with prototype $\boldsymbol{w}_j^q$. Moreover the input vector $\boldsymbol{x}^q$ is sufficiently close to the decision boundary between $\boldsymbol{w}_i^q$ and $\boldsymbol{w}_j^q$. This last condition is described by $\boldsymbol{x}^q$ falling in a window of relative width s, that is if:
$$\min\left(\frac{d(\boldsymbol{x}^q, \boldsymbol{w}_i^q)}{d(\boldsymbol{x}^q, \boldsymbol{w}_j^q)}, \frac{d(\boldsymbol{x}^q, \boldsymbol{w}_j^q)}{d(\boldsymbol{x}^q, \boldsymbol{w}_i^q)}\right) > \frac{1-s}{1+s} \tag{8.59}$$
where $d(\boldsymbol{x}^q, \boldsymbol{w}_k^q)$ represents the Euclidean distance between input pattern $\boldsymbol{x}^q$ and prototype $\boldsymbol{w}_k^q$. $s \in [0.2, 0.3]$ is recommendable. If all these conditions

are satisfied, the updating rule at the training step u is as follows [311]:

$$\begin{cases} \Delta \boldsymbol{w}_i^q(u) = -\eta(u)(\boldsymbol{x}^q - \boldsymbol{w}_i^q(u)) & \text{for the winner unit} \\ \Delta \boldsymbol{w}_j^q(u) = +\eta(u)(\boldsymbol{x}^q - \boldsymbol{w}_j^q(u)) & \text{for the closest neighbor} \\ \Delta \boldsymbol{w}_k^q(u) = 0 & \text{if } k \neq i,j \end{cases} \quad (8.60)$$

2. If $\boldsymbol{w}_i^q$, $\boldsymbol{w}_j^q$, and $\boldsymbol{x}^q$ all belong to the same class, the updating rule becomes as in eq. 8.61, with $\epsilon > 0$.

$$\begin{cases} \Delta \boldsymbol{w}_h^q(u) = +\epsilon\, \eta(u)\, (\boldsymbol{x}^q - \boldsymbol{w}_h^q(u)) & \text{with } h = i,j \\ \Delta \boldsymbol{w}_k^q(u) = 0 & \text{if } k \neq i,j \end{cases}$$

8.6.3 Kohonen's Self-Organizing Maps

Another important unsupervised neural network paradigm for data analysis is the *Kohonen map* [311]. The learning algorithm still follows the competitive model, but the updating rule produces an output layer, where the topology of the patterns in the input space is preserved. That means that if patterns $\boldsymbol{x}^r$ and $\boldsymbol{x}^s$ are close in the input space – close on the basis of the similarity measure adopted in the winner-take-all rule (eq. 8.54) – the corresponding firing neural units are topologically close in the network layer. A network that performs such a mapping is called a *feature map*. Feature maps not only group input patterns into clusters, but also visually describe the relationships among these clusters in the input space.

A Kohonen map consists usually of a two-dimensional array of neurons fully connected with the input vector, without lateral connections (fig. 8.16), arranged on a squared or hexagonal lattice. The latter is generally advised [311], because at the end of the learning process provides an easier visualization of the input space structure.

The topology preserving property is obtained by a learning rule that involves the winner unit and its neighbors in the weight updating process. As a consequence, close neurons in the output layer learn to fire for input vectors with similar characteristics. During training the network assigns to firing neurons a position on the map, based on the dominant feature of the activating input vector. For this reason Kohonen maps are also called *Self-Organizing Maps* (*SOM*).

The LVQ's learning set of rules (eq. 8.58) is changed into:

$$\Delta \boldsymbol{w}_k^q(u) = \eta(u)\; \Lambda(k, i^q, u)(\boldsymbol{x}^q - \boldsymbol{w}_k^q(u)) \quad \text{for all units } k \qquad (8.61)$$

where $\boldsymbol{x}^q$ represents the current training pattern, i^q the corresponding winner unit, $\boldsymbol{w}_k^q$ the weight vector to any unit k of the network, and $\Lambda(k, i^q, u)$ the *neighborhood function* involving the neighbor neurons of the winner i^q into the learning process. $\Lambda(k, i^q, u)$ has its maximum value for $k = i^q$ and decreases with the geometrical distance $\| \boldsymbol{r}_k - \boldsymbol{r}_{i^q} \|$ of unit k from the winner unit i^q on the map. Thus the winner unit i^q still receives the biggest reward, while other units receive a reward proportional to their physical distance from the

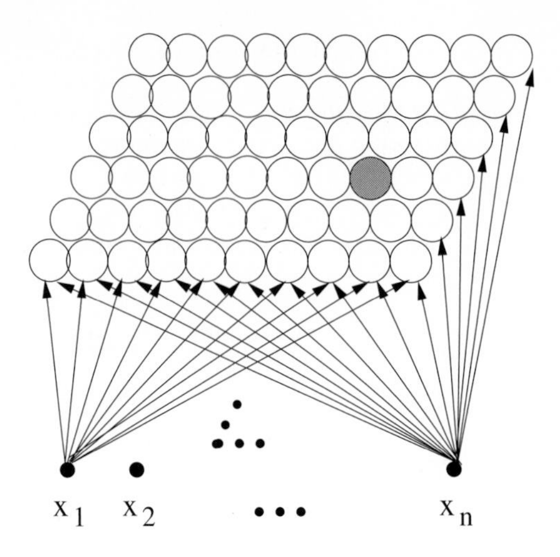

Fig. 8.16. A two-dimensional Kohonen map, with the winning unit in gray.

winner. Units far away from the winner on the map produce a small $\Lambda(k, i^q, u)$ and consequently experience little effect from the presentation of the training pattern $\boldsymbol{x}^q$. The reward consists of moving the prototype $\boldsymbol{w}_i^q$ of the winner unit, as in the winner-take-all learning strategy, as well as the prototype of the closest units towards the input vector $\boldsymbol{x}^q$.

Usually the algorithm starts ($u = 0$) with high values of $\Lambda(k, i^q, u)$ and $\eta(u)$ and reduces them gradually as learning proceeds, for the usual reason of allowing a rapid approaching to the solution. If $\eta(u) \to 0$ the algorithm eventually stops. $\eta(u)$ is usually chosen as $\approx u^{-\alpha}$ with $0 < \alpha < 1$ [311]. Several functions are possible for the $\Lambda(k, i^q, u)$. The most commonly adopted is a Gaussian function (eq. 8.62), where the variance $\sigma(u)$ is some monotonically decreasing function of training step u.

$$\Lambda(k, i^q, u) = e^{\frac{-\|\boldsymbol{r}_k - \boldsymbol{r}_{i^q}\|^2}{2\sigma(u)^2}} . \tag{8.62}$$

Even though a theoretical investigation has not been performed yet, there are several possible equilibrium points for the Kohonen maps. An optimal mapping would be the one that maps the probability density function of the input patterns in the most faithful way on the neural layer. A measure of "optimality" in this sense has not been defined yet, but an average quantization error Q (eq. 8.63) or an average distortion measure D (eq. 8.64) can be calculated, where $\boldsymbol{w}_i^q$ represents the weight to the winner unit i^q for the input pattern $\boldsymbol{x}^q$. Several networks can be trained and eventually the map with the smallest quantization error Q or average distortion D is selected.

$$Q = \frac{1}{m} \sum_{q=1}^{m} \| \boldsymbol{x}^q - \boldsymbol{w}_i^q \|^2 \tag{8.63}$$

$$D = \frac{1}{m} \sum_{q=1}^{m} \Lambda(i^q, i^q, q) \| \boldsymbol{x}^q - \boldsymbol{w}_i^q \|^2 \tag{8.64}$$

In a high dimensional input space, Kohonen maps can be interpreted as nonlinear projectors of the input data onto a two-dimensional neurons array, that takes into account the data probability density and does not change the original topology of the input patterns.

A very interesting example that illustrates the capability of the Kohonen maps to preserve the relationships among the training patterns in the input space is reported in [270]. Words from sentences of the Grimm brothers' tales are encoded and presented to a Kohonen map. During the learning process, the network automatically discovers the underlying grammatical rules and groups the different words on the basis of their grammatical role on the neurons layer. Thus all nouns, verbs, and adverbs excite different regions of the Kohonen map, as illustrated in figure 8.17.

8.7. Principal Components Analysis and Neural Networks

8.7.1 Hebbian Learning

Another unsupervised learning rule, borrowed directly from Hebb's theory [256], is the so called *plain Hebbian rule*. This rule is based on the observation that the more frequently an input pattern is presented to a given neuron, the stronger the corresponding answer is. Meaning, the most frequent pattern defines the neuron's answer.

For a neuron with linear activation function, that is $o = \sum_{k=1}^{n} w_k x_k = \boldsymbol{w}^T \boldsymbol{x}$, the previous physiological observation can be translated into the learning rule in eq. 8.65. The weight vector $\boldsymbol{w}^q$ is forced to follow the current input vector $\boldsymbol{x}^q$ according to the strength of the current answer o^q and the learning rate η.

$$\Delta w_k^q = \eta \; o^q \; x_k^q \tag{8.65}$$

A modified version of the plain Hebbian rule (eq. 8.66) introduces a weight decay proportional to $(o^q)^2$, that constrains the weight vector inside a sphere of radius $\| \boldsymbol{w} \| = 1$ preventing its uncontrolled growth.

$$\Delta w_k^q = \eta \; o^q \; (x_k^q - o^q \; w_k^q) \tag{8.66}$$

If this single neuron is adopted, the learning rule in eq. 8.66 converges to a vector solution $\boldsymbol{w}^*$ approaching the unit length and lying on the direction of maximum variance of the training set [261, 152]. Considering zero-mean data,

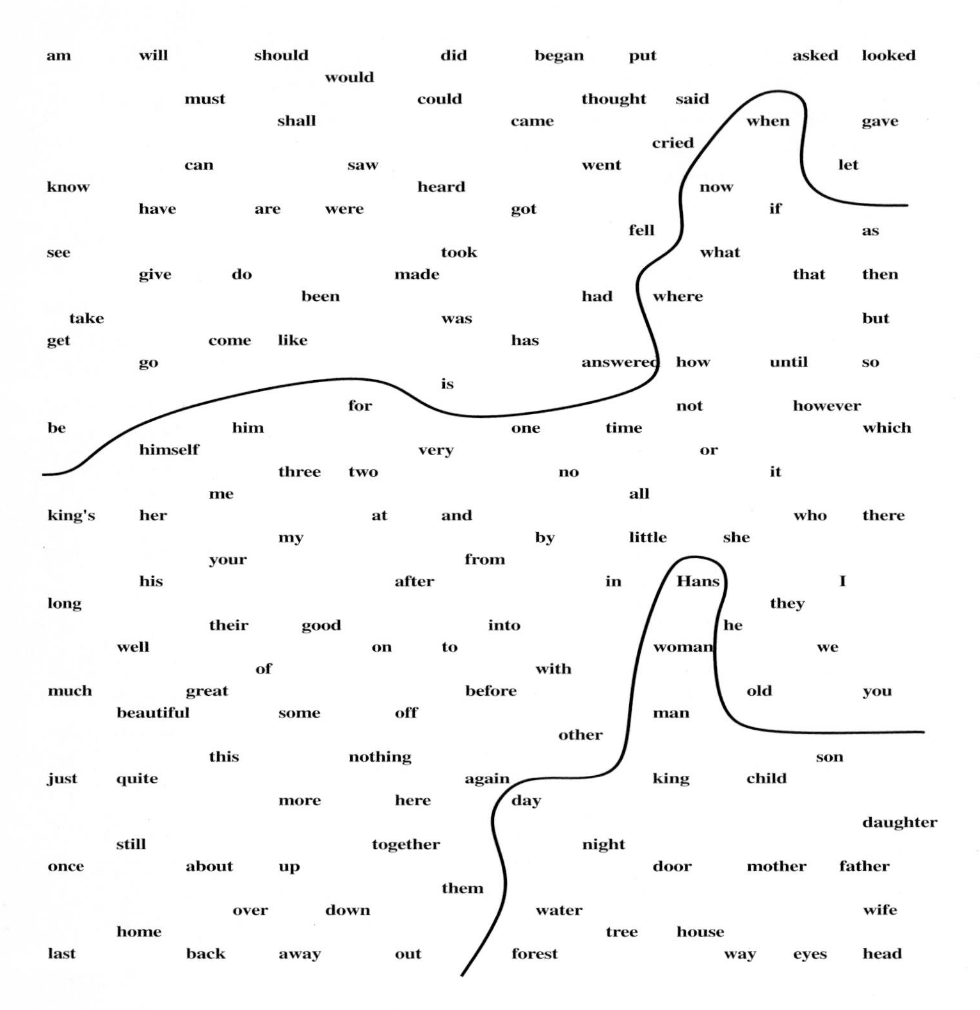

Fig. 8.17. A Kohonen map organizes words according to their grammatical role (reproduced from [270]).

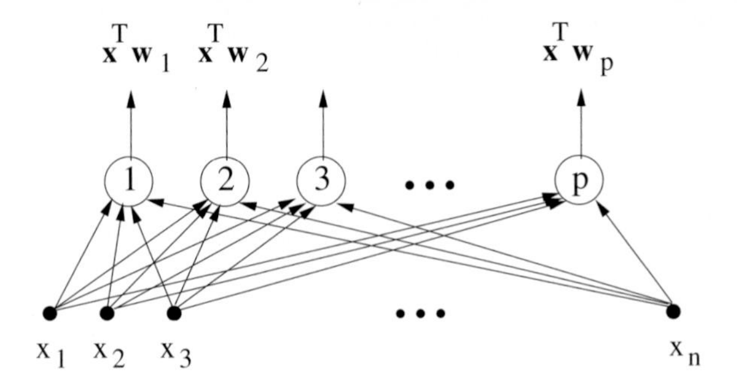

Fig. 8.18. Projection of the input vector onto the first p principal components.

the maximum variance direction is given by the eigenvector associated with the maximum eigenvalue of the covariance matrix, built on the training set (chapter 3). Such maximal eigenvector represents the first principal component direction in the input space. Thus, at the end of the training process, the neuron projects the input vector $\boldsymbol{x}$ on the direction of the maximum eigenvalue of the covariance matrix, i.e. it produces as output the first principal component of vector $\boldsymbol{x}$.

Introducing one layer of p such artificial neurons and appropriately modifying the learning rule, it is possible to project the input vector on the first p principal component directions of the training set. In figure 8.18 the resulting architecture is shown. $\boldsymbol{w}_i$ is the weight vector to unit i, that at the end of the learning process will represent the direction of the i-th principal component of the training set.

The learning rule for this network is called *Sanger's rule* and generalizes the weight update of eq. 8.66 as follows:

$$\Delta w_{ik}^q = \eta\ o_i^q \left(x_k^q - \sum_{j=1}^{i} o_j^q\ w_{jk}^q \right) . \tag{8.67}$$

Starting from the learning rule described in eq. 8.66 and proceeding by induction, it is possible to prove that, by applying Sanger's rule, the weight vectors $\boldsymbol{w}_i$, for $i = 1, \ldots, p$, converge exactly to the first p principal component directions of the training set, ranked according to the corresponding eigenvalues of the covariance matrix [261, 457]. Hence this network performs exactly the *Principal Component Analysis* (*PCA*) transformation described in chapter 3.

A similar learning rule (0ja's rule) extends the sum in eq. 8.67 up to all p neurons. In this case, however, the p weight vectors $\boldsymbol{w}_i$ form a different basis of the subspace spanned by the first p principal component directions, and they do not any longer represent the principal component directions themselves.

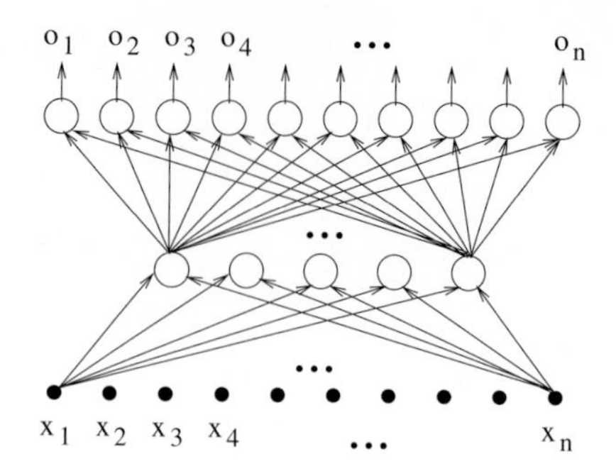

Fig. 8.19. The autoassociator architecture.

8.7.2 The Autoassociator Scheme

An interesting neural architecture, based on the Back-Propagation learning algorithm, is the *autoassociator* scheme (fig. 8.19). The autoassociator structure presents an equal number $n = p$ of input parameters and output units, and a hidden layer usually with fewer H units than input parameters ($H < n$). During the training procedure, whose mathematical properties are investigated in [66, 77], the network is trained to reproduce the input pattern on the output layer, that is to implement the identity mapping with target answer $\boldsymbol{d}^q = \boldsymbol{x}^q$. The network is said to operate in an *autoassociative* mode.

Such structure can be easily used for a two-class discrimination problem. The training set is built with examples of only one of the two output classes. During the test phase a distance between the input and the output vector is calculated. A small distance means that the network successfully reproduced the input pattern on the output layer. Therefore we can assume that the input pattern belongs to the output class included in the training set. A large distance means that the network was not able to reproduce the current input pattern on the output layer. Thus it is very likely that it belongs to the output class not included in the training set. Moving to a multiclass discrimination task, the same autoassociator scheme, appropriately modified, can be used for uncertainty characterization [482].

The autoassociator scheme can also be used to compress or reduce the input space dimension. If the number H of hidden units is lower than the number of input parameters n ($H < n$), the hidden layer is forced to code the input vector with fewer parameters than n. This indeed performs a dimensionality reduction of the input space. After training, the network can split in two parts: 1) the input and the hidden layer; 2) the hidden and the output layer. The first part performs a dimensionality reduction $f : \mathcal{R}^n \rightarrow \mathcal{R}^H$ of the input space, with reduction rate n/H. The second part implements the inverse function $g \approx f^{-1} : \mathcal{R}^H \rightarrow \mathcal{R}^n$, in order to rebuild the original input vector $\boldsymbol{x} \in \mathcal{R}^n$ from the output vector of the hidden layer $\boldsymbol{y} \in \mathcal{R}^H$.

If the autoassociator in figure 8.19 is trained to minimize the sum-of-squares error (eq. 8.8), the Back-Propagation learning procedure is equivalent to the minimization of the quadratic error of the input space reconstruction in the PCA transformation (chapter 3). It can be shown indeed [77] that a MLP, structured as an autoassociator, with linear activation functions and trained by using the sum-of-squares error, implements on its hidden layer a linear mapping $f : \mathcal{R}^n \rightarrow \mathcal{R}^H$, projecting the input data onto an H-dimensional subspace, spanned by the first H directions of maximum variance of the training set, i.e. by the first H principal components. The same result can be shown [77] also in the case of hidden units with nonlinear activation (sigmoidal) functions, since a one-layer Perceptron, with either linear or nonlinear activation functions, can only implement linearly separable mappings (section 8.2.5).

However there are two main differences between the standard technique for PCA, based on singular value decomposition (chapter 3), and the autoassociator scheme. The standard PCA transformation produces an ordered set of eigenvalues, each one representing the variance along the corresponding eigenvector of the covariance matrix of the training set, that is along the corresponding principal component direction. In the autoassociator structure the obtained first H principal components are not ordered, because the vectors of weights from the input to the hidden layer do not need to be orthogonal or normalized. Moreover standard PCA techniques are guaranteed to produce a solution in a finite time, which might not be true for a MLP, if its learning process does not converge.

8.7.3 Non-Linear Principal Components Analysis

In the PCA procedure the two functions – $f()$ for the dimensionality reduction and $g() \approx f^{-1}()$ for the input space reconstruction – are assumed to be linear. In some cases however a linear mapping may not be sufficient to detect the most efficient data representation [152]. In the most general case both the coding function $f : \mathcal{R}^n \rightarrow \mathcal{R}^H$ and the decoding function $g : \mathcal{R}^H \rightarrow \mathcal{R}^n$ with $H < n$ ought to be nonlinear. The minimization of the PCA quadratic reconstruction error (chapter 3), with $f()$ and $g()$ nonlinear functions, generates what is called the *Nonlinear Principal Components Analysis (NLPCA)* procedure.

In order to implement a NLPCA transform, a one-layer feedforward neural network, as the first part of the autoassociator scheme, is no more sufficient because one-layer MLPs can only implement linear input/output mappings. It becomes necessary to introduce a new hidden layer between the input vector and the hidden layer and between the hidden and the output layer of the autoassociator scheme in figure 8.19, to implement a nonlinear coding function. The resulting network is shown in figure 8.20.

This four-layer architecture is still trained on the basis of the sum-of-squares error (eq. 8.8) as an autoassociator, that is to reproduce the input vector on the output layer. The output units in layer 4 are linear, while units in layers 1 and 3 have sigmoidal nonlinear activation functions. The H units in layer 2 can either be linear or nonlinear. Layers 1 and 3 have the same number of units.

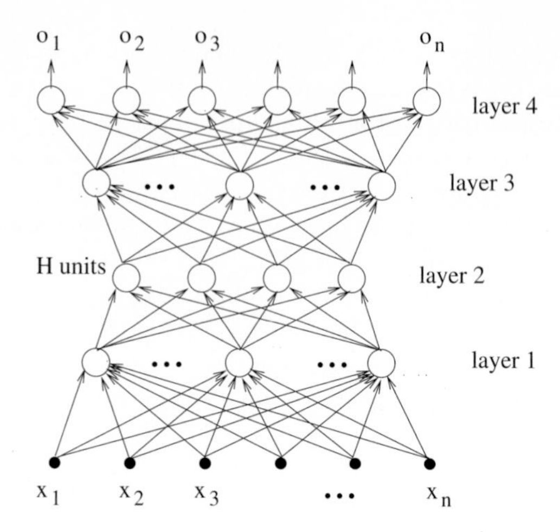

Fig. 8.20. The four-layer autoassociator architecture for NLPCA.

After training, the first two layers of the network perform a *nonlinear* transformation of the input data, projecting the original n-dimensional input vector into an H-dimensional subspace. Layers 3 and 4 of the network define an inverse mapping from the H-dimensional space into the original n-dimensional input space. Because of the sum-of-squares error function (eq. 8.8) adopted during the Back-Propagation training procedure, such nonlinear mappings define exactly the NLPCA transform [152, 69].

The NLPCA transform is able to produce an effective dimensionality reduction also in those problems, where PCA is unable to detect the most informative directions, due to the linearity of its two mapping functions $f()$ and $g()$. On the other hand, the ANN structure for NLPCA is more complicate than the autoassociator in figure 8.19, involving a much higher number of parameters into the error minimization procedure. The training becomes computationally more expensive and the risk of getting stuck in suboptimal local minima increases. As in the previous case, the final nonlinear principal components as output of layer 2 are not ordered and the number H of nonlinear principal components has to be set in advance, with the risk of a too strong or too mild reduction of the input dimensionality.

8.8. Time Series Analysis

In the ANN paradigms described in the previous sections, the output vector is always a direct consequence of the input vector. This kind of neural paradigm is called *static*. The neural network calculates the output vector $\boldsymbol{o}(t+1)$ at time $t+1$ based *only* on the corresponding input vector $\boldsymbol{x}(t)$ at time t and does not keep memory of the input patterns $\boldsymbol{x}(t-\tau)$ previously presented to the network at times $t-\tau$ with $\tau > 0$. Such an approach is usually sufficient to solve static association problems, like classification, where the goal is to produce only one output pattern for each input pattern. In this case the network has only one possible equilibrium point, that is only one possible output vector, produced by the feedforward propagation of the input stimulus through the neural architecture to the output layer.

However, the evolution of the input pattern $\boldsymbol{x}(t)$ in the last F time steps can be more interesting than its state at the final time t. It can be more significant to know how a given state was reached than the final state itself. In this case the sequence $S_x = [\boldsymbol{x}(t), \boldsymbol{x}(t-\Delta), \ldots, \boldsymbol{x}(t-(F-1)*\Delta)]^T$ becomes the subject of the investigation. F is the length and $\boldsymbol{x}(t-\tau)|_{\tau=0,\ldots,F-1}$ the frames of sequence S_x ($\boldsymbol{x}(t-\tau) \in \mathcal{R}^n$). In the following part of this section, $\Delta = 1$ will be considered without loss of generality.

In this section some ANN paradigms will be described, capable of generating an output pattern or an entire sequence of output patterns from a sequence of input patterns. The goal is to teach an appropriate neural network on one side to recognize sequences of input vectors and on the other side to generate the corresponding, possibly sequential, output. This kind of ANN paradigms are called *dynamic*.

In real world applications, several examples can be found, where the observation of the evolution of time sequences is necessary for the final action. Quite often in medicine the patient's clinical frame $\boldsymbol{x}(t)$ at a given time t does not say as much as the global evolution of its clinical conditions, in which the current frame at time t is included. In this case one particular output pattern (the diagnosis) is required at the end of the analyzed sequence. Such task is usually referred to as *sequence recognition*.

Another example could consist of a speech recognition problem. The recognition of a given phoneme $\boldsymbol{x}(t)$ is highly dependent on the entire sequence S_x of phonemes (the word), in which the currently analyzed phoneme is included. After the whole word utterance has been presented to the network, it could be required that the same clean word or another utterance is produced on the output layer. A particular output sequence is required to be associated with the presented input sequence. This represents a *temporal association* task [261].

The first task (sequence recognition) can also be performed by using static neural structures. In this case the whole sequence can be unfolded as a long input vector and any of the previously described supervised learning algorithms can be used to teach the network to perform a static association between such input vector and the corresponding target output pattern. In particular, derivations of the Back-Propagation algorithm applied to feedforward multilayer neural networks are usually adopted. Such methods, though, are computationally expensive and also miss the point of performing a dynamic input/output mapping, that is the recognition of the frame evolution inside the input sequence.

The second task is a bit more complex. For a given input the network has to go through a predetermined sequence of states, which produces a corresponding sequence of output vectors, instead of settling down immediately into an equilibrium point. To obtain such an evolution process of the network state, backward connections from units in a layer l to units in a previous layer $h < l$, crossward connections across units in the same layer, and autoconnections of a given unit with itself are introduced into the feed-forward structure. Back, auto, and crossward connections need some relaxation time before reaching the equilibrium state, which allows information to be held from previous input patterns. Neural architectures, which present cross and/or auto and/or backward connections, besides the traditional feedforward connections, are called *Recurrent Neural Networks* (*RNN*). RNNs have shown to be adequate to dynamic learning tasks and, in some cases, to appropriately characterize different dynamic evolutions of the input patterns [212].

The dynamical learning problem can be stated as the definition of a mapping function that associates the space of input sequences $\{S_x\}$ to the space of output sequences $\{S_o\}$, based on a training set T of m examples of pairs of input sequence, $S_x^q = \{\boldsymbol{x}(t)^q, \ldots, \boldsymbol{x}(t-(F-1))^q\}$ and target output sequence $S_d^q = \{\boldsymbol{d}(t)^q, \ldots, \boldsymbol{d}(t-(F-1))^q\}$, as defined in eq. 8.68. The length F of the input sequence S_x^q and of the corresponding output sequence S_d^q can be fixed or variable depending on the adopted learning algorithm. Frames $\boldsymbol{d}(t)$ in the target output sequence S_d^q can not be available for every $t-\tau = t-F+1, \ldots, t$. Indeed, it can be reasonable to supervise the training only in correspondence of some key points $t-\tau_j$ inside the input sequence S_x^q.

$$T = \{(S_x^q, S_d^q) \qquad \text{for } q = 1, \ldots, m\} \tag{8.68}$$

8.8.1 Time Delay Neural Networks

One possibility for performing sequence recognition consists of feeding a static neural network, usually a MLP, with an input vector with delayed values of the input signal, that is $[\boldsymbol{x}(t), \boldsymbol{x}(t-1), \ldots, \boldsymbol{x}(t-F+1)]^T$, as shown in figure 8.21. The entire input sequence S_x is presented as a single input pattern to the MLP by using delay units Δ to hold the past values of $\boldsymbol{x}(t)$. The MLP has to learn a mapping from the input pattern, representing the time sequence $S_x^q \in T$ to the desired output pattern $\boldsymbol{d}^q$, which is in this case not a sequence. The traditional Back-Propagation algorithm (section 8.3.2) can then be used to train neural architectures of this kind, called *Time Delay Neural Networks* (*TDNN*) [261, 212].

The length F of sequences S_x can not be variable and has to be defined before the TDNN design. To save storage space and to decrease computational expenses, an architecture where input delayed values close in time share the same weights can be used. The technique of sharing weights also allows the output values of the network to be shift-invariant for small time shifts in the input sequences.

TDNNs still perform a static association mapping of the input to the output pattern. In fact they do not exploit the time evolution of the input sequence, but for each S_x^q retrieve the output pattern associated with the prototype of the most frequent input vector configuration. Also the input sequence S_x can not be too long, because of the resulting unmanageable size of the TDNN and of the slow computation time.

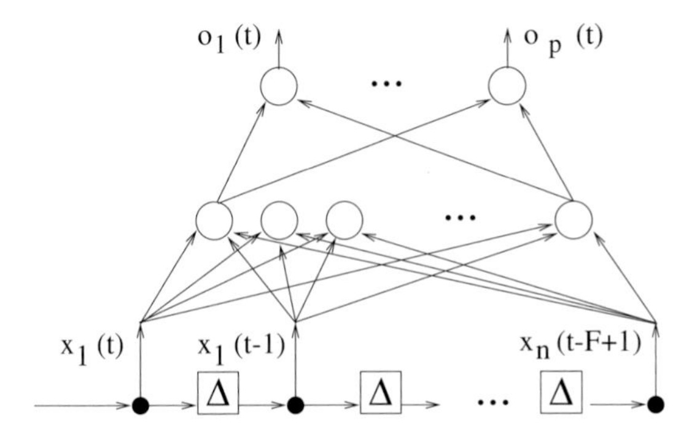

Fig. 8.21. The structure of a TDNN.

8.8.2 Back-Propagation Through Time

A more successful approach to the analysis of time series by using neural networks relies on Recurrent Neural Networks (RNN). A neural network is called recurrent, if cross, auto and backward connections are allowed. A very simple recurrent architecture is shown in figure 8.22.a with only one layer of neural units fully cross and autoconnected and no backward connections. For every input pattern $\boldsymbol{x}(t-\tau)$ of the sequence $S_x^q = \{\boldsymbol{x}^q(t), \boldsymbol{x}^q(t-1), \ldots, \boldsymbol{x}^q(t-F+1)\}$ the output value of each unit $o_i(t-\tau)$ evolves as:

$$o_i(t-\tau) = f\left(a_i(t-\tau-1)\right) = f\left(\sum_j w_{ij} o_j(t-\tau-1) + \chi_i(t-\tau-1)\right) \quad (8.69)$$

where the connection weights w_{ij} include also values for $i = j$ (autoconnections) and $\chi_i(t-\tau)$ represents that part of the input $a_i(t-\tau)$, if any, coming directly from the input vector $\boldsymbol{x}(t-\tau)$. The output vector $\boldsymbol{o}(t)$ then evolves in time according to the dynamics of the network and to the input sequence S_x.

An easy approach for dealing with RNNs, without introducing new learning algorithms, consists of turning an arbitrary RNN (fig. 8.22.a) into an equivalent static feedforward neural network (fig. 8.22.b), trainable with the Back-Propagation algorithm (section 8.3.2). This approach is called *Back-Propagation through Time*. The basic idea is to use F copies of the network to hold the output values at each delay τ, for $\tau = 0, \ldots, F-1$ [455]. That is for input sequences $S_x^q = \{\boldsymbol{x}^q(t), \ldots, \boldsymbol{x}^q(t-F+1)\}$ of length F, a copy of the network is produced for each time $t-F+1, t-F+2, \ldots, t$.

For example, considering input sequences of length $F = 3$, the RNN in figure 8.22.a has been translated into $F = 3$ consecutive feedforward neural networks, each one holding the network's outputs at step $t-\tau$, beginning with $\tau = 2$ (fig. 8.22.b). The weight connections w_{ij} are supposed to be the same in all the feedforward copies of the RNN and are independent of delay τ, because during learning they change only at the end of the presentation of sequence S_x^q.

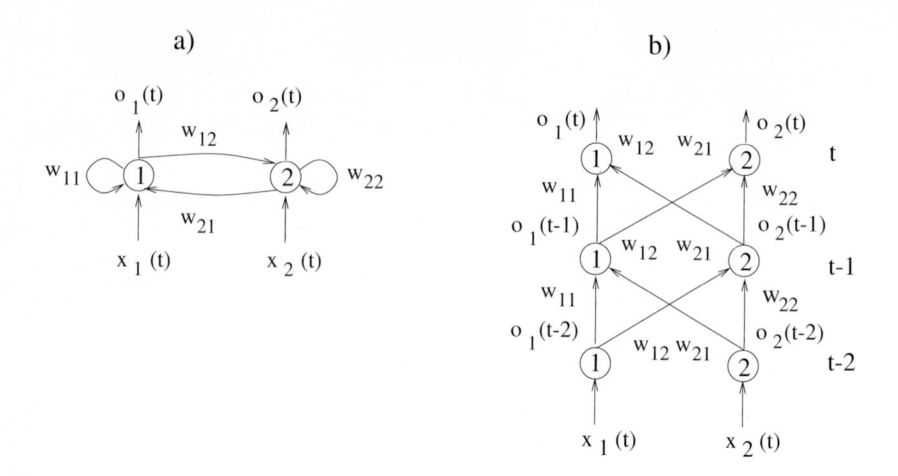

Fig. 8.22. The recurrent network in a) is equivalent to the feedforward network in b) from delay $\tau = 2$ to delay $\tau = 0$.

Under these constraints the two networks (fig. 8.22.a and 8.22.b) show the same behavior for $t - \tau = t - 2, t - 1$, and t.

The resulting unfolded feedforward network can be trained with Back-Propagation. In this case desired answers $d_i(t - \tau)$ can also be known for hidden units, as the supervision values inside the target output time sequence S_d^q. The error values from the hidden layers, as the error values from the output layer, are then back-propagated to the input vector during the learning process.

After the presentation of the input sequence S_x^q, in general Back-Propagation will supply different increments Δw_{ij}^{τ} for each w_{ij}^{τ} of the F instances of the unfolded network. However, after the weight update w_{ij}^{τ} have to be all the same through the feedforward copies of the original network. Usually a global increment $\sum_{\tau=1}^{F} \Delta w_{ij}^{\tau}$ is considered to update all the copies of weight w_{ij}.

Back-Propagation through time can only be used with very short input time sequences with fixed length F. For large values of F the transformation of the RNN requires a huge corresponding feedforward network, with consequent problems of data storage and algorithm's convergence speed.

8.8.3 Recurrent Learning

In the previous sections (sections 8.8.1 and 8.8.2), we have seen that going back to a static association problem and then applying the Back Propagation algorithm does not allow the investigation of input spaces with long sequences of data and does not really help in recognizing the dynamical properties of the sequence input space $\{S_x\}$. On the other hand, even though RNNs seem to represent the most appropriate tool for sequence analysis, their training algorithms are usually computationally expensive due to a large network topology.

If the sequence analysis problem is not too complicated, *Partially Recurrent Neural Networks* (PRNN) can be sufficient. In these networks the connections are mainly feedforward and only a subset of units include back, auto, and crossconnections. This subset of recurrent units, usually called *context units*, allows the network to memorize cues from the recent past and do not exceedingly complicate the structure of the network. In most cases [261] the recurrent connections are fixed, that is untrainable, and Back-Propagation (section 8.3.2) can be used as training algorithm for the remaining feedforward connections. An appropriate design of the PRNN structure and an appropriate initialization of the weights can also simplify the learning task [193].

However for real dynamical problems, where the observation of the dynamics of signal $\boldsymbol{x}(t)$ is necessary, dynamical learning algorithms for RNN structures need to be used [416, 422, 544, 212]. One of the first recurrent learning algorithms for RNN structures extends Back-Propagation to RNNs in continuous time [422]. This approach still uses RNNs for the solution of static problems, where the output of the system is considered at the fixed point. Other dynamical training algorithms for RNN architectures have been more specifically developed to solve dynamical problems. In this section two classic recurrent learning algorithms are described. For more details on the most recent developments in RNN learning algorithms see [212].

Real Time Recurrent Learning. This learning algorithm derives the gradient descent procedure (section 8.3.2) for recurrent neural architectures [544].

Let us suppose a RNN structure with n input values, p output units, and fed with training input sequences $S_x^q, q = 1, \ldots, m$ each of arbitrary length $F(q)$. The goal is to learn the association mapping between input sequences S_x^q and target sequences S_d^q. The error function in equation 8.70 describes how well the network implements this task for each training sequence $S_x^q \in T$.

$$E^q = \sum_{\tau=F(q)-1}^{0} E^q(t-\tau) = \sum_{\tau=F(q)-1}^{0} \left(\frac{1}{2} \sum_{i=1}^{p} \left(e_i^q(t-\tau)\right)^2 \right) \tag{8.70}$$

$$e_i^q(t-\tau) = \begin{cases} o_i^q(t-\tau) - d_i^q(t-\tau) & \text{if: } i \in U(\tau, q) \\ 0 & \text{otherwise} \end{cases} \tag{8.71}$$

where $U(\tau, q)$ represents the set of output units with a specified target $d_i^q(t-\tau)$ at delay τ for training example q. The goal of the learning procedure is the minimization of the error function E^q, by updating the weights of the network w_{ij} along the negative direction of its gradient $\nabla_{\boldsymbol{W}} E^q$, as follows:

$$\Delta w_{ij}^q = -\eta \frac{\partial E^q}{\partial w_{ij}} = \sum_{\tau=F(q)-1}^{0} -\eta \frac{\partial E^q(t-\tau)}{\partial w_{ij}} = \sum_{\tau=F(q)-1}^{0} \Delta w_{ij}^q(t-\tau) \tag{8.72}$$

The global updating Δw_{ij}^q is the result of the sum of the partial contributions $\Delta w_{ij}^q(t-\tau)$ collected during the presentation of the input sequence S_x^q to the network. η is the learning rate, as usual a positive parameter of small value.

By using eq. 8.69 as a description of the dynamics of the neural units, the recursive formula in eq. 8.73 can be obtained for the calculation of the derivative in eq. 8.72, as reported in [261].

$$\frac{\partial o_k(t)}{\partial w_{ij}} = f_k'(a_k(t)) \left[\delta_{ki}\, o_j(t) + \sum_h w_{kh} \frac{\partial o_h(t)}{\partial w_{ij}} \right] \tag{8.73}$$

$$\delta_{ki} = \begin{cases} 1 \text{ if } i = k \\ 0 \text{ if } i \neq k \end{cases} \tag{8.74}$$

where δ_{ki} in this case denotes the Kronecker delta (eq. 8.74), $f_k(a_k(t))$ the activation function of neural unit k, and index h in the sum in eq. 8.73 marks the units feeding unit k. In general $\frac{\partial o_k(t)}{\partial w_{ij}}$ depends on the weights w_{ij} at time t, because the network is supposed to generate a time sequence of events on the output layer. But we can reasonably assume that at time $t - (F(q) - 1)$ the network is in a stable condition and $\frac{\partial o_k(t-F(q)+1)}{\partial w_{ij}} = 0$. Starting from this initial condition and iterating eq. 8.73, it is possible to calculate the appropriate $\Delta w_{ij}^q(t-\tau)$ for each step $t - \tau$ of the q-th input sequence S_x^q. At the end of the input sequence presentation the global weight update is $\Delta w_{ij}^q = \sum_{\tau=F(q)-1}^{0} \Delta w_{ij}(t-\tau)$ as in eq. 8.72.

An important variation of this algorithm updates the network weight w_{ij} at each step $t - \tau$, by using $\Delta w_{ij}^q(t-\tau)$. This allows real time training, because weights can be updated while the network is running, but does not guarantee anymore that the weight update follows the gradient descent direction. However the practical differences between updating weights at the end of each sequence or at each delay τ inside each sequence become smaller and smaller as the learning rate η is made smaller [544]. This variation of the algorithm is called *Real Time Recurrent Learning.* This approach allows the storage of only as many values as the current length of the sequence $F(q)$. The algorithm though remains quite computationally expensive, due to the calculation of all the derivatives $\frac{\partial o_h(t)}{\partial w_{ij}}$ before each weight update.

The Forward-Backward Technique. This approach extends the Recurrent Back-Propagation learning algorithm from static association tasks to temporal sequence analysis [416]. Here the equation that describes the unit's evolution is taken as:

$$\gamma_i \frac{do_i}{dt} = -o_i + f\left(\sum_j w_{ij} o_j\right) + \chi_i \tag{8.75}$$

where $f()$ is the activation function, o_i the output value, χ_i the external input, and γ_i is the relaxation time scale of neuron i. The error function in eq. 8.8 is integrated through the time sequence length $F(q)$, as in eq. 8.76, in order to perform sequence analysis in continuous time.

$$E^q = \frac{1}{2} \int_{t-F(q)}^{t} \sum_{k=1}^{p} (o_k(t) - d_k(t))^2 dt \tag{8.76}$$

By minimizing the error measure with a gradient descent technique, the weight update in eq. 8.77 is obtained after each training sequence S_x^q presentation.

$$\Delta w_{ij}^q = -\eta \frac{\partial E^q}{\partial w_{ij}} \tag{8.77}$$

The functional derivatives $\frac{\partial E}{\partial w_{ij}}$ for the descent gradient procedure can be derived through the solution of the dynamical equation [261, 212] in eq. 8.78.

$$\frac{\partial E^q}{\partial w_{ij}} = \frac{1}{\gamma_i} \int_{t-F(q)}^{t} Y_i f' \left(\sum_j w_{ij} o_j \right) o_j d\tau \tag{8.78}$$

Y_i can be obtained as the fixed point of an equation with similar form and same parameters as eq. 8.78.

$$\gamma_i \frac{dY_i}{dt} = -\frac{1}{\gamma_i} Y_i + \sum_{k=1}^{p} \frac{1}{\gamma_k} f' \left(\sum_j w_{kj} o_j \right) w_{ki} Y_k + e_i(t-\tau) \tag{8.79}$$

where $e_i(t-\tau) = d_i(t-\tau) - o_i(t-\tau)$ and $Y_i(t) = 0$ for all output units i.

Thus the learning procedure consists of two steps, to be performed after the presentation of each training example (S_x^q, S_d^q).

1. Eq. 8.75 is integrated *forward* between $t - F(q)$, and t and the resulting output functions $o_i(t)$ are obtained.
2. The error functions $e_i(t-\tau)$ are calculated $\forall \tau = 0, \ldots, F(q)$ and for all output neurons i.
3. Eq. 8.79 is integrated *backward* between t and $t - F(q)$, to obtain $Y_i(t)$ for each network unit.
4. The partial derivatives $\frac{\partial E^q}{\partial w_{ij}}$ are calculated according to eq. 8.78.
5. The network weights are updated as in eq. 8.77.

The weight updates can also be accumulated in $\sum_q \Delta w_{ij}^q$, and the final weight update can be performed after the whole training set has been presented to the network.

This procedure, called the *forward-backward* technique, requires less computational expenses than the Real Time Recurrent Learning algorithm, but can not be performed in real time.

8.9. Conclusion

In this chapter a brief overview of the most commonly used ANN paradigms for data analysis is reported. Starting with the description of the first and still most commonly used ANN architecture, the Multilayer Perceptron, other ANN paradigms are also described, to provide solutions for MLP's main drawbacks. Radial Basis Function networks are introduced to allow an easier interpretation of the system's decisions. Some of the unsupervised learning algorithms proposed in the literature are reported, to discover unknown relationships among input patterns and in some case, as in the Kohonen maps, to produce an intuitive representation of them on the network output layer. Finally cross, auto, and backward connections transform the ANN structure from static to dynamic, allowing the recognition and the production of time sequences with different dynamics evolution (chapter 6).

ANN paradigms are all roughly based on the simulation of some specific feature of biological nervous systems. Nevertheless, in the last years an equivalence has been established between many ANN paradigms and statistical analysis techniques (chapters 2 and 3). For example a two-layer Perceptron implements a particular case of nonlinear regression. The autoassociator structure projects the input data onto the space spanned by the Principal Components of the training set. Radial Basis Function networks, if appropriately trained, can be interpreted as building statistical boundaries among the input patterns. Also unsupervised learning techniques can approximate Bayesian decision borders, as in the LVQ's case.

One of the main drawbacks of ANN paradigms consists of the lack of criteria for the a priori definition of the optimal network size for a given task. The space generated by all possible ANN structures with different size for a selected ANN paradigm can then become the object of other data analysis techniques. Genetic algorithms (chapter 10), for example, have been recently applied to this problem, to build a population of good ANN architectures with respect to a given task.

ANNs' decision processes remain still quite opaque and a translation into meaningful symbolic knowledge hard to perform. On the contrary, fuzzy systems (chapter 9) are usually appreciated for the transparency of their decisional algorithms. Neural networks which implement fuzzy rules could avoid the opacity of the ANN results. The combination of the ANN approach and of fuzzy logic has produced hybrid architectures, called neuro-fuzzy networks. Learning rules are no more constrained into the traditional crisp logic, but exploit the linguistic power of fuzzy logic.

Chapter 9
Fuzzy Logic

Michael R. Berthold
Data Analysis Research Lab, Tripos Inc., USA

9.1. Introduction

In the previous chapters a number of different methodologies for the analysis of datasets have been discussed. Most of the approaches presented, however, assume precise data. That is, they assume that we deal with exact measurements. But in most, if not all real-world scenarios, we will never have a precise measurement. There is always going to be a degree of uncertainty. Even if we are able to measure a temperature of 32.42 degrees with two significant numbers, we will never know the exact temperature. The only thing we can really say is that a measurement is somewhere in a certain range, in this case $(32.41, 32.43)$ degrees. In effect, all recorded data are really intervals, with a width depending on the accuracy of the measurement. It is important to stress that this is different from probability, where we deal with the likelihood that a certain crisp measurement is being obtained [558]. In the context of uncertainty we are interested in the range into which our measurement falls. Several approaches to handle information about uncertainty have already been proposed, for example interval arithmetic allows us to deal and compute with intervals rather than crisp numbers [388], and also numerical analysis offers ways to propagate errors along with the normal computation [34].

This chapter will concentrate on presenting an approach to deal with imprecise concepts based on *fuzzy logic*. This type of logic enables us to handle uncertainty in a very intuitive and natural manner. In addition to making it possible to formalize imprecise numbers, it also enables us to do arithmetic using such *fuzzy numbers*. Classical set theory can be extended to handle partial memberships, thus making it possible to express vague human concepts using *fuzzy sets* and also describe the corresponding inference systems based on *fuzzy rules*.

Another intriguing feature of using fuzzy systems is the ability to granulate information. Using fuzzy clusters of similarity we can hide unwanted or useless information, ultimately leading to systems where the granulation can be used to focus the analysis on aspects of interest to the user.

The chapter will start out by explaining the basic ideas behind fuzzy logic and fuzzy sets, followed by a brief discussion of fuzzy numbers. We will then concentrate on fuzzy rules and how we can generate sets of fuzzy rules from data. We will close with a discussion of Fuzzy Information Theory, linking this chapter to Appendix B by showing how Fuzzy Decision Trees can be constructed.

9.2. Basics of Fuzzy Sets and Fuzzy Logic

Before introducing the concept of fuzzy sets it is beneficial to recall classical sets using a slightly different point of view. Consider for example the set of "young people", assuming that our perception of a young person is someone with an age of no more than 20 years:

$$\text{young} = \{x \in P \mid \text{age}(x) \leq 20\}$$

over some domain P of all people and using a function *age* that returns the age of some person $x \in P$ in years. We can also define a characteristic function:

$$m_{\text{young}}(x) = \begin{cases} 1 & : \text{age}(x) \leq 20 \\ 0 & : \text{age}(x) > 20 \end{cases}$$

which assigns to elements of P a value of 1 whenever this element belongs to the set of young people, and 0 otherwise. This characteristic function can be seen as a *membership function* for our set *young*, defining the set *young* on P.

Someone could then argue with us that he, being just barely over 20 years old, still considers himself young to a very high degree. Defining our set *young* using such a sharp boundary seems therefore not very appropriate. The fundamental idea behind fuzzy set theory is now a variable notion of membership; that is, elements can belong to sets to a certain degree. For our example we could then specify that a person with an age of, let's say, 21 years, still belongs to the set of *young* people, but only to a degree of less than one, maybe 0.9. The corresponding membership function would look slightly different:

$$\mu_{\text{young}}(x) = \begin{cases} 1 & : \text{age}(x) \leq 20 \\ 1 - \frac{(\text{age}(x)-20)}{10} & : 20 < \text{age}(x) \leq 30 \\ 0 & : 30 < \text{age}(x) \end{cases}$$

Now our set *young* contains people with ages between 20 and 30 with a linearly decreasing degree of membership, that is, the closer someone's age approaches 30, the closer his degree of membership to the set of young people approaches zero (see Figure 9.1).

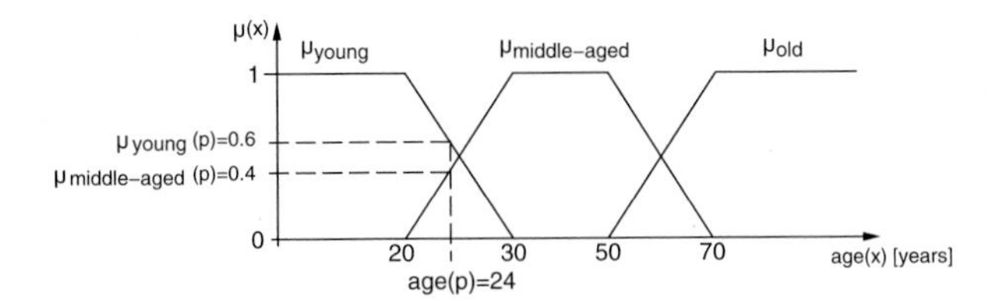

Fig. 9.1. A linguistic variable *age* with three fuzzy sets, and degrees of memberships for a certain age a.

The above is a very commonly used example for a *fuzzy set.* In contrast to classical sets, where an element can either belong to a set or lies completely outside of this set, fuzzy sets allow also partial memberships. A fuzzy set A is thus defined through specification of a *membership function* μ_A that assigns each element x a *degree of membership* to A: $\mu_A(x) \in [0, 1]$. Classical sets only allow values 1 (entirely contained) or 0 (not contained), whereas fuzzy set theory also deals with values in between 0 and 1. This idea was introduced in 1965 by Lotfi A. Zadeh [556].

9.2.1 Linguistic Variables and Fuzzy Sets

Covering the domain of a variable with several such fuzzy sets together with a corresponding semantic results in *linguistic variables*, allowing the computation with words. For our example this could mean that we define two more membership functions for middle-aged and old people, covering the entire domain of the variable *age.* This type of representation is especially appropriate for many real-world applications, where certain concepts are inherently vague in nature, either due to imprecise measurements or subjectivity.

The above example for a linguistic variable is shown in Figure 9.1. People are distinguished using their age (a function defined for all people) in groups of *young*, *middle-aged* and *old* people. Using fuzzy sets allows us to incorporate the fact that no sharp boundary between these groups exists. Figure 9.1 illustrates how the corresponding fuzzy sets overlap in these areas, forming non-crisp or fuzzy boundaries. Elements at the border between two sets belong to both. For example some person p with an age of $\text{age}(p) = 24$ years belongs to both groups *young* and *middle-aged* to a degree of 0.6 and 0.4 resp.; that is, $\mu_{\text{young}}(p) = 0.6$ and $\mu_{\text{middle-aged}}(p) = 0.4$. With an increase in age, the degree of membership to the group of young people will decrease whereas $\mu_{\text{middle-aged}}$ increases. The linguistic variable *age* is therefore described through three *linguistic values*, namely *young*, *middle-aged*, and *old.* The overlap between the membership functions reflects the imprecise nature of the underlying concept. We should keep in mind, however, that most concepts depend on the respective context. An old student can easily be a young professor.

This way of defining fuzzy sets over the domain of a variable is often referred to as *granulation*, in contrast to the division into crisp sets (quantization) which is

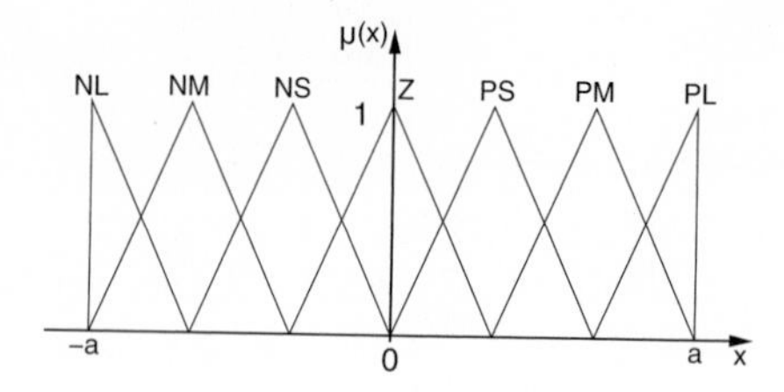

Fig. 9.2. The standard granulation using an odd number (here seven) of membership functions.

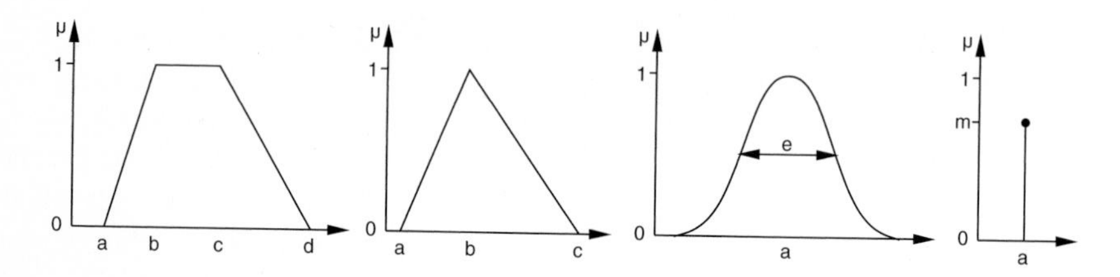

Fig. 9.3. Most commonly used shapes for membership functions (trapezoidal, triangular, Gaussian, singleton).

used by classical sets. Granulation results in a grouping of objects into imprecise clusters or *fuzzy granules*, with the objects forming a granule drawn together by similarity. Thus fuzzy quantization or granulation could also be seen as a form of fuzzy data compression. Often the granulation for some or all variables is obtained manually through expert interviews. If such expert knowledge is not available or the usage of a predefined granulation seems harmful, it is also possible to find a suitable granulation automatically. In the context of data analysis both approaches are used, depending on the focus of analysis. Later in this chapter we will discuss ways to use predefined granulations as well as algorithms that automatically build fuzzy clusters from data.

If no real semantic about the variable is known, a commonly used approach to label fuzzy sets is illustrated in Figure 9.2. For a symmetrical domain $[-a, a]$ of the variable often an odd number of membership functions is used, usually five or seven. The membership functions are then labeled `NL` (for "negative large"), `NM` ("negative medium"), `NS` ("negative small"), `Z` ("zero"), and the corresponding labels `PS`, `PM`, and `PL` for the positive side.

In real applications the shape of membership functions is usually restricted to a certain class of functions that can be specified with only few parameters. Figure 9.3 shows the most commonly used shapes for membership functions. On the left a trapezoidal function is depicted which can be specified through the four corner-points $<a, b, c, d>$. The triangular membership function can be seen as a special case of this trapezoidal function. Often used is also a Gaussian membership function which can be simply specified through two parameters a and e and offers nice mathematical properties such as continuity and differentiability. This is an often required property when membership functions are to be fine-tuned automatically during a training stage, as will be demonstrated later. Finally the singleton $<a|m>$ on the right can be used to define a fuzzy set containing only one element to a certain degree $m \leq 1$. The choice of membership function is mostly driven by the application. The Gaussian membership functions are usually used when the resulting system has to be adapted through gradient-descent methods. Knowledge retrieved from expert interviews will usually be modeled through triangular or trapezoidal membership functions, since the three resp. four parameters used to define these functions are intuitively easier to understand. An expert will prefer to define his notion of a fuzzy set by specifying the area where the degree of membership should be 1 ($[b, c]$ for the trapezoid) and where it should be zero (outside of (a, d)), rather than giving mean a and standard deviation e of a Gaussian. The resulting fuzzy system will not be affected drastically. Changing from one form of the membership function to another will affect the system only within the boundaries of its granulation.

The following parameters can be defined and are often used to characterize any fuzzy membership function:

- *support:* $s_A := \{x : \mu_A(x) > 0\}$, the area where the membership function is greater than 0.
- *core:* $c_A := \{x : \mu_A(x) = 1\}$, the area for which elements have maximum degree of membership to the fuzzy set A. Note that the core can be empty when the membership function stays below 1 over the entire domain.
- *α-cut:* $A_\alpha := \{x : \mu_A(x) \geq \alpha\}$, the cut through the membership function of A at height α.
- *height:* $h_A := \max_x\{\mu_A(x)\}$, the maximum value of the membership function of A.

Before discussing operators on fuzzy sets, the following section will discuss how fuzzy numbers can be treated.

9.2.2 Fuzzy Numbers

The motivation for using fuzzy numbers again stems from real-world applications. Real-world measurements, besides counting, are always imprecise in nature and a crisp number can not describe this fact adequately. Usually such measurements are modeled through a crisp number x for the most typical value together with an interval describing the amount of imprecision. In a linguistic sense this could be expressed as *about x*. Using fuzzy sets we can incorporate this additional information directly. This results in *fuzzy numbers* which are simply a special type of fuzzy sets restricting the possible types of membership functions:

- the membership function must be *normalized* (i.e. the core is non-empty, $c_A \neq \emptyset$) and *singular*. This results in precisely one point which lies inside the core modeling the typical value of our fuzzy number. This point is often also called the *modal value* of the corresponding fuzzy number.

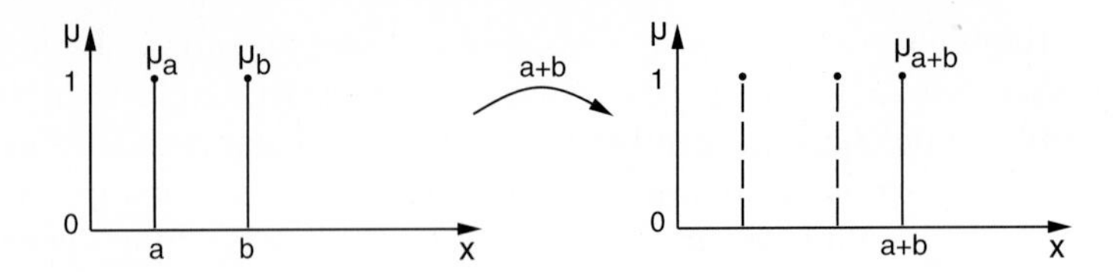

Fig. 9.4. An example for the membership functions of two crisp numbers and the result of adding the two.

- in addition μ_A has to be monotonically increasing left of the core and monotonically decreasing on the right. This makes sure that only one peak and therefore only one typical value exists. The spread of the support (i.e. the non-zero area of the fuzzy set) describes the degree of imprecision.

Typically a triangular membership-function is chosen, making it possible to specify a fuzzy number through only three parameters $< a, b, c >$ (Fig. 9.3).

Of course, we now want to use these fuzzy numbers for normal calculations, for example we want to add or multiply two fuzzy numbers or apply a certain function to one or more fuzzy arguments.

For clarification let us first consider the classical, crisp version. Here one would, for example, add two numbers a and b, resulting in $c = a + b$. This can also be illustrated as shown in Figure 9.4 using membership functions to represent the two numbers. The two singletons for a and b are then combined, resulting in a new singleton for c. Let us spend a moment to formulate this using membership functions for a and b. Obviously these functions are simple singletons, that is, $\mu_a(x) = 1$ for $x = a$ and $\mu_a(x) = 0$ everywhere else (similarly for $\mu_b(x)$). We can now construct the membership function for μ_{a+b} from here:

$$\mu_{a+b}(x) = \begin{cases} 1 & \text{if} \quad \exists y, z \in \mathbb{R} : y + z = x \wedge \mu_a(y) = 1 \wedge \mu_b(z) = 1 \\ 0 & \text{else} \end{cases}$$

In effect we define a function that assigns a degree of membership of 1 only to such points $x \in \mathbb{R}$ for which we can find points $y, z \in \mathbb{R}$ which also have a degree of membership of 1 using the membership functions for a and b and, in addition, who satisfy $y + z = x$. As expected for the crisp case, this results in another singleton, this time at the point $a + b$. Extending classical operators to their fuzzy counterparts can now be achieved similarly to the above, using the so-called extension principle [557]. We only have to extend the above method to also handle intermediate degrees of membership. For an arbitrary binary operator $\star$, this would be interpreted as follows:

$$\mu_{A \star B}(x) = \max_{y,z \in \mathbb{R}} \{\min\{\mu_A(y), \mu_B(z)\} \mid y \star z = x\}$$

that is, for a value $x \in \mathbb{R}$ a degree of membership is derived which is the maximum of $\min\{\mu_A(y), \mu_B(z)\}$ over all possible pairs of $y, z \in \mathbb{R}$ for which $y \star z = x$ holds.

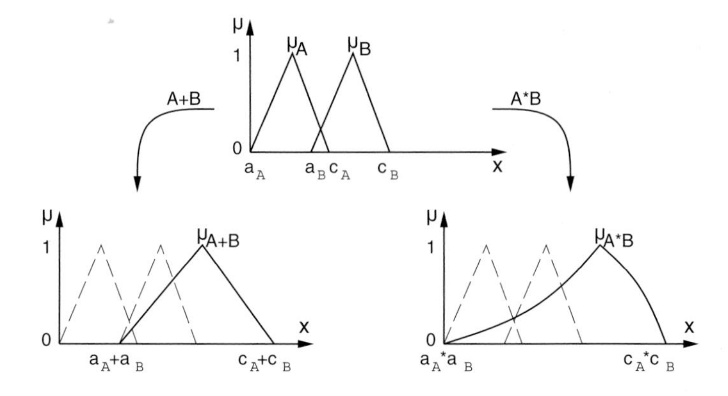

Fig. 9.5. An example for adding (left) and multiplying (right) two fuzzy numbers.

In other words, the new membership function assigns the maximum degree of membership which can be achieved by finding the best combination of parameters on the domain $\mathbb{R}$ of the involved variables. In effect, such operators convert fuzzy sets into a new fuzzy set, describing the result of the operation. Summation of two fuzzy numbers *about a* and *about b* would then result in a somewhat "fuzzier" version *(very) about a+b*. (It is interesting to note, however, that the result of 2·*about a* - *about a* and *about a* might not be the same when using the extension principle sequentially.) This extension principle can easily be extended to binary and also higher dimensional operators. Figure 9.5 demonstrates the result for addition and multiplication of two triangular fuzzy numbers. It is obvious how the resulting fuzzy numbers maintain their triangular shape after addition and become more complex after multiplication. They still maintain all properties required for fuzzy numbers but loose their triangular shape. Subtraction and division can be handled similarly, but for the latter special care has to be taken for cases where $\mu_B(0) \neq 0$, that is, where 0 is part of the quotient because then the resulting fuzzy set does not exist.

Whenever a non-linear function is applied to such fuzzy numbers, the triangular shape is not maintained. Similar to above, the extension principle is used again, resulting in:

$$\mu_{f(A)}(y) = \max\{\mu_A(x) \mid \forall x : f(x) = y\}$$

Figure 9.6 demonstrates this effect (In section 9.2.6 also the concept of fuzzy functions in contrast to such crisp functions will be discussed).

The fact that triangular fuzzy numbers are not closed under all operators, that is, their original (i.e. triangular) representation is not maintained through all operations, results in difficulties implementing calculations on fuzzy numbers, since the error of such a triangular approximation can possibly grow extremely fast [210]. A possible ad-hoc solution through an approximate representation uses a certain number of α-cuts for computation. On each of these α-cuts an exact computation can be done, but the linear approximation between two neighboring

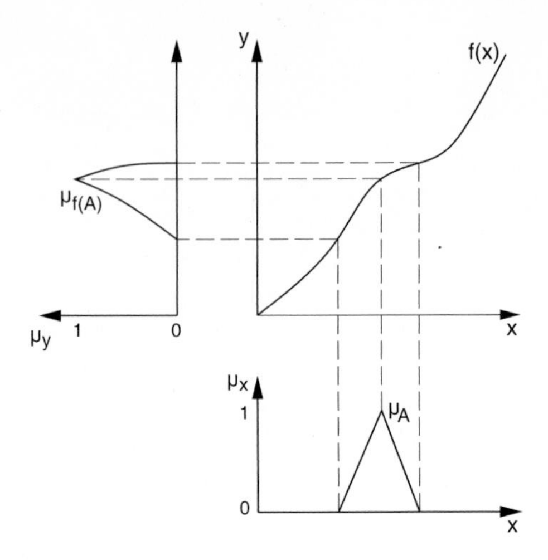

Fig. 9.6. An example for applying a function to a fuzzy number.

α-cuts again leads to a possibly large approximation error. The more different α-cuts are used, the smaller is the loss in accuracy, but also the complexity grows. Another approach is based on using a polynomial representation with respect to α. This representation is closed under addition and multiplication, that is, the result of adding or multiplying such fuzzy numbers will again be a polynom in α. This makes it possible to find an internal (though not necessarily finite) representation for such fuzzy numbers. Several other parametric representations of fuzzy numbers have been proposed, a good overview can be found in [210]. A more extensive discussion of fuzzy numbers can be found in [365].

9.2.3 Fuzzy Sets and Fuzzy Logic

Similar to the arithmetic operators defined on fuzzy numbers, classical operators from boolean logic like conjunction, disjunction, and complement can also be extended to fuzzy sets. Consider an example where we want to find the set of people that are both, young and tall. Using classical logic we would construct the new set using boolean conjunction:

$$m_{\text{tall } \mathbf{and} \text{ young}}(x) = \begin{cases} 1 & : \quad m_{\text{tall}}(x) = 1 \text{ and } m_{\text{young}}(x) = 1 \\ 0 & : \quad \text{else} \end{cases}$$

Using fuzzy sets defining such a strict condition is undesirable. Assume for example a person that belongs to the (fuzzy) set *young* to a degree of 0.5 and to the set *tall* to a degree of 0.8. What degree of membership should he or she have for the set of *young* **and** *tall* people? Should we use the minimum of the two degrees of membership? Another possibility would be the product, or one could even imagine more complicated arithmetic functions.

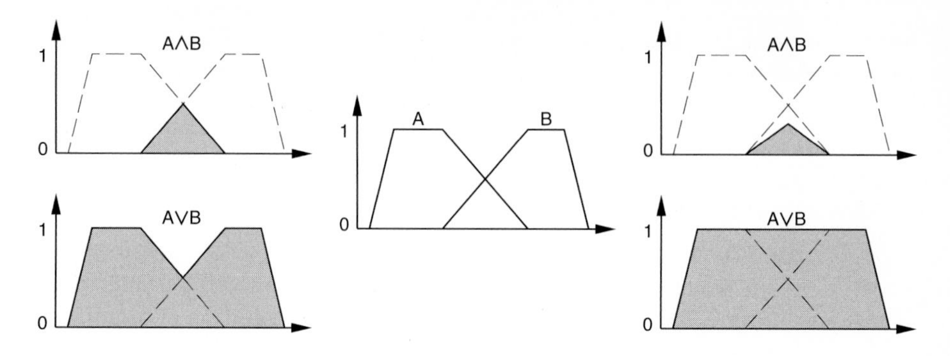

Fig. 9.7. Two variants for interpretation of fuzzy union and fuzzy intersection. Zadeh's min/max-interpretation (left) and the product/bounded sum-variant (right).

In effect, here the question arises how continuous truth-values can be processed. In contrast to the case of fuzzy numbers we are not interested in combining the entire fuzzy sets, but want to retrieve a new degree of membership resulting from an operation on one or more existing degrees of membership. Obviously an entire family of operators can be defined to derive the resulting membership functions. For the case of fuzzy logic, the most prominent examples were introduced by Lotfi Zadeh in [556], and are defined as follows:

- conjunction: $\mu_{A \wedge B}(x) := \min\{\mu_A(x), \mu_B(x)\}$
- disjunction: $\mu_{A \vee B}(x) := \max\{\mu_A(x), \mu_B(x)\}$
- complement: $\mu_{\neg A}(x) := 1 - \mu_A(x)$

For our example this would mean that the person belongs to the group of *young* and *tall* people to a degree of 0.5 ($= \min\{0.8, 0.5\}$).

Subsequently several other possible functions to interpret fuzzy set operators were proposed. Most of them can be formalized using the concept of T-norms (for the conjunction) and T-conorms (disjunction), the chapter by Gottwald in [404] and also [535] discuss these norms in more detail. The min/max-version above represents the most optimistic resp. most pessimistic version. Other common choices include the product for conjunction and the bounded sum for disjunction:

- conjunction II: $\mu_{A \wedge B}(x) := \mu_A(x) \cdot \mu_B(x)$
- disjunction II: $\mu_{A \vee B}(x) := \min\{\mu_A(x) + \mu_B(x), 1\}$

Figure 9.7 illustrates these two norms. As was pointed out in [327], however, using these definitions not all tautologies known from classical logic are true for the fuzzy case. For example consider the following (use $\mu_A(x) = 0.3$)[1]:

- $A \wedge \neg A \neq \emptyset$
- $A \vee \neg A \neq I$

[1] I and $\emptyset$ are defined according to classical logic, that is, $\mu_I \equiv 1$ and $\mu_\emptyset \equiv 0$.

Hence we can not simply use laws from classical logic to derive other operators. Most notably we have to define implication, rather than deriving it from disjunction and negation. For the min/max-interpretation the following definition for fuzzy implication is most often used (motivated by $A \to B = \neg A \vee (A \wedge B)$):

$$\mu_{A \to B}(x) := \max\{1 - \mu_A(x), \min\{\mu_A(x), \mu_B(x)\}\}$$

Several other norms have been defined, most of them motivated intuitively. Another set of norms are axiomatizable [327], and are all isomorphic to the Lukasiewicz-norm [358] which can be defined through a definition of implication as:

$$\mu_{A \to B}(x) := \min\{1, 1 - \mu_A(x) + \mu_B(x)\}$$

and negation as:

$$\mu_{\neg A} = 1 - \mu_A$$

and other operations being derived from there:

- disjunction III or strong-or, derived from $A \vee B = \neg A \to B$:

$$\mu_{A \vee B} = \min\{1, \mu_A + \mu_B\}$$

- conjunction III or strong-and, derived from $A \wedge B = \neg(\neg A \vee \neg B)$:

$$\mu_{A \wedge B} = 1 - \min\{1, 1 - \mu_A + 1 - \mu_B\} = \max\{0, \mu_A + \mu_B - 1\}$$

It should be noted, however, that also these definitions are not valid under all tautologies known from boolean logic. If the disjunction operator is, for example, derived through DeMorgan's law and $(A \to B) \to B$, the result will be the well-known maximum-operator.

In the next section it will be shown how these different definitions for fuzzy-implication affect the interpretation of inference in a fuzzy context.

9.2.4 Generalized Modus Ponens

In classical logic, conclusions can be drawn from known facts and implications based on these facts. In fuzzy logic this process of inference can be extended also to partial true facts, resulting in a generalized version of the classical Modus Ponens:

$$\frac{\begin{array}{l} x \texttt{ IS } A' \\ \texttt{IF } x \texttt{ IS } A \texttt{ THEN } y \texttt{ IS } B \end{array}}{y \texttt{ IS } B'}$$

where A' is the membership function for x, which does not have to be the same as in the antecedent of the implication (therefore the name *generalized* Modus

Ponens). In terms of fuzzy membership functions this can be expressed as follows:

$$\frac{\begin{array}{c}\mu_{A'}(x) \\ A \rightarrow B\end{array}}{\mu_{B'}(y)}$$

According to the previous different definitions of implication, different interpretations of such fuzzy inference can arise:

- Joint Constraint: using min-max norms, the implication can be seen as forming a constraint on a joint variable, that is, (x, y) is $A \times B$, where $\times$ denotes the Cartesian product $\mu_{A\times B}(x, y) = \min\{\mu_A(x), \mu_B(y)\}$. B' can now be obtained through $B' = A' \wedge (A \times B)$, or

$$\mu_{B'}(v) = \sup_u \{\min\{\mu_{A'}(u), \mu_{A\times B}(u, v)\}\}.$$

  Note that if A' is similar to A, also B' will be similar to B. On the other hand a very dissimilar A' will result in an empty set B', i.e. $\mu_{B'} \equiv 0$.
- Conditional Constraint: using Lukasiewicz's definition a possibility dependency arises: $\text{Poss}(x = u|y = v) = \min\{1, 1 - \mu_A(u) + \mu_B(v)\}$. Also here B' can be easily obtained through:

$$\mu_{B'}(v) = \sup_u \{\min\{\mu_{A'}(u), 1 - \mu_A(u) + \mu_B(v)\}\}.$$

  If A' is very dissimilar to A the resulting B' will be generic, i.e. $\mu_{B'} \equiv 1$.

Both definitions lead to a very different interpretation of fuzzy implication. In most rule-based applications the interpretation of fuzzy implication as a constraint on joint variables is used, because this variant is easier to implement and computationally less expensive. The remainder of this chapter will focus on this interpretation, but it should always be kept in mind that other interpretations exist (for a more detailed discussion see [405]). This lack of a generally accepted interpretation of fuzzy implication is often cited as a disadvantage of fuzzy logic, but if the user is aware of the potential of both interpretations they can provide interesting alternatives for different foci of analysis.

9.2.5 Fuzzy Rules

Fuzzy rules can be used to characterize imprecise dependencies between different variables. Consider for example a classical rule:

$$\texttt{IF } \text{age}(x) \leq 25 \texttt{ THEN } \text{risk}(x) > 60\%$$

describing the risk-factor for a car-insurance company. Obviously using linguistic variables can make such a rule much more readable:

$$\texttt{IF } \text{age}(x) \texttt{ IS } \text{young} \texttt{ THEN } \text{risk}(x) \texttt{ IS } \text{high}$$

Fuzzy rules are therefore of interest whenever a dependency is either imprecise or a high level of precision is not desired in order to maintain a high level of interpretability. A basic type of a categorical rule[2] which is widely used in control and other applications has the following form:

$$\texttt{IF } x_1 \texttt{ IS } A_1 \texttt{ AND } \ldots \texttt{ AND } x_n \texttt{ IS } A_n \texttt{ THEN } y \texttt{ IS } B$$

where the A_i in the antecedent and the B in the consequent are linguistic values of the input vector $\boldsymbol{x}$ and the output variable y, respectively. These types of rules are called *Mamdani rules* [363]. Most often, the consequent only consists of one variable and multi-dimensional cases are modeled through separate rules for each output.

In many modeling applications, also rules that assign simple (crisp) equations to the output variable are used. Most commonly these are either linear or quadratic dependencies on one or more input variables (first order or second order *Takagi-Sugeno models* [507]). Note that in this case the input $\boldsymbol{x}$ is required to consist of scalar values:

$$\texttt{IF } x_1 \texttt{ IS } A_1 \texttt{ AND } \ldots \texttt{ AND } x_n \texttt{ IS } A_n \texttt{ THEN } y = f(\boldsymbol{x})$$

Note that by using this expression, the output is no fuzzy set, but rather a singleton $< y, w >$. The degree of membership w of this singleton is equal to the degree of fulfillment of the antecedent:

$$w = \min\{\mu_{A_1}(x_1), \cdots, \mu_{A_n}(x_n)\}.$$

If several such rules exist, the corresponding output-singletons are combined using a fuzzy aggregation operator. Most often, the product is used to compute $\wedge$ and a weighted sum is used for aggregation (assuming r rules):

$$y_{\text{final}} = \frac{\sum_{i=1}^{r} w_i \cdot y_i}{\sum_{i=1}^{r} w_i}.$$

The resulting system is then similar to Radial Basis Function Networks, as discussed in chapter 8. Rules of this type are commonly used in control applications. In the context of data analysis applications, however, the above-mentioned Mamdani rules with fuzzy consequent "$\cdots$ THEN y IS B" are usually given preference because of their better interpretability.

Using these types of rules, our car-insurance agency might now describe a particular group of young drivers with high-output engines in their cars as follows:

IF *age* IS *young* AND *car power* IS *high* THEN *risk* IS *high*

As mentioned in section 9.2.4, such an implication is usually interpreted as a constraint on a joint variable. If a min-norm is used, this results in a "clipping"

[2] Qualified rules that assign additional constraints to rules (such as "very likely" or "not very true"), will not be dealt with in this chapter since they are not widely used.

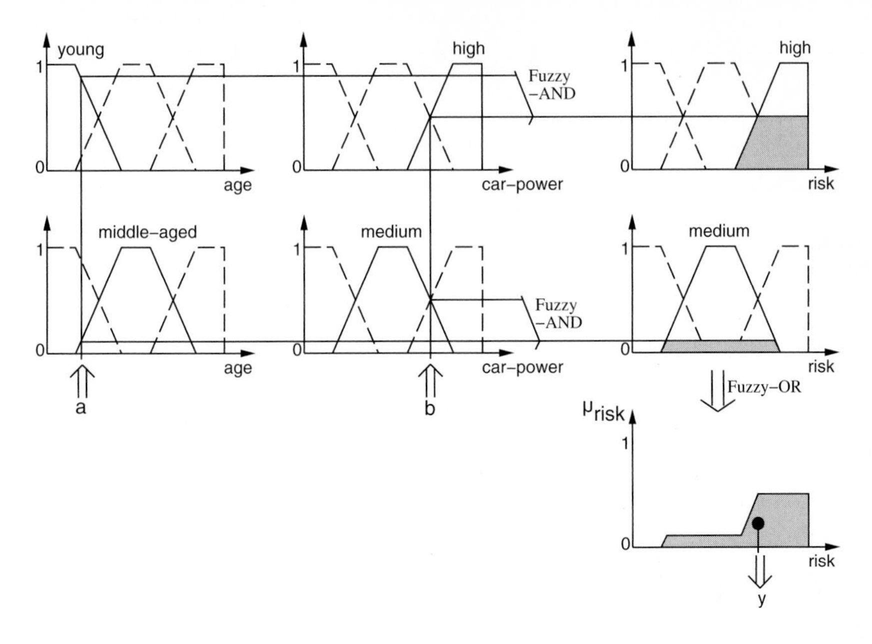

Fig. 9.8. An example for a fuzzy rule base in action. Receiving the crisp inputs age=a and car-power=b the defuzzified output will be risk=y.

of the fuzzy set of the consequent. Sometimes also the product-norm is used, resulting in a scaling of the consequent's membership function.

A fuzzy rule based system is a set of such fuzzy rules. The process of combining the individual results is often done through composition. An example for a fuzzy rule set in action using min/max-norms is shown in Figure 9.8. The rule-set contains two fuzzy `if-then`-rules for the example car-insurance risk-predictor:

`IF` *age* `IS` *young* `AND` *car power* `IS` *high* `THEN` *risk* `IS` *high*

`IF` *age* `IS` *middle aged* `AND` *car power* `IS` *medium* `THEN` *risk* `IS` *medium*

The first rule describes the fact that young drivers with high-power cars pose high risks for a car-insurance. Similarly the second rule describes middle-aged drivers with normal (i.e. medium power) cars. Figure 9.8 now shows how – using min-max-inference – a crisp input of *age=a* and *car-power=b* is being processed. For both rules the degrees of membership for the corresponding fuzzy terms are computed using Fuzzy-AND. This results in a clipped membership function of the rule-conclusion and together they produce the resulting membership function for the linguistic variable *risk* using Fuzzy-OR. A number of methods to determine a crisp output value are available. The most common one is shown in Figure 9.8. Here the crisp value y is determined from the generated membership function μ_{risk} through:

$$y = \int y \cdot \mu_{\text{risk}}(y) \mathrm{d}y$$

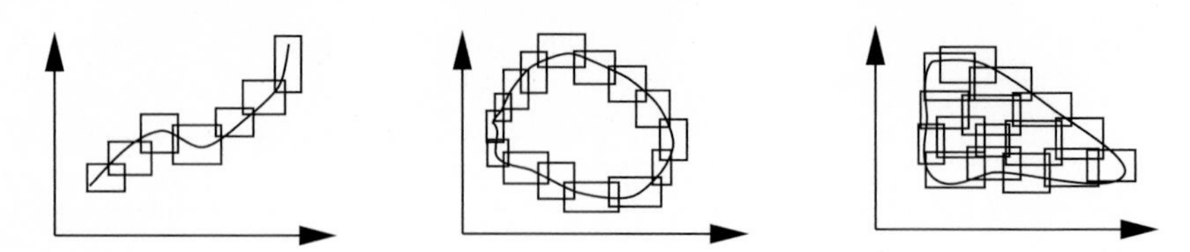

Fig. 9.9. Approximate representations of functions, contours, and relations (after Zadeh [559]).

In practical applications it is often ignored that the membership functions of the linguistic values can overlap and the crisp output is then easily computable through:

$$y = \frac{\sum_{j=1}^{r} \mu_j \cdot s_j}{\sum_{j=1}^{r} \mu_j}$$

where r indicates the number of rules, μ_j the corresponding degree of membership for each linguistic value and s_j denotes the center of gravity of the membership function. This method of *defuzzification* is known as COG- or *Center of Gravity*-method. It is worthwhile to note that using such rule sets it is possible to provide proofs for certain properties of the resulting system. The most notable is based on the Stone-Weierstrass theorem and shows that on the basis of overlapping, triangular membership functions, which cover the entire (compact) domain of all attributes, a set of rules can always be constructed that approximates an arbitrary real-valued, continuous function with an arbitrary small error [532, 98].

Unfortunately, such global granulation of the input space will result in an exploding number of rules for high-dimensional spaces. Hence, different ways to cover the input space using fuzzy rules were developed which will be discussed in the next section.

9.2.6 Fuzzy Graphs

To avoid a global granulation of the feature space the concept of fuzzy graphs can be used [559]. In contrast to a global grid, specified through linguistic variables as described above, fuzzy graphs employ individual membership functions for each fuzzy rule; that is, the membership function of the constraint on the joint variable is defined individually for each rule. Based on the interpretation of implication as a constraint on a joint variable, a set of individual fuzzy rules can also be seen as forming a fuzzy graph. Through the concept of such fuzzy graphs approximate representations of functions, contours, and sets can be derived (Figure 9.9).

In the following we will concentrate on fuzzy graphs for the representation of functions which approximate only one output variable[3] y depending on n input variables x_i $(1 \leq i \leq n)$.

[3] Multidimensional output variables can easily be modeled through a set of fuzzy graphs.

The typical fuzzy graphs build upon Mamdani rules similar to the ones mentioned in the previous section:

$$R: \texttt{ IF } x_1 \texttt{ IS } A_1 \texttt{ AND } \ldots \texttt{ AND } x_n \texttt{ IS } A_n \texttt{ THEN } y \texttt{ IS } B$$

Some of the input membership functions A_i can be constant ($\mu_{A_i} \equiv 1$). Thus different fuzzy rules may only depend on a subset of input variables. This is extremely beneficial in applications in high-dimensional spaces, where such rules will be much easier to interpret. Simplifying the above equation, using $\boldsymbol{A} = A_1 \times \cdots \times A_n$ (again $\times$ denotes the Cartesian product), leads to:

$$\texttt{IF } \boldsymbol{x} \texttt{ IS } \boldsymbol{A} \texttt{ THEN } y \texttt{ IS } B.$$

And, as mentioned in section 9.2.4, such a rule can also be seen as a fuzzy constraint on a joint variable $(\boldsymbol{x}, y)$, that is,

$$(\boldsymbol{x}, y) \texttt{ IS } \boldsymbol{A} \times B$$

In this representation $\boldsymbol{A} \times B$ is often called a *fuzzy point*. Using Zadeh's min/max-based interpretation of fuzzy operators, the membership function of $\boldsymbol{A} \times B$ can be computed as discussed above:

$$\mu_{\boldsymbol{A}\times B}(\boldsymbol{x}, y) = \min\left\{\mu_{\boldsymbol{A}}(\boldsymbol{x}), \mu_B(y)\right\}$$

or, more precisely

$$\mu_{\boldsymbol{A}\times B}(\boldsymbol{x}, y) = \min\left\{\mu_{A_1}(x_1), \cdots, \mu_{A_n}(x_n), \mu_B(y)\right\}$$

A collection of such rules can now be regarded as forming a superposition of r fuzzy points:

$$(\boldsymbol{x}, y) \texttt{ IS } (\boldsymbol{A}_1 \times B_1 + \cdots + \boldsymbol{A}_r \times B_r)$$

where $+$ denotes the disjunction operator (here defined as maximum). Zadeh calls this characterization of a dependency *fuzzy graph*, because the collection of rules can be seen as a coarse representation of a functional dependency f^* of y on $\boldsymbol{x}$. This fuzzy graph f^* can thus be defined as:

$$f^* = \sum_{j=1}^{r} (\boldsymbol{A}_j \times B_j)$$

The task of interpolation, that is, deriving a linguistic value B for y given an arbitrary linguistic value $\boldsymbol{A}$ for $\boldsymbol{x}$ and a fuzzy graph f^*:

$$\begin{array}{lll} \boldsymbol{x} & \texttt{IS} & \boldsymbol{A} \\ f^* & \texttt{IS} & \sum_{j=1}^{r} \boldsymbol{A}_j \times B_j \\ \hline y & \texttt{IS} & B \end{array}$$

results in an intersection of the fuzzy graph f^* with a cylindrical extension of the input fuzzy set $\boldsymbol{A}$:

$$\text{proj}_Y\{(\boldsymbol{A} \times I) \cap f^*\}$$

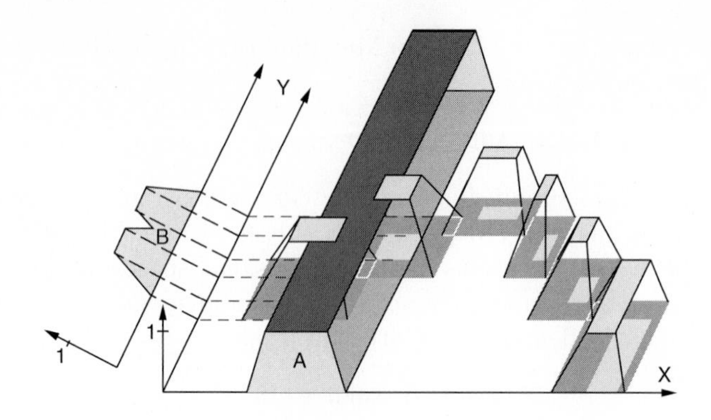

Fig. 9.10. Interpolation as the intersection of a fuzzy graph with a cylindrical extension of A.

with $\boldsymbol{A} \times I$ denoting the cylindrical extension along y and proj_Y the projection on Y. Figure 9.10 shows an example. This functional dependency can be computed through:

$$\begin{aligned}\mu_B(y) &= \mu_{f^*(\boldsymbol{A})}(y) = \sup_{\boldsymbol{x}} \{\min\{\mu_{f^*}(\boldsymbol{x}, y), \mu_{\boldsymbol{A}}(\boldsymbol{x})\}\} \\ &= \sup_{\boldsymbol{x}} \{\min\{\max\{\mu_{\boldsymbol{A}_1 \times B_1}(\boldsymbol{x}, y), \cdots, \mu_{\boldsymbol{A}_r \times B_r}(\boldsymbol{x}, y)\}, \mu_{\boldsymbol{A}}(\boldsymbol{x})\}\}.\end{aligned}$$

9.3. Extracting Fuzzy Models from Data

In the context of intelligent data analysis it is of great interest how such fuzzy models can automatically be derived from example data. Since, besides prediction, understandability is of prime concern, the resulting fuzzy model should offer insights into the underlying system. To achieve this, different approaches exist that construct grid-based rule sets defining a global granulation of the input space, as well as fuzzy graph based structures. Approaches that produce such models will be presented in the next two sections. Afterwards we will also briefly discuss a fuzzy clustering algorithm, similar to the one described in chapter 8.

9.3.1 Extracting Grid-Based Fuzzy Models from Data

Grid-based rule sets model each input variable through a usually small set of linguistic values. The resulting rule base uses all or a subset of all possible combinations of these linguistic values for each variable, resulting in a global granulation of the feature space into "tiles":

$$\begin{array}{lll} R_{1,\cdots,1}: & \texttt{IF } x_1 \texttt{ IS } A_{1,1} \texttt{ AND } \ldots \texttt{ AND } x_n \texttt{ IS } A_{1,n} & \texttt{THEN } \ldots \\ & \vdots \qquad\qquad\qquad \vdots & \\ R_{l_1,\cdots,l_n}: & \texttt{IF } x_1 \texttt{ IS } A_{l_1,1} \texttt{ AND } \ldots \texttt{ AND } x_n \texttt{ IS } A_{l_n,n} & \texttt{THEN } \ldots \end{array}$$

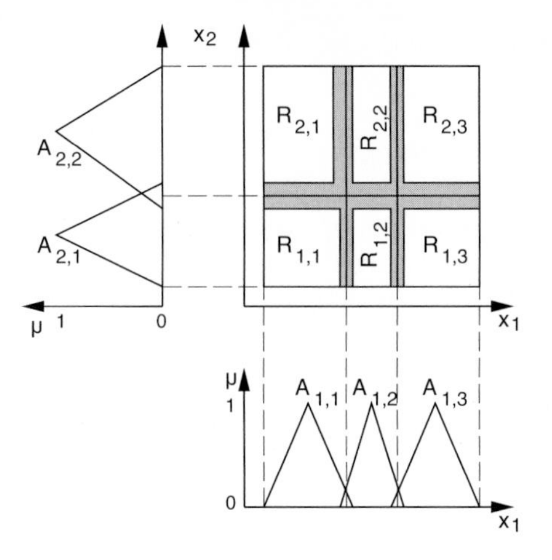

Fig. 9.11. A global granulation of the input space using three membership functions for x_1 and two for x_2.

where l_i $(1 \leq i \leq n)$ indicates the numbers of linguistic values for variable i in the n-dimensional feature space. Figure 9.11 illustrates this approach in two dimensions with $l_1 = 3$ and $l_2 = 2$. Extracting grid-based fuzzy models from data is straightforward when the input granulation is fixed, that is, the antecedents of all rules are predefined. Then only a matching consequent for each rule needs to be found.

Wang&Mendel [531] presented a straightforward approach to learning fixed-grid Mamdani models. After predefinition of the granulation of all input variables and also the output variable, one sweep through the entire dataset determines the closest example to the geometrical center of each rule, assigning the closest output fuzzy value to the corresponding rule:

1. Granulate the input and output space. Divide each variable x_i into l_i equidistant triangular membership functions. Similarly the granulation into l_y membership functions for the output variable y is determined, resulting in the typical overlapping distribution of triangular membership functions. Figure 9.11 illustrates this approach in two dimensions with 3 resp. 2 membership functions, resulting in six tiles.
2. Generate fuzzy rules from given data. For the example in Figure 9.11, this means that we have to determine the best consequence for each rule. For each example pattern $(\boldsymbol{x}, y)$ the degree of membership to each of the possible tiles is determined:

$$\min\left\{\mu_{\mathrm{msx}_{j_1,1}}(x_1), \cdots, \mu_{\mathrm{msx}_{j_n,n}}(x_n), \mu_{\mathrm{msy}_{j_y}}(y)\right\}$$

with $1 \leq j_i \leq l_i$ and $1 \leq j_y \leq l_y$. Then $\mathrm{msx}_{j_i,i}$ indicates the membership function of the j_i-th linguistic value of input variable i and similar for msy for the output variable y. Next the tile resulting in the maximum degree of membership is used to generate one rule:

$$R_{(j_1,\cdots,j_n)}: \texttt{ IF } x_1 \texttt{ IS } \text{msx}_{j_1,1} \cdots \texttt{ AND } x_n \texttt{ IS } \text{msx}_{j_n,n}$$
$$\texttt{THEN } y \texttt{ IS } \text{msy}_{j_y}$$

assuming that tile $(j_1, \cdots, j_n, j_y)$ resulted in the highest degree of membership for the corresponding training pattern.

3. Assign a rule weight to each rule. The degree of membership will in addition be assigned to each rule as rule-weight $\beta_{(j_1,\cdots,j_n)}$.
4. Determine an output based on an input-vector. Given an input $\boldsymbol{x}$ the resulting rule-base can be used to compute a crisp output $\hat{y}$. First the degree of fulfillment for each rule is computed:

$$\mu_{(j_1,\cdots,j_n)}(\boldsymbol{x}) = \min\left\{\mu_{\text{msx}_{j_1,1}}(x_1), \cdots, \mu_{\text{msx}_{j_n,n}}(x_n)\right\}$$

then the output $\hat{y}$ is combined through a centroid defuzzification formula:

$$\hat{y} = \sum_{j_1=1,\cdots,j_n=1}^{l_1,\cdots,l_n} \frac{\beta_{(j_1,\cdots,j_n)} \cdot \mu_{(j_1,\cdots,j_n)}(\boldsymbol{x}) \cdot \overline{y}_{(j_1,\cdots,j_n)}}{\beta_{(j_1,\cdots,j_n)} \cdot \mu_{(j_1,\cdots,j_n)}(\boldsymbol{x})}$$

where $\overline{y}_{(j_1,\cdots,j_n)}$ denotes the center of the output region of the corresponding rule with index $(j_1, \cdots, j_n)$.

Figure 9.12 demonstrates this algorithm using four membership functions for the input variable x and output y. The graph in the top left corner shows the example points used to generate the fuzzy model. On the right the used granulation is shown, and the data points which lie closest to the centers of the respective rules are marked (thick circles). At the bottom left the generated set of rules is shown, represented by their $\alpha = 0.5$-cuts:

$$\begin{array}{llll} R_1: & \texttt{IF } x \texttt{ IS } \textit{zero}_x & \texttt{THEN } y \texttt{ IS } \textit{medium}_y \\ R_2: & \texttt{IF } x \texttt{ IS } \textit{small}_x & \texttt{THEN } y \texttt{ IS } \textit{medium}_y \\ R_3: & \texttt{IF } x \texttt{ IS } \textit{medium}_x & \texttt{THEN } y \texttt{ IS } \textit{large}_y \\ R_4: & \texttt{IF } x \texttt{ IS } \textit{large}_x & \texttt{THEN } y \texttt{ IS } \textit{medium}_y \end{array}$$

and on the right the resulting crisp approximation is shown (thick line). Note how the generated model misses extrema that lie far from existing rule-centers. Intuitively rule R_2 should probably be used to describe the minimum of the function:

$$R_2': \texttt{ IF } x \texttt{ IS } \textit{small}_x \texttt{ THEN } y \texttt{ IS } \textit{small}_y$$

This behavior is due to the fact that only one pattern per rule is used to determine the outcome of this rule. Even a combined approach would very much depend on the predefined granulation. If the function to be modeled has a high variance inside one rule, the resulting rule model will fail to model this behavior. One interesting aspect of this approach is the availability of a proof that arbitrary real continuous functions over a compact set can be approximated to

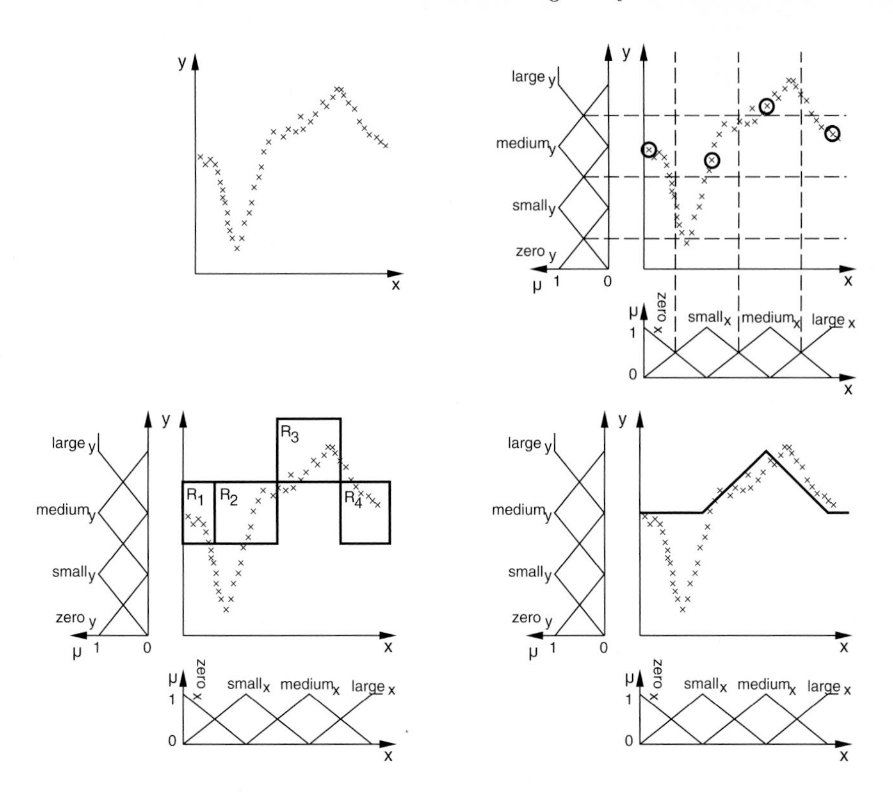

Fig. 9.12. An example for a fixed-grid fuzzy rule set produced by the Wang&Mendel-algorithm [531]. On the top the data points (left) and the used granulation (right) are shown. The figures at the bottom show the generated rules (left) and the resulting crisp approximation (right).

arbitrary accuracy. This proof is based on the Stone-Weierstrass theorem and can be found in [531].

For practical applications it is obvious, however, that using such a predefined, fixed grid results in a fuzzy model that will either not fit the underlying functions very well or consist of a large number of rules. This is why approaches are of more interest that fine-tune or even automatically determine the granulations of both input and output variables.

An approach that builds upon Wang&Mendel's fixed-grid algorithm was proposed by Higgins&Goodman in [262]. Initially only one membership function is used to model each of the input variables as well as the output variable, resulting in one large rule covering the entire feature space. Subsequently new membership functions are introduced at points of maximum error. This is done until a maximum number of divisions is reached or the approximation error remains below a certain threshold. Figure 9.13 demonstrates this algorithm for the same set of data points as used before to demonstrate the Wang&Mendel algorithm in Figure 9.12. Here training was stopped after a maximum of four membership

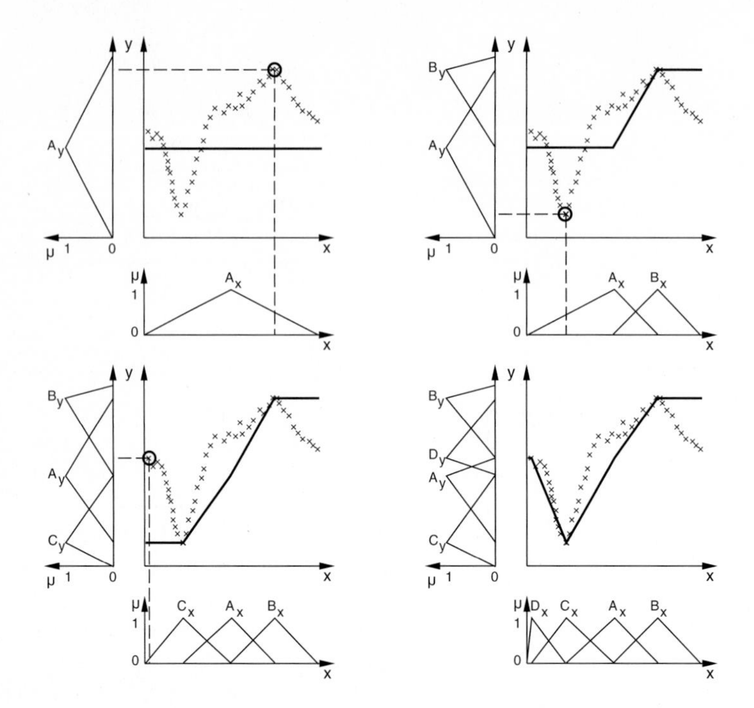

Fig. 9.13. An example how a grid-based fuzzy rule set will be produced by the Higgins&Goodman-algorithm [262] in four steps. The thick line shows the approximation and the circle indicates the point resulting in maximum error at this instance.

functions was generated for each variable and the graphs show all four steps. Each graph shows the generated membership functions and the corresponding approximation for the corresponding instance. The circles indicate the point of maximum error at each step. In the top left corner one large rule covers the entire input space:

IF x IS A_x THEN y IS A_y

The next step generates a second rule, describing the maximum of the example data points:

IF x IS B_x THEN y IS B_y

During the third step another rule is generated:

IF x IS C_x THEN y IS C_y

The final step generates yet another rule, but note how in the final set of four rules the consequent of the second rule is changed from B_y to C_y:

IF x IS A_x THEN y IS A_y
IF x IS B_x THEN y IS C_y
IF x IS C_x THEN y IS C_y
IF x IS D_x THEN y IS D_y

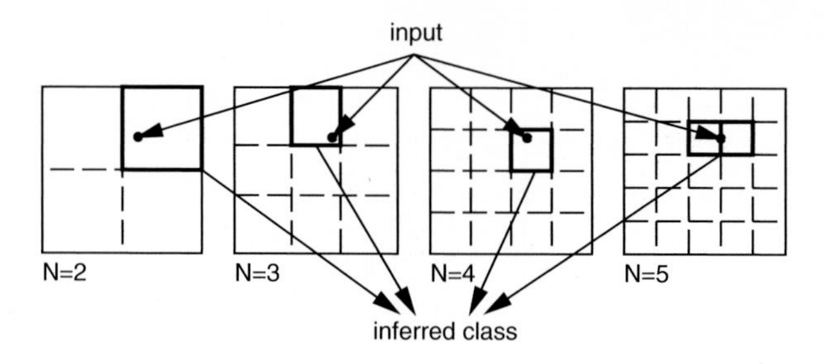

Fig. 9.14. An example for the multi-rule approach presented in [406]. Several rule sets are initialized, using different levels of granulation.

This is due to the underlying Wang&Mendel algorithm, which is used to generate the intermediate set of rules. At each step, new membership functions for all variables are introduced and therefore different choices can be made for the rules' output, effectively moving the output closer to the output value of the particular training pattern also in other regions then the one that caused the creation of new membership functions. Obviously this approach is able to model extrema much better than the Wang&Mendel algorithm alone, but has a definite preference to favor extrema and therefore a strong tendency to concentrate on outliers.

A hierarchical algorithm was presented in [406]. Here an ensemble of rule sets with different granulation is built at the beginning. Starting with a coarse granulation (usually having only two membership functions for each attribute), the remaining rule sets have increasingly finer granulation. Figure 9.14 illustrates this method. In the resulting multi-rule table the grade of certainty for each rule is then used for pruning; that is, rules with a low grade of certainty will be removed from the rule set.

Several other approaches still require a predefined grid but fine-tune the initial position of the grid-lines to better match the data. Similar to the Error Backpropagation algorithm for Neural Networks described in chapter 8, a heuristic version of gradient descent is used in [401]. The position and the shape of the membership functions is altered iteratively to reduce the overall approximation error.

Other approaches convert the fuzzy model into a neural network and use conventional training algorithms. In [4] an approach based on Radial Basis Function Networks with elliptical regions is described. The used training algorithm, however, needs to be restrained to maintain a possibility to extract a meaningful fuzzy model after training has finished. Also Genetic Algorithms (see chapter 10) can be used to fine tune an existing rule set. In [282] such an approach was presented. An algorithm which generates positive and negative rules using evolutionary concepts can be found in [322].

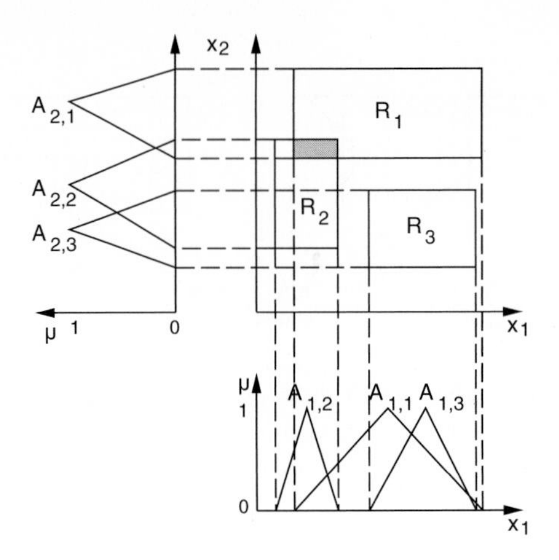

Fig. 9.15. A granulation of the input space using individual membership functions for each rule.

9.3.2 Extracting Fuzzy Graphs from Data

We mentioned before that especially in high-dimensional feature spaces a global granulation results in a large number of rules. For these tasks a fuzzy graph based approach is more suitable. Here the individual fuzzy rules are defined through independent membership functions in the feature space:

$$R_1: \texttt{IF } x_1 \texttt{ IS } A_{1,1} \texttt{ AND } \ldots \texttt{ AND } x_n \texttt{ IS } A_{1,n} \texttt{ THEN } \ldots$$

$$\vdots \qquad \vdots$$

$$R_r: \texttt{IF } x_1 \texttt{ IS } A_{r,1} \texttt{ AND } \ldots \texttt{ AND } x_n \texttt{ IS } A_{r,n} \texttt{ THEN } \ldots$$

Figure 9.15 shows an example in two dimensions using three rules. For comparison, Figure 9.11 in the previous section shows a global, grid-based granulation. A possible disadvantage of the individual membership functions is the potential loss of interpretation. Projecting all membership functions onto one variable will usually not lead to meaningful linguistic values. In many data analysis applications, however, such a meaningful granulation of all attributes is either not available or hard to determine automatically.

An algorithm that constructs such fuzzy graphs based on example data was presented in [62]. The only parameter specified by the user is the granulation of the output variable y; that is, the number and shape of the membership functions of y have to be determined manually[4]. In most applications this is no disadvantage, because it enables the user to define foci of attention or areas of interest where a finer granulation is desired. Thus c fuzzy sets are defined through membership functions μ_y^k (y indicating that this membership function

[4] In case of multi-dimensional output variables, several fuzzy graphs can be built independently.

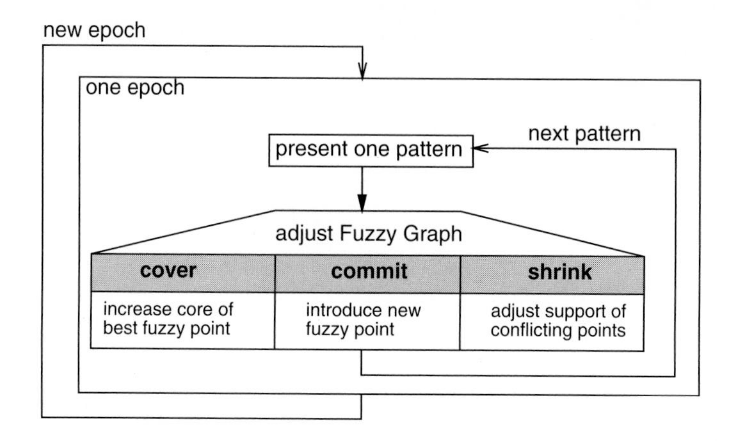

Fig. 9.16. The algorithm to construct a fuzzy graph based on example data [62].

is defined over the domain of the output variable y) for each output region k, with $1 \leq k \leq c$. The algorithm then iterates over the example data, and subsequently fine-tunes the evolving model.

The resulting fuzzy graph consists of a set of fuzzy points or fuzzy rules R_j^k, $1 \leq j \leq r_k$, where r_k indicates the number of rules for the output region k. A rule's activity $\mu_j^k(\boldsymbol{x})$ indicates the degree of membership of the pattern $\boldsymbol{x}$ to the corresponding rule R_j^k (note the difference between μ_j^k defined on input x_j and μ_y^k, defined on the output y). Using the normal notation of fuzzy rules, each rule can be decomposed into n individual, one-dimensional membership functions:

$$\mu_j^k(\boldsymbol{x}) = \min_{i=1,\ldots,n} \{\mu_{j,i}^k(x_i)\}$$

where $\mu_{j,i}^k$ indicates the projection of μ_j^k onto the i-th attribute. The degree of membership for output-region k is then computed through:

$$\mu^k(\boldsymbol{x}) = \max_{1 \leq j \leq r_k} \{\mu_j^k(\boldsymbol{x})\}$$

The algorithm relies on trapezoidal membership functions, so each rule can be described through four parameters for each dimension:

$$\texttt{IF } x_1 \texttt{ IS } < a_1, b_1, c_1, d_1 > \texttt{ AND } \cdots x_n \texttt{ IS } < a_n, b_n, c_n, d_n > \texttt{ THEN } y \texttt{ IS } \mu_y^k$$

It is important to note, however, that some trapezoids can cover the entire domain of the attribute, making the rule's degree of fulfillment independent of this dimension.

The flow of the underlying algorithm is illustrated in Figure 9.16. All existing example patterns are subsequently presented. During such an epoch, three different steps are executed. The algorithm ensures that each example pattern is covered by a fuzzy rule of the region to which it belongs and that rules of

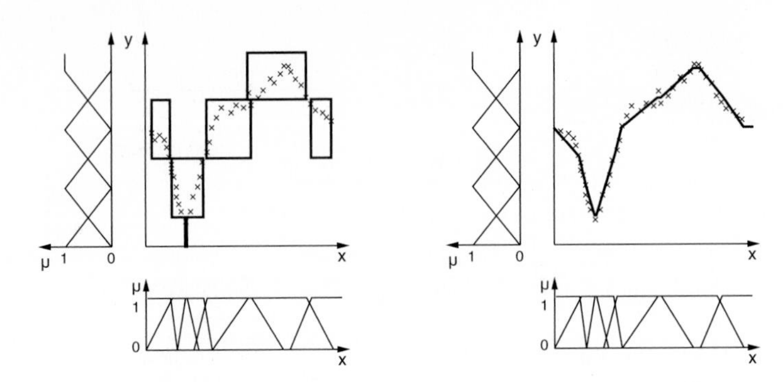

Fig. 9.17. An example for a fuzzy graph produced by the algorithm in [62] (left) along with the corresponding approximation (right).

conflicting regions do not cover the pattern. This enables the fuzzy graph to tolerate moderately noisy patterns or small oscillations along boundaries (see [62] for an example). The three main steps of the algorithm introduce new fuzzy points when necessary and adjust the core- and support-regions of existing ones in case of conflict:

- **covered**: if the new training pattern lies inside the support-region of an already existing fuzzy point and it belongs to the correct output-region, the core-region of this fuzzy point is extended to cover the new pattern.
- **commit**: if a new pattern is not covered, a new fuzzy point belonging to the correct output-region will be introduced. The new example pattern is assigned to its core, whereas the support-region is initialized "infinite"; that is, the new fuzzy point covers the entire domain.
- **shrink**: if a new pattern is incorrectly covered by an already existing fuzzy point of a conflicting region this fuzzy point's support-area will be reduced (e.g. shrunk) so that the conflict is solved. The underlying heuristic how this shrink is conducted aims to maximize the remaining volume. Details can be found in [62].

The algorithm clearly terminates after only few iterations over the set of example patterns. The final set of fuzzy points forms a fuzzy graph, where each fuzzy point is associated with one output region and is used to compute a membership value for a certain input pattern. The maximum degree of membership of all fuzzy points for one region determines the overall degree of membership. Fuzzy inference then produces a soft value for the output and using the well-known center-of-gravity method a final crisp output value can be obtained, if so desired. Approximation of the above used example-function produces the result shown in Figure 9.17. Here the output variable was divided into four soft regions. The fuzzy graphs tend to ignore oscillations that lie inside output regions, an effect often desirable in real applications. The input variable is divided into many membership functions, some of them extremely thin. In scenarios with multi-dimensional inputs these membership functions may even overlap each other.

Fuzzy graphs do not have a natural linguistic interpretation of the granulation of their input space. The main advantage is the low dimensionality of the individual rules. The algorithm only introduces restriction on few of the available input variables, thus making the extracted rules easier to interpret.

Other approaches that build fuzzy graphs or independent fuzzy rule sets based on example data were presented in [485,486]. In [3] an algorithm was presented that also generates negative fuzzy rules, thus allowing to handle conflicting regions where rules of different classes overlap. Several of the rule learning concepts discussed in chapter 7 and appendix B have also been extended to the fuzzy case. In section 9.4 an algorithm that builds a Fuzzy ID-3 decision tree will be discussed, along with the necessary extension of information theoretical concepts to the fuzzy case.

9.3.3 Fuzzy Clustering

In strong similarity to the clustering algorithm discussed in Chapter 8, a fuzzy version can also be derived. This is especially interesting in cases where different clusters overlap or noisy patterns interfere with the cluster building process. In contrast to classical clustering techniques, a fuzzy cluster algorithm does not try to assign each pattern to exactly one cluster. Instead, a degree of membership to each cluster is derived as well. Hence each pattern has a degree of compatibility with each cluster, rather then belonging to only one cluster.

As usual the clustering algorithm starts with a set of m patterns $\boldsymbol{x}_i$ $(1 \leq i \leq m)$ and a predefined number of clusters c. The result of the clustering will be a set of prototypes, or cluster-centers, $\boldsymbol{u}_j$ $(1 \leq j \leq c)$. A fuzzy clustering algorithm will then, in addition, produce a degree of membership $\mu_{i,j}$ to each of these clusters j for all patterns i.

A first example of such an algorithm, called fuzzy c-means, was explained in [63]. Here we will sketch a slightly different version which was presented in [321] and tends to be more stable against outliers and noise examples. The algorithm aims to minimize the following objective function (as usual n indicates the dimensionality of the feature space):

$$\sum_{j=1}^{c}\sum_{i=1}^{m}\mu_{i,j}^{n}\cdot \mathrm{d}(\boldsymbol{x}_i,\boldsymbol{u}_j)+\sum_{j=1}^{c}\eta_j\sum_{i=1}^{m}(1-\mu_{i,j})^{n}.$$

The first part of this objective function demands that the distances $\mathrm{d}(\boldsymbol{x}_i,\boldsymbol{u}_j)$ from the prototype vectors $\boldsymbol{u}_j$ to the patterns $\boldsymbol{x}_i$ is as low as possible, whereas the second term forces the degrees of membership to be as large as possible, thus avoiding the trivial solution with all $\mu_{i,j}$ being equal to zero. The η_j are factors to normalize and control the influence of this second term.

The resulting partition of the feature space can be regarded as a possibilistic partition, and the degrees of membership can be seen as degrees of compatibility with respect to the cluster centers. We will not go into detail about the formalism to construct these clusters since this is very similar to the one presented in Chapter 8. A detailed discussion of different approaches for the fuzzy case can be found for example in [321] and [140].

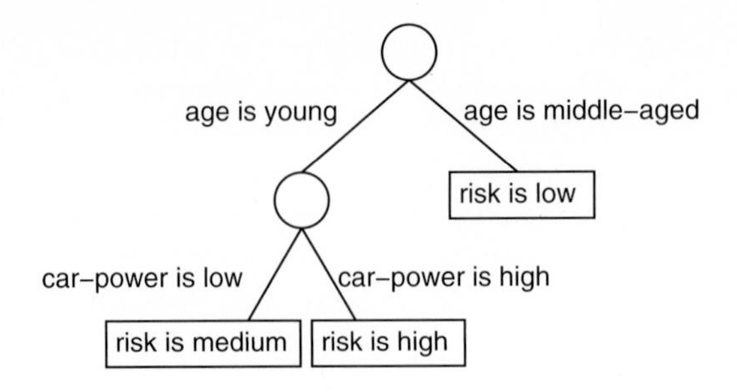

Fig. 9.18. An example fuzzy decision tree for a simple insurance risk predictor.

9.4. Fuzzy Decision Trees

In strong similarity to Appendix B, an extension to decision trees based on fuzzy logic can be derived. Different branches of the tree are then distinguished by fuzzy queries, an example for such a fuzzy decision tree is shown in Figure 9.18. Note how the decisions on each edge are fuzzy constraints rather than crisp conditions. In the following one possible extension to the concepts of information theory will be presented based on [288], followed by an example of a fuzzy ID3-algorithm to build decision trees.

9.4.1 Fuzzy Information Theory

Similar to probability a measure of a-priori membership can be defined for each class k $(1 \leq k \leq c)$ based on a data sample $T = \{\boldsymbol{x}_j : 1 \leq j \leq m\}$ and a degree of membership $\mu^k(\boldsymbol{x})$ of $\boldsymbol{x}$ to class k:

$$\mathcal{M}^k(T) = \frac{1}{|T|} \sum_{\boldsymbol{x} \in T} \mu^k(\boldsymbol{x})$$

(of course initially $|T| = m$, but later we will also have to deal with subsets of T, making this notation more consistent). Because fuzzy sets are not required to sum to one over the entire domain (as is the case for probabilities), the a-priori measure of membership for all classes is also not always equal to 1:

$$\mathcal{M}(T) = \sum_{k=1}^{c} \mathcal{M}^k(T)$$

Based on these measures we can now define a standard information content:

$$\mathcal{I}(T) = -\sum_{k=1}^{c} \frac{\mathcal{M}^k(T)}{\mathcal{M}(T)} \cdot \log \frac{\mathcal{M}^k(T)}{\mathcal{M}(T)}.$$

As also discussed in appendix B, this measure will reflect the amount of information contained in this soft assignment to various classes. If only one class has a degree of membership > 0, then $\mathcal{I}(T)$ will be zero. On the other hand, if several classes have a non-zero degree of membership, $\mathcal{I}(T)$ will be closer to one.

For the construction of decision trees, it is now of interest how much information we can gain by asking specific questions. Let us, therefore, assume that we can impose a constraint on the data sample, that is, define a fuzzy condition C that assigns a degree of fulfillment to each sample in T. In most cases this would represent a one-dimensional fuzzy condition, similar to a rule IF x_i IS C THEN $\cdots$. Now we can compute the conditional information:

$$\mathcal{M}^k(T|C) = \frac{1}{|T_C|} \sum_{\boldsymbol{x} \in T_C} \min\{\mu^k(\boldsymbol{x}), \mu_C(\boldsymbol{x})\}$$

$$\mathcal{M}(T|C) = \sum_{k=1}^{c} \mathcal{M}^k(T_C)$$

From there it is possible to derive a conditional information content:

$$\mathcal{I}(T|C) = -\sum_{k=1}^{c} \frac{\mathcal{M}^k(T|C)}{\mathcal{M}(T|C)} \cdot \log \frac{\mathcal{M}^k(T|C)}{\mathcal{M}(T|C)}$$

and we can easily derive how much information was obtained by applying the constraint C. The information gain computes to:

$$\mathcal{I}(T;C) = \mathcal{I}(T) - (\mathcal{I}(T|C) + \mathcal{I}(T|\neg C)).$$

Here we assume that we asked a question with only two possible outcomes, C and $\neg C$. Of course we could also have more than two outcomes (for example if we have several linguistic values for an attribute), then the information gain computes to:

$$\mathcal{I}(T;C) = \mathcal{I}(T) - \sum_{\text{cond} \in C} \mathcal{I}(T|\,\text{cond}).$$

when we have a set C of fuzzy constraints.

Using these measures we can now generate fuzzy decision trees, using an algorithm very similar to ID3 which is described in appendix B.

9.4.2 Fuzzy ID3

The following procedure to build a fuzzy decision tree is described in more detail in [288] and was applied to the recognition of handwritten numericals in [116]. In the following we will concentrate on the main algorithm which generates the decision tree itself.

We assume that in addition to the set of data examples T, a granulation for the input and output variables is available, dividing each attribute into a finite set of (soft) regions[5]. For each input variable x_i $(1 \leq i \leq n)$, this set of fuzzy terms is denoted by $D_i = \{v_p^i\}$. For each node N of the decision tree

[5] In case of a classification task this granulation needs only to be done for the input variables since the output already consists of a finite number of classes.

- F^N denotes the set of fuzzy restrictions on the path from F^{ROOT} leading to N ($F^{ROOT} = \emptyset$).
- V^N is the set of attributes appearing on the path from F^{ROOT} leading to N ($V^{ROOT} = \emptyset$). This assures that an attribute is only used once to come to a decision, analogous to the original ID3.
- χ assigns a membership to each training example at N ($\forall \boldsymbol{x} \in T : \chi^{ROOT}(\boldsymbol{x}) = 1$). If the training examples bear initial weights $\neq 1$ this can be taken into account as well. χ models the distribution of training patterns from T at different nodes. In contrast to classical decision trees where T would be split into disjunctive subsets, here only the degrees of memberships are adjusted, resulting in different values of χ for different nodes.
- $N|v_p^j$ denotes the child of N following the edge v_p^j which was created by using the test "V_j IS v_p^j" to split N.

The procedure to build the decision tree then operates iteratively.

1. At any node N still to be expanded compute the degree of membership for each class
$$\mathcal{M}^k(N) = \frac{1}{\sum_{\boldsymbol{x} \in T} \chi(\boldsymbol{x})} \sum_{\boldsymbol{x} \in T} \min\left\{\mu^k(\boldsymbol{x}), \chi(\boldsymbol{x})\right\}$$
and the entire node:
$$\mathcal{M}(N) = \sum_{k=1}^{c} \mathcal{M}^k(N)$$
2. Determine the information content of this node:
$$\mathcal{I}(N) = -\sum_{k=1}^{c} \frac{\mathcal{M}^k(N)}{\mathcal{M}(N)} \cdot \log \frac{\mathcal{M}^k(N)}{\mathcal{M}(N)}.$$
3. At each node we search the remaining attributes $V^{ROOT} - V^N$ to split this node:
   - compute the conditional information content, considering a test on v_p^j for attribute $V_j \in V^{ROOT} - V^N$:
$$\mathcal{M}^k(N|V_j \text{ IS } v_p^j) = \frac{1}{\sum_{\boldsymbol{x}} \chi(\boldsymbol{x})} \sum_{\boldsymbol{x} \in T} \min\left\{\mu^k(\boldsymbol{x}), \mu_{v_p^j}(\boldsymbol{x}), \chi(\boldsymbol{x})\right\}$$
$$\mathcal{M}(N|V_j \texttt{ IS } v_p^j) = \sum_{k=1}^{c} \mathcal{M}^k(N|V_j \texttt{ IS } v_p^j)$$
   and the conditional information content:
$$\mathcal{I}(N|V_j \texttt{ IS } v_p^j) = -\sum_{k=1}^{c} \frac{\mathcal{M}^k(N|V_j \texttt{ IS } v_p^j)}{\mathcal{M}(N|V_j \texttt{ IS } v_p^j)} \cdot \log \frac{\mathcal{M}^k(N|V_j \texttt{ IS } v_p^j)}{\mathcal{M}(N|V_j \texttt{ IS } v_p^j)}.$$

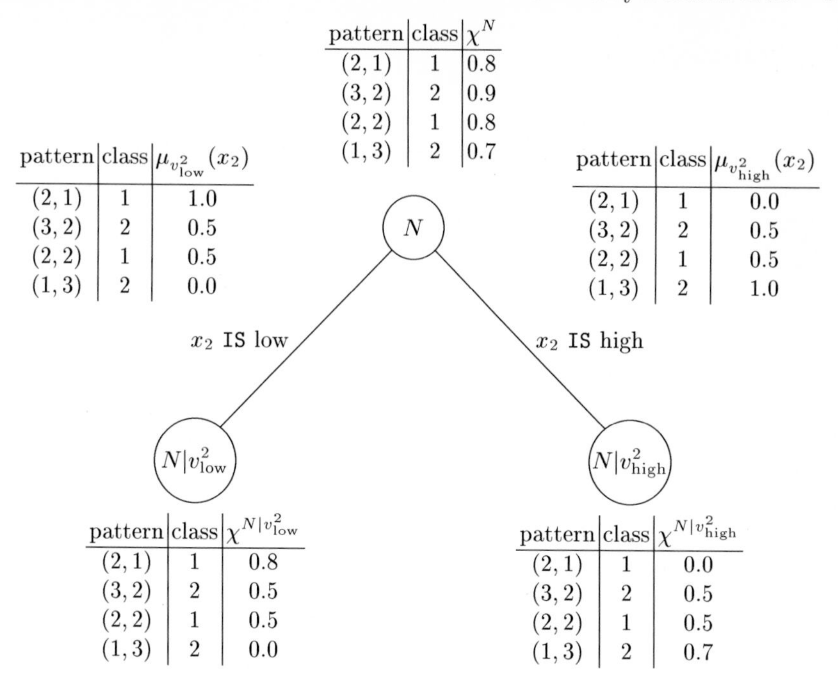

pattern	class	χ^N
(2, 1)	1	0.8
(3, 2)	2	0.9
(2, 2)	1	0.8
(1, 3)	2	0.7

pattern	class	$\mu_{v^2_{\text{low}}}(x_2)$
(2, 1)	1	1.0
(3, 2)	2	0.5
(2, 2)	1	0.5
(1, 3)	2	0.0

pattern	class	$\mu_{v^2_{\text{high}}}(x_2)$
(2, 1)	1	0.0
(3, 2)	2	0.5
(2, 2)	1	0.5
(1, 3)	2	1.0

pattern	class	$\chi^{N\|v^2_{\text{low}}}$
(2, 1)	1	0.8
(3, 2)	2	0.5
(2, 2)	1	0.5
(1, 3)	2	0.0

pattern	class	$\chi^{N\|v^2_{\text{high}}}$
(2, 1)	1	0.0
(3, 2)	2	0.5
(2, 2)	1	0.5
(1, 3)	2	0.7

Fig. 9.19. An example for a split of one node of a Fuzzy ID3 decision tree into two children. The tables show the new degrees of membership $\chi^{N|v^2_*}$ computed from the original node N and the degrees of membership $\mu_{v^2_*}$ for each pattern.

– select attribute V_j such that the information gain

$$\mathcal{I}(N) - \sum_{v^j_p \in V_j} \mathcal{I}(N|V_j \texttt{ IS } v^j_p)$$

is maximized. In the example in Figure 9.19 this attribute would be x_2.

4. Split N into $|V_j|$ subnodes $N|v^j_p$. The new memberships are computed using the fuzzy restriction leading to this node through

$$\chi^{N|v^j_p}(\boldsymbol{x}) = \min\left\{\chi^N(\boldsymbol{x}), \mu_{v^j_p}(\boldsymbol{x})\right\}.$$

In Figure 9.19 two new nodes are created and the tables show the new values for χ. Note how two patterns have nonzero degrees of membership for both nodes. Here a crisp decision using x_2 can not be made, therefore these elements maintain two memberships.

5. Continue until a certain depth of the tree is reached or node N has an overall membership $\mathcal{M}(N)$ below a certain threshold.

The described training procedure is the same as for ID3. The major difference comes from the fact that training examples can be found in more than one node to a certain degree of membership $\chi(\boldsymbol{x})$.

9.5. Conclusion

This chapter has discussed how fuzzy systems can be used for the analysis of data sets. In the scope of this book, however, only basic ideas were presented. Obviously more sophisticated methodologies have been proposed, chapter 10, for example, touches upon the subject of fine-tuning fuzzy rules by using genetic algorithms. We have also completely ignored how fuzzy rules can be used to inject expert knowledge into neural networks and how training algorithms from this area can then be used to adjust the membership functions of the corresponding rules. In [401] such Neuro-Fuzzy Systems are described in more detail.

In fact, most of the topics discussed in this book have in one way or another links with fuzzy logic, even probability – often seen as a rival to fuzzy logic. It has been merged with fuzzy logic, resulting in Fuzzy Probability and such interesting notions as the probability of a fuzzy event. The interested reader is referred to books on fuzzy logic [401, 404] and also to the relevant journals in this area, for example [562, 564]. Of additional interest are certainly two collections of Lotfi Zadeh's most influential papers [309, 553].

Chapter 10
Stochastic Search Methods

Christian Jacob
University of Calgary, Canada

10.1. Introduction: Exploring and Searching

Sophisticated search techniques form the backbone of modern machine learning and data analysis. Computer systems that are able to extract information from huge data sets (data mining), to recognize patterns, to do classification, or to suggest diagnoses, in short, systems that are adaptive and — to some extent — able to learn, fundamentally rely on effective and efficient search techniques. The ability of organisms to learn and adapt to signals from their environment is one of the core features of life. Technically, any adaptive system needs some kind of search operator in order to explore a feature space which describes all possible configurations of the system. Usually, one is interested in "optimal" or at least close to "optimal" configurations defined with respect to a specific application domain: the weight settings of a neural network for correct classification of some data, parameters that describe the body shape of an airplane with minimum drag, a sequence of jobs assigned to a flexible production line in a factory resulting in minimum idle time for the machine park, the configuration for a stable bridge with minimum weight or minimum cost to build and maintain, or a set of computer programs that implement a robot control task with a minimum number of commands.

These few examples show that learning means exploration of high-dimensional and multi-modal search spaces. The many dimensions make visualization as well as data analysis extremely difficult, hence, designing an appropriate search technique is a complex job. Multi-modality means that there is no single global maximum (optimum). There are many local maxima and several "interesting"

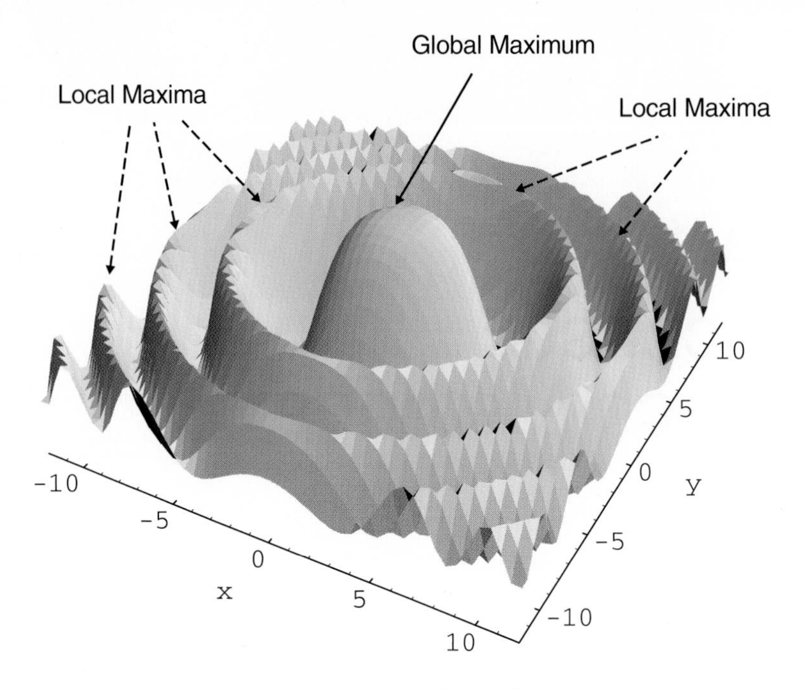

Fig. 10.1. Example of a multi-modal function.

(global) maxima, of which at least one has to be discovered by the search algorithm. Figure 10.1 shows a simple example of a multi-modal function for two dimensions.

In general, finding the global maximum (or minimum[1]) of an objective function that has many degrees of freedom and is subject to conflicting constraints is an NP-complete problem, since the objective function will tend to have many local extrema [144, 65]. Therefore, any procedure for solving hard optimization problems — these tend to be the practically relevant ones — should sample the multi-modal search space in such a way that there is a high probability of discovering near-optimal solutions. This technique should also lend itself to efficient implementation.

10.1.1 Categories of Search Techniques

Many search techniques that are more or less tailored to solve specific optimization problems have been developped: from calculus-based methods such as gradient strategies [443] in the versions of Gauss-Seidel, extrapolating gradient or simplex [409], [468, pp. 58] to enumerative techniques such as exhaustive search and dynamic programming [245], as well as stochastic methods such as

[1] Please note that searching for the maximum of a function $f(x)$ is the same as searching for the minimum of $-f(x)$.

Monte Carlo search [95], tabu search [468, pp. 162], [217], stochastic programming [170], or the recent approach of population-based incremental learning [40], [224], where randomness plays a major role in the search process.

Some kind of adaptability to the problem domain is part of most of these search methods. Any search algorithm for an adaptive (learning) system may be characterized by taking the following aspects into account [42]:

- How are solutions (parameters, hypotheses) represented? What data structures are used?
- What search operators are used to move from one configuration of solutions to the next? How is a single learning step defined?
- What type of search is conducted by applying the search operators iteratively? How is the search space explored and exploited?
- Is the adaptive system supervised or unsupervised? How is problem-specific knowledge incorporated into the learning system?

10.1.2 Nature's Paradigms of Stochastic, Adaptive Search

Search methods basically differ in their data structures encoding the search space and in their strategies of steping from one solution to another. Surprisingly, a certain degree of randomness may help tremendously in improving the ability of a search procedure to discover near-optimal solutions. Physical and biological processes in nature can teach us how to solve extremely complex optimization problems. Nature is able to arrange molecules as regular, crystal structures by careful temperature reduction or build highly adaptive, fractal structures by accretive growth [298]. Natural evolution has also designed adaptive, learning organisms with emergent competitive and cooperative strategies of interacting agents. This chapter will focus in particular on two categories of stochastic optimization techniques inspired by natural processes — simulated annealing and evolutionary algorithms.

Evolution in nature provides fascinating mechanisms for adapting living species to their environments. It is remarkable that such complex adaptations, taking place over huge time spans (from hundreds to millions of years) are essentially driven by a strikingly simple principle: iterated *selection* and *variation*, originally formulated by Charles Darwin [139]. All organisms that are able to reproduce, transmitting their specific genes to their descendants, have won nature's implicit *struggle-of-survival* test. The genes are passed on to the next generation, however, with some minor mutations, which are either due to imperfect copying of the genetic information or to recombination effects that are observable among sexually reproducing individuals.

A good overview of state-of-the-art optimization methods via evolutionary processes in the light of parallel programming techniques is provided by [436].

10.1.3 Chapter Overview

In the following sections we start with the stochastic optimization technique of simulated annealing inspired by the physical process of carefully cooling a molten substance. The derived algorithmic schemes used for solving optimization problems by probabilistic search bear many relations to evolution-based algorithms. The discussion of evolutionary search methods will comprise most part of this chapter, because evolutionary algorithms are playing a major role in the field of stochastic optimization.

Trying to integrate evolution into algorithmic schemes, is what many researchers, propagating the field known today as *evolutionary computation* (EC), have been doing for quite a long time (see, e.g., [192, 267, 443]). Consequently, many different versions and applications of evolutionary algorithms (EAs) have been developed. Most of them belong to one of the following three major EA-classes:

- ES: *Evolution Strategies* [445],
- GA: *Genetic Algorithms* [219, 268, 380], and
- GP: *Genetic Programming* [315–317, 42].

The following sections describe these three EA categories in detail. In Section 10.4, we will first explain *Evolution Strategies*, their typical representation of individuals as parameter vectors, mutations and recombinations, and their evolution schemes. The typical dynamics of Evolution Strategies are demonstrated by example of a parameter optimization task using multiple sub-populations.

Using parameter encodings by a discrete alphabet is the focus of *Genetic Algorithms* in Section 10.5. We will highlight examples for visualizing the dynamics of binary genotypic structures, comparing the GA operators of bit-mutation and crossover. Further genetic operators like inversion, deletion, and duplication are also discussed. Genetic Algorithms use several kinds of selection functions which will be introduced together with the typical GA evolution scheme. Finally, Genetic Algorithms will demonstrate their adaptability by example of a parameter optimization task in a changing environment.

In Section 10.6, we extend the GA scheme from parameter optimization to automatic evolution of computer programs known as *Genetic Programming.* Individuals are represented as symbolic expressions, that is, hierarchical structures which are evolving. How to breed and animate the solutions of a Genetic Programming robot control task will be demonstrated in this section.

10.2. Stochastic Search by Simulated Annealing

The simulated annealing approach is in the realm of problem solving methods that make use of paradigms found in nature. Annealing denotes the process of cooling a molten substance and is, for example, used to harden steel. One major effect of "careful" annealing is the condensing of matter into a crystalline solid. The hardening of steel is achieved by first raising the temperature close to the transition to its liquid phase, then cooling the steel slowly to allow the molecules to arrange in an ordered lattice pattern. Hence, annealing can be regarded as an adaptation process optimizing the stability of the final crystalline solid. Whether a state of minimum free energy is reached, depends very much on the actual speed with which the temperature is being decreased.

If the correspondence between macro variables like density, temperature, and entropy is described statistically, as first done by Boltzmann, the probability $p_T(\boldsymbol{s})$ for the particle system to be in state $\boldsymbol{s}$ for a certain temperature T depends on the free energy $E(\boldsymbol{s})$ and is described by the *Boltzmann distribution* [1]:

$$p_T(\boldsymbol{s}) = \frac{1}{n} \cdot \exp\left(\frac{-E(\boldsymbol{s})}{k \cdot T}\right) \quad \text{with} \quad n = \sum_{\boldsymbol{s} \in S} \exp\left(\frac{-E(\boldsymbol{s})}{k \cdot T}\right). \tag{10.1}$$

Here n is used to normalize the probability distribution, where S denotes the set of all possible system states, and k is the Boltzmann constant. A stochastic simulation algorithm has been proposed by Metropolis et al. in order to simulate the structural evolution of a molten substance for a given temperature [374].

10.2.1 The Metropolis Algorithm

The Metropolis procedure describes the probabilistic step in the evolution of a particle system from one configuration to another. The system is assumed to be in state $\boldsymbol{s}$ with free energy $E(\boldsymbol{s})$ at time t and temperature T. By means of a Monte-Carlo method a new configuration $\boldsymbol{s}_{\text{new}}$ is generated for time step $t+1$. Whether this new state $\boldsymbol{s}_{\text{new}}$ is actually accepted depends on the energy difference $\Delta E = E(\boldsymbol{s}_{\text{new}}) - E(\boldsymbol{s}))$. If $\Delta E < 0$, the new configuration with lower energy is accepted. Otherwise, $\boldsymbol{s}_{\text{new}}$ is rejected with probability

$$1 - \exp\left(\frac{\Delta E}{k \cdot T}\right). \tag{10.2}$$

The Metropolis algorithm is described in more detail in Figure 10.2. The procedure is parametrized by three arguments: The current system state is again denoted as $\boldsymbol{s}$, the current temperature is T, and the parameter m controls how long the equilibration phase should last.

Furthermore, two functions incorporate the stochastic elements: $Perturb(\boldsymbol{s})$ returns a modified system configuration most probably in the "vicinity" of the current system state $\boldsymbol{s}$. The function $\chi_{[0,1]}$ gives a uniformly distributed random number in the interval from 0 to 1.

10.2.2 Simulated Annealing

Starting from a configuration $\boldsymbol{s}$, the procedure $Metropolis(\boldsymbol{s}, T, m)$ simulates an equilibration process for a fixed temperature T over a (usually large) number of m time steps. Hence, for simulated cooling the Metropolis procedure simply has to be repeated for decreasing temperatures (Figure 10.3),

$$T_{\text{init}} = T_0 > T_1 > \cdots > T_{\text{final}}, \tag{10.3}$$

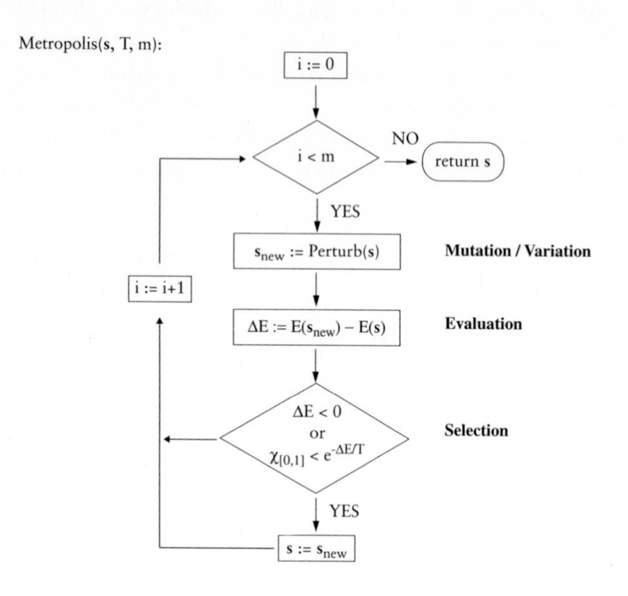

Fig. 10.2. Metropolis algorithm.

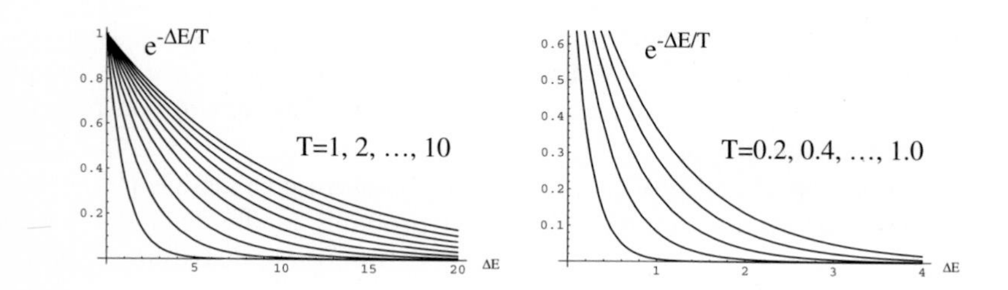

Fig. 10.3. The Metropolis selection (Figure 10.2) depends on ΔE as well as the temperature T which parametrizes the shape of the function $\exp(-\Delta E/T)$.

resulting in a sequence of annealing configurations

$$s_{\text{init}} = s_0; \tag{10.4}$$

$$s_1 = Metropolis(s_0, T_0, m); \tag{10.5}$$

$$s_2 = Metropolis(s_1, T_1, m); \ldots \tag{10.6}$$

with gradually decreasing free energies

$$E(s_0) \geq \cdots \geq E(s_i) \geq \cdots \geq E(s_{\text{final}}). \tag{10.7}$$

This methodology basically describes an optimization process where a sequence of solutions gradually evolves towards an equilibrium configuration for a minimum temperature (Figure 10.4). An optimization problem can be considered as a simulated cooling process using the following analogies to physical annealing:

- Solutions for the optimization problem correspond to system configurations.
- The system "energy" is described by the optimization function.
- Searching for a good or optimal problem solution is like finding a system configuration with minimum free energy.
- The temperature and equilibration time steps are two parameters (among others) for controlling the optimization process.

For the optimization process to be successful, i.e., not to be prone to premature convergence, a major influencing factor is the *annealing schedule* which describes how temperature T will be decreased and how many iterations will be used during each equilibration phase. In Figure 10.2 we use a simple cooling plan $T = \alpha \cdot T$, with $0 < \alpha < 1$, and a fixed number m of equilibration time steps. If the inital temperature T_0 is high, practically any new solution is accepted, thus premature convergence towards a specific region within the search space can be avoided. Careful cooling with $\alpha = 0.8 \ldots 0.99$ will lead to an asymptotic drift towards the minimum temperature T_{final}. On its search for a minimum or maximum, the algorithm may be able to escape from local extrema and finally reach an even better region, which is due to the fact that — with certain probability — good intermediate positions are "forgotten".

Annealing schedules such as these used to find minimum-cost solutions to large optimization problems, based on Metropolis' simulation algorithms, were first published by Kirkpatrick et al. [307]. They demonstrated the application of this stochastic computational technique to problems of combinatorial optimization, like wire routing and component placement in VLSI design.

10.2.3 Variants and Applications of Simulated Annealing

Since the first appearance of simulated annealing (SA) and its successful applications, many more SA variants and annealing schedules with special focus on specific optimization problems have been proposed. Davis et al. provide a good

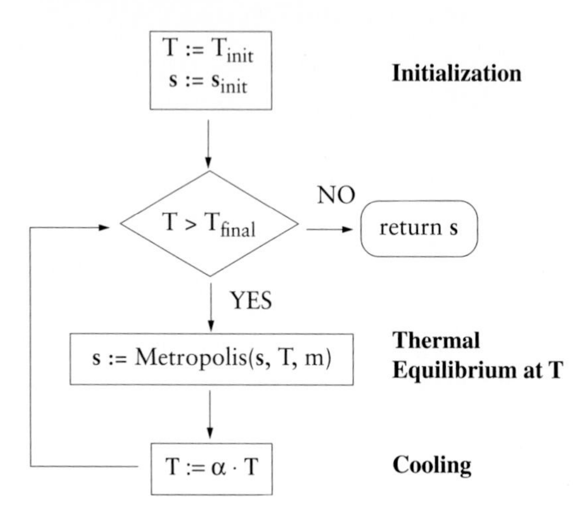

Fig. 10.4. Simulated Annealing (SA): The basic algorithmic SA scheme applies the Metropolis algorithm for iteratively reduced temperatures T, simulating a cooling process.

overview of SA applications at that time and partly compare simulated annealing with the evolutionary methods of genetic algorithms [142, 144]. Books by Azencott and Vidal provide more up-to-date overviews of simulated annealing and its applications [35, 528]. Collins et al. prepared an annotated bibliography for simulated annealing and its applications in the 1980s [125]. A recent bibliography of simulated annealing (as well as genetic algorithms) is available as an updated report and on the internet [12].

Recently, optimization techniques based on simulated annealing have been used for designing the page layout of newspapers [328], for evolving strategies to play the *MasterMind* game, which can be reduced to a dynamic constrained optimization problem in a string search space [58], for evolving symbolic expressions encoding computer programs [411], or for spacecraft design under performance uncertainty [431].

Mori et al. suggest a *thermodynamical genetic algorithm* (GA) as a hybrid optimization system combining evolutionary search with simulated annealing [390]. Most of the articles mentioned above are comparisons of SA and GA-based optimization techniques [411, 431, 58]. We will return to these references in Section 10.7 when we compare SA- and evolution-based search methods. A good overview of state-of-the-art optimization methods via evolutionary processes can be found in [436].

Threshold Accepting (TA) is an SA variant which accepts any solution whenever it is not much worse than the current solution. The allowed difference between these two solutions is controlled by a threshold value which becomes part of the annealing schedule and is gradually decreased to zero during the optimization process. Therefore, the degree of acceptable deterioration is reduced step by step until, finally, only improvements are allowed [160], [403, p. 223-228].

Another variant of the TA algorithm is *Record-to-Record Travel* using a slightly different rule for replacing the current solution by a new one: the best-so-far solution s_{best} is always updated; each new solution is compared to s_{best} and is only accepted if it is not worse than $s_{\text{best}} - d$ for some $d > 0$. The parameter d has to be adjusted during the optimization process, with d finally approaching zero [159].

The *Deluge Algorithm*, another promising SA-related optimization technique, is probably best explained by the following metaphor. The search algorithm is represented by a hiker who tries to climb up to the highest peak in a mountainous area. However, the hiker is short-sighted and only able to examine local height differences in her vicinity. The hiker has to rely on local gradient information. During her search, it starts raining, slowly flooding the surrounding landscape. Because the hiker wants to keep her feet dry, she tries to avoid the gradually rising water and is forced to continuously move uphill [159], [403, p. 228-230].

10.2.4 Parallel Simulated Annealing

Due to the inherent sequential nature of the Metropolis algorithm as well as the iterated cooling procedure, it is not an obvious task how to parallelize an optimization technique based on simulated annealing [226]. One suggestion is parallelization at the data level, which results in rather problem-dependent strategies, as shown, e.g., for combinatorial problems [14] and for computer aided design tasks [107]. An alternative and popular approach for SA parallelization is the *division strategy* where a set of independent annealings are performed, augmented by a periodic "coordinating" process. A common strategy is to periodically select the best current solution as the new starting point for subsequent independent annealings [154].

An even more flexible parallelization approach, using selection and migration, is derived from the *island model* also used for distributed genetic algorithm implementations which was itself inspired by the concept of co-evolving subpopulations in evolution theory [338]. The SA is implemented on an MIMD-type machine with p processors. On each processor, k annealings are performed in parallel, i.e., a subpopulation resides on each processor, the whole annealing process thus covering $k \cdot p$ annealings. All the processors perform n SA iterations on each of the k members of its population. After this annealing epoch, each processor exchanges individuals with some neighbouring processor(s). After this *migration* each processor performs a selection among its current together with its newly acquired solutions (e.g., by picking the k best individuals). Finally, another annealing epoch is started, etc. The only major decisions one has to take into account are how to actually perform migration and selection.

LIVERPOOL
JOHN MOORES UNIVERSITY
AVRIL ROBARTS LRC

This variant of a parallel simulated annealing method is already incorporating basic concepts that are typical of evolutionary algorithms: populations of individuals perform a parallel search for better solutions, the search process being directed by selection, and the exploration of new territory induced by some kind of mutation. Therefore, when we look closer at search algorithms based on principles of evolution, we will see that there are even more commonalities between SA and evolutionary algorithm (EA) techniques than the fact that both techniques are inspired by processes in nature — one drawn from physical, the other from biological paradigms.

10.3. Stochastic, Adaptive Search by Evolution

Computer algorithms modeling the search processes of natural evolution are crude simplifications of biological reality. However, during nearly three decades of research and application, they have turned out to yield robust, flexible and efficient algorithms for solving a wide range of optimization problems.

In this section a basic scheme of an evolutionary algorithm (EA) is presented to which we will refer in the later sections when the major EA variants (Evolution Stategies, Genetic Algorithms, and Genetic Programming) are introduced. We will also briefly discuss the role of randomness in evolutionary learning.

10.3.1 Variation and Selection: An Evolutionary Algorithm Scheme

Evolutionary algorithms (EAs) simulate a collective learning process within a population of individuals. More advanced EAs even rely on competition, cooperation and learning among several populations. Each individual represents a point (structure) in the search space of potential solutions to a specific machine learning problem. After arbitrary initialization of the population, the set of individuals evolves toward better and better regions of the search space by means of partly stochastic processes while the environment provides feedback information (quality, fitness) about the search points [38]:

Selection, which is deterministic in some algorithms, favors those individuals of better fitness to reproduce more often than those of lower fitness.

Mutation introduces innovation by random variation of the individual structures.

Recombination, which is omitted in some EA realizations, allows the mixing of parental genetic information while passing it to their descendants.

Figure 10.5 shows the basic scheme of an evolutionary algorithm. Let G denote the search space and let $\eta : G \to \Re$ be the fitness function that assigns a real value $\eta(\sigma_i) \in \Re$ to each individual structure encoded by a "genome" $\boldsymbol{g}_i \in G$. A population $G(t) = \{\boldsymbol{g}_1(t), \ldots, \boldsymbol{g}_\mu(t)\}$ at generation t is described by a (multi) set[2] of μ individuals. From a parent population of size $\mu \geq 1$ an offspring

[2] In a multi set the same element can appear more than just once.

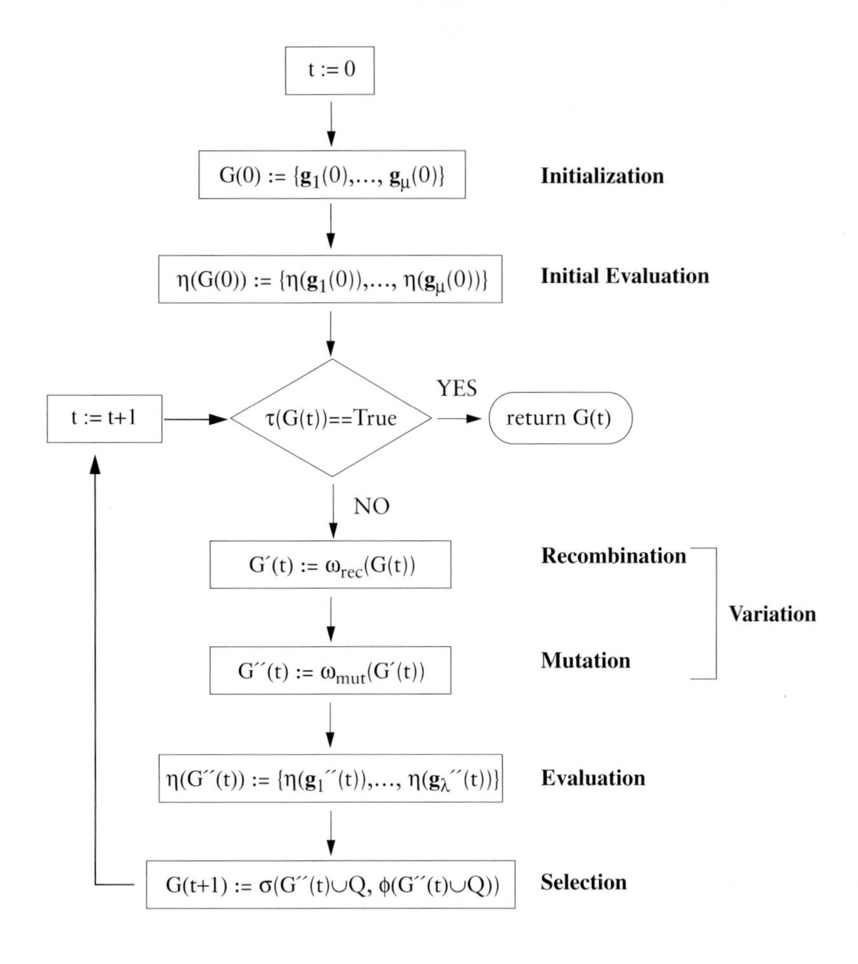

Fig. 10.5. An Evolutionary Algorithm Scheme.

population of size $\lambda \geq 1$ is created by means of recombination and mutation at each generation. A recombination operator $\omega_{\text{rec}} : G^\mu \to G^\lambda$, controlled by additional parameters $\Theta(\omega_{\text{rec}})$, generates λ structures as a result of combining the genetic information of μ individuals (for sexual recombination: $\mu = 2, \lambda = 1$).

A mutation operator $\omega_{\text{mut}} : G^\lambda \to G^\lambda$ modifies a subpopulation of λ individuals, again being controlled by parameters $\Theta(\omega_{\text{mut}})$. All the newly created solutions of the population set G'' are evaluated. The selection operator σ chooses the parent population for the next generation. The selection pool consists of the λ generated offspring. In some cases, the parents are also part of the selection pool, denoted by the extension set $Q = G(t)$. Otherwise, the extended selection set Q is empty, $Q = \emptyset$. Therefore, the selection operator is defined as $\sigma : G^\lambda \times \Re^\lambda \to G^\mu$ or $\sigma : G^{\lambda+\mu} \times \Re^{\lambda+\mu} \to G^\mu$. The termination criterion τ either makes the evolution loop stop after a predefined number of generations, or when an individual exceeding a maximum fitness value has been found.

10.3.2 The Role of Randomness in Evolutionary Learning

The principle dynamic elements that evolutionary algorithms simulate are *innovation* caused by mutation combined with natural *selection*. Driving a search process by randomness is, in some way, the most general procedure that can be designed. Increasing order among the population of solutions and causing learning by randomly changing the solution encoding structures may look counter-intuitive. However, random changes in combination with fitness-based selection make a formidable and more flexible search paradigm as demonstrated by nature's optimization methods [168], [42]. On average, it is better to explore a search space non-deterministically. This is independent of whether the search space is small enough to allow exhaustive search or whether it is so large that only sampling can reasonably cover it. At each location in a non-deterministic search, the algorithm has the choice of where to go next. Meaning that stochastic search — in combination with a selection procedure which might also be non-deterministic, at least to some degree — is used as a major tool not only to explore but also to exploit the search space.

In simulated annealing the Metropolis algorithm also relies on the stochastic search of the *Perturb* procedure which generates a new solution that competes with the current (best) solution. Evolutionary algorithms use operators like mutation and recombination (in some EA variants even more "genetic" operators are used) to produce new variants of already achieved solutions. These operators rely heavily on randomness: mutation may randomly change parameter settings of a solution encoding vector, and also the mixing of genes by recombination of two solutions is totally non-deterministic. EA-based search algorithms rely on evolution's "creative potential" which is largely due to a controlled degree of randomness.

10.4. Evolution Strategies

Evolution Strategies (ES) were invented during the 60s and 70s as a technique of evolutionary experimentation for solving complex optimization problems, mainly within engineering domains [442–445]. The preferred ES data structures are vectors of real numbers. Specifically tailored mutation operators are used to produce slight variations on the vector elements by adding normally distributed numbers (with mean zero) to each of the components. Recombination operators have been designed for the interchange or mixing of "genes" between two or more vectors. These recombinations range from simply swapping respective components among two vectors to component-wise computation of means. Evolution Strategies have developed sophisticated methods of selection, which are especially important when the evolution scheme involves several subpopulations.

10.4.1 Representation of Individuals

The basic data structures today's Evolution Strategies deal with, and around which most of the ES theory is built, are *vectors of real numbers* representing a set of parameters to be optimized. Therefore, an ES chromosome $\boldsymbol{g}$, as we prefer to call those vectors, can be simply defined as follows:

$$\boldsymbol{g} = (p_1, p_2, \ldots, p_n) \quad \text{with} \quad p_i \in \Re. \tag{10.8}$$

Usually the p_i are referred to as the *object parameters*. In the most elementary versions of Evolution Strategies only these parameter vectors are subject to evolution. However, in order to be able to solve more complex optimization tasks it turns out to be advantageous to keep a second vector of *strategy parameters*, which are used as variances to control the range of mutations on the object parameters. Thus we can extend our definition for an ES chromosome $\boldsymbol{g}$ the following way:

$$\boldsymbol{g} = (\boldsymbol{p}, \boldsymbol{s}) = ((p_1, p_2, \ldots, p_n), (s_1, s_2, \ldots, s_n)) \quad \text{with} \quad p_i, s_i \in \Re. \tag{10.9}$$

Now the ES chromosome vector is two-fold, with each s_i representing the mutation variance for the corresponding p_i. The additional control parameters $\boldsymbol{s}$ may be considered as an *internal model* of the optimization environment [467, 468]. As these parameters are also in the scope of the mutation and recombination operators, this has the effect of evolving a strategy for adaptation in conjunction with the object parameters being adapted.

10.4.2 ES Mutations

In order to demonstrate the dynamics of iterated selection and mutation, imagine the following scenario. We would like to find our way to the top of a hill as depicted in Figure 10.6. We start our journey at some distance from the top plateau with one individual, depicted as a gray spot. From the genotype, an ES chromosome containing the x- and y-coordinates of this individual, we generate five mutated offspring (children, the black spots), the locations of which are somewhat close to the parent individual. This set of points comprises our initial population on which we will perform a selection-mutation procedure as follows. We choose the individual currently with the highest location, from which we generate another five children, and continue the same way by selecting the best, generating mutants, etc.

Repeatedly choosing the best individual and generating mutated offspring, gradually makes the population move upwards, like following a gradient to the top plateau. A closer look at Figure 10.6 reveals that smaller mutations are prevalent, i.e. the majority of mutants (children) are located near to the wildtype (parent). This effect is mainly due to the Gaussian distributions used to create offspring genotypes, as we will see in the following subsections.

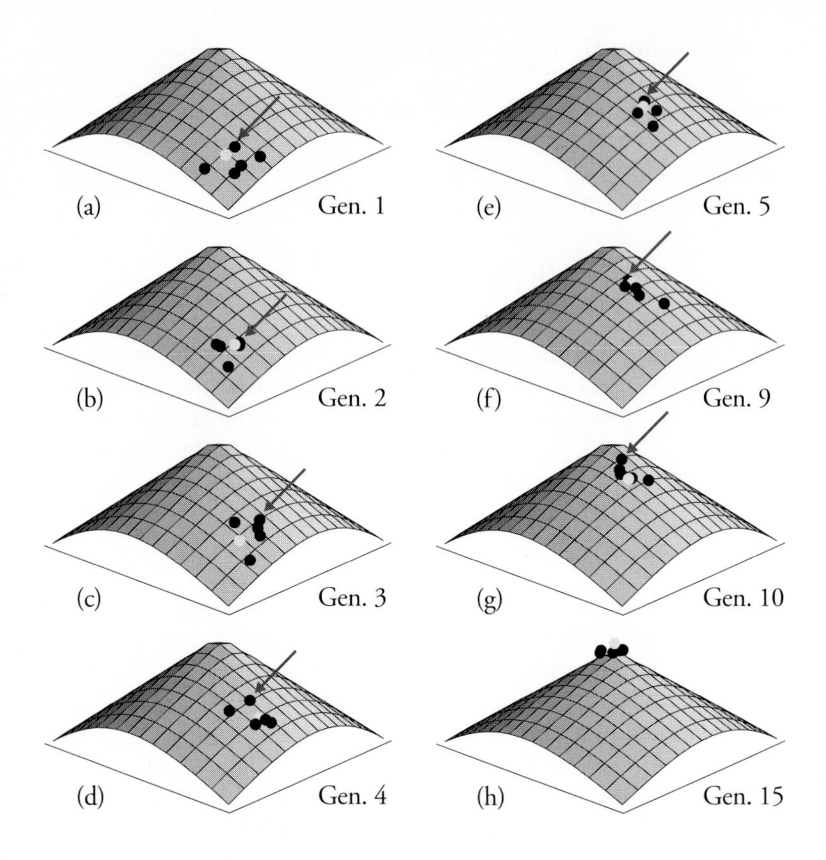

Fig. 10.6. An ES population consisting of 6 individuals climbing to the top by iterated elitist selection and mutation. The best individual of each generation is marked by an arrow.

Not only the object parameters, the coordinates in our example, but also the strategy parameters, controlling the mutation step sizes, are subject to change. This effect is also visible in the figure. Each strategy parameter controls the mutation step size, hence the distance of the children from their parent. With the collection of points moving uphill it is visible that the "flock radius" changes as well and becomes smaller when the population approaches the top. Towards the end of the optimization task, smaller mutations turn out to be more advantageous.

Mutating the Object Parameters Mutation is considered *the* major ES operator for variations on the chromosomes. Mutations of the object and strategy parameters are accomplished in different ways. Basically, ES mutation on a chromosome $\boldsymbol{g} = (\boldsymbol{p}, \boldsymbol{s})$ can be described as follows:

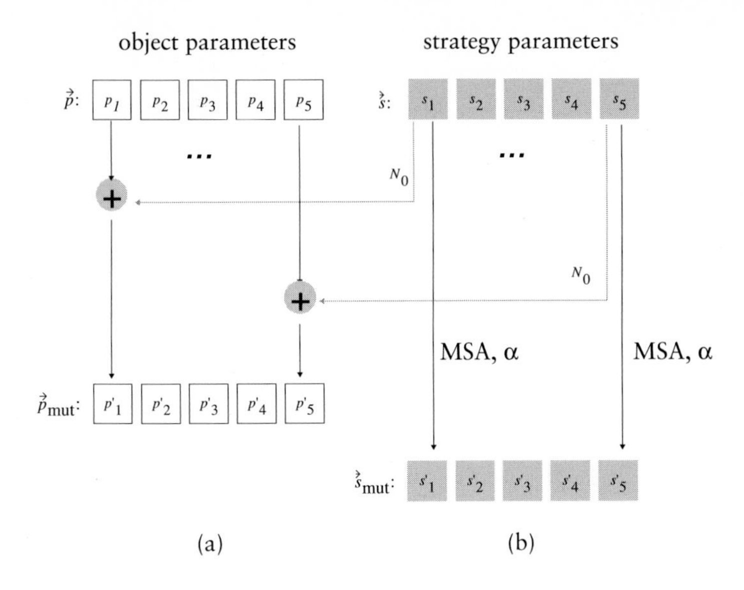

Fig. 10.7. Mutation scheme of an ES chromosome: (a) Mutation stepsizes of the object parameters are controlled by the strategy parameters. (b) Strategy parameters are adjusted by a heuristic function α (MSA).

$$\boldsymbol{g}_{\text{mut}} = (\boldsymbol{p}_{\text{mut}}, \boldsymbol{s}_{\text{mut}}) = (\boldsymbol{p} + \mathcal{N}_0(\boldsymbol{s}), \alpha(\boldsymbol{s})). \tag{10.10}$$

Here $\mathcal{N}_0(\boldsymbol{s})$ denotes normal distribution with mean zero and the vector of variances $\boldsymbol{s}$. α defines a function for adapting the strategy parameters. The variations on the object parameters are to be applied to each vector element separately:

$$\boldsymbol{p}_{\text{mut}} = (p_1 + \mathcal{N}_0(s_1), \ldots, p_n + \mathcal{N}_0(s_n)). \tag{10.11}$$

The strategy parameters are adjusted in an analogous way as

$$\boldsymbol{s}_{\text{mut}} = (\alpha(s_1), \ldots, \alpha(s_n)). \tag{10.12}$$

These adaptations of the mutation stepsizes will be discussed in the following subsection. For the rest of this section, we will assume that strategy parameters remain unchanged, i.e., we use the identity mapping $\alpha(x) = x$. Figure 10.7 graphically depicts this basic mutation scheme. Each strategy parameter controls the mutation range for its respective object parameter.

In order to have mutations prefer smaller changes to larger ones, Evolution Strategies use normal (Gaussian) distributed random numbers that are added to the object parameters. The characteristics of Gauss distributions, $\mathcal{N}_m(d)$ with mean m and standard deviation $\sqrt{d}$,

$$\mathcal{N}(m,d) = \frac{1}{\sqrt{2\pi d}} \cdot e^{-\frac{(x-m)^2}{2d}} \tag{10.13}$$

are their preference for numbers around mean m. Thus, in the case of $m = 0$, smaller values close to zero are selected more often than values with greater distance to m.

Mutating the Strategy Parameters In the previous subsection we learnt about mutations of the object parameters. We will discuss two of the heuristics generally used for adapting the strategy parameters $\boldsymbol{s}$ of an ES chromosome $\boldsymbol{g} = (\boldsymbol{p}, \boldsymbol{s})$:

$$\boldsymbol{g}_{\text{mut}} = (\boldsymbol{p}_{\text{mut}}, \boldsymbol{s}_{\text{mut}}) = (\boldsymbol{p} + \mathcal{N}_0(\boldsymbol{s}), \alpha(\boldsymbol{s})). \tag{10.14}$$

Here α defines a function for adapting the strategy parameters:

$$\boldsymbol{s}_{\text{mut}} = (\alpha(s_1), \ldots, \alpha(s_n)). \tag{10.15}$$

Each strategy parameter controls the mutation range of its respective object parameter, defining the variance for the normal distributed random values added to the object parameters.

There are several heuristics for defining α which work fairly well in changing the strategy parameters; this adaptation is usually referrred to as *mutative stepsize adaptation* (MSA). One suggestion for adjusting the standard deviations comes from Rechenberg [445]:

$$\boldsymbol{s}_{\text{mut}} = (s_1\zeta_1, \ldots, s_n\zeta_n) \tag{10.16}$$

$$\zeta_i = \begin{cases} \beta & : \ \chi < 0.5 \\ \frac{1}{\beta} & : \ \chi \geq 0.5 \end{cases} \tag{10.17}$$

Here χ denotes a uniformly distributed random variable from the interval $[0, 1]$. For $n < 100$ parameters Rechenberg suggests values for β between 1.3 and 1.5. For $\beta = 1.3$ this means that, on average, half of the strategy parameters are multiplied by 1.3, the rest is scaled by $0.77 = \frac{1}{1.3}$.

Another adaptation scheme for strategy parameters has been proposed by Hoffmeister and Bäck [265]:

$$\boldsymbol{s}_{\text{mut}} = (s_1\zeta_1, \ldots, s_n\zeta_n) \tag{10.18}$$

$$\zeta_i = e^{\mathcal{N}_0(\delta)}. \tag{10.19}$$

Here the parameter δ controls the range for the stepsize adjustment. Using an exponential function avoids negative values ($\zeta_i > 0$) and favours smaller changes by multiplication with values around 1. The strategy parameter is reduced for negative values of $\mathcal{N}_0(\delta)$, that is, for $0 < \zeta_i < 1$. For positive $\mathcal{N}_0(\delta)$ the parameter is scaled by a factor $\zeta_i > 1$.

10.4.3 ES Recombinations

Recombination operators create new chromosomes by composing corresponding parts of two or more chromosomes, thus mimicking sexual recombination mechanisms of cellular genomes. For the binary case, where two ES chromosomes, $\boldsymbol{g}_a = (\boldsymbol{p}_a, \boldsymbol{s}_a)$ and $\boldsymbol{g}_b = (\boldsymbol{p}_b, \boldsymbol{s}_b)$, are to be recombined by an operator ω_{rec}, we can describe the composition of a new chromosome as follows:

$$\omega_{\text{rec}}(\boldsymbol{g}_a, \boldsymbol{g}_b) = (\boldsymbol{p}', \boldsymbol{s}') = ((p'_1, \ldots, p'_n), (s'_1, \ldots, s'_n)). \tag{10.20}$$

Each element of the object and strategy parameter vector is a combination of the respective entries of $\boldsymbol{g}_a$ and $\boldsymbol{g}_b$ by two functions ρ_p and ρ_s:

$$p'_i = \rho_p(p_{ai}, p_{bi}) \quad \text{and} \quad s'_i = \rho_s(s_{ai}, s_{bi}). \tag{10.21}$$

Here the functions ρ_p and ρ_s define the component-wise recombination mapping for the object and strategy parameters, respectively. In order to keep the formulas simpler, we will assume an identical recombination mapping for both the object and strategy parameters, $\rho = \rho_p = \rho_s$.

Two recombination mappings are very common in Evolution Strategies: *discrete* and *intermediate* recombination. The following sections will discuss these two functions $\rho = \rho_{\text{dis}}$ and $\rho = \rho_{\text{int}}$ in detail.

Discrete Recombination With a *discrete recombination* function, ρ_{dis}, one of the two vector components is chosen at random and declared to be the new vector entry. For the case of binary recombination this means:

$$\rho_{\text{dis}}(x_a, x_b) = \begin{cases} x_a & : \quad \chi < 0.5 \\ x_b & : \quad \chi \geq 0.5 \end{cases} \tag{10.22}$$

Here χ computes a uniformly distributed random number from the interval $[0, 1]$. Each component x_a or x_b is selected with a 50-percent probability. In general, for μ values $x_1, \ldots, x_\mu$ to be recombined in discrete manner, $\rho_{\text{dis}}(x_1, \ldots, x_\mu)$, the probability to choose parameter x_i is $1/\mu$. Figure 10.8 illustrates discrete recombination of three ES chromosomes $(\boldsymbol{p}_1, \boldsymbol{s}_1)$, $(\boldsymbol{p}_3, \boldsymbol{s}_3)$ and $(\boldsymbol{p}_6, \boldsymbol{s}_6)$ into a new chromosome $(\boldsymbol{p}', \boldsymbol{s}')$. In Evolution Strategies discrete recombinations are mainly used for interchanging strategy parameters, i.e., usually $\rho_s = \rho_{\text{dis}}$.

Intermediate Recombination For many ES application domains dealing with real numbers that represent some control parameter settings, taking the mean value of corresponding elements turns out to be a sensible and natural operator. This is exactly what *intermediate* recombination does. Recombining μ chromosomes intermediately means that the following mapping, ρ_{int}, is applied to each set of corresponding vector components:

$$\rho_{\text{int}}(x_1, \ldots, x_\mu) = \frac{1}{\mu} \cdot \sum_{i=1}^{\mu} x_i \tag{10.23}$$

Although the choice of recombination functions is very much dependent on the requirements of a concrete application domain, for real-valued parameter optimization with Evolution Strategies intermediate recombination is the prevalent inter-chromosomal operator used for object parameters, i.e., in many cases $\rho_p = \rho_{\text{int}}$ turns out to be a natural choice.

Local and Global Recombinations We will extend our previous definitions of recombinations to a more flexible recombination scheme on populations of ES chromosomes. Basically, *local* recombinations work on a subpopulation of chromosomes, whereas for *global* recombinations each component can be selected from the set of corresponding entries among all chromosomes in a population. This results in an increased mixing of genotypic information [467]. In the following sections, with the term *recombination* we will always refer to a multi-recombination on a (sub-)population $G = (\boldsymbol{x}_1, \ldots, \boldsymbol{x}_\mu)$ of μ ES chromosomes, each being of the form

$$\boldsymbol{x_i} = (x_{i1}, \ldots, x_{in}) \quad \text{with} \quad 1 \le i \le \mu. \tag{10.24}$$

A multi-recombination operator ω_{rec} working on r chromosomes will be formalized as:

$$\omega_{\text{rec}}(\boldsymbol{x}_{i_1}, \ldots, \boldsymbol{x}_{i_r}) = \boldsymbol{x'} = (x'_1, \ldots, x'_n). \tag{10.25}$$

The indices $i_1, \ldots, i_r \in \{1, \ldots, N\}$, with $i_j \neq i_k$ for $1 \le j < k \le r$, refer to the chromosomes that are selected from G as recombination partners, none of which appears in more than one copy. However, this does not exclude the case that several chromosomes within the population may be identical.

We have already had a more detailed look at *local* intermediate and discrete recombinations in the previous sections (Figure 10.8). Here we will denote these recombinations by $\omega_{\text{rec}}^{\text{local}}$, i.e., for r recombinants the new chromosome is computed as follows:

$$\omega_{\text{rec}}^{\text{local}}(\boldsymbol{x}_{i_1}, \ldots, \boldsymbol{x}_{i_r}) = \boldsymbol{x'} = (x'_1, \ldots, x'_n) \tag{10.26}$$

$$x'_k = \begin{cases} x_{k,i_1} \text{ or} \ldots \text{or } x_{k,i_r} & : \quad \rho = \rho_{\text{dis}} \\ \frac{1}{r} \cdot \sum_{i=1}^{r} x_{k,i_m} & : \quad \rho = \rho_{\text{int}} \end{cases} \tag{10.27}$$

Again, ρ is the per-component recombination mapping. Figure 10.8 illustrates this local recombination scheme for a population of seven chromosomes from which three are chosen as recombinants: component-wise recombination is performed for chromosomes 1, 3 and 6, all marked by white squares. For discrete recombination, one element is selected (gray circle) within each column. This mode of recombination is called *local* because the recombination partners are selected prior to applying the combination mapping ρ. Thus, local recombinations work on subpopulations which are considered as a local domain.

For *global* recombination there is no preselection of vectors from which the recombining elements have to be chosen. Instead, all chromosomes of a population

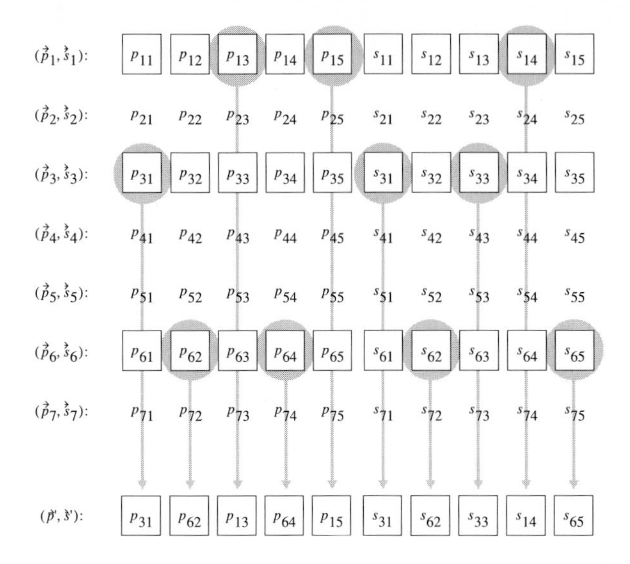

Fig. 10.8. Local, discrete multi-recombination: example with three recombined, locally preselected vectors.

are taken into account as possible recombinants. Figure 10.9 illustrates the basic selection scheme of global recombination on a population of seven parameter vectors. Within each column three elements are selected to which the recombination mapping is applied. For the discrete version, e.g., this means that one entry is chosen at random among the respective three marked column components.

We formalize global recombination analogously to local recombinations. The only difference is that the domain for selecting the components to be recombined is extended to the whole population. We define a global recombination operator, $\omega_{\text{rec},r}^{\text{global}}$, that combines elements "per column" as follows:

$$\omega_{\text{rec},r}^{\text{global}}(\boldsymbol{x}_1, \ldots, \boldsymbol{x}_N) = \boldsymbol{x}' = (x'_1, \ldots, x'_N) \tag{10.28}$$

$$x'_k = \rho(x_{i_{1k}k}, \ldots, x_{i_{rk}k}). \tag{10.29}$$

Note that all of the N population members $G = (\boldsymbol{x}_1, \ldots, \boldsymbol{x}_N)$ are in the scope of this recombination operator. The indices $i_{1k}, \ldots, i_{rk} \in \{1, \ldots, N\}$ with $1 \leq k \leq n$, refer to the selected vector elements per column. For the Figure 10.9 example, we have the following settings: $N = 7$, $n = 5$, $r = 3$, resulting in the following recombinations $p'_1 = \rho(p_{31}, p_{41}, p_{61})$, $p'_2 = \rho(p_{22}, p_{32}, p_{62})$, etc. Depending on the recombination mapping ρ, whether intermediate ($\rho = \rho_{\text{int}}$) or discrete ($\rho = \rho_{dis}$), the final elements are computed as in Equations 10.23 and 10.22, respectively.

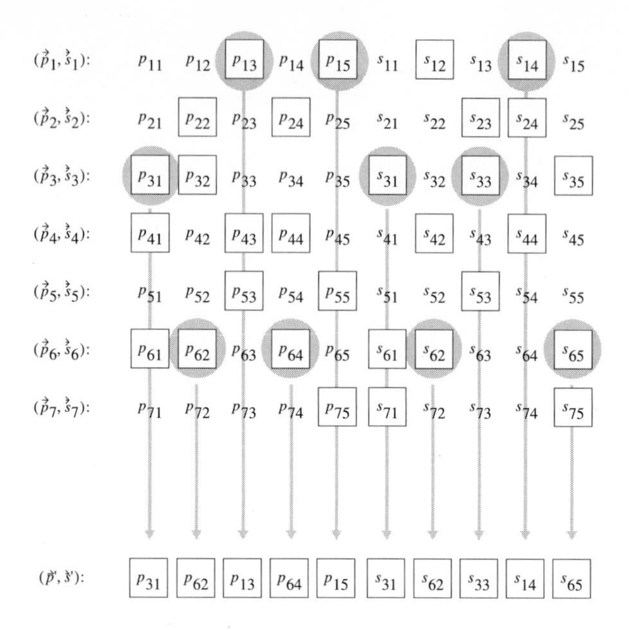

Fig. 10.9. Global, discrete multi-recombination for three recombining parameters.

10.4.4 ES Evolution Schemes

Having explained how to perform mutations and recombinations on ES chromosomes, we must now incorporate these operators into a selection scheme. This section will discuss the basic ES schemes, known as *comma* and *plus* strategies. Both of which will be illustrated with the help of a graphical notation introduced by Rechenberg [443]. We will explain the original Evolution Strategies of a single parent individual that produces one or several mutated offspring. We continue by extending this scheme to a more general one with μ parents producing λ offspring by mutation. Recombination operators will be included, and, finally, we will have a closer look at Meta^n-Evolution Strategies describing evolution among sub-populations at different levels.

$(1 \overset{+}{,} \lambda)$ Evolution Schemes The most simple and original ES scheme is known as a $(1+\lambda)$-Evolution Strategy. In Figure 10.10 a single parent individual produces λ offspring by mutation. Each individual's genotype is represented by a card on which its object and strategy parameters are written. The phenotype, corresponding to the parameter settings noted on the card, usually refers to an experimental setup. The zig-zag shape below the card should be a reminder of the very first ES test experiment with flexibly connected board-shaped strips, for which a configuration of minimum drag had to be developed by an evolutionary strategy [443, 445]. A set of cards forms a population (Figure 10.11). Additional signs on or around the cards symbolize genetic operators, such as a simple *copy* or *duplication* operator, *mutation*, and *recombination*.

In Figure 10.10a, the parent genotype is duplicated λ times and all copies are subsequently mutated. The offsprings' phenotypes are evaluated and both sets of individuals, the parent and offsprings, find themselves in a selection pool. Only the best of these individuals will survive and serve as the parent for the next generation loop. The abbreviating notation for this kind of reproduction process is $(1 + \lambda)$-ES, where the 1 refers to the number of parents and λ is the number of offspring. The '+' sign is used to describe the composition of the final selection pool (the encircled "cards" in Figure 10.10) which contains the parent as well as its children.

Two different selection functions are used with ES. First, there is *random* selection among μ individuals (in the figures indicated by the operator σ_{random}). Secondly, the *best* μ individuals of the selection pool are the designated parents for the following generation (according to their evaluation or fitness).

With a $(1+\lambda)$-scheme the single parent survives into the following generation if all its offsprings' fitnesses are worse. Thus the parent can only be replaced by a superior offspring individual, which means that the fitness of the best-so-far individual either remains the same or increases.

However, this selection scheme often lends itself to premature convergence of the search process. A simple remedy for this is to exclude the parent individual from the final selection pool using a $(1, \lambda)$-strategy (Figure 10.10b). By excluding the parent from the selection pool, the best individual among the offspring becomes the new parent individual for the next generation. This selection scheme is referred to as a $(1, \lambda)$-ES, the comma symbolizing the parent's exclusion from the selection pool.

$(\mu/\rho \overset{+}{,} \lambda)$ Evolution Schemes Taking into account that several (μ) parents produce a population of (λ) offspring, we arrive at a $(\mu + \lambda)$-ES or a (μ, λ)-ES, respectively (Figure 10.11). Instead of selecting a single individual from the selection pool as the designated parent, now the μ best individuals among the selection pool of λ individuals will survive into the next generation. In the case of a (μ, λ)-strategy, we must ensure that there are enough individuals from which to select, that is, $\lambda \geq \mu$, as illustrated in Figure 10.11.

Up to now we have not taken recombination into account. All offsprings are produced solely by mutation. Here we will also introduce recombination. Instead of producing λ copies from the set of μ parents we now repeat duplication for $\lambda \cdot \rho$ times, where ρ is the number of individuals that will be recombined. The ES notation is extended to a $(\mu/\rho + \lambda)$ or $(\mu/\rho, \lambda)$ scheme. Figure 10.11 shows these extensions for binary recombination, mutation, and a comma-strategy, $(\mu/2, \lambda)$-ES.

In general, from each of the λ subsets of ρ chromosomes one recombined genotype is composed. Finally, each of the resulting λ individuals are mutated and evaluated. From these (possibly together with the parents) the best individuals are selected as the new parent population.

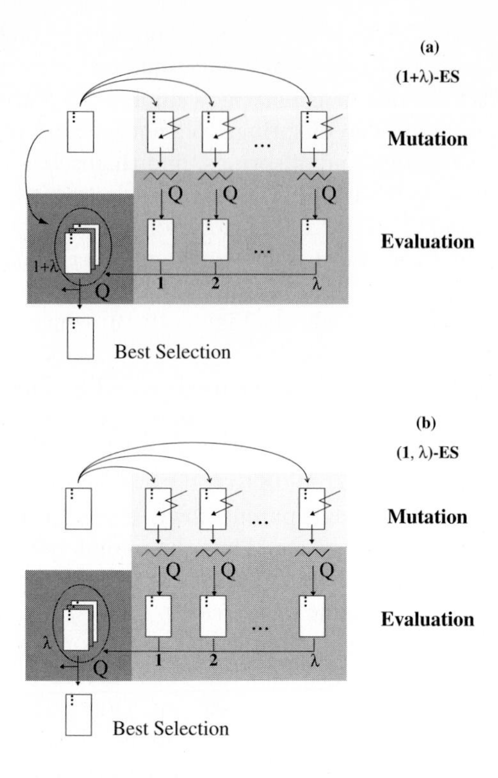

Fig. 10.10. Evolution strategies with descendants from a single parent individual: (a) Plus-strategy, the parent is a member of the selection pool. (b) Comma-strategy, the parent is excluded from the selection for the next generation.

Meta-Evolution Strategies The ES notation we have considered up to now tells us about the number of parents, the size of recombination pools, the number of offspring and the parent selection strategy. One might, however, also be interested in more detailed information about the evolution strategy as, e.g., in the number of generations or further control parameters for the genetic operators. These aspects will be taken into account with the following extended ES notation [287]:

$$\lambda_1((\mu_0/\rho_0) \overset{+}{,} \rho_0 \mid \Theta_0)^{\gamma_0} - \text{ES}. \tag{10.30}$$

This characterizes λ_1 independent evolution experiments each of which perform a $(\mu_0/\rho_0) \overset{+}{,} \rho_0 \mid \Theta_0)$-strategy over γ_0 generations. The additional Θ_0 entry denotes a list of rules or options describing parameter settings of, e.g., the recombination mapping, the mutative stepsize adaptation function or further control parameters as the following example shows:

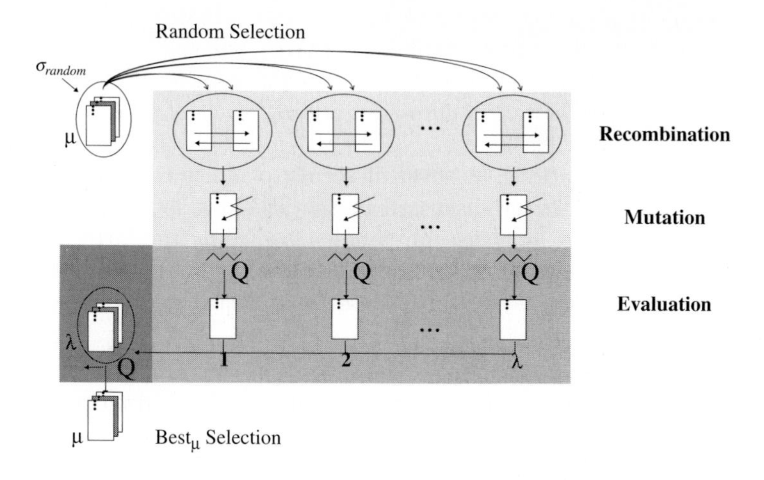

Fig. 10.11. Evolution strategy extended by binary recombination: $(\mu/2, \lambda)$-ES.

$$\Theta_0 = \{\rho = \rho_{\text{int}}, \alpha = e^{N_0(\delta)}\}. \tag{10.31}$$

After λ_1 independent evolution experiments we end up with a set of λ_1 populations, each of which might have explored a different area within the problem search space. Therefore we may compare the performance of these populations by calculating their mean fitnesses. The best of these populations could then serve as the parent population for another λ_1 independent runs, etc.

The described scenario is basically a meta-evolutionary strategy where subpopulations take the role of meta-individuals that adapt to their environment (by a plus or comma ES strategy). These may be considered meta-mutations, and are then evaluated and selected for survival in competition with other subpopulations. With this developed formalism in mind it is not difficult at all to describe meta-evolutions. This can be accomplished by extending the bracket hierarchy as follows:

$$(\mu_1/\rho_1 \overset{+}{,} \lambda_1 \cdot (\mu_0/\rho_0) \overset{+}{,} \rho_0 \mid \Theta_0)^{\gamma_0} \mid \Theta_1)^{\gamma_1} - \text{ES}. \tag{10.32}$$

This formula is a general description of a meta-evolution strategy with subpopulations (for $\rho_1 = 1$, i.e. without recombination). We begin with a set of μ_1 subpopulations. From this collection we randomly select and duplicate $\lambda_1 \cdot \rho_1$ many times. On each of the λ_1 groups of ρ_1 subpopulations we perform recombination, that is interchange of individuals. On the resulting subpopulations we perform meta-mutations in terms of a $((\mu_0/\rho_0) \overset{+}{,} \lambda_0|\Theta_0)$-strategy over γ_0 generations. Among those evolved subpopulations (possibly together with the initial parent populations) we select the μ_1 best populations. This meta-scheme is repeated over γ_1 meta-generations. The described meta-evolution finds its parallel in natural isolated populations (demes) among which genetic information is only scarcely exchanged. Meta-evolution strategies can be extended to Meta^n-strategies, because this notation allows for adding levels of evolution ad lib (for further details see [445, 286]).

For practical applications, however, it seems to suffice to stick to Meta^0-ES, including all the schemes we discussed in the previous sections, or to Meta^1-ES with subpopulations as meta-individuals. Regarding the latter ES scheme, some first theoretical steps towards meta-evolution have been formulated by Rechenberg [445, pp. 158]. We will discuss some Meta^1 evolution examples in the next section.

10.4.5 ES in Action: Multimodal Parameter Optimization

Evolution Strategies have developed sophisticated methods of selection, especially when several subpopulations are involved in the evolution scheme. Figure 10.12 illustrates a typical example. Three populations, marked as white, gray and black spots, have to independently search for locations of maximum heights within their "environment", which is defined by the function $f(x, y)$:

$$f(x, y) = g(x + 11, y + 9) + g(x - 11, y - 3) + g(x + 6, y - 9) \qquad (10.33)$$

$$g(x, y) = \frac{50 \cdot \sin(\sqrt{x^2 + y^2})}{\sqrt{x^2 + y^2}} - \sqrt{x^2 + y^2}. \qquad (10.34)$$

The x- and y-coordinates are encoded on the ES chromosomes, with each population consisting of 10 members. Initially, the individuals are randomly scattered within the lower right segment.

After 10 generations of iterative selection, mutation and recombination according to $(10/2, 10)$ strategies, all three populations have independently evolved towards higher regions. After another 10 generations, some individuals are even on their way up the peaks. Finally, at generation 30, all three populations have obviously converged to one of the major viewpoints of the "mountain region", showing that Evolution Strategies are capable of solving multimodal problems. For the last 10 generations $(10/2 + 10)$ strategies were used.

One should certainly bear in mind, that this is a simple example where specifically tailored ES strategies are used. It is only meant to illustrate the basic dynamics of ES stochastic search. Extensive experiments with Evolution Strategies including comparisons with other search strategies and empirical comparisons with Genetic Algorithms, which will be discussed in the next section, can be found in [468], [265], and [37, Chapter 4].

10.5. Genetic Algorithms

Genetic Algorithms (GAs), originally introduced by John Holland [267, 268]) have been popularized as universal optimization algorithms based on evolutionary methods (see also [219, 37, 364, 250]). The most remarkable difference

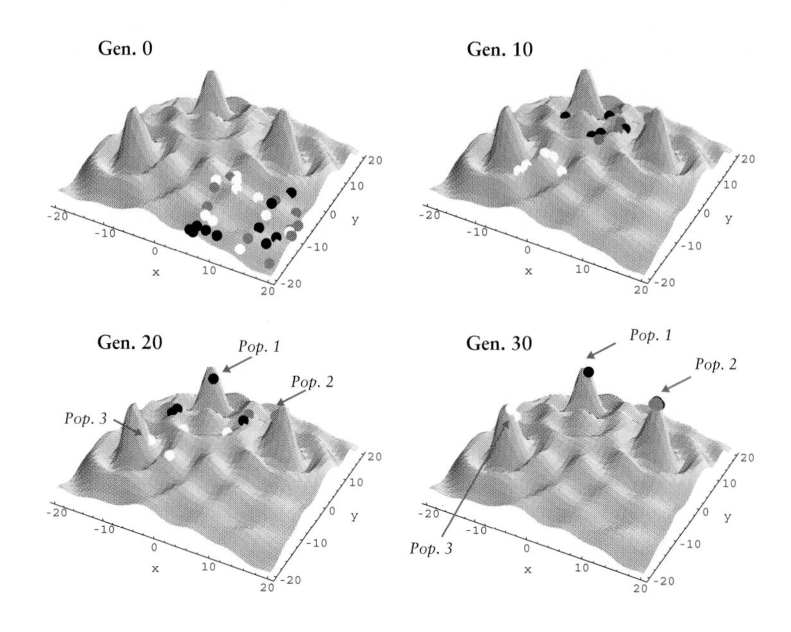

Fig. 10.12. Evolution Strategies solving a multi-modal optimization task: Three independent populations are approaching locally and globally optimal locations by a mixed (10/2, 10) and (10/2 + 10) strategy.

between Evolution Strategies and Genetic Algorithms is that GAs use a string-based, usually binary parameter encoding, resembling the discrete nucleotide coding scheme on cellular chromosomes. Therefore, GA genotypes can be defined as bit-vectors (bit-strings), on which *point mutations* are defined by switching bits with a certain probability. A recombinational operator of *crossover* models the breaking of two chromosomes and subsequent crosswise restituation. This is observed on natural genomes during sexual reproduction. Each individual is assigned a fitness value depending on the optimization task to be solved. Selection within the population is performed in a fitness-proportionate way: The more fit an individual, the more likely it is to be chosen for reproduction into the following generation.

In the following sections, we will explore Genetic Algorithms in more detail. We will start with the data structures of polyploid GA chromosomes explaining haploid and m-ploid chromosomes with a brief glimpse on the related topic of dominant and recessive alleles. Moreover, we will have a closer look at the two major GA operators already mentioned: *recombination* and *mutation*. Mutation will be demonstrated by a few examples, from simple point-mutations on haploid chromosomes to mutations on multiple strands. Examples of recombinations range from simple combinations of haploid GA chromosomes to meiotic recombination among diploid chromosomes. In addition to these two major GA operators, we will discuss the more specialized GA operators, such as *inversion, deletion, duplication*, and crossover between *non-homologous*, i.e., unequal-length, chromosomes, which is the recombination mechanism found in nature. Another section will explain how the basic GA evolution scheme works. The graphical notation introduced for Evolution Strategies will be used to characterize the basic GA evolution schemes. We will conclude this section with a GA experiment in the domain of parameter optimization. The evolution example will show the flexibility and adaptability of the Genetic Algorithm approach for solving optimization tasks in changing environments.

10.5.1 Representation of Genotypes

Nature encodes "growth programs", which implicitly describe the development of organisms, on the basis of a four-element alphabet, called the *nucleotide bases* A, C, G, and T on the DNA (*deoxyribo nucleic acid*) or A, C, G, and U on the RNA (*ribonucleic acid*) strands comprising the cellular genome. Binary alphabets are the preferred encoding used for Genetic Algorithm chromosomes. In the following sections, however, we will introduce GA chromosomes in the more general form of strings or vectors over an arbitrary, discrete alphabet.

Haploid GA Chromosomes In its simplest form a GA chromosome can be described as a haploid string, i.e., a single strand with alleles over a k-element alphabet $A = \{a_1, \ldots, a_k\}$. Therefore, we define a single GA chromosome of length n as a vector $\boldsymbol{g}$ of the form

$$\boldsymbol{g} = (g_1, \ldots, g_n) \quad \text{with} \quad g_i \in A \quad \text{for} \quad 1 \leq i \leq n \tag{10.35}$$

Such a GA chromosome can be interpreted as a sequence of *genes*. Each gene is represented by its *allele* which exhibits a specific value from a discrete pool of possible settings A. For the sake of simplicity, each gene locus i does not have its own alphabet A_i. Instead, all genes take their values from the same allele pool A.

Diploid and m-ploid GA Chromosomes The notion of a single chromosome strand is easily extended to a polyploid chromosome with several strands. We may, for example, represent a *diploid* pair of chromosomes $\boldsymbol{g}_1 = (g_{11}, \ldots, g_{1n})$ and $\boldsymbol{g}_2 = (g_{21}, \ldots, g_{2n})$ by the following structure:

$$\boldsymbol{g} = (\boldsymbol{g}_1, \boldsymbol{g}_2)^T = ((1, (g_{11}, g_{21})), \ldots, (n, (g_{1n}, g_{2n}))). \tag{10.36}$$

For each locus i the pair $(i, (g_{1i}, g_{2i}))$ contains the locus index as its first component, the second entry is the list of respective alleles of two chromosome strands g_1 and g_2, each with n components. More generally, we may define polyploid or m-ploid chromosomes with m homologous strands, i.e., strands of equal length n, as follows:

Parent chromosome:	1 1 0 0 0 1 0 0 0 1
Mutant 1, p_m = 1.0:	0 0 1 1 1 0 1 1 1 0
Mutant 2, p_m = 0.5:	0 1 1 0 0 1 1 1 1 0
Mutant 3, p_m = 0.2:	1 1 1 0 1 1 0 0 0 1
Mutant 4, p_m = 0.1:	1 1 0 0 0 1 0 1 0 1

Fig. 10.13. Example mutations for different mutation probabilities.

$$g = ((1, (g_{11}, \ldots, g_{m1})), \ldots, (n, (g_{1n}, \ldots, g_{mn}))). \tag{10.37}$$

Regarding the interpretation of polyploid chromosomes it has to be decided which of the "competing" alleles will be identified as the one to be expressed in the phenotype. This leads us to the question of dominant and recessive alleles which we will not discuss further. For additional details see, e.g., [286, Chapter 6.1], [287, Chapter 3.1] or [219, pp. 148].

10.5.2 Mutations on GA Chromosomes

Gene variations are the core ingredient for evolution to explore new territory of phenotypic structures and their functionality. The survival and reproductive capability of newly created traits by mutations on the genotypes decide whether new gene settings are passed on to future generations. Mutations in combination with recombinations are the driving forces of evolution.

We first consider simple point-mutations on a GA chromosome $\boldsymbol{g} = (g_1, \ldots, g_n)$ as introduced above with alleles g_i taken from a discrete alphabet $A = \{a_1, \ldots, a_k\}$. Point mutations change the settings of genes at randomly selected gene locations. For each gene, its allele is replaced by a new value from A with mutation probability p_m. The mutation operator $\omega_{\text{mut}} : G_A \longrightarrow G_A$, with G_A denoting the set of all GA chromosomes over alphabet A, generates a new chromosome $\boldsymbol{g'} = \omega_{\text{mut}}(\boldsymbol{g}) = \omega_{\text{mut}}((g_1, \ldots, g_n)) = (g'_1, \ldots, g'_n)$ as follows:

$$g'_i = \begin{cases} x \in A - \{g_i\} & : \ \chi \leq p_m \\ g_i & : \ \text{otherwise} \end{cases} \tag{10.38}$$

Here χ is a uniformly distributed random variable from the interval $[0, 1]$. The probability for a mutation per gene locus is denoted by p_m. Figure 10.13 shows a few examples of mutated chromosomes for different mutation probabilities.

Generally, we may define mutations on a polyploid set of single-strand chromosomes such as

$$\boldsymbol{g}^{(m)} = ((g_{11}, \ldots, g_{m1}), \ldots, (g_{1n}, \ldots, g_{mn})) \tag{10.39}$$

consisting of m homologous single chromosome strands of the form

$$\boldsymbol{g}_i = (g_{i1}, \ldots, g_{in}) \quad \text{with} \quad 1 \leq i \leq m. \tag{10.40}$$

Point mutation on the chromosome set $g^{(m)}$ is defined analogously as the mutated set $\boldsymbol{g}'^{(m)} = \omega_{\text{mut}}(\boldsymbol{g}^{(m)})$ where

$$\boldsymbol{g}'^{(m)} = ((g'_{11}, \ldots, g'_{m1}), \ldots, (g'_{1n}, \ldots, g'_{mn})) \tag{10.41}$$

The point mutations are performed independently for each gene locus g_{ij}:

$$g'_{ij} = \begin{cases} x \in A - \{g_{ij}\} & : \ \chi \leq p_m \\ g_{ij} & : \ \text{otherwise} \end{cases} \quad \text{for} 1 \leq i \leq m, 1 \leq j \leq n. \tag{10.42}$$

Again, p_m is the mutation probability for each haploid gene index.

What is the influence of point mutations on the search strategy? We explore the effects of mutation in combination with selection on the pheno- and genotypes of a population of GA chromosomes. An example on finding the minimum of a simple two-dimensional function will demonstrate that mutation is mainly an operator that drives populations towards *exploration* of new regions by randomly adding new allele settings to the genepool.

Figure 10.14 shows the results of five independent GA evolution runs, each according to a $(10 + 10)$-GA strategy over 10 generations. Each chromosome is 10 bits in length. The first five bits encode the x- and the second half encodes the y-coordinate. We search for the global minimum of the parabolic function $f(x, y) = x^2 + y^2$ in the region $[-1, 1] \times [-1, 1]$. Mutation is the only genetic operator which is applied to all GA chromosomes per generation. The point-mutation probability is set to $p_m = 0.5$.

As can be clearly seen in Figure 10.14, all five populations have made their way towards the location of the global minimum (marked by a flag). However, this effect is mostly due to selection where only the 10 individuals closest to the minimum (among the 20 individuals which comprise the selection pool) survive into the following generation. Mutation, on the other hand, constantly introduces new genetic material into the genepool thus preventing the population from converging too fast towards a set of individuals with nearly identical genotypes. The spreading effect of mutation is clearly visible; the populations are gathering around the optimum, but none of the populations has "collapsed" into a single point.

The influence of this mutation-selection scheme on the genotypes, the binary chromosomes, is depicted in Figure 10.15. Here each chromosome vector with its binary entries of 0's and 1's is represented as a pattern of black and white dots. The figure shows each of the five populations for generations 0, 5, and 10. Each population's chromosomes are sorted lexicographically and arranged as a black-and-white matrix. The resulting *building block* patterns show a slight tendency

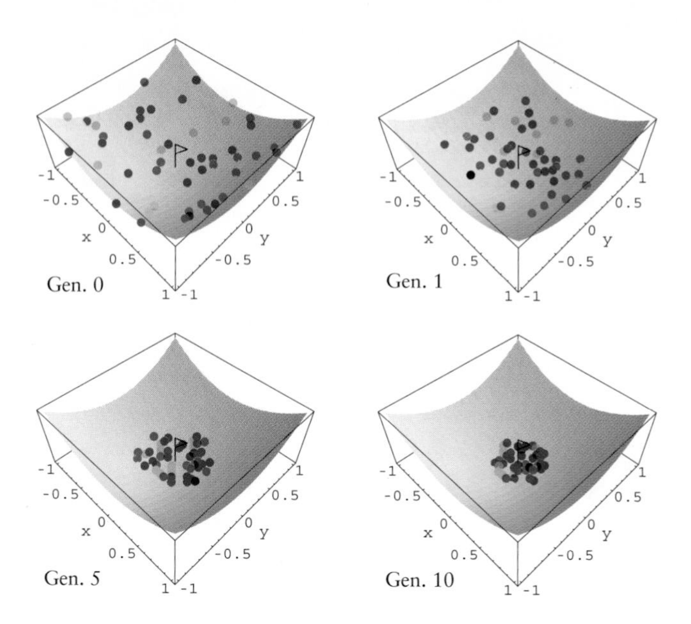

Fig. 10.14. GA population dynamics over 10 generations under the influence of mutation and selection only.

towards convergence, which would be visible by a larger number of vertically oriented stripes (compare Figure 10.18).

10.5.3 Recombinations among GA Chromosomes

As GA chromosomes are mostly defined over a discrete allele alphabet, all the discrete variants of recombination operators for Evolution Strategies (Section 10.4.3) can be used here as well. For m homologous, haploid GA chromosomes

$$\boldsymbol{g_1} = (g_{11}, \ldots, g_{1n}), \ldots, \boldsymbol{g_m} = (g_{m1}, \ldots, g_{mn}) \tag{10.43}$$

a recombined GA chromosome $\boldsymbol{g_{\text{rec}}}$ can be composed with the help of a recombination mask $\mu = (\mu_1, \ldots, \mu_n)$:

$$\boldsymbol{g_{\text{rec}}} = (g_{\mu_1 1}, \ldots, g_{\mu_n n}) \quad \text{with} \quad \mu_i \in \{1, \ldots, m\}, 1 \leq i \leq n. \tag{10.44}$$

The i-th component of g_{rec} is therefore the μ_i-th element from the set of "genes" $\{g_{1i}, \ldots, g_{mi}\}$.

A still widely used GA recombination operator know as binary GA *crossover* deserves special attention, because it played an important role in the early formulations [267] and implementations of Genetic Algorithms. If the recombina-

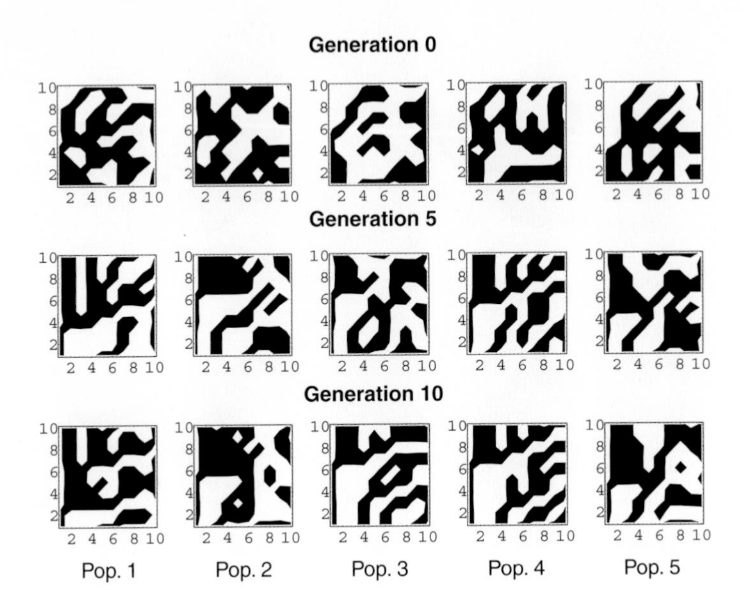

Fig. 10.15. Dynamics of the GA population genotypes of Figure 10.14 over 10 generations under the influence of mutation and selection only.

tion mask μ has the special property that it contains only two different elements, i_1 and i_2, from the index set $\{1, \ldots, m\}$ and there is exactly one index $k \in \{1, \ldots, n-1\}$ such that $\mu_1, \ldots, \mu_k = i_1$ and $\mu_{k+1}, \ldots, \mu_n = i_2$, then this recombination is termed a *one-point crossover*. An example is depicted in Figure 10.16a. Of course, the idea of one-point crossover can be easily extended to multi-point crossover among two (Figure 10.16b) or more GA chromosomes.

Single-point and two-point crossover for binary GA chromosomes are most widely used in GA applications. There are a great number of variants of recombination operators on GA chromosomes. Goldberg discusses crossover operators especially apt when dealing with combinatorial optimization problems, such as the travelling salesman problem [219, pp. 170]. *Parametrized uniform* crossover operators, similar to global ES recombination schemes are described in [490]. *Segmented* and *shuffle* crossover [171] work in a way similar to multi-point crossover. *Punctuated* crossover is one of the very few attempts to introduce self-adaptive strategy parameters into Genetic Algorithms [463]. Short discussions of GA crossovers in theory as well as practical applications are presented in [380, pp. 171] and [37, pp. 114]. Extensions for polyploid chromosomes are discussed in [286].

Let us explore the effects of one-point crossover in combination with selection on the pheno- and genotypes of a population of GA chromosomes. An example of finding the minimum of a simple two-dimensional function will show that

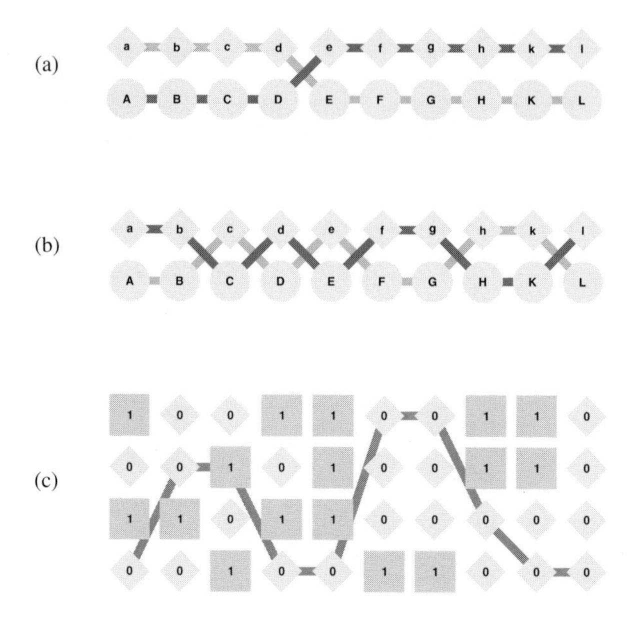

Fig. 10.16. Examples of recombinations on GA chromosomes: (a) one-point crossover, (b) 6-point crossover, (c) multi-point crossover of four GA chromosomes.

the basic effect of interchanging genetic information between chromosomes is convergence with respect to the genepool.

We perform five independent GA evolution runs each according to a $(10/2 + 10)^{10}$-GA. The only genetic operator is recombination which is applied to each of the 10 pairs of GA chromosomes. We use single-point crossover. The mutation operator is switched off. Figures 10.17 and 10.18 show graphical representations of the effects of recombination and selection on the pheno- and genotypes of the evolving populations. As we can see, each evolution run has led to strict convergence of the phenotypes within the fitness landscape. All populations finally gather at a very small region around the global minimum.

We can see that the chromosomes have converged to a high degree. Note the inverse patterns due to the symmetry of the problem domain. Comparing these graphics to the mutation results in Figure 10.15 reveals the interplay of recombination and mutation operators for genetic algorithms: Mutations introduce new alleles and gene combinations into the genepool, whereas recombinations, by interchanging and mixing genetic information, imply a tendency towards more similar chromosomes within a population, resulting in a convergent effect.

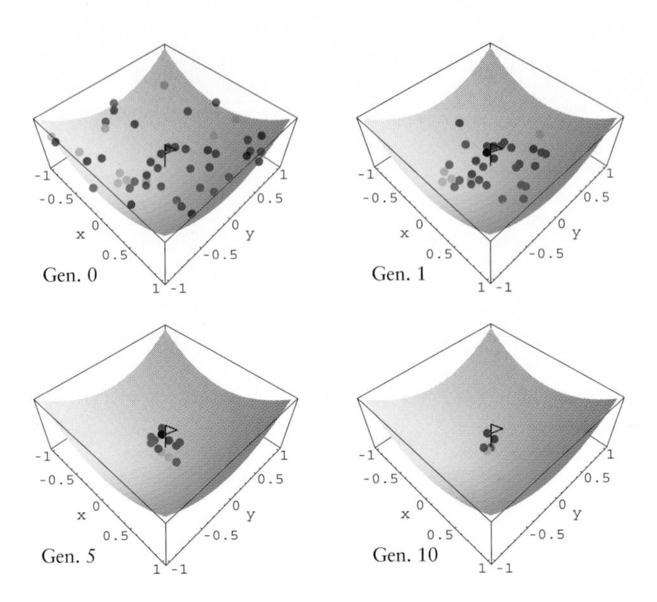

Fig. 10.17. The convergence effect of recombination and selection on the phenotypes of five independent GA populations.

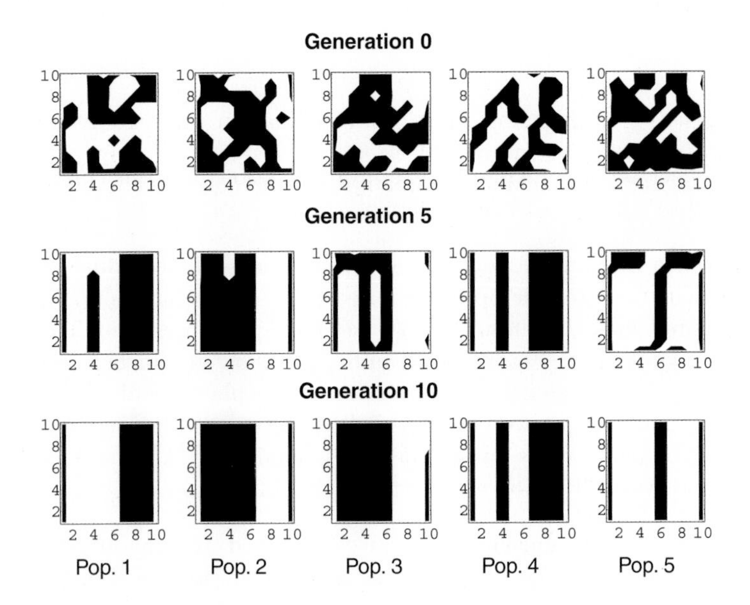

Fig. 10.18. The convergence effect of recombination and selection made visible by the dominance of column patterns on the genotypes of the five independent GA populations of Figure 10.17.

10.5.4 Further GA Operators

Beyond the two major GA operators, recombination as the primary and mutation as the secondary operator, some GA systems extend their repertoire of techniques for varying the contents of the chromosomes by further biologically inspired operators, such as *inversion*, *duplication*, and *deletion*. We will briefly present these operators and discuss application domains for which they are particularly useful. For the following definitions, we assume a haploid GA chromosome of the form

$$\boldsymbol{g} = (g_1, \ldots, g_{i_1}, \ldots, g_{i_2}, \ldots, g_n) \quad \text{with} \quad 1 \leq i_1 < i_2 \leq n, \tag{10.45}$$

where each gene locus, with an allele $g_i \in A$ from the allele alphabet A is identified by an index $1 \leq i \leq n$.

Inversion On natural genomes, an inversion occurs when a chromosome is split twice, after locus $i_1 - 1$ and before $i_2 + 1$, and the resulting middle section $(g_{i_1}, \ldots, g_{i_2})$ is reinserted into the strand in reverse order:

$$\boldsymbol{g}_{\text{inv}} = (g_1, \ldots, g_{i_2}, g_{i_2-1}, \ldots, g_{i_1+1}, g_{i_1}, \ldots, g_n) \tag{10.46}$$

In fact, the inversion function can be defined as any *permutation* π on the $i_2 - i_1 + 1$ elements:

$$\boldsymbol{g}_{\text{inv}} = (g_1, \ldots, \pi(g_{i_1}, \ldots, g_{i_2}), \ldots, g_n) \tag{10.47}$$

The operation of inversion, mainly in combination with extended versions of multi-point crossover, turns out to be advantageous for solving combinatorial optimization problems, such as the Travelling Salesman Problem [219, pp. 166] or job shop scheduling tasks [68].

Duplication A side effect of crosswise recombination of two homologous chromosomes is the duplication and deletion of subsections on the genome. Formally, the insertion of a copy of a gene sequence $(g_{i_1}, \ldots, g_{i_2})$ can be defined as follows:

$$\boldsymbol{g}_{\text{dup}} = (g_1, \ldots, g_{i_1}, \ldots, g_{i_2}, g_{i_1}, \ldots, g_{i_2}, \ldots, g_n). \tag{10.48}$$

This operation leads either to a redundant encoding of the genes $g_{i_1}, \ldots, g_{i_2}$ — the one being expressed depends on the interpretation function — or, if the genes do not carry specific information, the resolution of the encoding is enhanced (for example, this is the case if the chromosome encodes binary numbers). An increase (or decrease) of genes can also be used as an effective method to adaptively control mutation rates, as has been suggested by Holland [267]. With a constant mutation rate, the effective rate of mutation for a redundantly encoded gene increases with the number of its copies, because there is an increased probability that at least one copy undergoes mutation [219, p. 180], [287].

Deletion The duplication operator is mainly applied in conjunction with its counterpart, the deletion operator. Here a chromosome looses a gene sequence $(g_{i_1}, \ldots, g_{i_2})$:

$$\boldsymbol{g}_{\text{del}} = (g_1, \ldots, g_{i_1-1}, g_{i_2+1}, \ldots, g_n). \tag{10.49}$$

Both the deletion and duplication operator work best with *messy codings* which permit underspecification (with missing genes), variable lengths, and redundancy [219, p. 181], [287].

10.5.5 Selection Functions

A major influencing factor for an evolutionary algorithm to be successful, or not, is its manner of selecting individuals to serve as parents for the various kinds of mutation operators or to survive into the following generation. The selection functions define the "who shall live and who shall die" filters that all individuals pass through from one generation to the next.

In contrast to the Evolution Strategy's traditional selection procedure of *survival of the best*, Genetic Algorithms apply a more "natural" selection scheme. In nature, an individual's probability of survival is influenced by an abundance of factors. However, it is not at all the case that only "the most fit" individuals survive. In fact, even less adapted, hence less fit, individuals have a chance to reproduce and transfer their genes to their progeny. Their genes survive into the next generations with certain probability.

Fitness Proportionate Selection The idea of differentiated survival probabilities is incorporated into a probabilistic GA selection scheme which takes the individual fitnesses into account. For a population $G = \{g_1, \ldots, g_\mu\}$ of size μ with each individual g_i assigned a non-negative fitness value $\eta(g_i) \in \Re_0^+$, the probability $P_{\text{sel}}(\sigma_{\text{prop}}, g_i)$ for an individual to be selected fitness-proportionately is defined as

$$P_{\text{sel}}(\sigma_{\text{prop}}, g_i) = \frac{\eta(g_i)}{\eta_\Sigma(G)} \quad \text{with} \quad \eta_\Sigma(G) = \sum_{g \in G} \eta(g). \tag{10.50}$$

Here $\eta_\Sigma(G)$ denotes the sum of all fitnesses of the population. The probability of an individual to be reproduced into the next generation is directly proportionate to its fitness, hence the notion *fitness proportionate* selection. This selection scheme is also known as the *roulette wheel* selection, because the fitnesses can be represented as respective segments on an imaginary roulette wheel. The probability for a hit of one segment by an imaginary bowl directly depends on its size, that is, on the relative fitness value.

Rank-Based Selection The fitness proportionate selection scheme is problematic if there are a few "super-individuals" in a population, which have extremely high fitness values compared to their competitors. In addition, if the population size is somewhat small (much less than 100 individuals), this situation leads to premature convergence of the population which eventually comprises only of these few super-individuals. Thus the gene pool looses its heterogenuity and is reduced to only a small set of search points, causing the explorative power to be much more constrained. One remedy for this situation is to reduce the fitness differences among the individuals by assigning *ranks* to the individuals instead of using their actual fitness values. Just as in a tournament of μ competitors, the winner receives rank 1, the second best is assigned rank 2, etc., until the least fit individual ranks at μ.

The μ individuals are sorted in increasing fitness order such that $\eta(g_i) \leq \eta(g_j)$ for all $1 \leq i < j \leq \mu$. If $\rho(\boldsymbol{g})$ denotes the rank of individual $\boldsymbol{g} \in G$ within this sorted sequence, its fitness $\eta_{\text{rank}}(\boldsymbol{g})$ is defined by

$$\eta_{\text{rank}}(\boldsymbol{g}) = \mu - \rho(\boldsymbol{g}) + 1 \tag{10.51}$$

so that the rank-based selection probability is

$$P_{\text{sel}}(\sigma_{\text{rank}}, \boldsymbol{g}) = \frac{\eta_{\text{rank}}(\boldsymbol{g})}{\sum_{i=1}^{\mu} i} \tag{10.52}$$

This selection scheme has a normalizing effect on the distribution of fitnesses, especially when their mean variation is large .

Elitist Selection The elitist selection scheme is the one used for selection with Evolution Strategies and can be used for Genetic Algorithms, too. For the (μ, λ) or $(\mu+\lambda)$ strategies, the best μ individuals from the set of mutants in the pool of λ or $\mu+\lambda$ individuals are selected as the parents of the next generation (compare Figure 10.10). The GA elitist selection scheme implements exactly this selection method.

$$P_{\text{sel}}(\sigma_{\text{elitist}}, \boldsymbol{g}) = \begin{cases} \frac{1}{\delta} & : \ \boldsymbol{g} \in \text{Best}_{\delta}(G) \\ 0 & : \ \text{otherwise} \end{cases} \tag{10.53}$$

The best δ individuals (usually $\delta \ll \mu$) are selected, each is assigned the same fitness of $1/\delta$, and fitness proportionate selection is performed, which results in a random selection among these δ individuals.

Random Selection Sometimes, individuals $g \in G$ are selected by pure random choice,

$$P_{\text{sel}}(\sigma_{\text{random}}, \boldsymbol{g}) = \frac{1}{\mu}, \tag{10.54}$$

Here μ denotes the number of individuals in generation G. This is, for example, the preferred method of selection of Evolution Strategies in combination with an elitist selection *after* the individuals have been mutated and recombined (compare Section 10.4.4).

Further Selection Schemes Here, we can only discuss a very small collection of selection schemes. Examples, such as deterministic and stochastic sampling or tournament strategies are described in [219, p. 121-125], [93] and the classical reference of [295]. A recent comparison on GA selection schemes, such as tournament selection, truncation selection, linear and exponential rank-based selection, and fitness proportionate selection, is reported in [71].

10.5.6 GA Evolution Scheme

According to the general scheme for evolutionary algorithms (Figure 10.5), GA evolution starts with a randomly generated initial population of μ genotypic structures, usually over a discrete allele alphabet. After interpretation and evaluation, the population enters a selection-variation cycle which is iterated for either a maximum number of generations or until some termination criterion (τ in Figure 10.5) is met. In Figure 10.19, a single GA selection-variation step is depicted schematically according to the notation introduced in Section 10.4.

The canonical Genetic Algorithm performs a $(\mu/2, \mu)$ strategy. From the pool of μ parents $\lambda = \mu$ pairs of individuals are selected for recombination (crossover) and subsequent mutation. The resulting individuals represent the parents of the next generation. The major difference to the ES scheme (Figure 10.11) is that the individuals are not selected at random from the parent pool but according to one of the GA selection functions σ introduced in the previous section.

The probability of an individual remaining unchanged after recombination and mutation is rather low. Practically no individual survives more than one generation, hence the analogy to a comma ES survival scheme. In canonical GAs, the parent population is completely replaced by their offsprings.

If one does not want to loose the best-so-far individuals, the reproduction scheme can be extended as illustrated in Figure 10.19. The best $\mu - \lambda$ parents survive into the following generation, while λ mutants are added to this selection pool which already comprises all the parents of the next generation. In many GA implementations, only the best parent individual is kept and $\lambda = \mu - 1$ individuals are generated anew by recombination and mutation. In the *Steady State* GA, only the worst individual is replaced by a new mutant, that is, $\lambda = 1$ and the $\mu - 1$ best individuals survive [506].

Of course, this evolution scheme can also be extended by further operators, such as inversion, duplication, or deletion, which are omitted here to keep the figure simple. The operators are usually attributed with application probabilities. Not all operators are then sequentially applied to each individual. An alternative scheme, where for each individual only a single operator is applied (with repeated operator selection), is also quite popular (compare, e.g., the GA extension of Genetic Programming described in Section 10.6). Further GA evolution schemes and alternative GA reproduction techniques are discussed in [143, 219, 375, 380] and [395, 396].

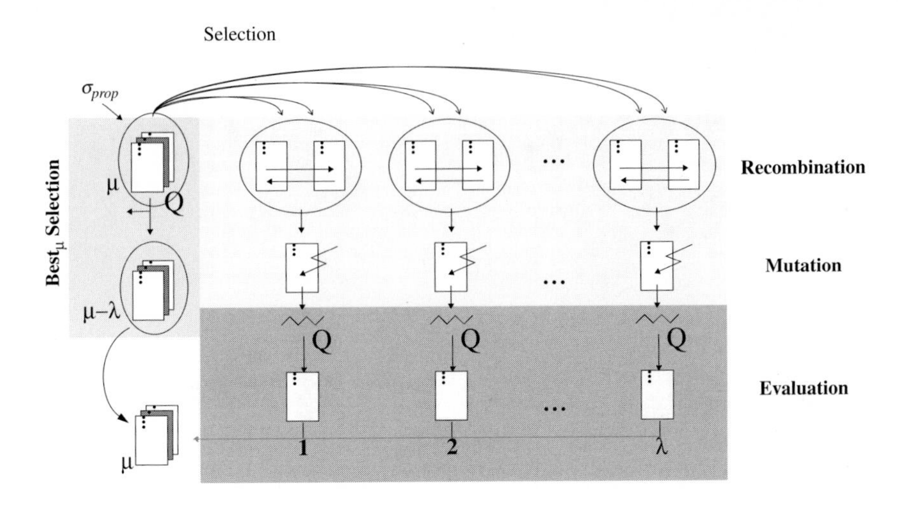

Fig. 10.19. Evolution scheme of an extended genetic algorithm with elitist selection. For a canonical GA we have $\lambda = \mu$.

10.5.7 GA in Action: Parameter Optimization in a Changing Environment

In order to show the capability of Genetic Algorithms to solve multi-modal optimization problems, we demonstrate an example of a parameter optimization task with the same fitness function used in Section 10.4.5 for Evolution Strategies. Furthermore, we show the flexibility of Genetic Algorithms to adapt to a changing environment by switching to a different fitness function during the evolution process.

We start three independent evolution runs to get a better idea of the GA's general behaviour (Figure 10.20). The task is to find the location, i.e., the coordinates, of the global maximum which is the peak at the back and center. The chromosomes consist of 12 binary bits, where the first 6 bits encode the x-coordinate and the remaining bits encode the y-coordinate. As genetic operators we use mutation (bit mutation rate $p_{\mathrm{mut}} = 0.2$), one-point crossover, and inversion, with equal selection probabilities for all three operators. The individuals are selected fitness-proportionately.

According to Figure 10.20, the three initial populations of generation 0 cover the search space fairly well. In generation 2, the search concentrates on the higher regions. This concentration effect is mainly due to fitness-proportionate selection. In generation 6, even all three peaks are occupied by a few individuals, and, after another 4 generations, the tips of the peaks, including the global maximum, are located even more precisely (of course, the precision is always constrained by the bit resolution given by the chromosome length).

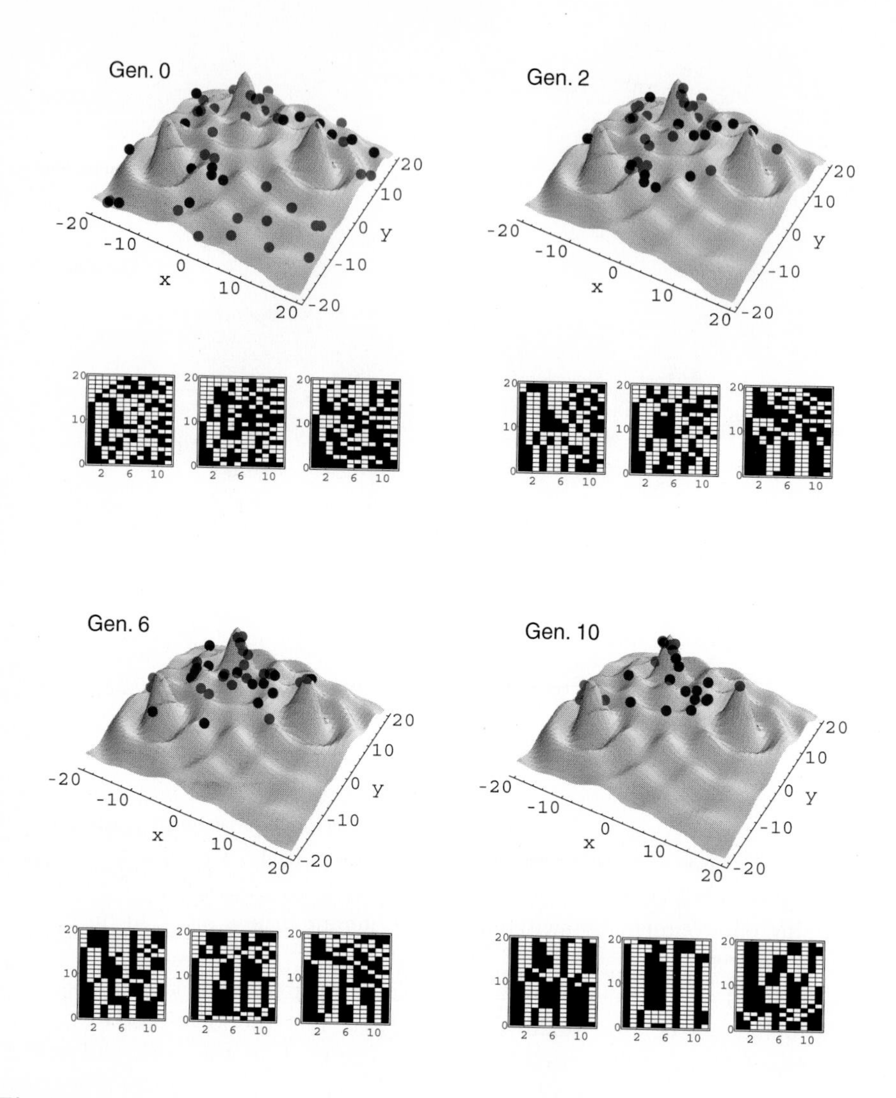

Fig. 10.20. Parameter optimization with three independent runs of a (20/2 + 10) Genetic Algorithm with fitness proportionate selection (Part 1).

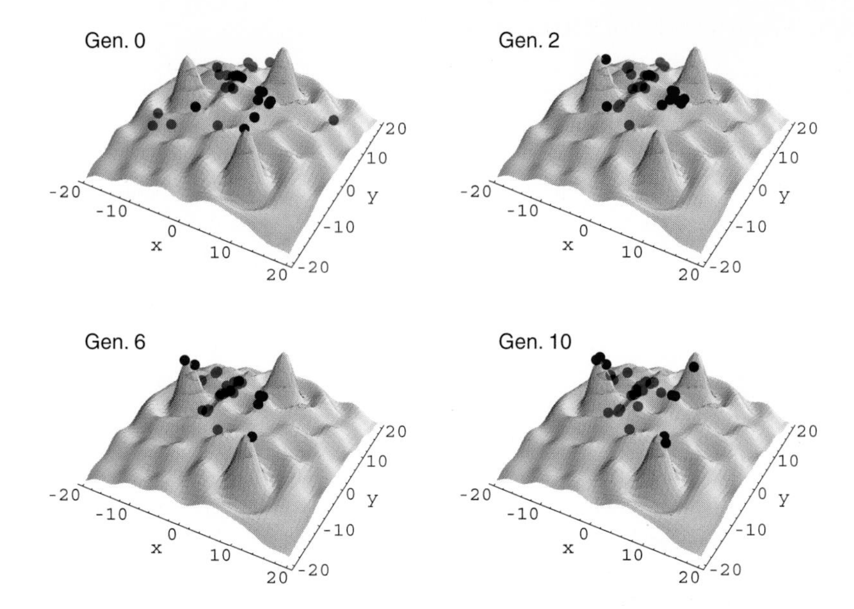

Fig. 10.21. Parameter optimization in a changed environment with three independent runs of a $(20/2 + 10)$ Genetic Algorithm with fitness proportionate selection (Part 2). The fitness function has been changed by rearranging the peaks (compare Figure 10.20).

Figure 10.20 also shows the evolution dynamics of the genepools. The black and white patterns are matrix representations of the 20 binary chromosomes of length 12 of each of the three populations. Within each population, the genotypic structures are sorted lexicographically.[3] For the initial generation, the patterns look rather irregular, whereas towards the end vertical stripes appear which indicate identical alleles in the genepools.

Now we can change the fitness function: the peaks are "rotated" to the right, so that most of the individuals loose their highly fit positions (Figure 10.21). What will happen if we continue the evolution process? Will the GA be able to discover the new peak locations? How long will it take the GA to react to this change of its fitness landscape?

The initial populations in Figure 10.21 are identical to the last generation of Figure 10.20. The adaptation to the new environment is successful and takes only another 10 generations. This flexibility is only possible because the populations' genepools remain rather heterogeneous, an effect also due to the fitness-proportionate selection scheme which is — compared to Evolution Strategies — inherently less prone to premature convergence.

[3] Lexicographic ordering "$\prec$" on pairs is defined as $(n, m) \prec (n', m') :\Leftrightarrow (n < n') \vee ((n = n') \wedge (m < m'))$

10.6. Genetic Programming

Since the first steps of computer programming, many auxiliary tools and paradigms have been developed in order to support the (still) tedious task of program and code generation. Editors, syntax checkers, code generators, compiler-compilers, etc. are used to support structured programming in different flavors — such as procedural, functional, rule-based, or object-oriented —, to generate object libraries or to do rapid prototyping. Despite these improvements, computers are not yet able to *learn to program themselves*. Instead, programming still remains a "craftsman" approach, although computer scientists made a number of attempts during the last four decades to meet the primary idea of *machine learning* [456], namely to make computers program themselves (compare [42, p. 4] for a more detailed discussion on this subject; see also Chapter 8 in this book about neural network learning paradigms).

Genetic Programming (GP) is certainly one of the major steps towards automatic programming using evolutionary principles. The term *genetic programming* was coined by John Koza, who initially introduced his evolutionary programming approach as a "hierarchical genetic algorithm" [313] and later on switched to a symbolic representation for evolving LISP programs [314–316]. This approach had and still has an immense influence on the field of automatic programming. The main contribution of Koza's research group is documented by a three-volume treatise [315–317], demonstrating that Genetic Programming can successfully be applied to induce computer programs in a wide range of areas, such as symbolic regression or learning boolean parity functions, the evolution of emergent behavior, the evolution of robot control programs, the evolution of classifiers for prediction of transmembrane domains and omega loops in proteins, or the recent work on evolution of analog electrical circuits [317].

The successful use of tree structures to represent data and program instructions has led to an immediate association of Genetic Programming with symbolic expressions, although many more encoding schemes can be used and actually are used for automatic programming by evolution. The following items characterize the field of *Genetic Programming* in a broader sense [42]:

Program Induction: Genetic Programming deals with the induction of computer programs using evolutionary principles (partly incorporating evolutionary algorithms, such as GAs). The programs can be either directly executable (machine) code or expressions — data and/or instructions — of any programming language.

Learning Algorithms: Not all learning algorithms are explicitly represented as programs in a strict sense, such as neural networks and learning fuzzy systems. Algorithms for adapting these data structures (neuron weights, activation functions, fuzzy rule parameters, etc.) are also in the domain of Genetic Programming.

Representation: There are many different ways of representing programs, for example, by linear, string-based structures, by tree-like symbolic expressions, by growth grammars, or by graphs.

Operators: Selection operators are used to designate the survivors among the population of competing programs. Reproduction operators are used in conjunction with recombination operators, generating new variants by, primarily, mutation and crossover, or further operators, such as permutation, deletion, duplication, or encapsulation.

Keeping this more general characterization of GP in mind, we will focus on the original Genetic Programming approach, using symbolic expressions to encode program structures demonstrating the approach of programming by evolution in the following sections.

10.6.1 Representation of Computer Programs by Symbolic Expressions

Programs evolved by a tree-based GP system are typically composed from a finite set F of building blocks, the *functions*

$$F = \{f_1, \ldots, f_N\}, \quad \text{with} \quad \nu(F) = \{\nu(f_1), \ldots, \nu(f_N)\}, \tag{10.55}$$

where $\nu(F)$ denotes the arities of the function symbols, and the *terminals*

$$T = \{t_1, \ldots, t_M\}. \tag{10.56}$$

Considering the terminals as function symbols of arity zero, both sets can be merged into a single set of elementary building blocks:

$$S = F \cup T. \tag{10.57}$$

The set $GPterms_S$ of program trees, representing the GP search space, can be defined as follows:

1. Each terminal $t \in S$ with $\nu(t) = 0$ is an element of $GPterms_S$.
2. For each $f \in S$ with $\nu(f) = n$ and $g_1, \ldots, g_n \in GPterms_S$, the term $f(g_1, \ldots, g_n)$ is also an element of $GPterms_S$.

The set of problem-specific elementary components has to be predefined for each problem domain. This turns out to be one of the major problems with Genetic Programming. However, finding suitable encodings of data structures is *the* core problem for search algorithms in general. Similar to Genetic Algorithms, the GP encoding of the structures to be evolved has an important impact on whether the evolutionary approach will succeed or not. The choice of functions and terminals also influences the potential of the genetic operators to create innovative and optimized program structures.

Given the set of building blocks

$$S = \{\text{Mult}_2, \text{Add}_2, \text{If-Then-Else}_3, \text{Equal}_2, A_0, B_0, C_0, D_0\}, \tag{10.58}$$

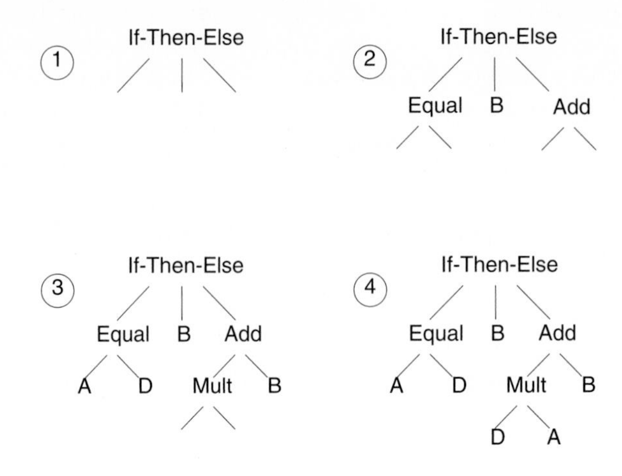

Fig. 10.22. Stochastic, step-by-step generation of a symbolic expression by selecting functions and terminals from a problemspecific expression pool as described in the text.

where the indices in S denote the arities of the symbols. Figure 10.22 illustrates the recursive, step-by-step construction of a random program term, an element of $GPterms_S$. A symbol from S is randomly selected (If-Then-Else) as the root of the expression tree. At each of the branches, further subterms have to be composed by further random selection from S. This procedure is repeated until all leaf nodes are labeled with terminals.

In principle, the depth of the generated expression tree is not constrained, although one usually defines a maximum tree depth and width in order to reduce memory requirements for storage and evaluation time of the program term. Furthermore, the symbol set S has to be *closed* with respect to composition. Each symbol must be combineable with any other symbol, so that the final expression is always syntactically correct and can be interpreted as a proper program or data structure. For the example in Figure 10.22, the expression should still evaluate to a valid result, if the Equal_2 symbol is replaced by Add_2. For further restrictions on the composition of building blocks, a *Strongly Typed* GP approach can be used, where legal syntactic structures are obtained by attributing the arguments and return values of the functions as well as the terminals with type information [384, 254]. For our example, a typed version might look as follows:

$$S = \{\text{Mult}_{(\text{int},\text{int})\longrightarrow\text{int}}, \text{If-Then-Else}_{(\text{bool},\text{int},\text{int})\longrightarrow\text{int}}, \text{Equal}_{(\text{int},\text{int})\longrightarrow\text{bool}}, \ldots\}.$$

Further concepts of generating and maintaining constrained syntactic structures are also described in [315, Chapter 19] for symbolic expressions and in [231] and [284, 285, 287] for grammar-based approaches.

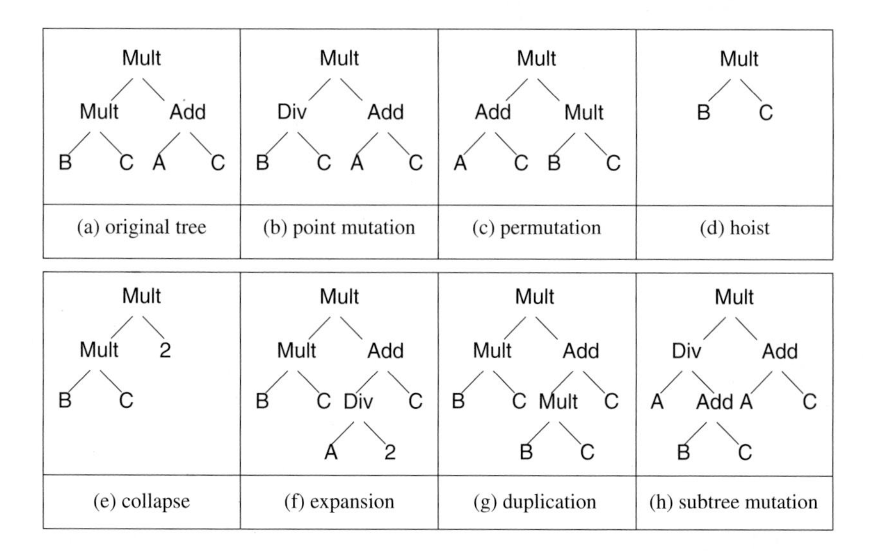

Fig. 10.23. GP mutations on tree-structured expressions.

10.6.2 Mutations on Symbolic Expressions

The GP mutation operators come in a great number of varieties. Basically, subterms or leaves of a tree structure are substituted by either newly generated or duplicated terms or terminals. Figure 10.23 gives a brief overview of mutation operators on tree structured expressions.

Point Mutation A *point mutation* (b) exchanges a single node by a random node of the same class. In the simplest case this means that terminals are substituted by terminals, and function symbols are substituted by function symbols with the same arity (and types, if applicable).

Permutation A *permutation* (c) merely permutes the sequence of arguments of a node. The *hoist* operator (d) substitutes the whole tree by a randomly selected proper subtree (terminals are not in the scope of selection).

Collapse Subtree Mutation With the *collapse subtree mutation* (e) a subtree is replaced by a random terminal.

Expansion Mutation The inverse operation is the *expansion mutation* (f) where a terminal is exchanged against a random, newly generated subtree.

Duplication The *duplication* operator (g) replaces a terminal node with a copy of a proper subtree.

Subtree Mutation The most general mutation operator is defined by *subtree mutation* (h), where a subtree is substituted with a newly generated subtree.

10.6.3 Crossover of Symbolic Expressions

Another important operator used in GP for generating new term structures is a variant of the one-point crossover known from Genetic Algorithms (Section 10.5.3). By crossing two linear GA chromosomes, substrings are exchanged between the chromosomes. An analogous recombination operator for GP terms is defined by interchanging (sub-)trees between two GP terms. Figure 10.24 shows the simplest crossover version known as *subtree exchange crossover*, which is performed as follows: For each term a node is chosen at random. Inner nodes (including the root) as well as leaves are selectable. In Figure 10.24 the selected subtrees are marked by triangular shapes. The two recombined terms result from a mutual exchange of the selected sub-expressions.

In comparison to the crossover or recombination operators of Genetic Algorithms or Evolution Strategies, an interesting aspect of GP crossover is the following. Even for a recombination among two identical trees, the GP crossover (*self crossover*) results in a pair of new structures whenever the two crossover nodes differ. The GP crossover operator is, by far, less prone to convergence of the GP structures within a population than the string-based GA chromosomes. GA crossover for identical chromosomes is reduced to a simple reproduction operator without changing the structures. As the GA crossover operator enhances the similarity among the strings, mutation operators are essentially needed to introduce new allele settings into the genepool. Obviously, this is not the case for GP crossover on symbolic expressions. Genetic Programming in its original proposal even refrains from using any of the described mutation operators. In [315] a set of "secondary" operators is described, but they are not used for the discussed experiments — with only few exceptions [315, p. 105-112].

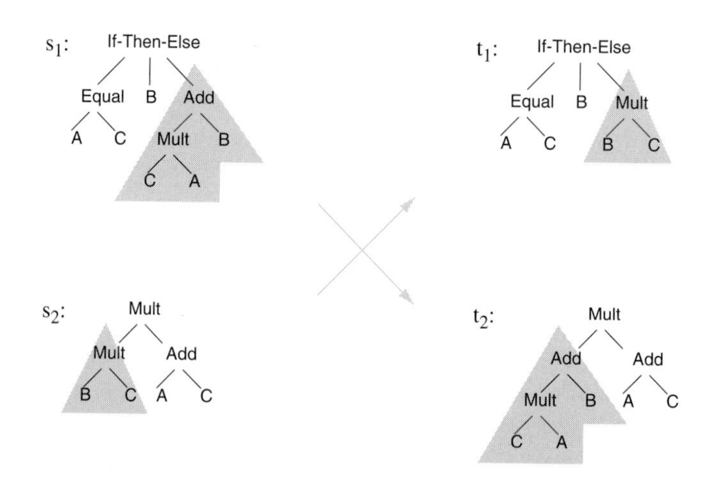

Fig. 10.24. Subtree exchange crossover for symbolic expressions.

Several variants for GP crossover have been developed, such as *context-preserving crossover* (CPC), where subtrees are exchanged only if either their node coordinates match exactly (*strong* CPC) or match approximately (*weak* CPC). With *module crossover*, parametrized sub-trees are exchanged between two individual structures. The modules are comparable to parametrized macros in programming languages like C. One variant of module crossover is known as *encapsulation* and *decapsulation* [315, 23]. An extension of this approach denoted as *macro extraction* is presented in [286].

10.6.4 GP Evolution Scheme

For canonical Genetic Programming a slightly modified GA evolution scheme is used. The scheme depicted in Figure 10.25 is a slight modification of the basic GP algorithm [315, p. 76] and is better compared to the GA and ES schemes already discussed (Sections 10.4.4 and 10.5.6). The comma as well as the plus reproduction scheme can be used. Figure 10.25 depicts the comma scheme where parents are not part of the selection pool.

A notable extension of the GP evolution scheme is the operator pool, which contains a set of genetic operators, each attributed by a selection probability. One reproduction cycle works as follows: First, a genetic operator is selected from the operator pool (in the example we have only a reproduction, a mutation, and a crossover operator). Secondly, depending on the arity n of the operator, n individuals are chosen by a fitness-proportionate selection function ($n = 1$ for reproduction and mutation, $n \geq 2$ for crossover). After applying the operators the new program structures are evaluated and constitute the selection pool for the parents of the following generation.

10.6.5 GP in Action: The AntTracker Task

The following brief example illustrates what Genetic Programming with symbolic expressions is about. Although this is only a "toy example", it has been widely used by many GP research groups as a testbed to understand the basic principles of genetic program evolution.

Suppose, an *artificial ant* is placed within a two-dimensional grid area as depicted in Figure 10.26. The dark black squares represent walls. The darker gray squares, spread out like a trace, are "food pieces" the ant has to collect. The lighter gray squares are "pheromones", which would lead the ant to the next food piece. However, the ant has to learn how to use this information. Initially, the ant sits at position (11, 17) heading east. The ant is able to sense the field it is facing and is equipped with sensors that can differentiate food squares from wall pieces. At one action step, the ant can either move forward to the next square (if it is not in front of a wall) or turn 90 degrees left or right.

The optimization task is to evolve control programs that enable the ant to reach all food squares within a given time limit. The set of building blocks used as elementary instructions is defined as follows. Four commands, implemented as terminals, control the ant's movement:

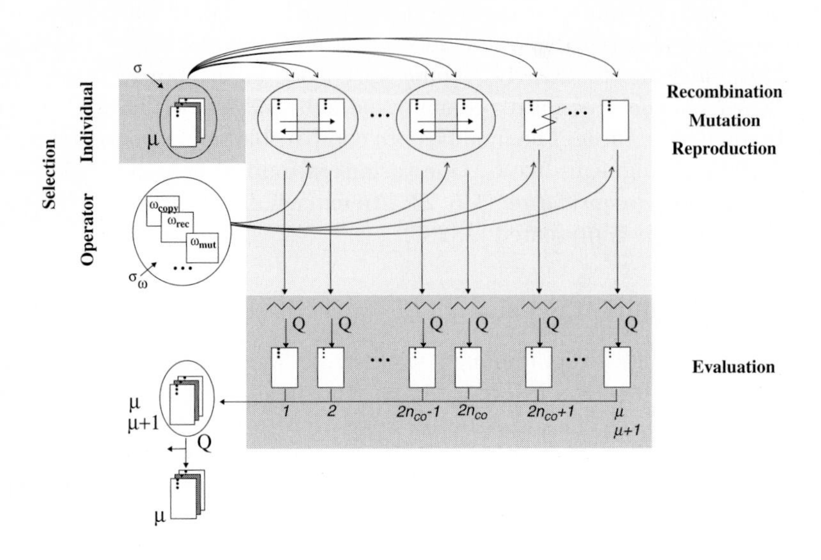

Fig. 10.25. GP selection, variation and reproduction scheme.

$$T_{\text{move}} = \{\text{advance}, \text{turnLeft}, \text{turnRight}, \text{nop}\}. \tag{10.59}$$

The function set defines how an ant can react on environmental signals:

$$F_{\text{sensor}} = \{\text{ifSensor}[\mathit{signal}]_1, \text{ifSensor}[\mathit{signal}]_2\}, \tag{10.60}$$

where

$$\mathit{signal} \in \{\text{food}, \text{wall}, \text{pheromone}, \text{dust}\} \tag{10.61}$$

describe the sensitivity of the ant to the four categories of squares. The parametrized 1- and 2-ary function *ifSensor* implements an *if-then* and an *if-then-else* conditional. The complete set of functions and terminals is defined as

$$S = T_{\text{move}} \cup F_{\text{sensor}} \cup \{\text{seq}_{1\dots m}, \text{again}\}, \tag{10.62}$$

where *seq* is a function with a varying number of arguments, defining sequences of instructions, and the *again* terminal reevaluates the last *ifSensor* condition, which leads to implicit instruction looping [286].

We use a population size of 100 individuals per generation. The time limit for the maximum number of action steps is set to 400. The following genetic operators are used for constructing and recombining new ant control programs: subtree mutation, crossover, permutation, duplication, and deletion. The operator ranks are automatically adjusted. Operators which lead to better-fit individuals are

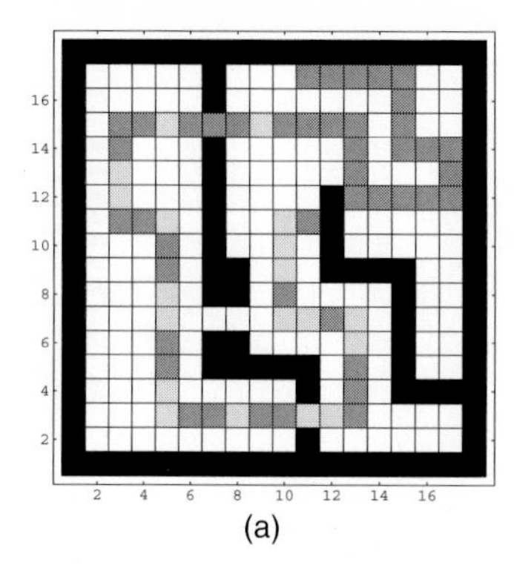

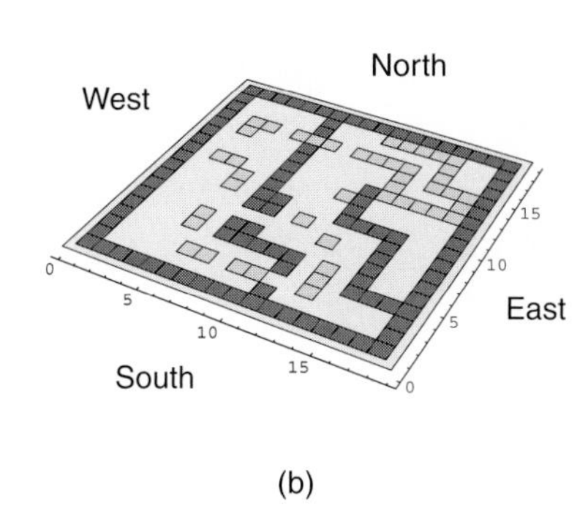

Fig. 10.26. Two views of the discrete environment in which each "ant" control program is evaluated. Black squares denote wall segments, dark grey squares represent food, light grey squares are pheromones that may be used by the control program to find the next food piece. The task is to move over all food squares, by possibly following the food path, starting from the top (north) at position $(11, 17)$.

used more often, whereas operators, which tend to produce offspring programs with lower fitness than their parent, receive decreased selection ranks (see [287, Chapter 8.3] for more details).

Figure 10.27 gives examples of control programs that have been evolved by Genetic Programming methods. Shown are the best-of-generation programs with their associated action diagrams visualizing the ant's movements in time. We again started with randomly generated program structures, 100 per population. For the first generation (Gen. 0) a very simple program turns out to be the best. The ant moves forward until it stops in front of the wall towards which it is heading. Some generations later, periodic, circular movements of the ant evolve (Gen. 9 and 16). In generation 22 the ant follows the food trail almost perfectly, except at the beginning, where it misses several food squares. Finally, after 59 generations, a program is evolved that implements an ant not only following the given trail but also able to actively search for food in off-trail regions.

This example is interesting because it gives an impression of the potential of Genetic Programming as a paradigm to automatically evolve algorithms for specific problems without having to do any explicit programming. The system starts from a set of basic building blocks, recombines them to compose a population of program structures which compete for better and better solutions to the given problem. Recombinations lead to new, slightly better program variants. Useful subroutines invented by earlier programs are built into later structures (by crossover), from which eventually a general trail follower program is evolved. A more detailed discussion of this example can be found in [286].

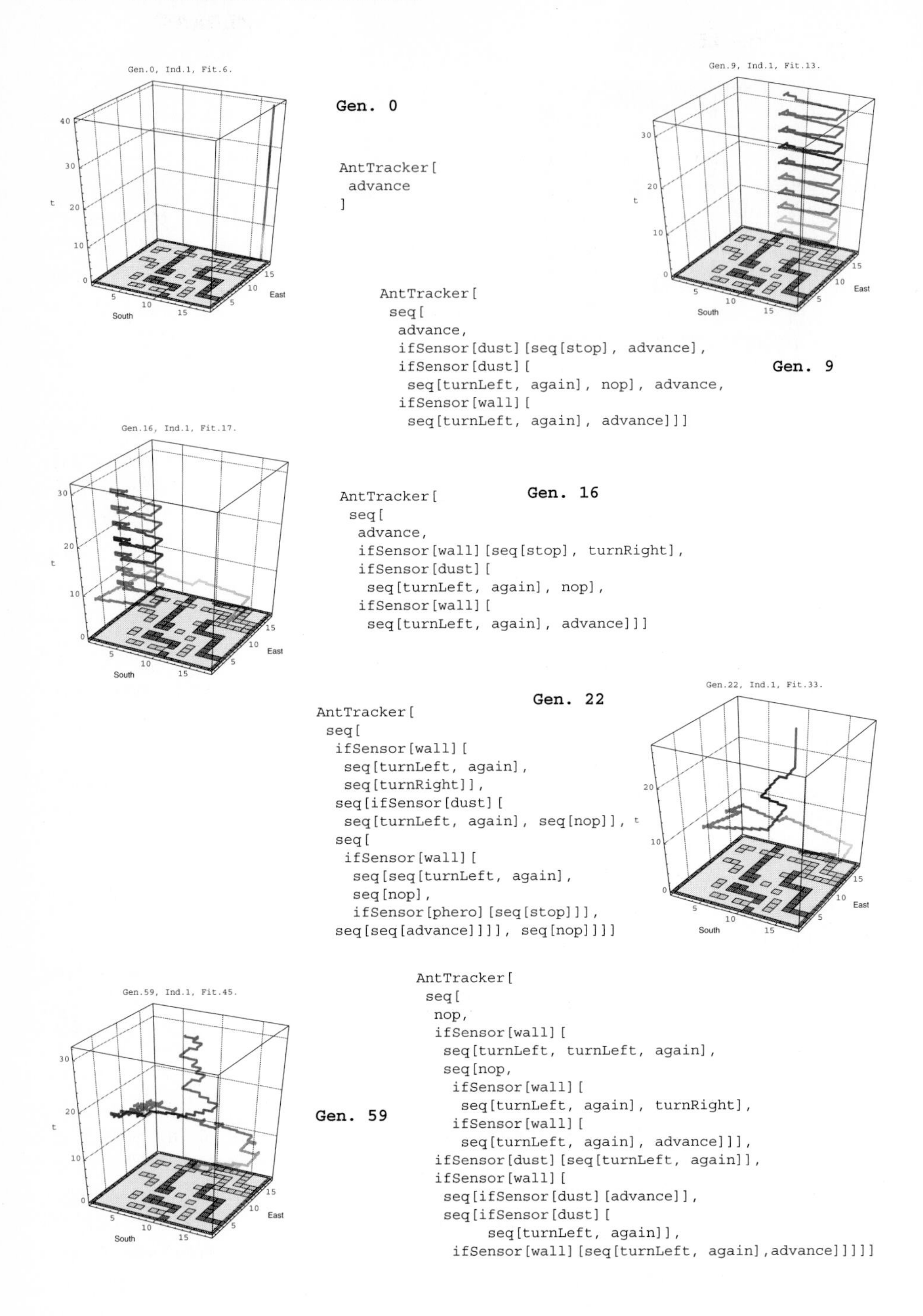

Fig. 10.27. Genetic Programming: Evolving food-foraging programs.

10.6.6 Further GP Applications and References

Genetic Programming has been applied to a wide variety of problem domains of which we can only mention a few here. The three-volume treatise of Koza's research group gives an excellent overview of GP at work [315–317]. A list of almost 200 GP applications is provided in [42, pp. 342] ranging from GP evolution of art images, jazz melodies and musical structures, to applications in biotechnology, computer graphics, computing and algorithms, process and robotics control, data analysis, electrical engineering and circuit design, image processing, signal processing and optimization in general. Further research results, theoretical issues and applications are collected in the *Advances in Genetic Programming* volumes [306], [24], and [331]. The proceedings of the annual GP conference and the Genetic and Evolutionary Computation Conference (GECCO) series give an overview of state-of-the-art genetic programming [320, 319, 318, 41, 539, 492, 102].

10.6.7 Further Remarks on Programming by Evolution

Programming by evolution has a long history. In the early 1950s, Friedberg et al. generated assembler programs for a single-bit register machine by a stochastic process, trying to evolve new programs by solely random mutations [195, 196]. Mainly due to the lack of any selection and crossover mechanism this simulated evolution scheme for automatic programming did not perform better than pure random search. The importance of recombination or crossover was only recognized later for optimizing vectors of discrete and continuous object variables in linear programming, where operators similar to the discussed ES and GA operators were used [91, 92].

A more general approach of evolution on the basis of an operational programming model is presented by Fogel et al. [192]. Here the individuals are descriptions of finite state automata, the behavior of which is adapted to problem-specific tasks by a mutation and reproduction scheme (similar to a $(1+1)$-ES scheme). Mutations are defined by deleting or inserting automaton states, changing transitions or changing the input/output signals of the states. This approach is known today as *Evolutionary Programming* (EP) although the evolution of programs, with respect to higher programming languages, do not play a prominent role in EP. Most of the EP systems either deal with the evolution of finite automata [188, 191] or with parameter optimization [188, 190]. However, there are also more general EP approaches. The proceedings of the annual Evolutionary Programming conferences give an up-to-date overview [25, 425]. For example, problem-specific data structures are adapted by specifically designed mutation operators combined with a general selection scheme. Angeline reports on a number of successful EP experiments in the areas of neural networks, game theory, induction of grammars, and robot control [23].

A more general objective followed by Holland — eventually leading to the development of Genetic Algorithms — was the design of a genetic code by which the structure of any computer program could be represented [268, p. 45]. The idea of applying an adaptation process to a population of programs describing complex behavioral strategies leads to the concept of *classifier systems* (CS). A classifier system is defined by a set of condition-action (if-then) rules. The adaptation of the rule set to specific environmental conditions is performed by a Genetic Algorithm. More detailed descriptions of this rule-based CS approach, which also allows complex interactions among the (sub-)programs (rules), can be found in [75], [219, Chapter 6 and pp. 219], and [268, p. 171-184]. Classifier systems have also been used to model interactions among gene complexes in cellular genomes.

There are numerous other approaches on using Genetic Algorithms for automatic program induction. We mention a few here: Compositions of binary operations are encoded as strings of constant length and evolved by a GA scheme [132]. De Jong shows how to use Genetic Algorithms to search program spaces [296]. A GA approach to generate LISP code for evolving strategies for the prisoner's dilemma is described in [203]. Another proposal on how to encode tree-structured terms with binary operators on bit strings by mapping complete binary trees can be found in [545].

10.7. Conclusion and Summary

In this chapter we have shown how stochastic search techniques can be used in the context of machine learning and data analysis. We have focused on two stochastic search paradigms: *simulated annealing* (SA) and *evolutionary algorithms* (EA). Comparing the basic schemes for the SA Metropolis algorithm (Figure 10.2) and the standard evolutionary algorithm (Figure 10.5), their similarities and differences can be summarized as follows:

Solution Variation In simulated annealing, variations of the set of solutions[4] are generated using the stochastic *Perturb* function. The current solutions may be considered the "parents" from which mutated "offspring" solutions are created. In the context of evolutionary algorithms, analogous concepts are used; here, mutations of the offspring data structures (genomes) lead to explorations of new territory in the solution space. Additionally, by mimicking sexual reproduction, recombination operators are applied. This introduces a novel approach into stochastic search techniques, because new combinations of "building blocks" are composed from already discovered partial solutions.

Solution Evaluation For the subsequent selection, the newly created solution set has to be evaluated. For simulated annealing, evaluation is reduced to calculating the fitness difference between the current best and the new solution. For evolutionary algorithms, all the offspring solutions receive a fitness value, on the basis of which they enter the selection filter.

[4] Only for parallel SA versions there is a *set* of solutions, otherwise only a single solution is considered. However, a single solution may also represent a "population" of individual solutions.

Solution Selection In the Metropolis algorithm, the new solution is selected if it has higher fitness (lower energy) than the current solution. However, worse solutions are also selectable, with a certain probability, which itself is controlled by the temperature. The lower the temperature, the less likely it is that solutions with lower fitness are accepted. This adaptive selection strategy is similar to what can be achieved by using a variable mutation stepsize for evolutionary algorithms (as defined for Evolution Strategies, Section 10.4.2).

For evolutionary algorithms we have discussed a broad range of selection schemes, ranging from fitness-proportionate, elitist, and rank-based selection (Section 10.5.5) to combinations of random and best selections (Section 10.4.4). Therefore, the EA selection functions allow for a broad range of solution selection with subtle control of search space exploration versus exploitation. For SA applications this can be controlled by the annealing schedule, where higher temperatures result in explorative (global) search, whereas lower temperatures result in more constrained, local and exploitative search.

We have discussed a number of variants of simulated annealing and their applications (Sections 10.2.3 and 10.2.4). For evolutionary algorithms, the major approaches, successfully applied to optimization, data analysis and machine learning problems in science and industry, have been introduced. Evolution Strategies have a broad range of application domains, especially in engineering sciences, when huge sets of parameters have to be adjusted and optimized. Genetic Algorithms are an alternative approach, with closer connections to natural genomes (discrete alphabet encoding, crossover operator, etc.). They also can be used in a wide range of application domains, and they can be especially useful for solving combinatorial optimization problems. The Genetic Programming approach, an extension of Genetic Algorithms, is particularly interesting for evolving complex, hierarchical data structures, such as computer programs, graphs, grammar and transformation systems, fuzzy rules, or neural network architectures [332, 333]. Further historical and up-to-date overviews of evolutionary computing and its applications can be found in [49, 138, 189] and [287].

Chapter 11
Visualization

Daniel Keim* and Matthew Ward**
*University of Konstanz, Germany and **Worcester Polytechnic Institute, USA

11.1. Introduction

The progress made in hardware technology allows today's computer systems to store very large amounts of data. Researchers from the University of Berkeley estimate that every year about 1.5 Exabytes (= 1.5 Million Terabytes) of data are generated, of which a large portion is available in digital form [359]. It is possible that in the next three years more data will be generated than in all of human history to date. The data are often automatically recorded via sensors and monitoring systems. Even simple transactions of everyday life, such as paying by credit card or using the telephone, are typically recorded by computers. Usually many variables are recorded, resulting in data with a high dimensionality. The data are collected because people believe that it is a potential source of valuable information, providing new insights or a competitive advantage (at some point). Finding valuable information hidden in the data, however, is a difficult task. With today's data management systems, it is only possible to examine quite small portions of the data. If the data are presented textually, the amount of data that can be displayed is in the range of some one hundred data items, but this is like a drop in the ocean when dealing with data sets containing millions of data items. Having no possibility to adequately explore the large amounts of data that have been collected because of their potential usefulness, the data becomes useless and the databases become data 'dumps'.

Information visualization and visual data analysis can help to deal with the flood of information. The advantage of visual data exploration is that the user is directly involved in the data analysis process. There are a large number of information visualization techniques that have been developed over the last two decades to support the exploration of large data sets. In this chapter, we present an overview of information visualization and visual exploration using a classification based on the *data type to be visualized*, the *visualization technique*, and the *interaction technique*. We illustrate the classification using a few examples, and indicate some directions for future work.

Benefits of Visual Data Exploration

For data analysis to be effective, it is important to include the human in the data exploration process and combine the flexibility, creativity, and general knowledge of the human with the enormous storage capacity and the computational power of today's computers. Visual data mining aims at integrating the human in the data analysis process, applying human perceptual abilities to the analysis of large data sets available in today's computer systems. The basic idea of visual data mining is to present the data in some visual form, allowing the user to gain insight into the data, draw conclusions, and directly interact with the data. Visual data analysis techniques have proven to be of high value in exploratory data analysis. Visual data mining is especially useful when little is known about the data and the exploration goals are vague. Since the user is directly involved in the exploration process, shifting and adjusting the exploration goals can be done in a continuous fashion as needed.

Visual data exploration can be seen as a hypothesis generation process; the visualizations of the data allow the user to gain insight into the data and come up with new hypotheses. The verification of the hypotheses can also be done via data visualization, but may also be accomplished by automatic techniques from statistics, pattern recognition, or machine learning, as discussed earlier in this volume. In addition to the direct involvement of the user, the main advantages of visual data exploration over automatic data analysis techniques are:

- Visual data exploration can easily deal with highly non-homogeneous and noisy data.
- Visual data exploration is intuitive and requires no understanding of complex mathematical or statistical algorithms or parameters.
- Visualization can provide a qualitative overview of the data, allowing data phenomena to be isolated for further quantitative analysis.

As a result, visual data exploration usually allows a faster data exploration and often provides more interesting results, especially in cases where automatic algorithms fail. In addition, visual data exploration techniques provide a much higher degree of confidence in the findings of the exploration. These facts lead to a high demand for visual exploration techniques and make them indispensable in conjunction with automatic exploration techniques.

Visual Exploration Paradigm

Visual Data Exploration usually follows a three step process: *Overview first, zoom and filter, and then details-on-demand.* According to Ben Shneiderman this process is called the Information Seeking Mantra [479]. In analyzing large data sets, the user first needs to get an overview of the data. In the overview, the user identifies interesting patterns or groups in the data and focuses on one or more of them. For analyzing the patterns, the user needs to drill-down and access details of the data. Visualization technology may be used for all three steps of the data exploration process. Visualization techniques are useful for showing an overview of the data, allowing the user to identify interesting subsets. In this step, it is important to keep the overview visualization while focusing on the subset using another visualization technique. An alternative is to distort the overview visualization in order to focus on the interesting subsets. This can be performed by dedicating a larger percentage of the display to the interesting subsets while decreasing screen utilization for currently uninteresting data. To further explore the interesting subsets, the user needs a drill-down capability in order to observe the details about the data. Note that visualization technology does not only provide the base visualization techniques for all three steps but also bridges the gaps between the steps.

11.2. Classification of Visual Data Analysis Techniques

Information visualization focuses on data sets lacking inherent 2D or 3D semantics and therefore also lacking a standard mapping of the abstract data onto the physical screen space. There are a number of well known techniques for visualizing such data sets, such as x-y plots, line plots, and histograms. These techniques are useful for data exploration but are limited to relatively small and low dimensional data sets. In the last decade, a large number of novel information visualization techniques have been developed, allowing visualizations of multi-dimensional data sets without inherent two- or three-dimensional semantics. Nice overviews of the approaches can be found in a number of recent books [103, 534, 493, 466]. The techniques can be classified based on three criteria (see Figure 11.1) [300]: The data type to be visualized, the visualization technique, and the interaction technique used.
The **data type to be visualized** [479] may be:

- One-dimensional data, such as temporal (time-series) data
- Two-dimensional data, such as geographical maps
- Multi-dimensional data, such as relational tables
- Text and hypertext, such as news articles and Web documents
- Hierarchies and graphs, such as telephone calls and Web documents
- Algorithms and software, such as debugging operations

The **visualization technique** used may be classified as:

- Standard 2D/3D displays, such as bar charts and x-y plots
- Geometrically-transformed displays, such as landscapes and parallel coordinates
- Icon-based displays, such as needle icons and star icons

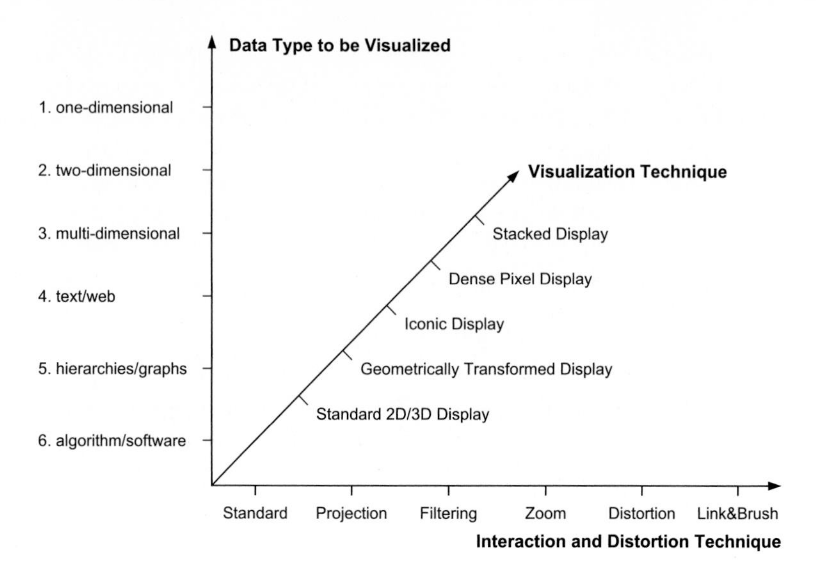

Fig. 11.1. Classification of information visualization techniques.

- Dense pixel displays, such as recursive patterns and circle segments
- Stacked displays, such as treemaps and dimensional stacking

The third dimension of the classification is the **interaction technique** used. Interaction techniques allow users to directly navigate and modify the visualizations, as well as select subsets of the data for further operations. Examples include:

- Dynamic Projection, that allows smooth navigations through the data space
- Interactive Filtering, to enable users to isolate subsets of data for focussed analysis
- Zooming, to enlarge data for detailed analysis
- Distortion, to increase the screen space allocated to areas of interest while preserving the context of the entire data set
- Linking and Brushing, to enable users to select data of interest in one view and see it highlighted in other views

Note that the three dimensions of our classification – data type to be visualized, visualization technique, and interaction technique – can be assumed to be orthogonal. Orthogonality means that any of the visualization techniques may be used in conjunction with any of the interaction techniques for any data type. Note also that a specific system may be designed to support different data types and that it may use a combination of visualization and interaction techniques.

11.3. Data Type to be Visualized

In information visualization, the data usually consists of a large number of records, each consisting of a number of variables or dimensions. Each record

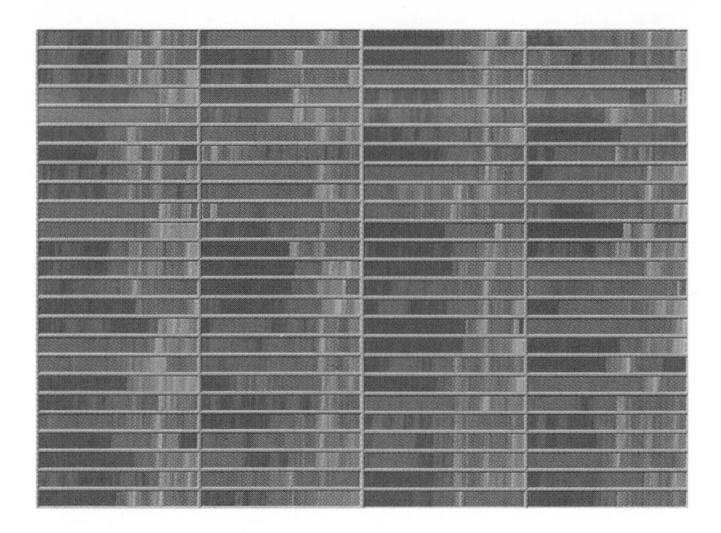

Fig. 11.2. The Recursive Pattern Technique [303] is a dense display technique mapping each data value to a colored pixel (high values correspond to bright colors), which is based on a recursive generalization of a line- and column-based arrangement of the pixels. The example shows about 20 years (Jan. 74 - Apr. 95) daily data of the 100 stocks of the FAZ index (Frankfurt stock index). *(from [303] ©IEEE)*

corresponds to an observation, measurement, or transaction. Examples are customer properties, e-commerce transactions, and sensor output from physical experiments. The number of attributes can differ from data set to data set; one particular physical experiment, for example, can be described by five variables, while another may need hundreds of variables. We call the number of variables the dimensionality of the data set. Data sets may be one-dimensional, two-dimensional, multi-dimensional or may have more complex data types such as text/hypertext or hierarchies/graphs. Sometimes, a distinction is made between dense (or grid) dimensions and the dimensions that may have arbitrary values. Depending on the number of dimensions with arbitrary values the data are sometimes also called univariate, bivariate, or multivariate.

One-dimensional data

One-dimensional data usually have one dense dimension. A typical example of one-dimensional data is temporal data. Note that one or multiple data values may be associated with each point in time. An example are time series of stock prices (see Figures 11.2 and 11.3) or time series of news data (see e.g. ThemeRiver [251]).

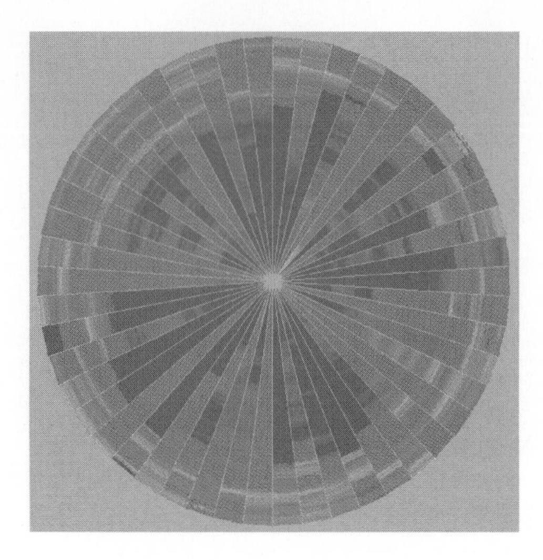

Fig. 11.3. The Circle Segments Technique [29] is a dense pixel display technique, mapping each data value to a colored pixel (high values correspond to bright colors). The different dimensions are mapped to the segments of the circle and the pixels are arranged in a back-and-forth fashion adjacent to the segment-halving line. The example shows about 20 years (Jan 74 - Apr 95) daily data of 50 stocks of the FAZ index (Frankfurt stock index). Note for example the three bright outer 'rings' which corresponds to different high price periods with subsequent lower price periods. *(from [29] ©IEEE)*

Two-dimensional data

A typical example of two-dimensional data is geographical data, where the two distinct dimensions are longitude and latitude. A standard method for visualizing two-dimensional data are x-y plots and maps are a special type of x-y plots for presenting two-dimensional geographical data. Examples are geographical maps. Although it seems easy to deal with temporal or geographical data, caution is advised. If the number of records to be visualized is large, temporal axes and maps get quickly cluttered – and may not help to understand the data.

Multi-dimensional data

Many data sets consist of more than three dimensions and therefore do not allow a simple visualization as 2-dimensional or 3-dimensional plots. Examples of multi-dimensional (or multivariate) data are tables from relational databases, which often have tens to hundreds of columns (or dimensions). Since there is no simple mapping of the data dimensions to the two dimensions of the screen, more sophisticated visualization techniques are needed. An example of a technique that allows the visualization of multi-dimensional data is the Parallel Coordinates Technique [280] (see Figure 11.4). Parallel Coordinates display each multi-dimensional data item as a set of line segments that intersect each of the parallel axes at the position corresponding to the data value for that dimension.

Fig. 11.4. The Parallel Coordinates Technique [280] belongs to the class of geometrically-transformed displays. In conjunction with similarity-based coloring, it displays each multi-dimensional data item as a polygonal line intersecting the dimension axes at the position corresponding to the data value for the dimension. *(from [301] ©IEEE)*

Text and Hypertext

Not all data types can be described in terms of dimensionality. In the age of the World Wide Web, important data types are text and hypertext, as well as multimedia web page contents. These data types differ in that they cannot be easily described by numbers, and therefore most of the standard visualization techniques cannot be applied. In most cases, a transformation of the data into description vectors is necessary before visualization techniques can be used. An example of a simple transformation is word counting (see ThemeRiver [251]) which is often combined with principal component analysis or multidimensional scaling [323] to reduce the dimensionality to two or three (for example, see [546]). See Chapter 3 for a description of these techniques.

Hierarchies and Graphs

Data records often have some relationship to other pieces of information. These relationships may be ordered, hierarchical, or arbitrary networks of relations. Graphs are widely used to represent such interdependencies. A graph consists of a set of objects, called nodes, and connections between these objects, called edges or links. Examples are the e-mail interrelationships among people, their shopping behavior, the file structure of the hard disk, or the hyperlinks in the World Wide Web. There are a number of specific visualization techniques that deal with hierarchical and graphical data. An example of an internet graph is shown in Figure 11.5. An overview of hierarchical information visualization techniques can be found in [113], an overview of Web visualization techniques is presented in [156], and an overview book on all aspects related to graph drawing is [47].

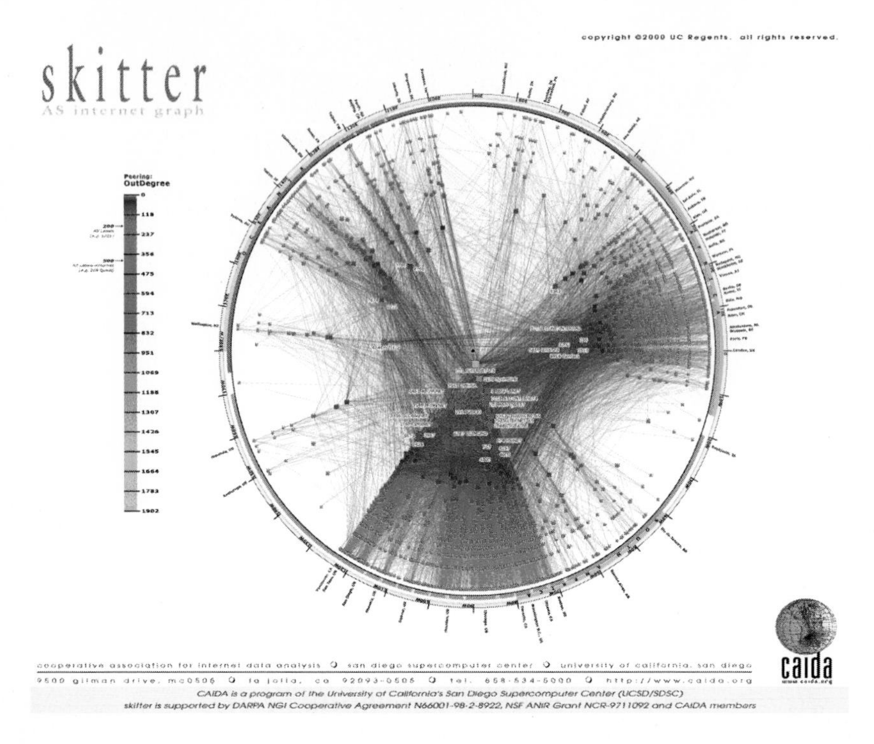

Fig. 11.5. The Skitter Graph Internet Map shows the global structure of the internet. The nodes represent autonomous systems which in general correspond to ISPs. The nodes are positioned according to their geographical longitude in polar coordinates. The more important nodes with a high number of connections are positioned towards the center and are colored accordingly. The visualization shows the high connectivity within North America and the strong connections between Europe (or Asia) and the US. In contrast, direct connections between Europe and Asia are rare. The data shown have been collected in two weeks of October 2000. *(used by permission of the Cooperative Association for Internet Data Analysis, "Skitter Graph" ©2000 UC Regents. Courtesy University of California)*

Algorithms & Software

Another class of data are algorithms and software. Coping with large software projects is a challenge. The goal of software visualization is to support software development by helping to understand algorithms (e.g., by showing the flow of information in a program), to enhance the understanding of written code (e.g., by representing the structure of thousands of source code lines as graphs), and to support the programmer in debugging the code (e.g., by visualizing errors). There are a large number of tools and systems that support these tasks. Overviews of software visualization can be found in [428] and [498].

11.4. Visualization Techniques

There are a large number of visualization techniques that can be used for visualizing data. In addition to standard 2D/3D-techniques such as x-y (x-y-z) plots, bar charts, line graphs, and maps, there are a number of more sophisticated classes of visualization techniques. The classes correspond to basic visualization principles that may be combined in order to implement a specific visualization system.

11.4.1 Geometrically-Transformed Displays

Geometrically-transformed display techniques aim at finding "interesting" transformations of multi-dimensional data sets. The class of geometric display methods includes techniques from exploratory statistics such as scatterplot matrices [22, 119] and techniques that can be subsumed under the term "projection pursuit" [274], a class of techniques that attempt to locate projections that satisfy some computable quality of interestingness. Other geometric projection techniques include Prosection Views, where only user-selected slices of the data are projected [205, 495], Hyperslice [524], and the well-known Parallel Coordinates visualization technique [280]. The Parallel Coordinate Technique maps the k-dimensional space onto the two display dimensions by using k axes that are parallel to each other (either horizontally or vertically oriented) and are evenly spaced across the display. The axes correspond to the dimensions and are linearly scaled from the minimum to the maximum value of the corresponding dimension. Each data item is presented as a chain of connected line segments, intersecting each of the axes at the location corresponding to the value of the dimension considered (see Figure 11.4).

11.4.2 Iconic Displays

Another class of visual data exploration techniques are the iconic display methods. The idea is to map the attribute values of a multi-dimensional data item to the features of an icon. Icons may be defined arbitrarily – for example as little faces [115], needle icons [5], star icons [533], stick figure icons [421], color icons [302, 349], or TileBars [255]. The visualization is generated by mapping the attribute values of each data record to the features of the icons. In case of the stick figure technique, for example, two dimensions are mapped to the display dimensions and the remaining dimensions are mapped to the angles and/or limb length of the stick figure icon. If the data items are relatively dense with respect to the two display dimensions, the resulting visualization presents texture patterns that vary according to the characteristics of the data. Transitions between two regions of different texture are readily detectable, as this is an innate pre-attentive perceptual ability of the human visual system. Figure 11.6 shows an example of this class of techniques. Each data point is represented by a star icon/glyph, where each data dimension controls the length of a ray emanating

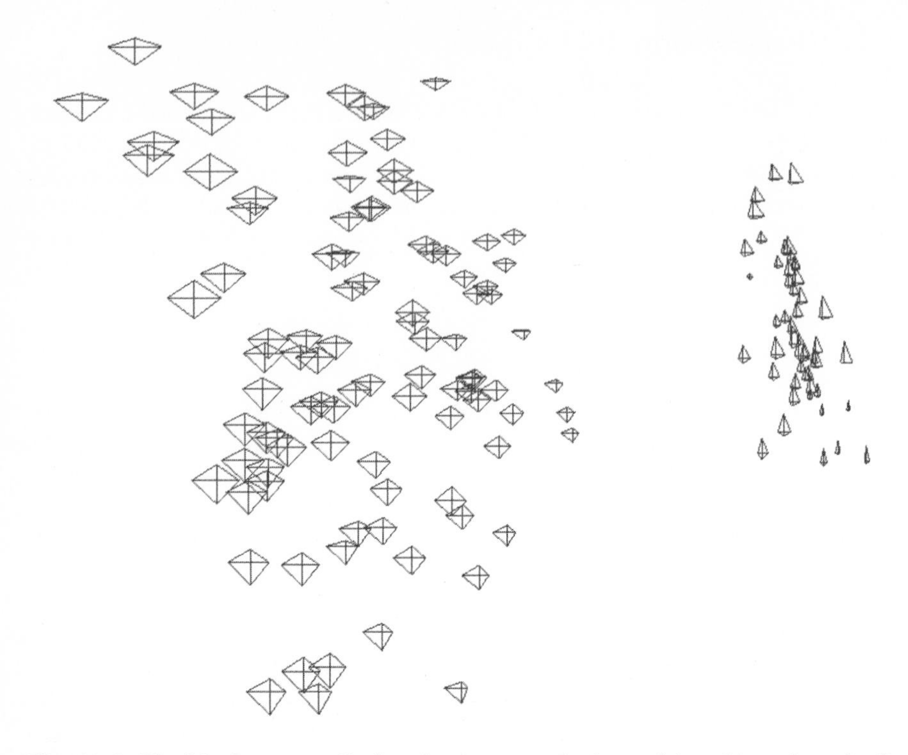

Fig. 11.6. The iris data set – displayed using star glyphs positioned based on the first two principal components. *(generated using XmdvTool [533])*

from the center of the icon. In this example, the positions of the icons are determined using principal component analysis (PCA) to convey more information about data relations. Other data attributes could also be mapped to the icon position.

11.4.3 Dense Pixel Displays

The basic idea of dense pixel techniques is to map each dimension value to a colored pixel and group the pixels belonging to each dimension into adjacent areas [299]. Since in general dense pixel displays use one pixel per data value, the techniques allow the visualization of the largest amount of data possible on current displays (up to about 1,000,000 data values). If each data value is represented by one pixel, the main question is how the pixels are arranged on the screen. Dense pixel displays use different arrangements to provide detailed information on local correlations, dependencies, and hot spots.

Well-known examples are the recursive pattern technique [303] and the circle segments technique [29]. The recursive pattern technique is based on a generic recursive back-and-forth arrangement of the pixels (see Figure 11.7) and is particularly aimed at representing data sets with a natural order according to one

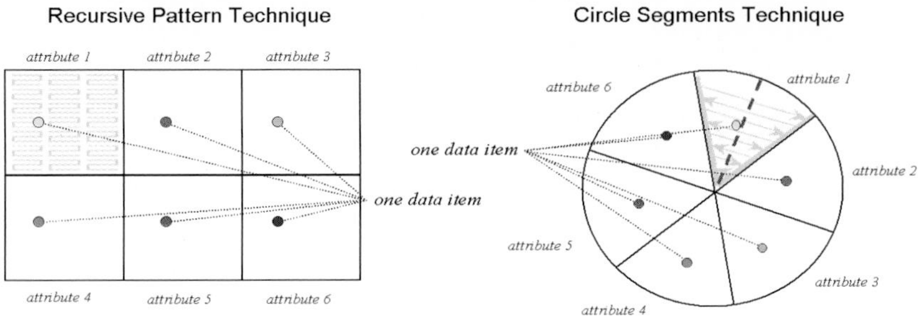

Fig. 11.7. The basic idea of Recursive Pattern and Circle Segments techniques.

attribute (e.g. time-series data). The user may specify parameters for each recursion level and thereby control the arrangement of the pixels to form semantically meaningful substructures. The basic element on each recursion level is a pattern of height h_i and width w_i as specified by the user. First, the elements correspond to single pixels that are arranged within a rectangle of height h_1 and width w_1 from left to right, then below backwards from right to left, then again forward from left to right, and so on. The same basic arrangement is done on all recursion levels with the only difference that the basic elements that are arranged on level i are the patterns resulting from level $(i-1)$. Figure 11.2 provides an example recursive pattern visualization of financial data. The visualization shows twenty years (January 1974 - April 1995) of daily prices of the 100 stocks contained in the Frankfurt stock index (FAZ). The idea of the circle segments technique [29] is to represent the data in a circle that is divided into segments, one for each attribute (see Figure 11.7). Within the segments each attribute value is again visualized by a single colored pixel. The arrangement of the pixels starts at the center of the circle and continues to the outside by plotting on a line orthogonal to the segment halving line in a back and forth manner. The rationale of this approach is that close to the center all attributes are close to each other enhancing the visual comparison of their values. Figure 11.3 shows an example of circle segment visualization using the same data (50 stocks) as shown in Figure 11.2.

11.4.4 Stacked Displays

Stacked display techniques are tailored to present data partitioned in a hierarchical fashion. In the case of multi-dimensional data, the data dimensions to be used for partitioning the data and building the hierarchy have to be selected appropriately. An example of a stacked display technique is *Dimensional Stacking* [346]. The basic idea is to embed one coordinate system inside another coordinate system, i.e. two attributes form the outer coordinate system, two other attributes are embedded into the outer coordinate system, and so on. The display is generated by dividing the outermost level coordinate system into rectangular cells. Within the cells, the next two attributes are used to span the

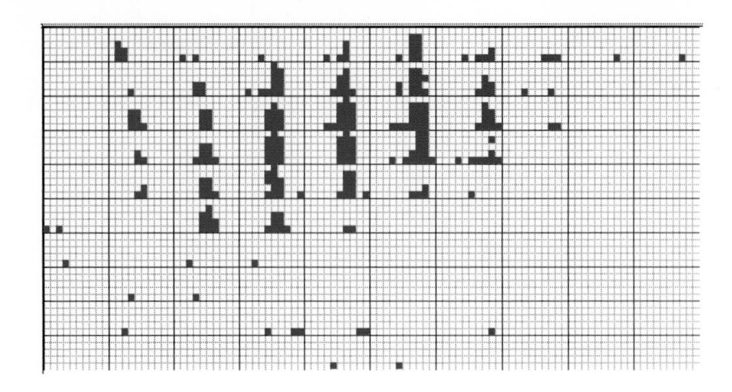

Fig. 11.8. Dimensional Stacking visualization of drill hole mining data. *(used by permission of M. Ward, Worcester Polytechnic Institute ©IEEE)*

second level coordinate system. This process may be repeated multiple times. The usefulness of the resulting visualization largely depends on the data distribution of the outer coordinates and therefore the dimensions that are used for defining the outer coordinate system have to be selected carefully. A rule of thumb is to choose the most important dimensions first. A Dimensional Stacking visualization of mining data with longitude and latitude mapped to the outer x and y axes, as well as ore grade and depth mapped to the inner x and y axes is shown in Figure 11.8. Other examples of stacked display techniques include Worlds-within-Worlds [178], Treemap [294, 478], and Cone Trees [450].

11.5. Interaction Techniques

In addition to the visualization technique, for an effective data exploration it is necessary to use one or more interaction techniques. *Interaction techniques* allow the data analyst to directly interact with the visualizations and dynamically change the visualizations according to the exploration objectives. In addition, they also make it possible to relate and combine multiple independent visualizations.

Interaction techniques can be categorized based on the effects they have on the display. *Navigation techniques* focus on modifying the projection of the data onto the screen, using either manual or automated methods. *View enhancement methods* allow users to adjust the level of detail on part or all of the visualization, or modify the mapping to emphasize some subset of the data. *Selection techniques* provide users with the ability to isolate a subset of the displayed data for operations such as highlighting, filtering, and quantitative analysis. Selection can be done directly on the visualization (*direct manipulation*) or via dialog boxes and other query mechanisms (*indirect manipulation*). Some examples of interaction techniques are described below.

Dynamic Projection

Dynamic projection is an automated navigation operation. The basic idea is to dynamically change the projections in order to explore a multi-dimensional data set. A well-known example is the GrandTour system [33] which tries to show all interesting – i.e., those exhibiting desireable properties such as well-separated clusters – two-dimensional projections of a multi-dimensional data set as a series of scatterplots. Note that the number of possible projections is exponential in the number of dimensions, i.e., it is intractable for large dimensionality. The sequence of projections shown can be random, manual, precomputed, or data driven. Systems supporting dynamic projection techniques include XGobi [99, 505], XLispStat [514], and ExplorN [105].

Interactive Filtering

Interactive filtering is a combination of selection and view enhancement. In exploring large data sets, it is important to interactively partition the data set into segments and focus on interesting subsets. This can be done by a direct selection of the desired subset (*browsing*) or by a specification of properties of the desired subset (*querying*). Browsing is difficult for very large data sets and querying often does not produce the desired results. Therefore, a number of interactive selection techniques have been developed to improve interactive filtering in data exploration. An example of a tool that can be used for interactive filtering is the Magic Lens [67, 181]. The basic idea of Magic Lenses is to use a tool similar to a magnifying glass to filter the data directly in the visualization. The data under the magnifying glass is processed by the filter and displayed in a different way than the remaining data set. Magic Lenses show a modified view of the selected region, while the rest of the visualization remains unaffected. Note that several lenses with different filters may be used; if the filter overlap, all filters are combined. Other examples of interactive filtering techniques and tools are InfoCrystal [497], Dynamic Queries [9, 167, 220], and Polaris [502].

Zooming

Zooming is a well known view modification technique that is widely used in a number of applications. In dealing with large amounts of data, it is important to present the data in a highly compressed form to provide an overview of the data but at the same time allow a variable display of the data at different resolutions. Zooming does not only mean displaying the data objects larger, but also that the data representation may automatically change to present more details on higher zoom levels. The objects may, for example, be represented as single pixels at a low zoom level, as icons at an intermediate zoom level, and as labeled objects at a high resolution. An interesting example applying the zooming idea to large tabular data sets is the TableLens approach [440]. Getting an overview of such data sets is difficult if the data are displayed in textual form. The basic idea of TableLens is to represent each numerical value by a small bar. All bars have a

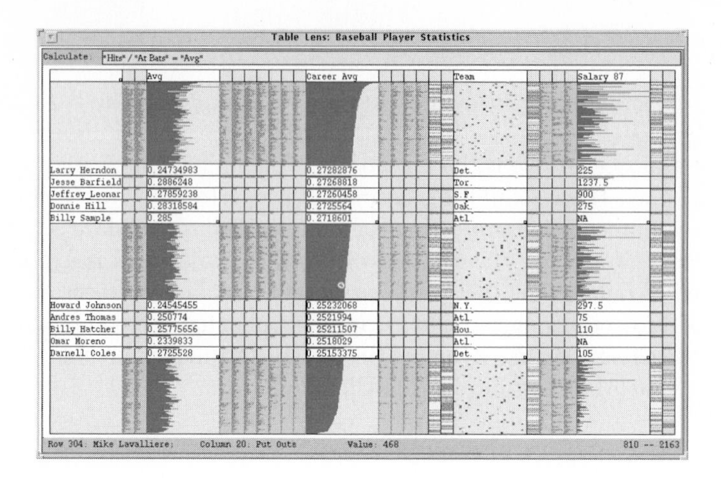

Fig. 11.9. The Table Lens approach. *(used by permission of R. Rao, Xerox PARC ©ACM)*

one-pixel height and the lengths are determined by the attribute values. This means that the number of rows on the display can be nearly as large as the vertical resolution and the number of columns depends on the maximum width of the bars for each attribute. The initial view allows the user to detect patterns, correlations, and outliers in the data set. In order to explore a region of interest the user can zoom in, with the result that the affected rows (or columns) are displayed in more detail, possibly even in textual form. Figure 11.9 shows an example of a baseball database with a few rows and columns being selected in full detail. Other examples of techniques and systems that use interactive zooming include PAD++ [52, 53, 418], IVEE/Spotfire [10], and DataSpace [30]. A comparison of fisheye and zooming techniques can be found in [461].

Distortion

Distortion is a view modification technique that supports the data exploration process by preserving an overview of the data during drill-down operations. The basic idea is to show portions of the data with a high level of detail while others are shown with a lower level of detail. Popular distortion techniques are hyperbolic and spherical distortions; these are often used on hierarchies or graphs but may also be applied to any other visualization technique. An overview of distortion techniques is provided in [348] and [104]. Examples of distortion techniques include Bifocal Displays [494], Perspective Wall [361], Graphical Fisheye Views [204, 458], Hyperbolic Visualization [330, 397], and Hyperbox [15]. Figure 11.10 shows the effect of distorting part of a scatterplot matrix to display more detail from one of the plots while preserving context from the rest of the display.

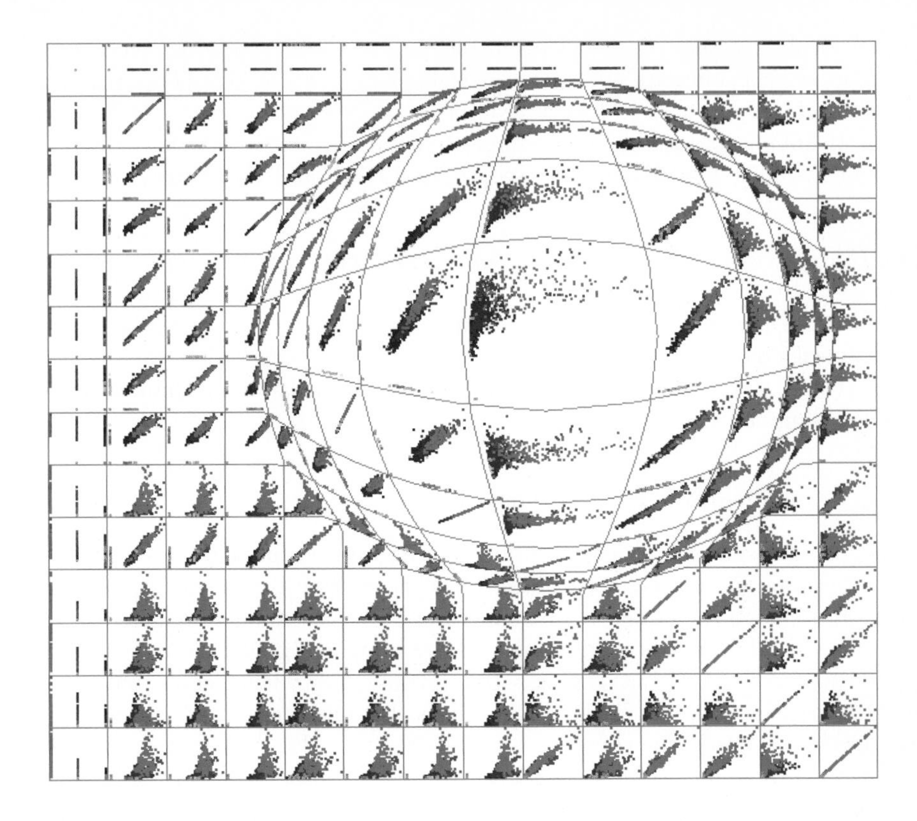

Fig. 11.10. A scatterplot matrix with part of the display distorted using a fisheye lens. *(used by permission of M. Ward, Worcester Polytechnic Institute)*

Brushing and Linking

Brushing is an interactive selection process that is often, but not always, combined with *linking*, a process for communicating the selected data to other views of the data set. There are many possibilities to visualize multi-dimensional data, each with their own strengths and weaknesses. The idea of linking and brushing is to combine different visualization methods to overcome the shortcomings of individual techniques. Scatterplots of different projections, for example, may be combined by coloring and linking subsets of points in all projections. In a similar fashion, linking and brushing can be applied to visualizations generated by all visualization techniques described above. As a result, the brushed points are highlighted in all visualizations, making it possible to detect dependencies and correlations. Interactive changes made in one visualization are automatically reflected in the other visualizations. Note that connecting multiple visualizations through interactive linking and brushing provides more information than considering the component visualizations independently.

Typical examples of visualization techniques that have been combined by linking and brushing are multiple scatterplots, bar charts, parallel coordinates, pixel displays, and maps. Most interactive data exploration systems allow some form of linking and brushing. Example tools and systems include S Plus [50], XGobi [51, 505], XmdvTool [533] (see Figure 11.11), and DataDesk [526, 542].

11.6. Specific Visual Data Analysis Techniques

There are a number of visualization techniques that have been developed to support specific data mining tasks such as association rule generation, classification, and clustering. In the following, we describe how visualization techniques can be used to support these tasks.

11.6.1 Association Rule Generation

The goal of association rule generation is to find interesting patterns and trends in transaction databases. Association rules are statistical relations between two or more items in the data set. In a supermarket basket application, associations express the relations between items that are bought together. It is for example interesting if we find out that in 70% of the cases when people buy bread, they also buy milk. Association rules tell us that the presence of some items in a transaction imply the presence of other items in the same transaction with a certain probability, called confidence. A second important parameter is the support of an association rule, which is defined as the percentage of transactions in which the items co-occur.

Let $I = \{i_1, ...i_n\}$ be a set of items and let D be a set of transactions, where each transaction T is a set of items such that $T \subseteq I$. An association rule is an implication of the form $X \Rightarrow Y$, where $X \subseteq I$, $Y \in I$, $X, Y \neq \emptyset$. The confidence c is defined as the percentage of transactions that contain Y, given X. The support is the percentage of transactions that contain both X and Y. For a given support and confidence level, there are efficient algorithms to determine all association rules [7]. A problem, however, is that the resulting set of association rules is usually very large, especially for low support and confidence levels. Using higher support and confidence levels may not be effective since then, useful rules may be overlooked.

Visualization techniques have been used to overcome this problem and to allow an interactive selection of good support and confidence levels. Figure 11.12 shows SGI MineSets *Rule Visualizer* [97] which maps the left and right hand sides of the rules to the x- and y-axes of the plot, respectively, and shows the confidence as the height of the bars and the support as the height of the discs. The color of the bars shows the interestingness of the rule (see Chapter 7 for a discussion of methods to compute interestingness). Using the visualization, the user is able to see groups of related rules and the impact of different confidence and support levels. The number of rules that can be visualized, however, is limited and the visualization does not suport combinations of items on the left

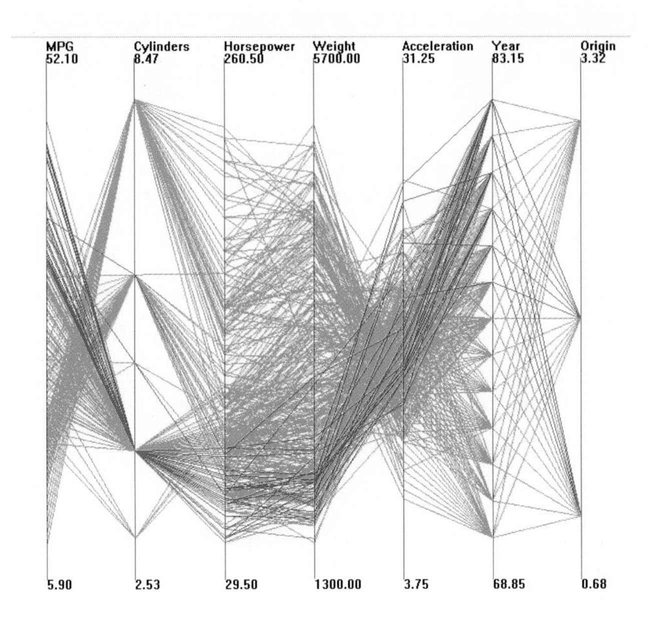

(a) Parallel Coordinates

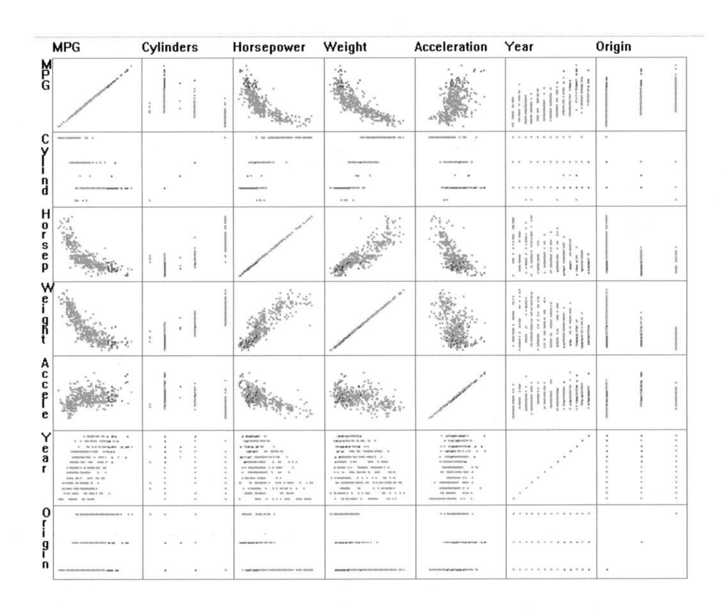

(b) Scatterplot Matrix

Fig. 11.11. Linked brushing between two multivariate visualization techniques in XmdvTool [533]. Highlighted data (in red) from one display is also highlighted in the other display. *(used by permission of M. Ward, Worcester Polytechnic Institute)*

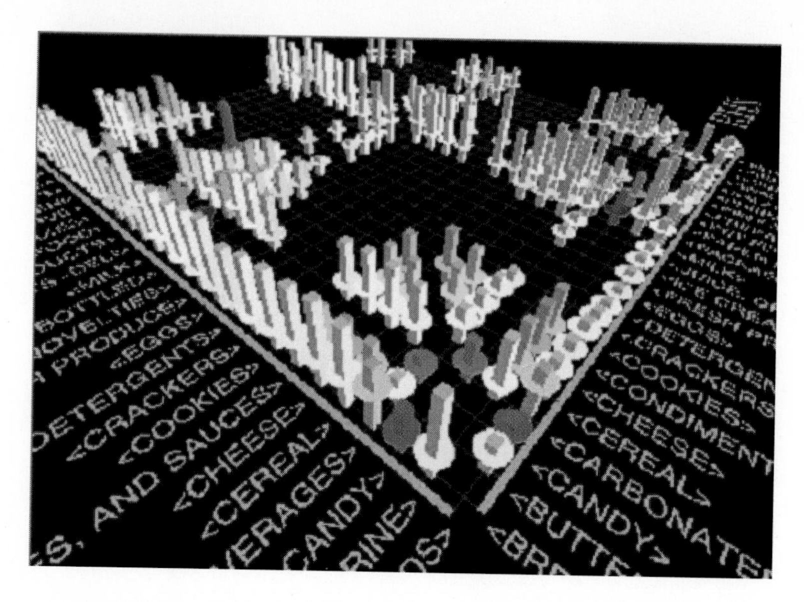

Fig. 11.12. MineSet's association rule visualizer [97]. *(©SGI)*

or right hand side of the association rules. Figure 11.13 shows two alternative visualizations called mosaic and double decker plots [266]. The basic idea is to partition a rectangle on the y-axis according to one attribute and make the size of the regions proportional to the sum of the corresponding data values. Compared to bar charts, mosaic plots use the height of the bars instead of their width to show the parameter value. Then each resulting area is split in the same way according to a second attribute. The coloring reflects the percentage of data items that fulfill a third attribute. The visualization shows the support and confidence values of all rules of the form $X_1, X_2 \Rightarrow Y$. Mosaic plots are restricted to two attributes on the left side of the association rule. Double decker plots can be used to show more than two attributes on the left side. The idea is to display a hierarchy of attributes on the bottom (heineken, coke, chicken in the example shown in Figure 11.13(b)) corresponding to the left hand side of the association rules. The bars on the top correspond to the number of items in the considered subset of the database and therefore visualize the support of the rule. The colored areas in the bars correspond to the percentage of data transactions that contain an additional item (sardines in Figure 11.13(b)) and therefore represent the support. Other approaches to association rule visualization include graphs with nodes corresponding to items and arrows corresponding to implications (as used in DBMiner [279]) and association matrix visualizations to cluster related rules [247].

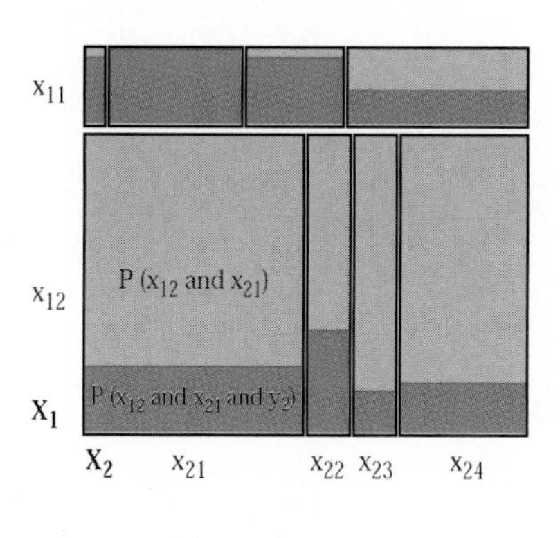

(a) Mosaic Plot

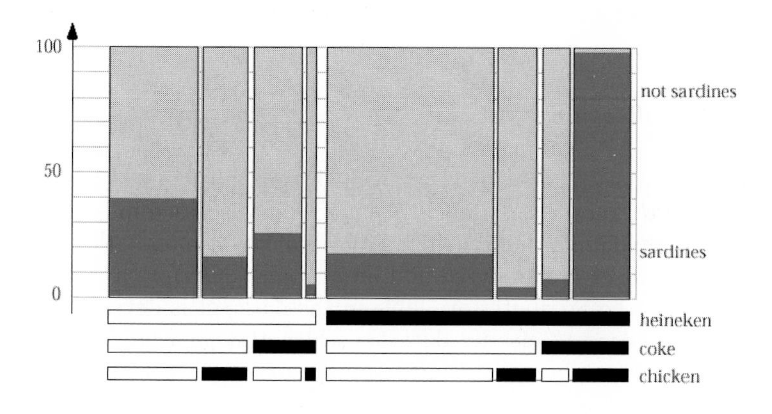

(b) Double Decker Plot

Fig. 11.13. Association rule visualizations. *(from [266] ©ACM)*

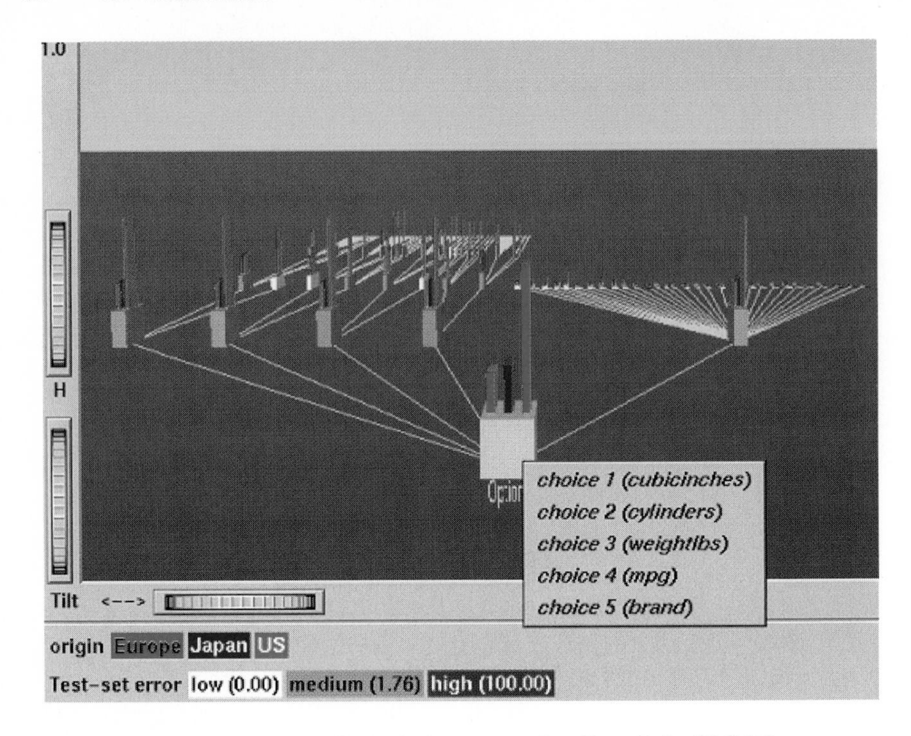

Fig. 11.14. MineSet's decision tree visualizer [97]. *(©SGI)*

11.6.2 Classification

Classification is the process of developing a classification model based on a training data set with known class labels. As has been discussed in several previous chapters in this volume, to construct the classification model, the attributes of the training data set are analyzed and an accurate description or model of the classes based on the attributes available in the data set is developed. The class descriptions are used to classify data for which the class labels are unknown. Classification is sometimes also called *supervised learning* because the training set is used to teach the system how to classify the data. There are a large number of algorithms for solving classification tasks. A popular class of approaches are algorithms that inductively construct decision trees (see Appendix B.Examples are ID3 [433], CART [90], ID5 [520, 521], C4.5 [435], SLIQ [372], and SPRINT [472]. In addition, there are approaches that use neural networks, genetic algorithms, or Bayesian networks (see Chapter 4) to solve the classification problem. Since most algorithms work as black box approaches, it is often difficult to understand and optimize the decision model. Problems such as overfitting or tree pruning are difficult to tackle.

Visualization techniques can help to overcome these problems. The decision tree visualizer in SGIs MineSet system [97] shows an overview of the decision tree together with important parameters such as the attribute value distributions

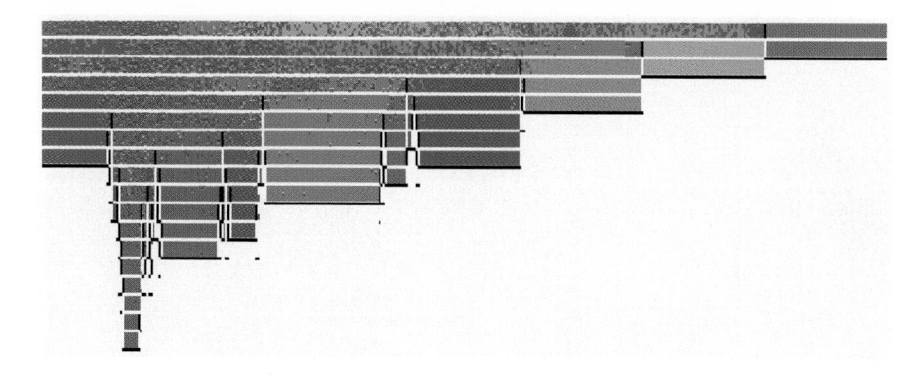

Fig. 11.15. Visualization of a decision tree [28] for the segment training data from the Statlog benchmark having 19 attributes *(used by permission of M. Ankerst [28] ©ACM)*

(see Figure 11.14). The system allows an interactive selection of the attributes shown and helps the user to understand the decision tree. A more sophisticated approach, which also helps in decision tree construction, is visual classification as proposed in [28]. The basic idea is to show each attribute value by a colored pixel and arrange them in bars – similar to the Dense Pixel Displays presented in subsection 11.4.3. The pixels of each attribute bar are sorted separately and the attribute with the purest value distribution is selected as the split attribute of the decision tree. The procedure is repeated until all leaves correspond to pure classes. An examplary decision tree resulting from this process is shown in Figure 11.15. Compared to a standard visualization of a decision tree, additional information is provided that is helpful for explaining and analyzing the decision tree, namely

- size of the nodes (number of training records corresponding to the node)
- quality of the split (purity of the resulting partitions)
- class distribution (frequency and location of the training instances of all classes).

Some of this information might also be provided by annotating the standard visualization of a decision tree (for example, annotating the nodes with the number of records or the gini-index), but this approach clearly fails for more complex information such as the class distribution. In general, visualizations can provide a better understanding of the classification models and they can help to interact more easily with the classification algorithms in order to optimize the model generation and classification process.

11.6.3 Clustering

Clustering is the process of finding a partitioning of the data set into homogeneous subsets called clusters (see Chapters 3 and 9 for more discussions on cluster

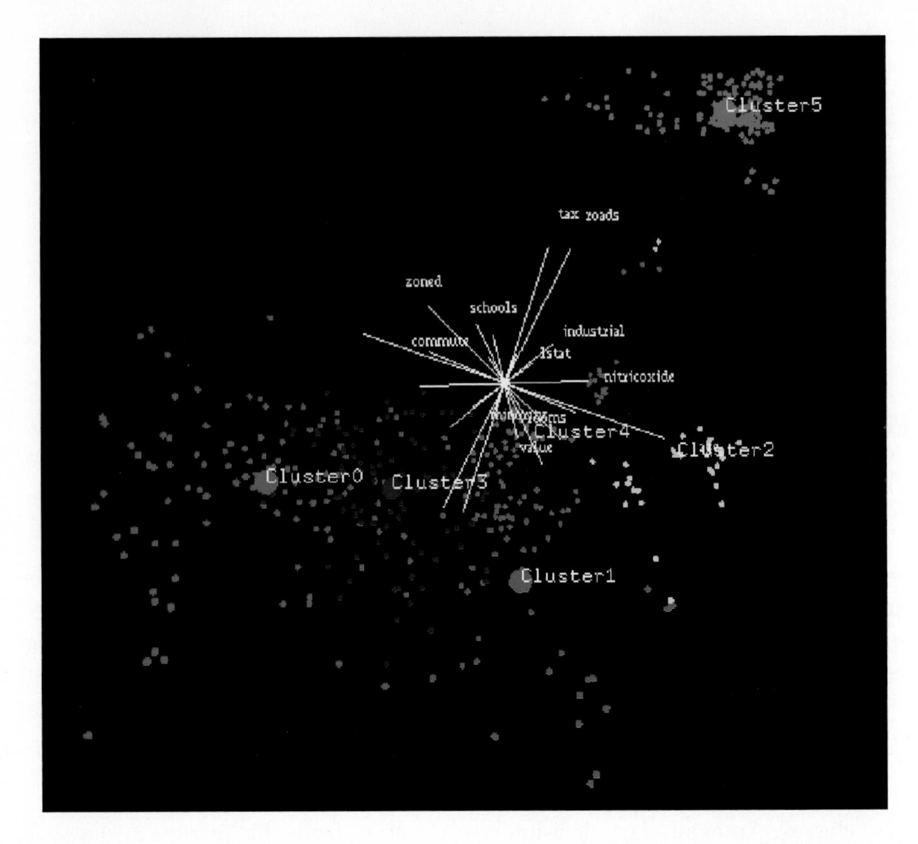

Fig. 11.16. Visualization based on a projection into 3D space. *(from [555] ©ACM)*

analysis). Unlike classification, clustering is often implemented as a form of *unsupervised learning.* This means that the classes are unknown and no training set with class labels is available. A wide range of clustering algorithms have been proposed in the literature including density-based methods such as KDE [469] and linkage-based methods [73]. Most algorithms use assumptions about the properties of the clusters that are either used as defaults or have to be given as input parameters. Depending on the parameter values, the user obtains different clustering results. In two- or three-dimensional space, the impact of different algorithms and parameter settings can be explored easily using simple visualizations of the resulting clusters (for example, x-y plots), but in higher dimensional space the impact is much more difficult to understand. Some higher-dimensional techniques try to determine two- or three-dimensional projections of the data that retain the properties of the high-dimensional clusters as much as possible [555]. Figure 11.16 shows a three-dimensional projection of a data set consisting of five clusters.

While this approach works well with low- to medium-dimensional data sets, it is difficult to apply it to large high-dimensional data sets, especially if the

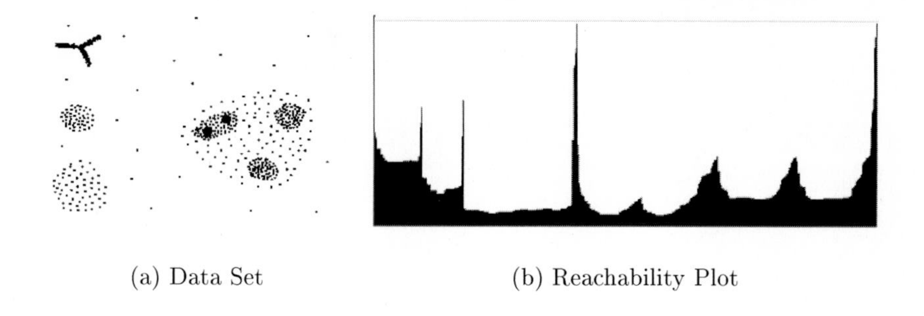

(a) Data Set (b) Reachability Plot

Fig. 11.17. OPTICS visual clustering. *(used by permission of M. Ankerst [27] ©ACM)*

clusters are not clearly separated and the data set also contains noise (data that does not belong to any cluster). In this case, more sophisticated visualization techniques are needed to guide the clustering process, select the right clustering model, and adjust the parameter values appropriately. An example of a system that uses visualization techniques to help in high-dimensional clustering is OPTICS (*Ordering Points To Identify the Clustering Structure*) [27]. The idea of OPTICS is to create a one-dimensional ordering of the database representing its density-based clustering structure. Figure 11.17 shows a two-dimensional examplary data set together with its reachability distance plot. Intuitively, points within a cluster are close in the generated one-dimensional ordering and their reachability distance (shown in Figure 11.17) is similar. Jumping to another cluster results in higher reachability distances. The idea works for data of arbitrary dimension. The reachability plot provides a visualization of the inherent clustering structure and is therefore valuable for understanding the clustering and guiding the clustering process.

Another interesting approach is the *HD-Eye* system [263]. The *HD-Eye* system considers the clustering problem as a partitioning problem and supports a tight integration of advanced clustering algorithms and state-of-the-art visualization techniques, allowing the user to directly interact in the crucial steps of the clustering process. The crucial steps are the selection of dimensions to be considered, the selection of the clustering paradigm, and the partitioning of the data set. Novel visualization techniques are employed to help the user identify the most interesting projections and subsets as well as the best separators for partitioning the data. Figure 11.18 shows an example screen shot of the *HD-Eye* system with its basic visual components for cluster separation. The separator tree represents the clustering model produced so far in the clustering process. The *abstract iconic displays* (top right and bottom middle in figure 11.18) visualize the partitioning potential of a large number of projections. The properties are based on histogram information of the point density in the projected space. The number of icons corresponds to the number of peaks in the projection and their color to the number of data points belonging to the maximum. The color follows a given color table ranging from dark colors for large maxima to bright

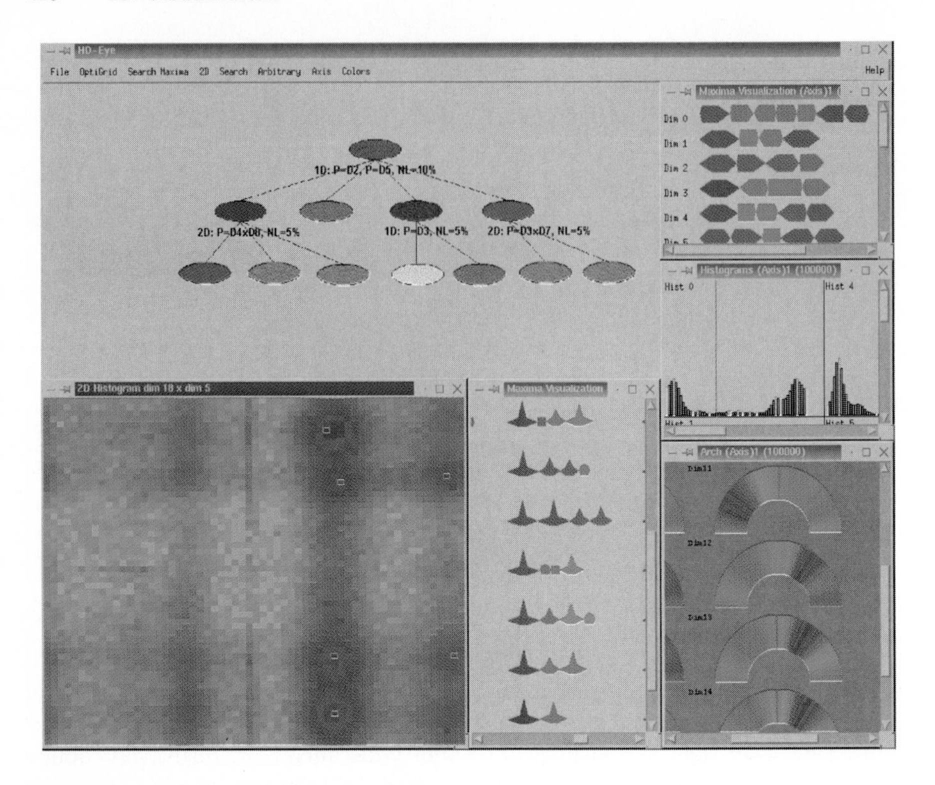

Fig. 11.18. *HD-Eye* A screen-shot [263] showing different visualizations of projections and the separator tree. Clockwise from the top: separator tree, iconic representation of 1D projections, 1D projection histogram, 1D color-based density plots, iconic representation of multi-dimensional projections and color-based 2D density plot. *(from [263] ©IEEE)*

colors for small maxima. The measure of how well a maximum is separated from the others is reflected by the shape of the icon and the degree of separation varies from sharp spikes for well-separated maxima to blunt spikes for weak-separated maxima. The *color- and curve-based point density displays* present the density of the data and allow a better understanding of the data distribution, which is crucial for an effective partitioning of the data. The visualizations are used to decide which dimensions are taken for the partitioning. In addition, the partitioning can be specified interactively directly within the visualizations, allowing the user to define non-linear partitionings (e.g., in the 2D density plots).

11.7. Conclusion

The exploration of large data sets is an important but difficult problem. Information visualization techniques can be useful in solving this problem. Visual data exploration has a high potential, and many applications such as fraud detection and data mining can use information visualization technology for improved data analysis.

Avenues for future work include the tight integration of visualization techniques with traditional techniques from such disciplines as statistics, machine learning, operations research, and simulation. Integration of visualization techniques and these more established methods would combine fast automatic data analysis algorithms with the intuitive power of the human mind, improving the quality and speed of the data analysis process. Visual data analysis techniques also need to be tightly integrated with the systems used to manage the vast amounts of relational and semistructured information, including database management and data warehouse systems. The ultimate goal is to bring the power of visualization technology to every desktop to allow a better, faster, and more intuitive exploration of very large data resources. This will not only be valuable in an economic sense but will also stimulate and delight the user.

LIVERPOOL JOHN MOORES UNIVERSITY
LEARNING & INFORMATION SERVICES

Chapter 12
Systems and Applications

Xiaohui Liu
Brunel University, United Kingdom

12.1. Introduction

Recent developments in computing have provided the basic infrastructure for fast access to vast amounts of online data; processing power and storage devices are continuing to be cheaper and more powerful, networks are providing more bandwidth and higher reliability, personal computers and workstations are widespread, and On-Line Analytic Processing (OLAP) allows rapid retrieval of data from data warehouses [120]. In addition, many of the advanced computational methods for extracting information from large quantities of data, or "data mining" methods, are beginning to mature, e.g. artificial neural networks, Bayesian networks, decision trees, genetic algorithms, statistical pattern recognition and support vector machines. These developments coincide with the desire from various communities for extracting useful knowledge from large amount of data that has become readily available. For example, new exciting applications such as bioinformatics and e-science have called for new ways of organising and managing data as well as novel ways of analysing these data. The quest in the business community to gain competitive advantage using the knowledge from corporate data has led to "business intelligence" – the gathering and management of data, and the analysis of that data to turn it into useful information to improve decision making.

The analysis of data, though obviously a crucial part of these relatively new applications such as bioinformatics, e-science and business intelligence, is currently lagging behind other aspects of the process; most of the data collected so far have not been analyzed, and there are few tools around that support the effective analysis of "big data". Despite these, many intelligent systems for data analysis have been developed and much progress has been made. This chapter aims to provide a snapshot of intelligent data analysis (IDA) applications and to look into a few important issues in the system development.

12.2. Diversity of IDA Applications

12.2.1 Business and Finance

There is a wide range of successful business applications reported, although the retrieval of technical details is not always easy, perhaps for obvious reasons. These applications include fraud detection, customer retention, cross selling, stock prediction, marketing and insurance.

Fraud is costing industries billions of pounds, so it is not surprising to see that systems have been developed to combat fraudulent activities in such areas as credit card [153], stock market dealing [392], and health care [131]. Worth particular mention is a system developed by the Financial Crimes Enforcement Network (FINCEN) of the US Treasury Department, called "FAIS": the FINCEN AI Systems [471]. FAIS detects potential money-laundering activities from a large number of big cash transactions. The Bank Secrecy Act in 1971 required the reporting of all cash transactions greater than $10,000, and these transactions, about 14 million a year, are the basis for detecting suspicious financial activities. By combining user expertise with the system's rule-based reasoner, visualization facility, and association-analysis module, FAIS uncovers previously unknown and potentially high-value leads for possible investigation. The reports generated by the FAIS application have helped FINCEN uncover more than 400 cases of money-laundering activities, involving more than $1 billion in potentially laundered funds. In addition FAIS is reported to be able to discover criminal activities that law enforcement in the field would otherwise miss, e.g. connections in cases involving nearly 300 individuals, more than 80 front operations, and thousands of cash transactions [110].

To many business organizations, how to retain their existing customers, especially those "vulnerable" ones, is a key marketing issue. For example, Mellon Bank at USA has used the data they have on existing credit card customers to characterize their behavior and then try to predict what they will do next. Using IBM's Intelligent Miner, Mellon developed a credit card attrition model to predict which customers will stop using Mellon's credit card in the next few months. Based on the prediction results, the bank can take marketing actions to retain these customers loyalty. Similarly Lockheed has developed customer vulnerability models for frozen orange juice and reported the results of applying the models to an A C Nielsen database containing the supermarket purchases of 15,000 households over a three-year period [484].

TV program schedulers would like to know the likely audience for a proposed program and the optimum time to show it. In audience prediction, the data are fairly complex. Factors which determine the audience share gained by a particular program include not only the characteristics of the program itself and the time at which is shown, but also the nature of the competing programs at other channels. Using Clementine, the Integral Solutions Limited developed a system to predict television audiences for the BBC. The prediction accuracy was reported to be the same as that achieved by the best performance of BBC's planners [182].

"Market basket analysis", examining associations between different products bought by customer, has been popular with organizations such as supermarkets and catalog selling companies. The underlying technology is the so-called association rule algorithm [8]. RS Components, a UK-based distributor of technical products such as electronic and electrical components and instrumentation, has used the IBM's Intelligent Miners to develop a system to do cross selling (suggesting related products on the phone when customers ask for one set of products), and in warehouse product allocation. The company had one warehouse in Corby before 1995, and decided to open another in Midlands due to business expansion. The problem was how to split the products into these two warehouses so that the number of partial orders and split shipments could be minimized. Remarkably, the percentage of split orders is just about 6% after using the patterns found by the system, much better than expected.

MITA, MetLife's Intelligent Text Analyzer, has been developed to help automate the underwriting of 260,000 life insurance applications received by the company every year. Automation is difficult because the applications include many free-form text fields. The use of keywords or simple parsing techniques to understand the text fields has proven to be inadequate, while the application of full semantic natural language processing was perceived to be too complex and unnecessary because the text often contains details that are not directly relevant to decision making. Instead the "information extraction" approach [127] where the input text is skimmed for specific information relevant to the particular application was seen as a good compromise and adopted by the system. MITA is currently processing 20,000 life insurance applications a month and it is reported that 89% of the text fields processed by the system exceed the established confidence-level threshold [216].

12.2.2 Science and Engineering

Enormous amounts of data have been generated in science and engineering, e.g. in cosmology, molecular biology and chemical engineering. In cosmology, advanced computational tools are needed to help astronomers understand the origin of large-scale cosmological structures as well as the formation and evolution of their astrophysical components (galaxies, quasars, and clusters) [412]. In molecular biology, recent technological advances in such areas as molecular genetics, protein sequencing and macro-molecular structure determination have created huge amounts of data [39]. Some exciting developments in bioinformatics will be discussed in the next section. In chemical engineering, mathematical models have been used to describe the interactions among various chemical processes occurring inside a plant. These models are typically very large systems of nonlinear algebraic or differential equations and a whole plant simulation may involve more than a million equations [45].

Obviously managing problems of the above-mentioned complexity would require the support of "high-performance" computers, which will be touched upon in section 12.3.4. Here let us have a brief look at a few cases.

Over three terabytes of image data have been collected by the Digital Palomar Observatory Sky Survey (POSS-II), which contain on the order of two billion sky objects. It has been a challenging task for astronomers to catalog the entire data set, i.e. a record of the sky location of each object and its corresponding classification such as star or galaxy. The Sky Image Cataloguing and Analysis Tool (SKICAT) [175] has been developed to automate this task. The SKICAT system integrates methods from machine learning, image processing, classification and database, and is reported to be able to classify objects too faint for visual classification with high accuracy [175].

Pavilion Technologies' Process Insights, an application development tool that combines neural networks, fuzzy logic and statistical methods, has been successfully used by Eastman Kodak and other companies to develop chemical manufacturing and control applications to reduce waste, improve product quality and increase plant throughput [179]. Typically, historical process data is used to build a predictive model of plant behavior and this model is then used to change the control setpoints in the plant for optimization.

DataEngine is another IDA tool which has been used in a wide range of applications. The basic components of the tool are neural networks, fuzzy logic and graphical user interfaces. One of the most successful application areas of DataEngine is data analysis in process industry. In particular, the tool has been applied to process analysis in chemical, steel and rubber industries, with savings in input materials and improvement in quality and productivity as a result [26].

The role of an intelligent data analyzer is discussed in a qualitative modeling tool called PRET and applied to a sensor-equipped driven pendulum [84]. PRET automates system identification, the process of finding a dynamic model of a black-box system. In particular, PRET performs both structural identification and parameter estimation by integrating several reasoning modes, including qualitative reasoning and simulation, numeric simulation, geometric reasoning, constraint satisfaction and meta-level control.

To improve its manufacturing processes, Boeing has successfully applied machine learning algorithms to the discovery of informative and useful rules from its plant data. In particular, it has been found that it is more beneficial to seek concise predictive rules which cover small subsets of the data, rather than full-blown decision trees. A variety of rules were extracted to predict such events as when a manufactured part is likely to fail inspection or when a delay will occur at a particular machine. These rules have been found to facilitate the identification of relatively rare, but potentially important anomalies [447].

12.2.3 Bioinformatics

Huge amount of data has been generated by genome-sequencing projects and other experimental efforts to determine the structures and functions of biological molecules and to understand the evolution of life. This has provided an unprecedented demand and opportunity for making sense of these data, whose outcome can have important impact on the discovery of new drugs and therapies as well as a better understanding of many disease processes [39, 17]. Bioinformatics is concerned with the development and application of computational and mathematical methods for organising, analysing and interpreting biological data, and requires an interdisciplinary research effort from biologists, computer scientists, statisticians and mathematicians in general.

One of the most significant developments in bioinformatics is the use of high-throughput technologies to collect huge amount of information such as gene expression data. The study of biological processes during health and disease has traditionally followed the "reductionist" method of identifying a single gene and encoded protein products that define an observed phenotype. In many situations this has led to a greater understanding of the processes involved within a cell and how these processes are dynamically linked. However, such reductionist approaches often over-simplify complex systems and are less useful when multi-gene/multi-protein processes are involved. New functional genomics technologies such as DNA microarray [464] allow the study of thousands of genes in a single experiment and provide a global view of the underlying biological process, for example by revealing which genes are responsible for a disease process, how they interact and are regulated, and which genes are being co-expressed under different experimental conditions and participating in common biochemical pathways.

The race has been on for a while to discover hidden knowledge from the DNA sequences and to understand the functions and roles of individual genes in different disease and health processes. Further pressure on the life science community was observed with the publication of the draft sequence of the human genome in 2001 [527, 281]. DNA microarray technology is now widely used to produce gene expression data in various areas. Although a considerable research effort is required for the effective analysis of such data especially in the area of modelling gene interaction and regulation, important progress has been made in using clustering and visualization techniques to profile gene expression. For example, a small group of genes whose expression patterns varied among differnt breast tumours has been identified, and these "molecular portraits" as revealed in the gene expression patterns have been found not only uncover similarities and differences among differnt tumours, but also in many cases point to a biological interpretation [419].

The work on gene expression profiling has also led to the call for a revised definition of what is deemed a "disease" [13]. It has been found that patients receiving the same diagnosis can have markedly different clinical courses and treatment responses. In particular, the current taxonomy of cancer appears to group together molecularly distinct diseases with distinct clinical phenotypes. For example, the diffuse large B-cell lymphoma has been found clinically heterogeneous: 40% of patients respond well to current therapy and have prolonged survival, while the remainder succumb to the disease [13]. The corresponding gene expression analysis demonstrated that this single diagnostic category of lymphoma masks at least two distinct diseases with the differential expression of hundreds of different genes. These molecular differences correspond well to the clinical differences (very different survival rates) between the two groups, suggesting that these two subgroups should be considered separate diseases. The work of this kind will provide a good perspective on the development of new cancer therapeutics for clinical trials since patients can be then classified into molecularly relevant categories.

12.2.4 Medicine and Health Care

Medical applications have inspired many methods and developments in AI, and they have also provided interesting platforms for developing IDA systems and applications. With the increasing development of electronic patient records and medical information systems, a large amount of clinical data is available online. Regularities, trends and surprising events extracted from these data by IDA methods are important in assisting clinicians to make informed decisions, thereby improving health services. Clinicians evaluate patient's condition over time. The analysis of large quantities of time-stamped data will provide doctors with important information regarding the progression of the disease. Therefore systems capable of performing temporal abstraction and reasoning become crucial in this context. Shahar has developed a general framework for this purpose [473]. Although the use of temporal abstraction and reasoning methods require an intensive knowledge acquisition effort, they have found many successful medical applications, including data validation in intensive care [271], the monitoring of children growth [233], and the monitoring of heart transplant patients [334], and intelligent anesthesia monitoring [146]. In particular, Bellazzi et al. discusses the use of temporal abstraction to transform longitudinal data into a new time series containing more meaningful information, whose features are then interpreted using statistical and probabilistic techniques. The methods were successfully applied to the analysis of diabetic patients' data [55].

Guardian is an intelligent autonomous agent for monitoring and diagnosing intensive care patients recovering from cardiac surgery [335]. The system has several algorithms cooperating to produce diagnoses and treatment plans under real-time conditions. These algorithms were developed for various purposes including the continuous acquisition and interpretation of data, diagnosis and warning humans of complications and diseases, selection of treatment plans and monitoring of their execution, and closed-loop control of physiological parameters. A standard data representation scheme was created and the blackboard architecture [252] was used to coordinate the activities of the system. Although it has not been used in a real hospital setting, Guardian has undergone several tests with the help of a patient-simulator system. The performance comparison with human physicians demonstrate that such a system can be a valuable companion to medical personnel, especially under difficult and stressful circumstances [335].

Visual field testing provides the eye care practitioner with essential information regarding the early detection of major blindness-causing diseases such as glaucoma. Testing of the visual field is normally performed using an expensive, specially designed instrument that is currently mostly restricted for use in eye hospitals. However it is of great importance that visual field testing be offered to subjects at the earliest possible stage. By the time a patient has displayed overt symptoms and has been referred to a hospital for eye examination, it is possible that the visual field loss is already at an advanced stage and cannot be easily treated. To remedy the situation, personal computers have been exploited as an affordable test machine. One of the major technical challenges is to obtain reliable test data for analysis, given that the data collected by PCs are going to be much more noisy than those collected by the instrument. In this connection, a software-based test system has been developed using machine learning techniques (e.g. neural networks and decision tree induction), an intelligent user interface and a pattern discovery model, and this system has been successfully used in several primary care settings [354].

12.2.5 Others

The Pilot's Associate program was a five-year effort to help pilots of advanced fighter aircraft [487]. The Associate was designed to provide the pilot with enhanced situational awareness by sorting and prioritizing data, analyzing sensor and aircraft system data, and turning the data into useful information. Based on this information, plans for achieving mission goals can be developed and presented to the pilot for approval and execution.

Despite the best efforts of web designers, we have all had the experience of not being able to find a certain page we want. A bad design for a commercial web site obviously means the loss of customers. One challenge for the AI community has been the creation of "adaptive web sites", web sites that automatically improve their organization and presentation by learning from user access patterns [417]. One early attempt is WebWatcher, an operational tour guide for the WWW [292]. It learns to predict what links users will follow on a particular page, highlight the links along the way, and learn from experience to improve its advice-giving skills. The prediction is based on many previous access patterns and the current user's stated interests. It has also been reported that Microsoft is to include in its electronic-commerce system a feature called Intelligent Cross Sell that can be used to analyze the activity of shoppers on a web site and automatically adapt the site to that user's preferences.

Advanced Scout was developed for the American National Basketball Association (NBA) coaching staffs to discover interesting patterns in basketball game data. These patterns are then used to assess the effectiveness of certain coaching decisions and to formulate game strategies for subsequent games. In the 1995-96 season, the software was distributed to 16 of the 29 NBA teams. Positive feedback from the coaching staffs was reported, including the improvement of the quality of play [64].

Another example is the real-time image processing system in an ambitious project called "No Hands Across America". This is a vision-based robotics system that automatically steered a vehicle, the CMU Navlab 5, across the whole of America - from Washington DC to San Diego - with an average speed of 63 mph. In only about 2% of the entire 2850-mile journey the steering was taken over by humans who were monitoring the vehicle, in situations such as driving towards direct sunshine, through a city, or on new roads which don't have lane markings. An early version of its image data analyzer was a single hidden-layer, back-propagation network. This was evolved into a model of road geometry and image reprojection to extract and adapt to the relevant features for driving [293].

12.3. Several Development Issues

12.3.1 Strategies

Data analysis in a problem-solving context is typically an *iterative* process involving problem formulation, model building, and interpretation and communication of the results. The question of how data analysis may be carried out effectively should lead us to having a close look not only at those individual components in the data analysis process, but also at the process as a whole, asking what would constitute a sensible data analysis *strategy*. This strategy should describe the steps, decisions and actions which are taken during the process of analyzing data to build a model or answer a question, and one might define a good strategy as being the hallmark of intelligent data analysis [242].

There has indeed been work on data analysis strategies [130], often under "statistical expert systems" [237]. For example, strategies for multivariate analysis of variance and for discriminant analysis were examined in [236], while strategies for linear regression were used to illustrate some of the strategic issues [427]. Although research in this area has not progressed as fast as Tukey predicted [519], a lot of insight has been gained into the complex, iterative, and interactive data analysis process.

Part of the problem is that there can be many different ways of performing analysis, given the same data/problem, and it is not always possible to judge which one is better in advance. Moreover, strategies emerge from experience in doing analysis and more experience is certainly needed to develop good strategies for performing large-scale data analysis tasks. The work reported so far in this area has very much been restricted to relatively small data sets.

12.3.2 Problem Formulation and Model Building

In problem formulation, various factors affecting the application development are analyzed, including the key requirements from the application, human and organizational constraints, and of course what data should be used or collected. The nature of the problem solving task is defined, e.g. classification, clustering, etc. In the general statistics context, it was found that establishing the mapping from an application to a statistical question is one of the most difficult parts of a statistical analysis, which is yet to receive sufficient amount of attention [238].

Once the problem is formulated, a problem-solving model will be built based on matching the key application factors identified in the formulation of the problem to the general characteristics of the models under consideration. This typically involves model selection, fitting and checking, in which an initial model is chosen, parameters of the model are estimated, and the underlying assumptions and adequacy of the model are checked. If it is decided that the model is not justified, or does not fit the data well enough, then a return to an earlier phase is called for, with a renewed search for alternative models [80]. Model fitting and checking have now become relatively straightforward, thanks to the availability of many software packages. However, model selection is much harder: how do you choose the initial model to be considered? This is arguably the most important and most difficult aspect of model building, and yet is the one where there is least help [111].

A notable exception in this aspect from the statistics community is the work of Leamer [345] where he highlighted the gap between statistical theory and statistical practice. He argued the case for "metastatistics" – the theory of inference *actually* drawn from data – which would emphasize how the investigator's motives and opinions influence her choice of model and her choice of data. This is in contrast to traditional statistical theory – a theory of inference *ideally* drawn from data – which takes as given the model, the data and the purity of the investigator' motives. He then went on to address the search for model specifications from nonexperimental data. By observing economists at work, he defined six logically distinct varieties of model specification searches, which were then discussed under a Bayesian decision analysis framework [345]. Other methods suggested in the statistics literature include the use of "computer-intensive methods" such as simulation and resampling to assess the size of model selection biases and to find ways to avoid them [166], the use of Bayesian approaches to address model uncertainty, and the use of Markov Chain Monte Carlo (MCMC) methods to perform the computations necessary for model averaging [214].

In the computing community, model selection has received much attention lately and this is largely triggered by the fact that there are so many different methods for model building in addition to traditional statistical methods. These methods include association rules algorithms, decision tree and rule induction, fuzzy and rough sets, genetic algorithms and genetic programming, heuristic search, inductive logic programming, knowledge-based systems and neural networks [382, 420].

There have been many comparative studies regarding competing methods in the literature, e.g. [378], with some reporting very little difference between methods, while others report sharp contrasts. This raises an issue which concerns many of the analysts working in practical problem solving: if there are many methods which appear to be applicable to the problem or sub-problem under consideration, which one should be used? One approach would be to try every method and choose the best. The assumption behind this approach is that there are sufficient resources available for the experiments, but this is not always realistic in practice. Could we anticipate the results of a particular comparative study so that an "optimal" or near-optimal method can be chosen without trying all of the possibilities? It would be necessary to have a deep understanding of the characteristics of the data and the properties of individual methods and problem solving contexts. Results from those comparative studies which have already been carried out should be able to help here.

12.3.3 Mapping Method to Application

So what are the important factors in choosing a method for problem solving, say, classification? The primary and most commonly used criterion is the prediction accuracy [537]. However, it is not always the only, or even the most important, criterion for evaluating competing methods. Credit scoring is one of the most quoted applications where misclassification cost is more important than predictive accuracy [398]. Other important factors in deciding that one method is preferable to another include the computational efficiency and the understandability/interpretability of the methods. Recently a study has been done to find a comprehensive list of such factors and to see how such a list may be used to map methods onto problems [150].

In this study, the factors cover four different dimensions: quality of model, engineering considerations, available resources, and logistical constraints. The "quality of model" is concerned with issues such as accuracy, explainability, speed and efficiency. Engineering considerations include flexibility - how flexible is the system in allowing the problem specifications to be changed; scalability (see section 12.3.4); embeddability - how easily can the system be embedded into a large system or the existing work flow of an organization; usability - how easy is the system to use. The third set of factors addresses issues relating to the resources available to attack the problem. Are there good, *high-quality* online data available? Are there *a lot* of data available? Is the organization far enough up the learning curve? And how subtle and easily understood are interactions between the problem variables? The final set of factors relates to the logistical constraints of an organization: the accessibility of the expert(s), the adequacy of the computing infrastructure for the problem, and the development time allowed.

These factors, referred to as "intelligence density", are a checklist of the objectives and constraints of the various parties involved in the application development: users, developers, and management. This provides a handle for bringing to the surface the important questions involved in imposing structure on an initially ill-defined problem. Each factor in the checklist is concerned with a separate aspect of the development process or of the organizational impact of the system, and each tries to influence the outcome of the final choice. Together these factors form a "committee of critics", that will help reduce the risk of getting trapped into developing a solution that uses a favorite technique or an *obvious* solution that could turn out to be a bad choice from a business standpoint [150]. Seven different methods were then evaluated under these factors and case studies were provided to illustrate how to map the appropriate technique to the problem in hand using the identified factors. Below is a brief description of one such case.

LBS is a management investment firm. In a field where even a slight advantage can mean millions and any additional market insight can be profitable, LBS has explored a variety of technologies in helping better predict the performance of the individual stocks in the firm's portfolios. But it is extremely difficult to predict the market movement in the short term; financial processes are generally characterized by high levels of non-linearity and complexity, making them hard to model. The amount of data generated daily is overwhelming. To develop a tool that will predict the direction and magnitude of price movements over time, the LBS team evaluated traditional statistical methods, rule-based systems and neural networks against nine criterion: accuracy, explainability, response speed, efficiency, flexibility, scalability, embeddability, tolerance for complexity, tolerance for noise in data, and independence from experts. It was found that the neural network satisfies these criteria better than the other two in general, especially on those aspects which matter most to the application, e.g. flexibility, tolerance for complexity and tolerance for noise in data. Consequently different neural network models have been developed for different stocks, and have been found to "consistently outperform the market" [150].

12.3.4 Scalability

Perhaps the first word that comes to mind when thinking of managing and analyzing "big data" is *scalability*: if a method works well for a task involving a dozen variables and 10,000 cases, is it still going to perform well for one with 100 variables and one million cases? Much research has been done in the development of efficient, approximate or heuristic algorithms that are able to scale well [122, 234]. Among many related methods are sampling, feature selection, restriction of model search space, and application of domain knowledge. A brief overview of issues and examples related to the scaling-up of inductive learning algorithms is provided in [430]. To deal with "big data" of gigabytes or even terabytes, however, some basic hardware support is essential.

Now let us look at one application in some depth. It's a project initiated by Mellon Bank in the USA, in collaboration with IBM Business Intelligence Solutions, on how to predict which customers will stop using Mellon's credit card in the next few months and switch to a competitor. The data regarding 2.7 million credit card customers and corresponding transaction data were already stored in a data warehouse. Stored data items included information related to current account, mortgage, credit card, investment, etc. Initially there were 29 gigabytes of raw data corresponding to 2.7 million credit card customers, in which 23 gigabytes were customer master data with 135 columns, and six gigabytes were transaction data with ten columns. Among these, data corresponding to 350,000 credit card customers (three gigabytes) were selected for preprocessing and transformation, which produced seven tables of two gigabytes of final data (565 columns) for model building. Run on IBM's S/390 G3 Enterprise Server with parallel processing capabilities (ten CPUs) and two gigabytes real storage space, a model of vulnerable customers was built using the Intelligent Miner. All the columns were selected and the tables were read five times during processing, thus comprising about 1.7 million rows or ten gigabytes of data. The turn-around time was about one hour and it is not difficult to imagine how difficult it would be for an ordinary PC to manage this application!

In this connection, it is pleasing to see that the High Performance Computing Community is taking initial steps in this direction, asking how a desktop user might be able to utilize the computer power offered by a set of supercomputers and massive data stores interconnected by a high-speed, high bandwidth communication network, and to use resources from such a network as though it were from a single machine [499]. An analyst then no longer needs to travel to a supercomputer center to get analysis done there, a typical scenario in the 80s. To achieve that, different kinds of software support are essential. At the lowest level are distributed operating systems that allow distributed scheduling and resource management. New high-performance programming languages such as HPC++ and HP Fortran have been developed, which allow the exploitation of parallelism in computations [304]. And new visualization tools based on virtual reality environments have been developed to allow multiple-gigabyte data sets to be explored interactively [136].

However, we are still some distance away from a truly distributed computing environment where we can manipulate and analyze "big data" with ease on some remote powerful machines in the network. For those who do manage to have direct access to powerful parallel machines or supercomputers at the present time, work has already been carried out to develop large-scale data mining systems. Quakefinder is one such system [501], developed to automatically detect and measure tectonic activity in the earth's crust by analyzing satellite data. Using methods from statistics, optimization and massive parallel computing, Quakefinder was implemented on a 256-node Cray T3D at the Jet Propulsion Laboratory. Each node is based on a DEC Alpha processor running at 150MHz. Quakefinder has been used to automatically map the direction and magnitude of ground displacements due to the 1992 Landers earthquake in South California [501].

12.3.5 Data Quality

Data are now viewed as a key organizational resource and the use of high-quality data for decision making has received increasing attention [509]. It is commonly accepted that one of the most difficult and costly tasks in large-scale data analysis is trying to extract clean and reliable data, and many have estimated that as much as 50% to 70% of a project's effort is typically spent on this part of the process. In response to this opportunity, "data cleaning" companies are being created, and "data quality groups" are being set up in corporations.

There are many dimensions of data quality: accuracy, completeness, consistency, timeliness etc and much work has been done on the management of noisy, missing, and outlying data [550, 438, 44]. The handling of outlying observations is a difficult but important topic for the following reasons. First, outlying observations can have a considerable influence on the results of an analysis. Second, although outliers are often measurement or recording errors, some of them can represent phenomena of interest, something significant from the viewpoint of the application domain. Third, for many applications, exceptions identified can often lead to the discovery of unexpected knowledge.

There two principal approaches to outlier management [44]. One is outlier *accommodation*, which is characterized by the development of a variety of statistical estimation or testing procedures which are *robust* against, or relatively unaffected by, outliers [273]. In these procedures, the analysis of the main body of data is the key objective and outliers themselves are not of prime concern. This approach is unsuitable for those applications where explicit identification of anomalous observations is an important consideration, e.g. suspicious credit card transactions.

The other approach is characterized by identifying outliers and deciding whether they should be retained or rejected. Many statistical methods have been proposed to detect outliers [44]. These approaches range from informal methods such as the ordering of multivariate data, the use of graphical and pictorial methods, and the application of simple test statistics, to some more formal approach in which a model for the data is provided and tests of hypotheses that certain observations are outliers are set up against the alternative that they are part of the main body of the data. The identification of outliers has also received much attention from the computing and pattern recognition communities [310, 367]. However, there appears to be much less work on how to decide whether outliers should be retained or rejected. In the statistical community, a commonly-adopted strategy when analyzing data is to carry out the analysis both including and excluding the suspicious values [16]. If there is little difference in the results obtained then the outliers had minimal effect, but if excluding them does have an effect, options need to be sought to analyze these outliers further.

In order to successfully distinguish between noisy outlying data and noise-free outliers, different kinds of information are normally needed. These should not only include various data characteristics and the context in which the outliers occur, but also relevant domain knowledge. The procedure for analyzing outliers has been experimentally shown to be subjective, depending on the above mentioned factors [124]. The analyst is normally given this task of judging which suspicious values are obviously impossible or which, while physically possible, should be viewed with caution. However in the context of data mining where a great many cases are normally involved, the number of suspicious cases would also be sizable and manual analysis would become too labor-intensive.

In this connection, two general strategies for distinguishing between phenomena of interest and measurement noise have been proposed. The first attempts to model "real measurements", namely how measurements should be distributed in a domain of interest, and rejects values that do not fall within the real measurements [353]. The other strategy uses knowledge regarding our understanding of noisy data points instead, so as to help reason about outliers. Noisy data points are modeled, and those outliers are accepted if they are not accounted for by a noise model [552]. Both strategies have been found effective for handling different types of medical application, depending on the availability of relevant domain knowledge.

Here are the key ideas behind the outlier analysis strategy using the knowledge of noisy data points [552]:

Assume there is a set of training data which can be labeled into two classes: "noisy" and "rest" by using relevant domain knowledge. Class "noisy" indicates that the corresponding outliers in this data set are noisy data points, while class "rest" says we have good reason to believe the outliers in the data set are not due to measurement noise. Given sufficient amount of training data, one can use any supervised learning techniques to construct a "noise model", which, after validation, can then be used to help distinguish between two types of outliers.

Note that the labeling of training instances is not always easy, especially with multi-dimensional data. To assist this process, the Self-Organizing Map [311] is used to visualize and compress data into a two-dimensional map. Data clusters and outliers then become easy to spot, and data can then be carefully interpreted using the meanings of the map and relevant domain knowledge.

So given a data set, outliers may be detected using appropriate detection algorithms, and these outliers can then be tested using the noise model. As a result, noisy outliers can be deleted, and the rest of outliers kept in the data set for further analysis. This strategy has been applied to a set of visual field data collected in a large urban general practice in North London for a glaucoma case findings study [552]. In that study, nearly 1000 patients were screened and the *discriminating power* of the visual field test in terms of its glaucoma detection rate versus false alarms was examined. The decision threshold used for discriminating between normal and abnormal eyes was the average percentage of positive responses within the test. It was found that the data with selected outliers using the above-mentioned strategy perform better than the original data set or the one obtained by deleting all the outliers, in terms of maximizing the detection rate and minimizing false alarms [552].

12.4. Conclusion

It seemed an impossible task to have a sensible overview of IDA systems and applications as there are so many of them and they have been developed in so many different fields. Moreover, there are currently not just one, but many different perspectives on IDA [60]. Despite these constraints I do hope that this chapter has provided a snapshot of possible IDA applications and a general awareness of IDA system development issues as well as the relevant literature.

Data analysis is performed for a variety of reasons by scientists, engineers, medical and government researchers, business communities and so on. Statistical methods have been the primary analysis tool, but many new computing developments have been applied to challenging real-world problems. This is a truly exciting time. Questions have been constantly raised: how can one perform data analysis most effectively –intelligently– to gain new scientific insights, to capture bigger portions of the market, to improve the quality of life, and so on? And is there a set of established guiding principles to enable one to do so? Statistics has laid some important foundations, but the evolution of computing technology and the ever-increasing size and variety of data sets have led to new challenges as well as new opportunities for both computing and statistical communities [352, 242]. The quest for crucial insights into the process of intelligent data analysis will require an interdisciplinary effort, as demonstrated by many of the real-world applications outlined in this chapter.

Inevitably we are going to see more applications as well as more analysis tools. A major problem is that, despite the best intentions of the user, a tool can be used without an essential understanding of what it can offer and how the results should be interpreted. The misuse of statistical packages or methods is common, but now the chance of misusing a variety of data mining tools is even greater [218, 275]. So how can we prevent the misuse? Codifying competent data analysis strategies and making them widely available in software packages was one early attempt [236, 427]. We have also started to see "collaborative systems", asking what the analyst is good at, what the computer is good at, and what is the best way of dividing the analysis work between the two [18, 481]. It would be helpful if some formal collaborative strategies emerge. Research from "human-computer collaboration" should shed some interesting light here [230].

Finally, a deep understanding of the IDA process will require much more experience in analyzing large, complex real-world data sets. Currently technical reports of analyzing "big data" are sketchy. A lot of heuristics and trial-and-error have been used in exploring and analyzing these data, especially the data collected opportunistically. As we gain more practical experience, important lessons will be learned, techniques and solutions will be refined, and principles will be better established. Analysts of the future will approach tough, complex problems on a sounder basis of theory and experience.

Appendix A: Tools

Luis Aburto[+], Mariano Silva[*], and Richard Weber[**]
[+]Data Analysis for Decision Support, [*]Egoland Consultores (both Chile), and [**]University of Chile

Intelligent Data Analysis (IDA) has emerged from a research area to a field, which is very active in various aspects such as pure and applied research, real-world applications, and development of commercial software tools. This chapter presents a selection of IDA-tools. It does not claim to be complete and can only show examples. As sources we have used Internet sites (e.g. www.kdnuggets.com) and descriptions from the respective software providers. We use the terms IDA-tools and Data Mining tools synonymously since most of the IDA techniques are applied in Data Mining. During the last years (since the first edition of this survey) several things concerning IDA-tools have changed. New and promising IDA-techniques have been presented (e.g. Support Vector Machines), existing techniques have been improved, the main players in the software market changed and new application areas appeared (e.g. Text and Web Mining).

These developments make it even harder than before to select a suitable tool given a certain application. The present survey aims at delivering an overview on existing IDA-tools and this way help the reader to find the right solution for his/her data analysis problem.

Matching the structure of this book the tools are subdivided into the categories statistical analysis, tools for modeling data using Bayesian techniques, neural networks, evolutionary computation, knowledge extraction tools, Support Vector Machines and tool suites for versatile IDA-applications. New trends like e.g. Text Mining and Web Mining are also added to this survey. Finally we point at current trends in the software market.

A.1. Tools for statistical analysis

All of the statistical packages in this section provide basic and advanced descriptive statistics. Depending on the tool, time series analysis or multivariate methods are integrated in the standard package or can be added as extensions. The scope of functions provided by the tools that fall into this category makes them an indispensable part of every data analyst's toolbox. Sampling, analysis of distributions or hypothesis testing are necessary in almost any data analysis project regardless of whatever modeling technique is applied to the data afterwards. The statistical packages combine all this basic functionality and thus serve as a uniform platform for the analyses carried out using other tools. Typical tasks for statistical packages include reduction of dimensionality (feature selection, principal component analysis, correlation analysis), summarization (describing larger data sets by their statistical descriptions or parameters of fitting functions).

S-Plus

Description: S is an interactive, object-oriented programming language for data analysis, developed by AT&T Bell Labs. S-Plus is the commercial version supported and marketed by MathSoft. S/S-Plus has proved to be a useful platform for both general-purpose data analysis and development of new methodologies. In addition to supporting a large number of modeling and visualization techniques in the base language, S and S-Plus have proved to be an active platform for research in new statistics methodologies. An interactive, pseudo-object-oriented language, S enjoys many of the features which make APL and Lisp attractive, but additionally has an easier to understand algebraic syntax and many built-in statistical primitives. The S language also allows the user to easily link in new code written in C or Fortran. Thus for developers of new methodology, S can provide data cleaning and visualization functionality allowing the developer to concentrate on algorithm development. The StatLib archive (http://lib.stat.cmu.edu/) contains a large collection of experimental code for various statistical methods, much of it written in S or S-Plus.
Data analysis functionalities: Clustering, Classification, Summarization, Visualization, Regression
Platform(s): Windows 98, NT and Unix
Contact: Insightful Headquarters, USA
E-mail: `info@insightful.com`
URL: www.insightful.com

SAS

Description: The SAS System is a modular, integrated, hardware-independent system of software for enterprise-wide information delivery. Distinguishing the software is its ability to:

- make enterprise data, regardless of source, a generalized resource available to any user or application that requires it,
- transform enterprise data into information for a broad range of applications,
- deliver that critical information through a variety of interfaces tailored to the needs and experience of the individual computer user,
- perform consistently across and cooperatively among a broad range of hardware environments while exploiting the particular advantages of each.

Data analysis functionalities: Statistics, Visualization
Platform(s): Windows 98, NT

Contact: SAS Institute Inc., USA
E-mail: `software@sas.com`
URL: www.sas.com

SPSS
Description: a powerful statistical package for business or research, featuring top quality statistics, high-resolution graphics and reporting and distributing capabilities. SPSS is easy to use with lots of help including a Statistics Coach for those whose statistical knowledge is rusty. With SPSS, data can be read from a variety of sources or entered in a familiar spreadsheet format. Using SPSS Smart Viewer, you can electronically distribute live reports (reports that recipients can work with by pivoting and reformatting tables). SPSS' statistical and graphing capabilities range from basic to sophisticated. Results can be presented with a wide range of report formats and charts, including business, statistical and quality control charts.
Data analysis functionalities: Classification, Clustering, Summarization, Statistics
Platform(s): Windows 98, NT
Contact: SPSS Inc., USA
E-mail: `sales@spss.com`
URL: www.spss.com

SPSS SYSTAT
Description: SYSTAT is an integrated desktop statistics and graphics software package for scientists, engineers, and statisticians. SYSTAT's core strength derives from the breadth, depth and robustness of its statistical procedures. Built upon this foundation is an extensive and interactive graphical tool library.
Data analysis functionalities: Clustering, Dimensional Analysis, Statistical and Scientific Visualization, Statistics
Platform(s): Windows 98, NT
Contact: SPSS Inc., USA
E-mail: `sales@spss.com`
URL: www.spsscience.com

A.2. Tools for exploration/modeling

In this section, tools are presented that provide a single-method approach to creating models by learning from data. They all focus on different paradigms like Bayesian methods, neural networks, evolutionary computation or clustering, but the limits are fluid as e.g. neural networks can as well be applied to clustering as traditional clustering techniques. The characteristic feature of the tools falling into this category is that they generally cover a wide range of different algorithms from their special area and allow a very detailed configuration of the methods. This in return usually requires quite a bit of expertise in order to find the right method and the right configuration.

A.2.1 IDA-tools using Bayesian techniques

Bayesian networks are a popular representation for encoding uncertain expert knowledge in expert systems (see Chapter 4). This section presents tools that apply Bayesian techniques.

BKD: Bayesian Knowledge Discoverer
Description: BKD is designed to extract Bayesian Belief Networks (BBNs) from (possibly incomplete) databases. It is based on a new estimation method called Bound and Collapse and its extensions to model selection. The aim of BKD is to provide a Knowledge Discovery tool able to extract knowledge from databases, using sounds and accountable statistical methods. The capabilities of BKD include: estimation of conditional probability from data, extraction of the graphical structure from data, goal oriented evidence propagation, missing data handling using Bound and Collapse discretization of continuous variables, automated definition of nodes from data, conversion from and to the proposed standard Bayesian Networks Interchange Format (BNIF), Graphic User Interface and a movie-based on-line help.
Data analysis functionalities: Dependency analysis, belief networks
Platform(s): Windows 98, NT, Mac and Solaris
Contact: Marco Ramoni
E-mail: `M.Ramoni@open.ac.uk`
URL: kmi.open.ac.uk/projects/bkd/

Business Navigator
Description: Business Navigator is a standalone package that can help analysts to build predictive models from data. It is useful for getting a rapid understanding of the predictive relationships within data with little or no special training. Business Navigator can be used to build extremely high-quality predictive models.
Data analysis functionalities: Bayesian networks, GUI, data preparation tools.
Platform(s): Windows 98, NT, 2000, XP
Contact: Data Digest
E-mail: `inquires@data-digest.com`
URL: www.data-digest.com

Hugin Explorer
Description: The Hugin Explorer package contains a flexible, user friendly and comprehensive graphical user interface. The user interface contains a compiler and a runtime system for construction, maintenance and usage of knowledge bases, based on Bayesian networks. The main features in the Hugin Explorer package are:

- Construction of knowledge bases using Bayesian networks and Influence diagrams technology
- Supports development of object oriented Bayesian networks
- Automated learning of knowledge bases from databases

– Wizard for generation of probability tables

Data analysis functionalities: Bayesian networks.
Platform(s): Windows 98, NT, 2000, XP, Solaris, Linux
Contact: Hugin, Denmark
E-mail: `info@hugin.com`
URL: www.hugin.com

A.2.2 IDA-tools using neural networks

This section lists tools implementing various neural network approaches. In general these tools are not particularly devoted to data analysis problems but rather general- purpose toolboxes that provide many kinds of neural networks without a special application background. Usually these tools provide a wide range of different neural network architectures and allow a very detailed configuration. They are tailored to the neural network specialist who can use their versatility in order to create very elaborate models.

Viscovery SOMine
Description: A data mining tool based on Self-organizing maps with unique capabilities for visualizing of multi-dimensional data. Viscovery SOMine is a powerful tool for exploratory data analysis and data mining. Employing an enhanced version of Self Organizing Maps [311] it puts complex data into order based on its similarity. The resulting map can be used to identify and evaluate the features hidden in the data. The result is presented in a graphical way, which allows the user to analyze non-linear relationships without requiring profound statistical knowledge. Through the implementation of new techniques - such as SOM scaling - the speed in creating maps is notably increased compared to the original SOM algorithm.
Data analysis functionalities: Clustering, Classification, and Visualization
Platform(s): Windows 98, NT
Contact: Eudaptics Software GmbH, Austria
E-mail: `office@eudaptics.co.at`
URL: www.eudaptics.co.at

MATLAB NN Toolbox
Description: An engineering environment for neural network research, design, and simulation. Offers over fifteen proven network architectures and learning rules. Includes backpropagation, perceptron, linear, recurrent, associative, and self- organizing networks. Fully open and customizable.
Data analysis functionalities: Classification, Function Approximation
Platform(s): Windows 98, NT, Unix; Mac
Contact: The Math Works, Inc., USA
E-mail: `info@mathworks.com`
URL: www.mathworks.com

NeuralWorks Professional II/PLUS
Description: NeuralWorks Professional II/PLUS is a complete and comprehensive multi-paradigm prototyping and development system. You can use it to design, build, train, test and deploy neural network models to solve complex real-world problems. Professional II/PLUS allows to easily generate models using over two dozen paradigm types. A wide variety of diagnostic tools and specialized instruments allow monitoring network performance and diagnose learning. Powerful options include Designer Pack, which enables you to incorporate dynamic models into the application and User- Defined Neuro-Dynamics for customizing existing network architectures. Discovery tasks: Classification, Clustering, and Prediction
Platform(s): Windows 98, NT, SGI IRIS, Sun, HP series, AIX and Mac
Contact: NeuralWare, Inc., USA
E-mail: info@neuralware.com
URL: www.neuralware.com

SPSS Neural Connection
Description: Neural network package for prediction, classification, time series analysis and data segmentation. Neural Connection can work with data from a large variety of formats. Also featuring an icon-based interface that makes it easy to build and access all models, Neural Connection is applicable for market research, database marketing, financial research, operational analysis and scientific research.
Data analysis functionalities: Classification, Prediction
Platform(s): Windows 98, NT
Contact: SPSS Inc., Chicago IL, USA
E-mail: sales@spss.com
URL: www.spss.com/software/Neuro/nc2info.html

NeuroShell2/NeuroWindows
Description: NeuroShell 2 combines neural network architectures, a Windows icon driven user interface, and sophisticated utilities for MS-Windows. Internal format is spreadsheet, and users can specify that NeuroShell 2 use their own spreadsheet when editing. Includes both Beginner's and Advanced systems, a Runtime capability, and a choice of 15 Backpropagation, Kohonen, PNN and GRNN architectures. Includes Rules, Symbol Translate, Graphics, File Import/Export modules (including MetaStock from Equis International) and NET-PERFECT to prevent overtraining. NeuroWindows is a programmer's tool in a Dynamic Link Library (DLL) that can create as many as 128 interactive nets in an application, each with 32 slabs in a single network, and 32K neurons in a slab. NeuroWindows can mix supervised and unsupervised nets. The DLL may be called from Visual Basic, Visual C, Access Basic, C, Pascal, and VBA/Excel 5.
Data analysis functionalities: Classification, Function Approximation
Platform(s): Windows 98, 2000, NT, XP
Contact: Ward Systems Group, Inc.

E-mail: `WardSystems@msn.com`
URL: www.wardsystems.com

STATISTICA: Neural Networks
Description: STATISTICA: Neural Networks is a comprehensive application capable of designing a wide range of neural network architectures, employing both widely- used and highly-specialized training algorithms. It offers features such as sophisticated training algorithms, an Automatic Network Designer, a Neuro-Genetic Input Selection facility, complete API (Application Programming Interface) support, and the ability to interface with STATISTICA data files and graphs. STATISTICA: Neural Networks includes traditional learning algorithms, such as back propagation and state-of-the-art training algorithms such as Conjugate Gradient Descent and Levenberg-Marquardt iterative procedures. STATISTICA: Neural Networks features an Automatic Network Designer that utilizes heuristics and sophisticated optimization strategies to determine the best network architecture. The Automatic Network Designer compares Linear, Radial Basis Function, and Multilayer Perceptron networks, determines the number of hidden units, and chooses the Smoothing factor for Radial Basis Function networks. The process of obtaining the right input variables also is facilitated by STATISTICA: Neural Networks. Neuro-Genetic Input Selection procedures aid in determining the input variables that should be used in training. It uses an optimization strategy to compare the possible combinations of input variables to determine which set is most effective. STATISTICA: Neural Networks offers complete API support so advanced users (or designers of corporate knowledge seeking or data mining systems) may be able to integrate the advanced computational engines of the Neural Networks module into their applications. STATISTICA: Neural Networks can be used as a stand-alone application or can interface with STATISTICA.
Data analysis functionalities: Classification, Function Approximation
Platform(s): Windows 98, NT
Contact: StatSoft, Inc. USA
E-mail: `info@statsoft.com`
URL: www.statsoftinc.com

A.2.3 IDA-tools using Evolutionary Algorithms

Evolutionary algorithms offer flexible techniques to solve optimization tasks (see Chapter 10). Some data analysis tasks like e.g. clustering require the optimization of a given objective function. That is where evolutionary algorithms can be used in the context of intelligent data analysis.

Evolver
Description: Evolver is an optimization add-in for Microsoft Excel. Evolver uses innovative genetic algorithm (GA) technology to quickly solve complex optimization problems in finance, distribution, scheduling, resource allocation, manufacturing, budgeting, engineering, and more. Virtually any type of problem that can

be modeled in Excel can be solved by Evolver. It is available in three versions: Standard, Professional, and Industrial. The Professional and Industrial versions have increased problem capacities and advanced features, including the Evolver Developer's Kit.
Data analysis functionalities: Optimization
Platform(s): Windows 98, NT, 2000, XP
Contact: Palisade Corporation
E-mail: `sales@palisade.com`
URL: www.palisade.com

A.2.4 IDA-tools for knowledge extraction

The tools presented in this section implement methods from machine learning: rule induction and decision trees. The advantage of these techniques lies in their interpretability. The result is usually a set of linguistic rules. The decision tree itself provides information about relevance of features and about decision boundaries. For this reason, decision trees are frequently used for feature selection.

KnowledgeSEEKER
Description: A data mining tool which automates the process of discovering significant relationships or patterns within your data, uncovering correlation within seemingly unrelated sets of data. KnowledgeSEEKER is fully scalable.
Data analysis functionalities: Classification, Rule Discovery
Platform(s): Windows 98, NT, Unix
Contact: ANGOSS International, Canada
E-mail: `info@angoss.com`
URL: www.angoss.com

Decisionhouse
Description: Decisionhouse is a tool that incorporates various aspects of the data mining process, OLAP, reporting, database query, data preparation and data visualization. Decisionhouse interfaces to standard relational databases and flat files through an intuitive but powerful point-and-click interface. The package provides powerful data preprocessing including cleansing, selection, transformation, sampling, aggregation and computation of statistics. It allows drill-down to the level of individual records and fields, as well as providing summary information over the selected data, and allowing arbitrary cross-tabulation of that data. The visualization tools in Decisionhouse offer arbitrary 3-dimensional views of the data (graphical cross-tabulations) over non-precomputed dimensions, giving unprecedented flexibility in *slicing and dicing.* A 3-dimensional map-based viewer also allows powerful insights into geo-demographic patterns, either over the complete database or for segments selected using other visualization tools. Decisionhouse not only supports manual data mining through visualization, selection and drill-down, but also using sophisticated automated analysis. The first two analysis engines available with Decisionhouse are TreeHouse for analysis based on decision trees (CART and ID3) for segmentation and targeting,

and ScoreHouse for building scorecards based on discriminant analysis, logistic and probit regression, Gini or Kolmogorov-Smirnov analysis for risk assessment and profitability management.
Data analysis functionalities: Classification, Visualization
Platform(s): Releases for Sun, NCR, Data General and SGI.
Contact: Quadstone Ltd, Scotland
E-mail: `info@quadstone.com`
URL: www.quadstone.com

CART
Description: CART is a data-mining tool that automatically searches for important patterns and relationships and quickly uncovers hidden structure even in highly complex data [90]. This discovered knowledge is then used to generate accurate and reliable predictive models for applications such as profiling customers, targeting direct mailings, detecting telecommunications and credit-card fraud, and managing credit risk. CART is a pre-processing complement to other data-mining packages. CART's outputs (predicted values) can also be used as inputs to improve predictive accuracy of neural net and logistic regression models.
Data analysis functionalities: Classification, Prediction
Platform(s): Windows 98, NT, Mac, UNIX, IBM MVS and CMS.
Contact: Salford Systems
E-mail: `info@salford-systems.com`
URL: www.salford-systems.com

WizWhy
Description: WizWhy is a program that reveals the rules in a given database and predicts the value of future cases on the basis of these rules. WizWhy reveals two kinds of rules:

1. If-then rules
2. Mathematical formulae rules

The search of rules is based on a new mathematical deterministic algorithm, which completes the revealing task.
Data analysis functionalities: Classification, Rule Discovery
Platform(s): Windows 98, NT
Contact: WizSoft, Israel
E-mail: `info@wizsoft.com`
URL: www.wizsoft.com

SPSS AnswerTree
Description: AnswerTree can detect segments and patterns in large databases. AnswerTree's four scalable decision-tree algorithms uncover this valuable information and solve business problems.
Data analysis functionalities: Classification with CART, CHAID.
Platform(s): Windows 98, NT, 2000, Solaris

Contact: SPSS Inc., USA
E-mail: `sales@spss.com`
URL: www.spss.com/spssbi/answertree

A.2.5 IDA-tools using Support Vector Machines

KXEN
Description: KXEN Analytic Framework is a powerful and easy-to-use suite of analytic components. It is the ideal environment for modeling your data as easily and quickly as possible, while maintaining relevant and readily interpretable results. KXEN Analytic Framework places latest Data Mining technology within reach of business decision makers and data mining professionals. The patented, state-of-the-art predictive and descriptive modeling engines are based upon Vladimir Vapnik's mathematical theory for scoring, classification, clustering, and variable contribution.
Data analysis functionalities: time series analysis, support vector machines, robust regression, segmentation.
Platform(s): Windows NT, 2000, Solaris and Linux.
Contact: Knowledge Extraction Engine
E-mail: `support@kxen.com`
URL: www.kxen.com

NeuroSolutions
Description: NeuroSolutions combines a modular, icon-based network design interface with an implementation of advanced learning procedures, such as Support Vector Machines, recurrent backpropagation and backpropagation through time. Some other notable features include C++ source code generation, customized components through DLLs, a comprehensive macro language, and Visual Basic accessibility through OLE Automation.
Data analysis functionalities: Classification, Prediction, Support Vector Machine
Platform(s): Windows 98, 2000, NT, XP
Contact: NeuroDimension, Inc.
E-mail: `info@nd.com`
URL: www.nd.com/products

A.3. Tools for Text and Web Mining

IBM Intelligent Miner for Text
Description: Intelligent Miner for Text turns unstructured information into business knowledge for organizations. The knowledge-discovery toolkit includes components for building advanced text mining and text search applications, a wide range of text analysis tools for feature extraction, clustering, categorization and summarization, and full-text retrieval components using the IBM Text Search Engine. With Intelligent Miner for Text, you can unlock the business information that is trapped in email, insurance claims, news feeds and Lotus Notes, as well

as analyze patent portfolios, customer complaint letters, and even competitors' Web pages.
Data analysis functionalities: clustering, categorization, and summarization
Platform(s): Windows NT, Solaris, AIX
Contact: IBM, USA
URL: www-3.ibm.com/software/data/iminer/fortext/index.html

InFact
Description: InFact improves the productivity and decision-making skills of knowledge workers while reducing the cost of knowledge retrieval and transfer. Question Answering uses natural language to retrieve direct answers, not just lists of documents. Search across text, images, tables and graphs. Summarization extracts the most relevant sentence for retrieval. Query using similar sentences, paragraphs or documents to save time and increase the drill down accuracy. Supports exploration of a concept or idea by automatically identifying topics that are related to a question, categorized and hyperlinked by their relationships to the key elements of the query.
Data analysis functionalities: text queries, summarization
Platform(s): Windows 98, NT
Contact: Insightful Headquarters, USA
E-mail: `info@insightful.com`
URL: www.insightful.com

OK Log
Description: OK Log is a commercial software tool that scans websites, generates screens considering frames that describes what the user sees on the screen at any given time, and generates enhanced log files.
Data analysis functionalities: path information, visitors' analysis
Platform(s): Windows 98, NT
Contact: Cuende Infometrics, Spain
E-mail: `oklog@cuende.com`
URL: www.oklog.biz

WLS XP
Description: WebMining Log Sessionizator XPert (WLS) is a software tool that discovers patterns of user activity on websites by generating association rules using the Apriori algorithm. WLS scans and processes log files and automatically generates user sessions, extended association rule reports and active server pages scripts to incorporate rules in a web site.
Data analysis functionalities: association rules, log analyzer
Platform(s): Windows 98, NT
Contact: Egoland Consultores, Chile
E-mail: `info@webmining.cl`
URL: www.webmining.cl

A.4. Data Analysis Suites

As [243] suggests, intelligent data analysis is characterized by "a strategy, describing steps, decisions and actions that are taken during the process of analyzing data to build a model or answer a question". The tools described in this section support to a certain degree this process. An industry- and tool-neutral process model has been developed by the DM-CRISP project (www.crisp-dm.org). The category of tools in this section combines multiple modeling techniques from machine learning or artificial intelligence and adds preprocessing (e.g. transformations, scaling, outlier and missing value treatment), visualization and statistics in order to provide support for a broad range of applications.

Alice
Description: ALICE is an On-line Interactive Data Mining tool that can bring understandable and explicit results. It is an interactive and complete segmentation tool. Worded for the final, business users, it allows discovering hidden trends and relationships in the data and makes reliable predictions thanks to Interactive Decision Trees.
Data analysis functionalities: prediction, Clustering and OLAP capabilities
Platform(s): Windows 98, NT, 2000, TSE
Contact: Isoft, France
E-mail: `info@isoft.fr`
URL: www.alice-soft.com

Clementine
Description: Clementine is a visual data mining environment providing decision trees and neural networks. Based on a visual programming interface which links data access, manipulation and visualization together with machine learning (decision tree induction and neural networks). Trained rules and networks can be exported as C source code. Uses a graphical *building block* approach to develop applications.
Data analysis functionalities: Data reduction, clustering and segmentation.
Platform(s): Windows, Sun, HP, SGI, VAX/VMS, Alpha
Contact: SPSS Inc., USA
E-mail: `sales@spss.com`
URL: www.spss.com

DBMiner
Description: DBMiner Insight is a new generation of business intelligence and performance management applications powered by data mining. With new and insightful business patterns and knowledge revealed by DBMiner, the user can present his/her products or services to customers.
Data analysis functionalities: Decision trees, clustering, OLAP summaries and association rules.
Platform(s): Windows 98, NT, 2000, XP
Contact: DBMiner Technology Inc.

E-mail: `sales@dbminer.com`
URL: www.dbminer.com

DataDetective
Description: DataDetective, Sentient Machine Research's product line for matching and mining, allows businesses to retrieve and analyze their data easy, fast and powerful. DataDetective features new ways of dealing with large complex databases, and is ideal for creating user-friendly vertical solutions.
Data analysis functionalities: segmentation, fuzzy queries and prediction models
Platform(s): Windows 98, NT, 2000
Contact: Sentient Machine Research, The Netherlands
E-mail: `info@smr.nl`
URL: www.smr.nl

DataEngine
Description: A software tool for intelligent data analysis, combining conventional data analysis methods with fuzzy technologies, neural networks, decision trees and statistical methods [373]. General description of DataEngine:

- It is an efficient tool for technical and management applications.
- Its 32-bit architecture, a powerful data visualization component and an easy operation of the user interface provide lots of conveniences for the user.
- Data can be accessed by the import and export of ASCII or MS-Excel* files as well as by the ODBC interface.
- DataEngine can be extended by PlugIns, such as: automatic feature selection, accessing data acquisition hardware, advanced clustering and interfaces with S-Plus.
- Applications of DataEngine: quality control, process analysis, forecasting, database marketing and diagnosis.

Data analysis functionalities: Classification, Clustering, Dependency analysis, Function Approximation
Platform(s): Windows 98, NT
Contact: MIT-Management Intelligenter Technologien, Germany
E-mail: `products@mitgmbh.de`
URL: www.dataengine.de

Data Prospector Suite
Description: This tool derives statistical data models (Synes Data Summaries) from databases in an unbiased and non-parametric fashion. There is no need for a priori knowledge of or assumptions about the relationships between variables.
Data analysis functionalities: Market basket analysis, regression, clustering.
Platform(s): Windows 98, 2000, NT
Contact: Synes NV
E-mail: `info@synes.com`
URL: www.synes.com

Enterprise Data-Miner
Description: The software is a collection of routines for efficient mining of large data sets. Both classical and the more computationally expensive state-of-the-art prediction methods are included. Using a standard spreadsheet data format, this kit implements many data-mining tasks.
Data analysis functionalities: neural nets, decision rules and trees
Platform(s): Windows 98, NT, Unix (Sun, DEC-Alpha, IBM RS/6000, HP, Linux)
Contact: Data-Miner Pty, Australia
E-mail: `comment@data-miner.com`
URL: www.data-miner.com

Genio Miner
Description: Hummingbird Genio Miner is a data mining suite that integrates data acquisition and cleansing with algorithmic models. This architecture provides more precise results by improving the quality of the source data and facilitating access from heterogeneous sources including flat files and direct database connections.
Data analysis functionalities: Bayesian, symbolic nearest mean, LVQ, MLP, Kohonen, k-means, Chaid, C4.5 and Gini trees.
Platform(s): Windows 98, NT, 2000, XP
Contact: Hummingbird
E-mail: `getinfo@hummingbird.com`
URL: www.hummingbird.com

Ghost Miner
Description: The GhostMiner Software is a general-purpose data mining system. Its goal is to discover hidden knowledge: facts, patterns, and subtle relationships in databases, and thus extract relevant information from the data. To achieve its goals the GhostMiner system uses a combination of neural networks, machine learning, statistical analysis, visualization techniques and database technology.
Data analysis functionalities: Statistics, Neural networks, Neurofuzzy system, Decision tree, and the k-nearest neighbor.
Platform(s): Windows NT, 2000, XP
Contact: FQS, Poland
E-mail: `info@fqspl.com.pl`
URL: www.fqspl.com.pl/ghostminer/generalInfo.asp

iData Analyzer
Description: As an MS Excel add-on, the iData Analyzer software tool has been used in a diverse set of industries from healthcare and medical informatics to finance and telecommunications. Specific applications include business intelligence, database cleansing, data warehouse development prototyping, domain and attribute correlation, manufacturing failure analysis, and stock price prediction. iData Analyzer is a business intelligence tool that can be easily used by

anyone with a PC and MS Excel. As an Excel add-on, iData Analyzer employs state-of-the-art artificial intelligence technology to unveil hidden knowledge from any spreadsheet of source data.
Data analysis functionalities: neural networks, supervised and unsupervised classification, decision rules
Platform(s): Windows 98, NT, 2000, XP
Contact: Information Acumen Corp.
E-mail: `info@infoacumen.com`
URL: www.infoacumen.com

Insightful Miner
Description: Insightful Miner Desktop Edition is a data mining workbench that gives new data miners and skilled modelers the ability to deploy predictive intelligence throughout the enterprise.
Data analysis functionalities: neural networks, decision trees, linear and logistic regression, naive Bayes classifier, k-means.
Platform(s): Windows 98, NT
Contact: Insightful Headquarters, USA
E-mail: `info@insightful.com`
URL: www.insightful.com

Intelligent Miner
Description: Scalable and with support for multiple platforms, the Intelligent Miner for Data provides a single framework for database mining using proven, parallel mining techniques. Business applications for this technology vary widely, and a variety of mining algorithms is provided.
Data analysis functionalities: clustering, association rules, classification and predictive algorithms
Platform(s): Windows NT, 2000, Solaris, OS/400, V4R3 and AIX
Contact: IBM, USA
URL: www-3.ibm.com/software/data/iminer/fordata/

K.wiz
Description: K.wiz is a fully integrated Business Analysis Automation platform incorporating knowledge discovery and data mining techniques designed to provide business users with intelligent analysis capabilities. It combines ease-of-use with enterprise scalability to uncover latent knowledge from massive amounts of data through host applications. As a result, it offers business users data mining benefits without the need to learn about specialized algorithms or techniques.
Data analysis functionalities: statistics, visualization, data mining techniques.
Platform(s): Windows NT, 2000, Solaris, HP, AIX and Linux
Contact: thinkAnalytics
E-mail: `info@thinkanalytics.com`
URL: www.thinkanalytics.com

Kensigton Discovery Edition
Description: Kensington Discovery Edition is an enterprise-wide discovery platform that supports entire processes of information based discovery works. It supports dynamic information integration of independently monitored databases and other information resources within an enterprise and provides an extensible layer of generic and scalable analytical components including functions for clustering, classification and dependency analysis for the analysis of data. It is a uniform framework of dynamic knowledge management for verifying and warehousing discovered knowledge and supports a wide range of deployment mechanisms for automatically generating analytical applications from well-verified discovery processes.
Data analysis functionalities: statistics, neural networks, k-means, EM clustering, association analysis, predictive models.
Platform(s): Windows NT, 2000, Solaris and Linux.
Contact: InforSense
E-mail: `Marketing@inforsense.com`
URL: www.inforSense.com

Knowledge Miner
Description: KnowledgeMiner is an artificial intelligence tool designed to easily extract hidden knowledge from data. It was built on the cybernetic principles of self- organization: Learning a completely unknown relationship between output and input of any given system in an evolutionary way from a very simple organization to an optimally complex one.
Data analysis functionalities:, neural networks, fuzzy rule induction, analog complexing.
Platform(s): Windows 98, NT, 2000, MacOS 7.5.
Contact: Script Software International
E-mail: `info@knowledgeminer.net`
URL: www.knowledgeminer.net

Microsoft SQL Server 2000
Description: Business today demands a different kind of database solution. Performance, scalability, and reliability are essential, and time to market is critical. Beyond these core enterprise qualities, SQL Server 2000 provides agility to your data management and analysis, allowing your organization to adapt quickly and gracefully to derive competitive advantage in a fast-changing environment. From a data management and analysis perspective, it is critical to turn raw data into business intelligence and take full advantage of the opportunities presented by the Web. A complete database and data analysis package, SQL Server 2000 opens the door to the rapid development of a new generation of enterprise-class business applications that can give your company a critical competitive advantage.
Data analysis functionalities: Clustering, decision trees, OLAP
Platform(s): Windows 98, 2000, NT

Contact: Microsoft, USA
URL: www.microsoft.com/sql/default.asp

Nuggets
Description: Nuggets is a powerful solution that sifts through data and uncovers hidden facts and relationships. The insights that Nuggets can discover will reveal which indicators most impact your business and help predict future results.
Data analysis functionalities: SiftAgent, machine learning based, decision rules.
Platform(s): Windows 98, 2000, NT
Contact: Data Mining Technologies Inc.
E-mail: `info@data-mine.com`
URL: www.data-mine.com

Oracle 9i Data Mining
Description: Oracle9i Data Mining, an option to Oracle9i Enterprise Edition, allows companies to build advanced business intelligence applications that mine corporate databases to discover new insights, and integrate those insights into business applications. Oracle9i Data Mining embeds data-mining functionality into the Oracle9i database, for making classifications, predictions, and associations. All model-building, scoring, and metadata management operations are initiated via a Java- based API and occur entirely within the relational database.
Data analysis functionalities: Classification and Prediction.
Platform(s): Client - Windows 95, NT; Server - Sun Solaris, IBM AIX , HP UX
Contact: Oracle, USA
URL: otn.oracle.com/products/bi/9idmining.html

Partek Pro
Description: Featuring the Analytical SpreadsheetTM and Pattern Visualization SystemTM, it is a comprehensive data processing, analysis, and visualization system. Some features are:
- No software imposed limit to number of rows or columns in Analytical SpreadsheetTM
- Visualize and interact with high-dimensional data using Pattern Visualization SystemTM
- Complete set of summary statistics
- Robust normalization and rescaling
- Interactive, programmable data filtering
- Pattern matching using over 20 built-in measures of similarity
- Customize and automate your work using the built-in Tcl scripting language

Data analysis functionalities: Statistics, neural networks, discriminant analysis, data reduction.
Platform(s): Windows 98, 2000, NT
Contact: Partek Inc.
E-mail: `info@partek.com`
URL: www.partek.com

PolyAnalyst

Description: PolyAnalyst is a comprehensive and versatile suite of advanced data mining tools. PolyAnalyst incorporates the latest achievements in automated knowledge discovery to analyze both structured and unstructured data. The PolyAnalyst platform offers a complete end-to-end analytical solution - from data importing, cleaning, manipulation, visualization, modeling, scoring, and reporting. The intuitive interface and an online tutorial smooth the learning curve to enable analysts to reach conclusions comfortably and confidently.
Data analysis functionalities: Market basket analysis, neural networks, regression, decision trees, clustering.
Platform(s): Windows 98, 2000, NT
Contact: Megaputer Intelligence Inc.
URL: www.megaputer.com

Prudsys Discoverer 2000

Description: Prudsys DISCOVERER 2000 is based on the newest technology to analyze your customer and enterprise data: non-linear decision trees. The system is particularly suitable for customer qualification, creation of canceler profiles, evaluation of target groups and optimization of marketing actions.
Data analysis functionalities: statistics, non linear decision trees, classification.
Platform(s): Windows 98, 2000, NT
Contact: Prudsys
E-mail: `info@prudsys.com`
URL: www.prudsys.com

SAS Enterprise Miner

Description: Enterprise Miner is an integrated software product that provides an end- to-end business solution for data mining. A graphical user interface (GUI) provides a user-friendly front-end to the SEMMA process (Sample, Explore, Modify, Model, Assess). All of the functionality needed to implement the SEMMA process is accessed through a single GUI. The SEMMA process is driven by a process flow diagram (pfd), which you can modify and save. SAS Enterprise Miner contains a collection of sophisticated analysis tools that have a common user-friendly interface. Statistical tools include clustering, decision trees, linear and logistic regression, and neural networks. Data preparation tools include outlier detection, variable transformations, random sampling, and the partitioning of data sets (into train, test, and validate data sets). Advanced visualization tools enable you to quickly and easily examine large amounts of data in multidimensional histograms, and to graphically compare modeling results.
Data analysis functionalities: Visualization, Exploration
Platform(s): Windows 98, NT
Contact: SAS Institute Inc., USA
E-mail: `software@sas.com`
URL: www.sas.com

Statistica Data Miner
Description: Statistica Data Miner is a system of user-friendly tools for the data mining process, from querying database to generating final reports.
Data analysis functionalities: statistics, OLAP analysis, neural networks, ARIMA, association rules, CHAID trees.
Platform(s): Windows 98, NT
Contact: StatSoft, Inc., USA
E-mail: info@statsoft.com
URL: www.statsoftinc.com

Storm
Description: Storm is a component toolkit that can be used to complete advanced data-mining tasks, such as e.g. customer profiling, scoring, and forecasting. Storm offers an extensive range of data-mining techniques, from classical methods (e.g. neural networks, decision trees) to highly innovative technologies (bayesian networks, non-linear principal component analysis).
Data analysis functionalities: neural networks, decision trees, cluster analysis, bayesian networks, PCA, k-nearest neighbor.
Platform(s): Windows 98, 2000, NT
Contact: Elseware, France
E-mail: storm@elseware.fr
URL: www.storm-central.com

Teradata Warehouse Mining
Description: Teradata has evolved Data Mining from the realm of raw algorithms to proven business tools that have significantly increased corporate revenue and reduced cost. Teradata Warehouse Mining provides the tools and services to put data mining in your hands enabling you to solve complex business problems and insert knowledge into your CRM and analytic applications.
Data analysis functionalities: statistics, data visualization, linear and logistic regression, decision trees, rule induction, clustering.
Platform(s): Windows 98, 2000, NT
Contact: NCR, USA
URL: www.ncr.com/products/software/teradata_mining.htm

XpertRule Miner
Description: XpertRule Miner is a tool evolved from the established Profiler scalable client-server data mining software. Using ActiveX technology, the Miner client can be deployed in a variety of ways. Solutions can be built as stand-alone mining systems or embedded in other vertical applications under MS-Windows. Deployment can also be over Intranets or the Internet. The ActiveX Miner client works with Attar's high performance data mining servers to provide multi-tier client- server data mining against large data bases. Mining can be performed either directly against the data in situ, or by mining against tokenised cache data tables.

Data analysis functionalities: decision trees, data visualization, market basket analysis.
Platform(s): Windows 98, 2000, NT and XP
Contact: Attar Software
E-mail: info@attar.com
URL: www.attar.com

A.5. Conclusion

The selection of tools presented in this survey confirms the current development of the IDA-software market: We have witnessed the evolution from tools focussing on single techniques towards IDA-suites that support the main steps of the process of Intelligent Data Analysis. This trend has been accompanied by the integration of IDA- functionality into main data storage technology (offered e.g. by IBM, Microsoft, NCR, and Oracle, among others) as well as by the development software standards such as the Predictive Model Markup Language (PMML). PMML is an XML mark up language to describe statistical and data mining models and contains the inputs to data mining models, the transformations used prior to prepare data for data mining, and the parameters which define the models themselves (for more information see www.dmg.org). Another trend, which is not covered by this survey, but very important for the future development of IDA-applications, is the increase of solutions instead of tools (for more details see e.g. www.kdnuggets.com). Worth mentioning is also a new business model that can be described as *Outsourcing of IDA-applications.* In this case, providers such as e.g. digiMine (www.digimine.com) analyze their customers' data using high-performance IDA-tools in-house and deliver solutions to their customers, which in turn do not have to invest in data storage, data analysis, updating of IDA-models and training.

Appendix B: Information-Theoretic Tree and Rule Induction

Gerard van den Eijkel
Telematica Institute, The Netherlands

This appendix outlines the information-theoretic approach to decision tree and rule induction. The goal is to arrive at a measure that can select some partition (or even a complete rule base) over other partitions (rule bases). In the information-theoretic approach, the measure used in both tree induction as well as in rule induction is based on mutual information. In order to explain this measure, it is necessary to clarify the notion of information, often referred to as entropy or uncertainty [522].

B.1. Information and Uncertainty

Suppose somebody picks randomly a number ξ out of the set $N = \{1, 2, 3, 4\}$. Suppose further that it is our task to find out which number this is with a minimum number of yes/no questions. As a first guess we may think that we can always find the answer with at most three questions, since there are only four possible numbers. If we are lucky we may have the answer right the first time, but if we are unlucky we need three questions. However, the answer is that we can always find the answer with at most two questions; see Figure B.1. The trick is to note that an answer to a question reduces the uncertainty we have with respect to ξ. If we do not ask any question, then there are four possible numbers, all being equally likely. The uncertainty $H(N)$ in this discrete case is now defined as the *discrete information* and is usually expressed in bit(s):

$$\begin{aligned} H(N) &= \sum_{n=1}^{4} -P(n) \log_2 P(n) \\ &= \sum_{n=1}^{4} -0.25 \log_2 0.25 = 2 \text{ bit} \end{aligned} \tag{B.1}$$

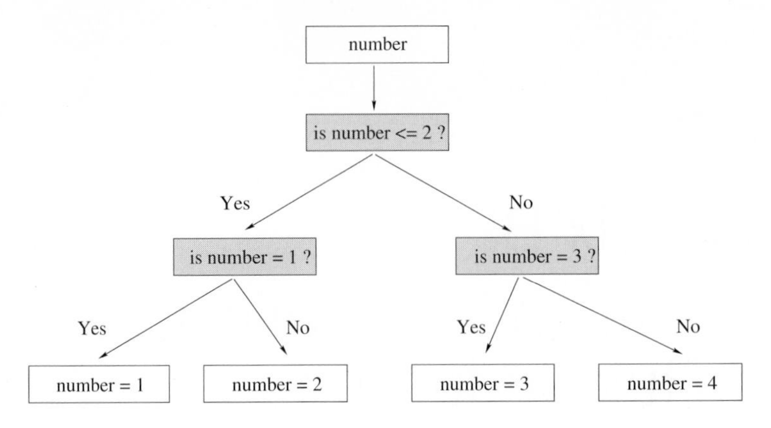

Fig. B.1. Example of finding a number between 1 and 4 with only two questions. Start from the top and work your way down to the leaves.

where $P(n)$ is the probability that ξ equals n. Suppose we ask the question: "Is ξ equal to three or four?", then the answer (either yes or no) leaves only two equally likely numbers. The uncertainty about ξ that we have after obtaining the answer "yes" to our question is then defined as the discrete *conditional information* $H(N|yes)$:

$$\begin{aligned} H(N|\text{yes}) &= \sum_{n=1}^{4} -P(n|\text{yes}) \log_2 P(n|\text{yes}) \\ &= 0 + 0 + 0.5 + 0.5 = 1 \text{ bit} \end{aligned} \tag{B.2}$$

where $P(n|yes)$ is the probability ξ equals n *given* that the answer is "yes"; a conditional probability. This quantity can be viewed as the uncertainty in the answer or as the information still necessary to find the final answer. Likewise for $H(N|no)$:

$$\begin{aligned} H(N|\text{no}) &= \sum_{n=1}^{4} -P(n|\text{no}) \log_2 P(n|\text{no}) \\ &= 0.5 + 0.5 + 0 + 0 = 1 \text{ bit} \end{aligned} \tag{B.3}$$

It is clear that discrete information nicely predicts the remaining number of yes/no questions which we still have to ask in order to obtain a *certain* answer. However, since we do not know the answer to the question beforehand, we can also calculate the *expectation* of the remaining uncertainty. This quantity is also known as the discrete *average conditional information* $H(N|Q)$, where $Q = \{\text{yes}, \text{no}\}$.

$$
\begin{aligned}
H(N|Q) = & \sum_{q \in \{\text{yes,no}\}} \sum_{n=1}^{4} P(q)\ P(n|q) \log_2 P(n|q) \\
= & \sum_{n=1}^{4} -P(\text{yes})\ P(n|\text{yes}) \log_2 P(n|\text{yes}) + \\
& + \sum_{n=1}^{4} -P(\text{no})\ P(n|\text{no}) \log_2 P(n|\text{no}) \\
= & \ 0.5 * 1 + 0.5 * 1 = 1 \text{ bit}
\end{aligned} \tag{B.4}
$$

We may now ask how much uncertainty the answer to this question may reduce, or what the *expected information gain* is upon receiving the answer. Since we started with an uncertainty of 2 bits and expect to have a single bit of uncertainty left having asked the question, we may expect that the reduction of uncertainty equals 1 bit. In general this quantity is known as the discrete *mutual information* and is defined as:

$$I(N;Q) = H(N) - H(N|Q) \tag{B.5}$$

Suppose we would like to immediately ask the question "is ξ equal to four" instead of asking is "is ξ equal to 3 or 4", then we can immediately calculate that the expected information gain of this question equals:

$$
\begin{aligned}
I(N;Q) &= H(N) - H(N|Q) \\
&= 2 - \sum_{n=1}^{4} -P(\text{yes})\ P(n|\text{yes}) \log_2 P(n|\text{yes}) + \\
& \quad + \sum_{n=1}^{4} -P(\text{no})\ P(n|\text{no}) \log_2 P(n|\text{no}) \\
&= 2 - (0 + 1.1996) = 0.8004 \text{ bit}
\end{aligned} \tag{B.6}
$$

Hence, the expected information gain of this question is less than the gain found for the question "is ξ equal to 3 or 4" (which was 1 bit), therefore *on average* we are better of with asking "is ξ equal to 3 or 4" (and only after we obtain a yes to this question it is useful to ask: "is ξ equal to four").

Mutual information helps in selecting the proper questions. In much the same way it can help in selecting the proper partition, since a partition is essentially formed by conditions put on the instance of a test (question). In the information-theoretic approach it is now assumed that a partition should provide an information gain, if it is to be useful. The partition that provides the highest gain is then selected over others.

Suppose we have a partition of R into n disjoint regions of the instance space; $R = \{R_1, \ldots, R_j, \ldots, R_n\}$, where each region can be associated with an event (class) C_i from the event space $C = \{C_1, \ldots, C_i, \ldots, C_k\}$ by a conditional probability $P(C_i|R_j)$. The discrete information of the event space is then defined

Table B.1. Examples of patients.

Patient no.	Heart Rate	Blood Pressure	Class
1	irregular	normal	ill
2	regular	normal	healthy
3	irregular	abnormal	ill
4	irregular	normal	ill
5	regular	normal	healthy
6	regular	abnormal	ill
7	regular	normal	healthy
8	regular	normal	healthy

as:

$$H(C) = E[-\log_2 P(C_i)] = \sum_{i=1}^{k} -P(C_i)\log_2 P(C_i) \tag{B.7}$$

the discrete conditional information as:

$$H(C|R) = E[-\log_2 P(C_i|R_j)] = \sum_{j=1}^{n}\sum_{i=1}^{k} -P(C_i, R_j)\log_2 P(C_i|R_j) \tag{B.8}$$

and the mutual information as:

$$I(C;R) = H(C) - H(C|R) \tag{B.9}$$

B.2. Decision Tree Induction

As an example of tree induction, consider the set of examples of eight patients classified as healthy or ill on the basis of their heart rate and mean blood pressure, shown in Table B.1. Suppose our problem is to find out, with a minimal number of questions, when a patient is ill or healthy. Or in other words, to divide these examples in groups of "ill" patients and "healthy" patients.

By using the mutual information, the first question leading to a partition is easily found. We note that the uncertainty with respect to the class without partitioning equals:

$$\begin{aligned} H(\text{Class}) &= -P(\text{healthy})\log_2 P(\text{healthy}) - P(\text{ill})\log_2 P(\text{ill}) \\ &= 0.5 + 0.5 = 1 \text{ bit} \end{aligned}$$

where the a priori probabilities are calculated from the examples (each being 0.5 since there are 4 ill and 4 healthy patients). Now we have several possible ways to partition the examples. We could partitioning the examples according to the heart rate, but we could also partition the examples according to the blood pressure. In order to choose between these possibilities, we simply calculate the information gain (mutual information) of each. For the heart rate, we note that it can take on the values “regular” and “irregular” and we obtain:

$$
\begin{aligned}
&H(\text{Class}|\text{Heart Rate}) = \\
&= -P(\text{irregular})\ P(\text{healthy}|\text{irregular}) \log_2 P(\text{healthy}|\text{irregular}) + \\
&\quad -P(\text{regular})\ P(\text{healthy}|\text{regular}) \log_2 P(\text{healthy}|\text{regular}) + \\
&\quad -P(\text{irregular})\ P(\text{ill}|\text{irregular}) \log_2 P(\text{ill}|\text{irregular}) + \\
&\quad -P(\text{regular})\ P(\text{ill}|\text{regular}) \log_2 P(\text{ill}|\text{regular}) \\
&= -0.375 * 0 - 0.625 * 0.8 \log_2 0.8 - 0.375 * 0 - 0.625 * 0.2 \log_2 0.2 \\
&= 0.45 \text{ bit}
\end{aligned}
$$

Whereas we have for the blood pressure:

$$
\begin{aligned}
&H(\text{Class}|\text{Blood Pressure}) = \\
&= 0.25 * 0 + 0.75 * 0.66 \log_2 0.66 + 0.25 * 0 + 0.75 * 0.33 \log_2 0.33 \\
&= 0.69 \text{ bit}
\end{aligned}
$$

hence the information gain for these partitions is:

$$
\begin{aligned}
I(\text{Class}; \text{Heart Rate}) &= 1 - 0.45 = 0.55 \text{ bit} \\
I(\text{Class}; \text{Blood Pressure}) &= 1 - 0.69 = 0.31 \text{ bit}
\end{aligned}
$$

Because it gives a larger information gain, we choose to partition according to the heart rate. Now the group for which the heart rate is “irregular” is perfect in the sense that the three patients that have an irregular heart rate are indeed all “ill”. Hence, we can decide with certainty that if the heart rate is irregular, then the patient is ill. However, the other group is less clear-cut: of the five patients that have a regular heart rate, four patients are healthy but one is ill. Therefore, we cannot decide without uncertainty what the patient class is if the heart rate is regular, so we need to refine this group. Normally, this refinement entails a recursion: generate possible (sub-)partitions for the group to be refined and select the one having highest information gain. In this example problem, however, we only have one possibility for refinement: the blood pressure. If it is normal, then the patients who have a regular heart rate are indeed healthy, but if the blood pressure is abnormal than the patient is “ill”. In this way we have arrived at a scheme depicted in Figure B.2, which can be used for other patients which where not in the data base. What we have arrived at is a decision tree, and the type of induction that we performed was identical to that introduced by Quinlan in his famous ID3 tree induction algorithm [432, 433].

In view of this powerful and general tree induction algorithm, the reader might wonder why so many other algorithms exist, even for the information theoretic approach. The reason is that real-world problems that are more complicated than our simple, noise-free database without overlapping classes, demand better methods. If noise is present the tree may start to fit the noise or become too specific to deal with the class overlap, a condition called “over-specification”

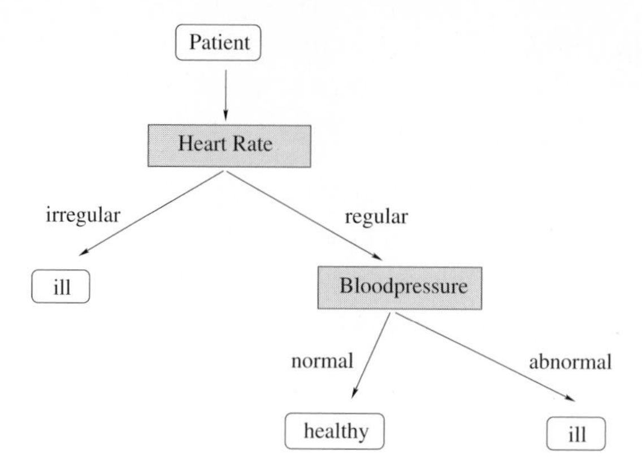

Fig. B.2. Example of a simple decision tree.

or "over-fitting". The solution to that problem is to build in an appropriate stopping criterion or a tree pruning algorithm, by which irrelevant sub-partitions can be prevented or removed, respectively. Irrelevance can also be tackled by information theory and many other statistical measures [399]. Other reasons for the diversity in algorithms stem from the type of discretization, hypothesis generation, search strategy etc.

B.3. Rule Induction

Although decision trees provide a powerful means for generating useful partitions, there is a different way of approaching the problem. If we take a good look at the examples in Table B.1, it is clear that the patient is ill if the heart rate is irregular or if the blood pressure is abnormal; otherwise he is healthy. Hence, we can form the following *rule base*, that gives the same decision as the decision tree in Figure B.2:

- **If** Heart Rate is irregular **Then** Patient is ill
- **If** Blood Pressure is abnormal **Then** Patient is ill
- **If** Heart Rate is normal **And** Blood Pressure is normal **Then** Patient is healthy

The construction of such a rule base is not trivial, even though it seems easy. A way to construct a rule base from the tree of Figure B.2 is by "walking along the path of the tree". We then get the following rule base:

- **If** Heart rate is irregular **Then** Patient is ill
- **If** Heart rate is regular **And** Blood Pressure is abnormal **Then** Patient is ill
- **If** Heart rate is regular **And** Blood Pressure is normal **Then** Patient is healthy

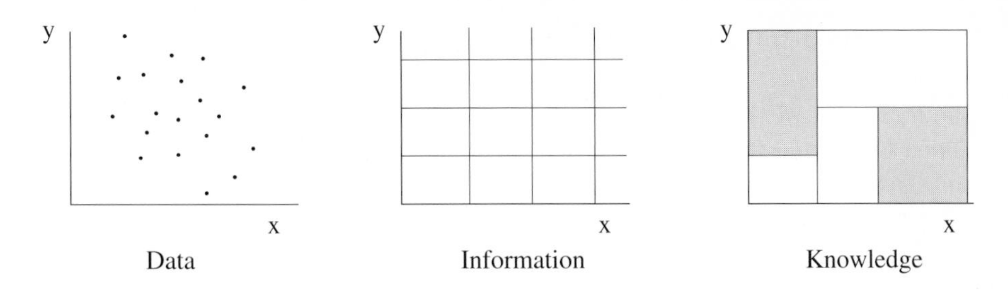

Fig. B.3. Figure depicts the Data-Information-Knowledge paradigm. The information is a discrete representation of the data, as indicated by the regions. The knowledge consists of general rules for each class (event), which are indicated by the shading.

As can be seen through comparison of the rule bases, the second rule is somewhat more specific than the second rule of the first rule base, and is therefore somewhat more complicated to explain. For more complex problems in high-dimensional feature spaces, explanations become much more complicated. In such cases a rule base with *overlapping* rules provides a simpler explanation than a decision tree or disjoint rule-base.

In the final section we will outline an information-theoretic approach to rule induction that uses the J-information measure, first introduced by Smyth and Goodman [489]. With this approach it is possible to construct rule bases with overlapping rules.

B.3.1 Data, Information, and Knowledge

A useful paradigm for rule induction is based on layers of data, information, and knowledge. Suppose we have a continuous n-dimensional domain which we would like to partition. The first step to be taken is to transform the continuous data in discrete information which consists of specific regions, formed by qualifications, and an associated decision. In the information layer, the examples are represented by discrete values or symbols. The search for possible rules is then performed on the information layer and the selected rules are stored in a rule base, called the knowledge layer. The knowledge layer is used for classification and explanation of new instances. In this way we have obtained three layers, a data layer, an information layer and a knowledge layer. The paradigm is depicted in Figure B.3.

Since the rules are expressed in qualifications obtained from a discretization, it can be advantageous to let an expert determine the discretization. In that case the qualifications can be viewed as a reference frame in which the classification problem should be cast. In decision-support systems, such a reference frame is useful for explaining results to an expert "in his own words" [146]. However, the transformation from data to information can be provided by any appropriate discretization method, such as k-means clustering [19], see Chapter 3, Section 3.4.4.

B.3.2 The J-information Measure

The formation of the knowledge layer from the information layer is essentially guided by mutual information, just as in tree induction. However, the main advantage of the J-measure over mutual information is that it allows the evaluation of a single rule (a region associated with a class) rather than a complete partition. To see this, we note that, according to [70], the mutual information $I(C;R)$ can be written as an expectation:

$$\begin{aligned} I(C;R) &= H(C) - H(C|R) \\ &= E_{C_i,R_j}[\log_2 \frac{P(C_i|R_j)}{P(C_i)}] \\ &= \sum_{j=1}^{n}\sum_{i=1}^{k} P(R_j)P(C_i|R_j)\log_2 \frac{P(C_i|R_j)}{P(C_i)} \end{aligned} \tag{B.10}$$

The j-information measure is defined as:

$$j(C;R_j) = \sum_{i=1}^{k} P(C_i|R_j)\log_2 \frac{P(C_i|R_j)}{P(C_i)} \tag{B.11}$$

such that:

$$I(C;R) = E_{R_j}[j(C;R_j)] \tag{B.12}$$

Here, $j(C|R_j)$ expresses the goodness-of-fit of the region R_j with respect to the classes; this a measure for the data fit of the rule that covers region R_j. The J-measure is an extension of the j-measure with a mental fit part:

$$J(C;R_j) = P(R_j)j(C;R_j) \tag{B.13}$$

The larger the probability $P(R_j)$, the better is the mental fit of the rule that covers region R_j. Since a rule essentially gives information over a single class C_m (the majority class) and its complement (not C_m), the J-measure according to:

$$\begin{aligned} J(C;R_j) = P(R_j)P(C_m|R_j)\log_2 \frac{P(C_m|R_j)}{P(C_m)} + \\ +P(R_j)(1 - P(C_m|R_j))\log_2 \frac{1 - P(C_m|R_j)}{1 - P(C_m)} \end{aligned} \tag{B.14}$$

is more suitable for the specific task of rule induction, see [399].

Since the J-measure compares the a priori probability with the a posteriori probability, it is also referred to as information gain. This J-measure has been introduced by Smyth and Goodman [489] and is considered to be one of the most promising measures [312] for rule induction.

B.3.3 Example

As an example of rule induction using the J-measure, we will use the previous problem of the patient database. If we allow overlapping rules then the following initial hypotheses can be formed either data-driven or model-driven (we have written the J-values behind the hypotheses and ordered the list for convenience):

- **If** Heart Rate is regular **And** Blood Pressure is normal **Then** Patient is healthy (0.5 bit)
- **If** Heart Rate is irregular **Then** Patient is ill (0.375 bit)
- **If** Blood Pressure is abnormal **Then** Patient is ill (0.25 bit)
- **If** Heart Rate is regular **Then** Patient is healthy (0.18 bit)
- **If** Blood Pressure is normal **Then** Patient is healthy (0.06 bit)
- ...

Clearly the first hypothesis has the highest J-value (0.5 bit) and we select this one for our rule base. All patients that are healthy are described (covered) by this rule. However, there are still patients left which are not yet covered by this rule base. Reviewing our list of hypotheses, we select the second hypothesis as the next rule to add to our rule base. Now only one patient remains uncovered, the ill patient with a regular heart rate and an abnormal blood pressure. For this patient we choose the third hypothesis to add to the rule base. Our rule base (or knowledge) does now cover all the examples and consists of the following rules:

- **If** Heart Rate is regular **And** Blood Pressure is normal **Then** Patient is healthy
- **If** Heart Rate is irregular **Then** Patient is ill
- **If** Blood Pressure is abnormal **Then** Patient is ill

Note that both the second and third rule cover patient number three of the database. It is said that the rules overlap or are non-disjoint. In this case they both have the same conclusion, and hence we have no conflict. However, conflicting rules make it sometimes necessary to form only disjoint rules. A disjoint scheme of rule induction essentially follows the same procedure. It iteratively generates hypotheses and selects the one having the highest J-value *and* that is disjoint with all existing rules in the rule base. In this case we would have obtained for the patients:

- **If** Heart Rate is regular **And** Blood Pressure is normal **Then** the Patient is healthy
- **If** Heart Rate is irregular **Then** the Patient is ill
- **If** Heart rate is regular **And** Blood Pressure is abnormal **Then** the Patient is ill

Note that these are exactly the same rules as we have obtained from the tree induction algorithm (after transformation to rules). In general, this is usually not the case; it is only due to the small number of examples and features present in this synthetic database.

References

1. E. H. Aarts and J. Korst. *Simulated Annealing and Boltzmann Machines*. John Wiley and Sons, Chichester, 1989.
2. H. D. I. Abarbanel. *Analysis of Observed Chaotic Data*. Springer, 1995.
3. S. Abe and M.-S. Lan. A method for fuzzy rules extraction directly from numerical data and its application to pattern classifiction. *IEEE Transactions on Fuzzy Systems*, 3(1):18–28, 1995.
4. S. Abe and R. Thawonmas. A fuzzy classifier with ellipsoidal regions. *IEEE Transactions on Fuzzy Systems*, 5(3):358–368, 1997.
5. J. Abello and J. Korn. MGV: A system for visualizing massive multi-digraphs. *Transactions on Visualization and Computer Graphics*, 8(1), 2002.
6. R. Agrawal, H. Mannila, R. Srikant, H. Toivonen, and A. Verkamo. Fast discovery of association rules. In U. Fayyad, G. Piatetski-Shapiro, P. Smyth, and R. Uthurusamy, editors, *Advances in Knowledge Discovery and Data Mining*, pages 307–328. AAAI Press, 1996.
7. R. Agrawal, H. Mannila, R. Srikant, H. Toivonen, and A. Verkamo. Fast discovery of association rules. *Advances in Knowledge Discovery and Data Mining*, pages 307–328, 1996.
8. R. Agrawal, H. Mannila, R. Srikant, H. Toivonen, and A. Verkamo. Fast discovery of association rules. *Advances in Knowledge Discovery and Data Mining*, pages 307–328, 1996.
9. C. Ahlberg and B. Shneiderman. Visual information seeking: Tight coupling of dynamic query filters with starfield displays. In *Proc. Human Factors in Computing Systems CHI '94 Conf., Boston, MA*, pages 313–317, 1994.
10. C. Ahlberg and E. Wistrand. IVEE: An information visualization and exploration environment. In *Proc. Int. Symp. on Information Visualization, Atlanta, GA*, pages 66–73, 1995.
11. M. Aizerman, E. Braverman, and L. Rozonoer. Theoretical foundations of the potential function method in pattern recognition learning. *Automations and Remote Control*, 25:821–837, 1964.
12. J. T. Alander. An indexed bibliography of genetic algorithms and simulated annealing: Hybrids and comparisons. Technical report, University of Vaasa, Vaasa, Finland, 1995.
13. A. A. Alizadeh et al. Distinct types of diffuse large b-cell lymphoma identified by gene expression profiling. *Nature*, pages 503–511, 2000.
14. J. R. A. Allwright and D. B. Carpenter. A distributed implementation of simulated annealing for the traveling salesman problem. *Parallel Computing*, 10:335–338, 1989.

15. B. Alpern and L. Carter. Hyperbox. In *Proc. Visualization '91, San Diego, CA*, pages 133–139, 1991.
16. D. Altman. *Practical Statistics for Medical Research.* Chapman and Hall, 1991.
17. R. B. Altman. Challenges for intelligent systems in biology. *IEEE Intelligent Systems*, 16(6):14–18, 2001.
18. R. S. Amant and P. Cohen. Interaction with a mixed-initiative system for exploratory data analysis. *Knowledge-Based Systems*, 10:265–273, 1998.
19. M. Anderberg. *Cluster Analysis for Applications.* Academic Press, New York, 1973.
20. E. Anderson. The irises of the Gaspe Peninsula. *Bulletin of the American Iris Society*, 59:2–5, 1935.
21. T. W. Anderson. *An Introduction to Multivariate Analysis.* John Wiley & Sons, New York, 1958.
22. D. F. Andrews. Plots of high-dimensional data. *Biometrics*, 29:125–136, 1972.
23. P. Angeline. *Evolutionary Algorithms and Emergent Intelligence.* PhD thesis, Ohio State University, Nov. 1993.
24. P. J. Angeline and K. E. Kinnear, Jr., editors. *Advances in Genetic Programming.* MIT Press, Cambridge, MA, 1996.
25. P. J. Angeline, R. G. Reynolds, J. R. McDonnell, and R. Eberhart, editors. *Evolutionary Programming VI, 6th International Conference, EP 97*, Berlin, 1997. Springer.
26. J. Angstenberger and R. Weber. Applications of intelligent techniques in process analysis. In D. Ruan, editor, *Intelligent Hybrid Systems: Fuzzy Logic, Neural Networks, and Genetic Algorithms*, pages 189–208. Kluwer, 1997.
27. M. Ankerst, M. Breunig, H. Kriegel, and J. Sander. OPTICS: Ordering Points To Identify the Clustering Structure. In *Proc. ACM SIGMOD'99, Int. Conf on Management of Data, Philadelphia, PA*, pages 49–60, 1999.
28. M. Ankerst, M. Ester, and H. Kriegel. Towards an effective cooperation of the computer and the user for classification. In *SIGKDD Int. Conf. On Knowledge Discovery & Data Mining (KDD 2000), Boston, MA*, pages 179–188, 2000.
29. M. Ankerst, D. A. Keim, and H.-P. Kriegel. Circle Segments: A Technique for Visually Exploring Large Multidimensional Data Sets. In *Proc. Visualization 96, Hot Topic Session, San Francisco, CA*, 1996.
30. V. Anupam, S. Dar, T. Leibfried, and E. Petajan. Dataspace: 3D visualization of large databases. In *Proc. Int. Symp. on Information Visualization, Atlanta, GA*, pages 82–88, 1995.
31. M. A. Arbib. *The Handbook of Brain Theory and Neural Networks.* MIT Press, Cambridge, MA, 1998.
32. V. I. Arnold. *Mathematical Methods of Classical Mechanics.* Springer, 2nd edition, 1989.
33. D. Asimov. The grand tour: A tool for viewing multidimensional data. *SIAM Journal of Science & Stat. Comp.*, 6:128–143, 1985.
34. K. Atkinson. *An introduction to numerical analysis.* Wiley, New York, 2nd edition, 1989.
35. R. Azencott. *Simulated Annealing.* Wiley, New York, 1992.
36. F. R. Bach and M. I. Jordan. Kernel independent component analysis. *Journal of Machine Learning Research*, 3:1–48, 2002.
37. T. Bäck. *Evolutionary Algorithms in Theory and Practice.* University Press, Oxford, 1996.
38. T. Bäck and H.-P. Schwefel. An overview of evolutionary algorithms for parameter optimization. *Evolutionary Computation*, 1(1):1–23, 1993. MIT Press.

39. P. Baldi and S. Brunak. *Bioinformatics: The Machine Learning Approach.* The MIT Press, 1998.
40. S. Baluja and R. Caruana. Removing the genetics from the standard genetic algorithm. In A. Prieditis and S. Russel, editors, *International Conference on Machine Learning (ML-95)*, pages 38–46, San Mateo, CA, 1995. Morgan Kaufmann.
41. W. Banzhaf, J. Daida, and A. E. Eiben, editors. *GECCO'99: Proceedings of the Genetic and Evolutionary Computation Conference.* Morgan Kaufmann, San Francisco, 1999.
42. W. Banzhaf, P. Nordin, R. Keller, and F. D. Francone. *Genetic Programming – An Introduction.* Morgan Kaufmann, San Francisco, CA, and dpunkt, Heidelberg, 1997.
43. H. B. Barlow. Unsupervised learning. *Neural Computation*, 1:151–160, 1989.
44. V. Barnet and T. Lewis. *Outliers in Statistical Data.* Wiley, 1994.
45. P. Barton and C. Pantelides. Modeling of combined discrete/continuous processes. *AIChE J.*, 40:966–979, 1994.
46. T. Bass. *The Eudaemonic Pie.* Penguin, New York, 1992.
47. G. D. Battista, P. Eades, R. Tamassia, and I. G. Tollis. *Graph Drawing.* Prentice Hall, 1999.
48. E. B. Baum and D. Haussler. What size net gives valid generalization? *Neural Computation*, 1:295–311, 1989.
49. T. Bäck, D. B. Fogel, and Z. Michalewicz, editors. *Handbook of Evolutionary Computation.* Oxford University Press, Oxford, 1997.
50. R. Becker, J. M. Chambers, and A. R. Wilks. *The New S Language.* Wadsworth & Brooks/Cole Advanced Books and Software, Pacific Grove, CA, 1988.
51. R. A. Becker, W. S. Cleveland, and M.-J. Shyu. The visual design and control of trellis display. *Journal of Computational and Graphical Statistics*, 5(2):123–155, 1996.
52. B. Bederson. PAD++: Advances in multiscale interfaces. In *Proc. Human Factors in Computing Systems CHI '94 Conf., Boston, MA*, page 315, 1994.
53. B. B. Bederson and J. D. Hollan. PAD++: A zooming graphical interface for exploring alternate interface physics. In *Proc. UIST*, pages 17–26, 1994.
54. R. K. Belew and L. B. Booker, editors. *Proceedings of the Fourth International Conference on Genetic Algorithms*, San Francisco, CA, 1991. Morgan Kaufmann.
55. R. Bellazzi, C. Larizza, and A. Riva. Interpreting longitudinal data through temporal abstractions: An application to diabetic patients monitoring. In *[355]*, pages 287–298, 1997.
56. J. Berger. *Statistical Decision Theory and Bayesian Analysis.* Springer-Verlag: New York, 2nd edition, 1985.
57. J. Bernardo and A. Smith. *Bayesian Theory.* Wiley, New York, 1994.
58. J. L. Bernier, C. I. Herráiz, J. J. Merel, S. Olmeda, and A. Prieto. Solving mastermind using gas and simulated annealing: A case of dynamic constraint optimization. In *[529]*, pages 554–563, 1996.
59. D. Berry. *Statistics: a Bayesian perspective.* Wadsworth, Belmont (CA), 1996.
60. M. R. Berthold, P. Cohen, and X. Liu. Intelligent data analysis: Reasoning about data. *AI Magazine*, 19(3):131–134, 1999.
61. M. R. Berthold and J. Diamond. Constructive training of probabilistic neural networks. *Neurocomputing*, 19:167–183, 1998.
62. M. R. Berthold and K.-P. Huber. Constructing fuzzy graphs from examples. *Intelligent Data Analysis*, 3(1), 1999. (http://www.elsevier.nl/locate/ida).
63. J. C. Bezdek, R. Ehrlich, and W. Full. FCM: the fuzzy c-means clustering algorithm. *Computer & Geosciences*, 10(2-3):191–203, 1984.

64. I. Bhandari, E. Colet, J. Parker, Z. Pines, and R. Pratap. Advanced scout: Data mining and knowledge discovery in NBA data. *Data Mining and Knowledge Discovery*, 1:121–125, 1997.
65. M. A. Bhatti. *Practical Optimization Methods With Mathematica Applications.* Springer-TELOS, New York, 1998.
66. M. Bianchini, P. Frasconi, and M. Gori. Learning in multilayered networks used as autoassociators. *IEEE Transactions on Neural Networks*, 6(2):512–515, 1995.
67. E. A. Bier, M. C. Stone, K. Pier, W. Buxton, and T. DeRose. Toolglass and magic lenses: The see-through interface. In *Proc. SIGGRAPH '93, Anaheim, CA*, pages 73–80, 1993.
68. C. Bierwirth, D. C. Mattfeld, and H. Kopfer. On permutation representations for scheduling problems. In *[529]*, pages 310–318, 1996.
69. C. M. Bishop. *Neural networks for pattern recognition.* Oxford University Press, New York, 1995.
70. N. Blachman. The amount information that y gives about x. *IEEE Transactions on Information Theory*, 14:27–31, 1968.
71. T. Blickle and L. Thiele. A comparison of selection schemes used in genetic algorithms. TIK Report 11, ETH Technical University of Zurich, Switzerland, 1995.
72. H. Blockeel and L. De Raedt. Top-down induction of first order logical decision trees. *Artificial Intelligence*, 101(1-2):285–297, 1998.
73. H. H. Bock. *Automatic Classification.* Vandenhoeck and Ruprecht, Göttingen, 1974.
74. K. A. Bollen. *Structural Equations with Latent Variables.* Wiley, New York, 1989.
75. L. B. Booker, D. E. Goldberg, and J. H. Holland. Classifier systems and genetic algorithms. In J. G. Carbonell, editor, *Machine Learning: Paradigms and Methods*, pages 235–282. The MIT Press / Elsevier, 1989.
76. B. E. Boser, I. M. Guyon, and V. N. Vapnik. A training algorithm for optimal margin classifiers. In D. Haussler, editor, *5th Annual ACM Workshop on COLT*, pages 144–152, Pittsburgh, PA, 1992. ACM Press.
77. H. Bourlard and Y. Kamp. Autoassociation by multilayer perceptrons and singular value decomposition. *Biological Cybernetics*, 59:291 – 294, 1988.
78. A. Bowers, C. Giraud-Carrier, and J. Lloyd. Classification of individuals with complex structure. In *Proceedings of the 17th International Conference on Machine Learning*, pages 81–88. Morgan Kaufmann, 2000.
79. G. Box and D. Cox. An analysis of transformations. *J. R. Statist. Soc. B*, 26:211–252, 1964.
80. G. E. Box. Science and statistics. *Journal of the American Statistical Association*, 71:791–799, 1976.
81. G. E. Box and W. Hunter. The experimental study of physical mechanisms. *Technometrics*, 7:57–71, 1965.
82. G. E. Box and G. C. Tiao. *Bayesian Inference in Statistical Analysis.* Wiley, New York, 1973.
83. G. E. P. Box and F. M. Jenkins. *Time Series Analysis: Forecasting and Control.* Holden Day, 2nd edition, 1976.
84. E. Bradley and M. Easley. Reasoning about sensor data for automated system identification. In *[355]*, pages 561–572, 1997.
85. E. Bradley, M. Easley, and R. Stolle. Reasoning about nonlinear system identification. *Artificial Intelligence*, 133:139–188, December 2001.
86. E. Bradley and R. Mantilla. Recurrence plots and unstable periodic orbits. *Chaos*, 12:596–600, 2002.

87. J. L. Breeden, F. Dinkelacker, and A. Hübler. Noise in the modeling and control of dynamical systems. *Physical Review A*, 42(10):5827–5836, 1990.
88. L. Breiman. Bagging predictors. *Machine Learning*, 26(2):123–140, 1996.
89. L. Breiman. Bias, variance, and arcing classifiers. Technical Report 460, University of California, 1996.
90. L. Breiman, J. H. Friedman, R. A. Olshen, and C. J. Stone. *Classification and Regression Trees.* Wadsworth & Brooks, Monterey, 1984.
91. H. J. Bremermann. Optimization through evolution and recombination. In M. C. Yovits, G. T. Jacobi, and D. G. Goldstein, editors, *Self-organizing systems*, pages 93–106. Spartan Books, Washington D.C., 1962.
92. H. J. Bremermann, M. Rogson, and S. Salaff. Search by evolution. In M. Maxfield, A. Callahan, and L. J. Fogel, editors, *Biophysics and Cybernetic Systems - Proceedings of the 2nd Cybernetic Sciences Symposium*, pages 157–167. Spartan Books, Washington, D.C., 1965.
93. A. Brindle. *Genetic algorithms for function optimization.* PhD thesis, University of Alberta, Edmonton, Nov. 1981.
94. P. J. Brockwell and R. A. Davis. *Time Series: Theory and Methods.* Springer Verlag, 2nd edition, 1991.
95. S. H. Brooks. A discussion of random methods for seeking maxima. *Operations Research*, 6:244–251, 1958.
96. R. Brown, P. Bryant, and H. D. I. Abarbanel. Computing the Lyapunov spectrum of a dynamical system from an observed time series. *Physical Review A*, 43:2787–2806, 1991.
97. C. Brunk, J. Kelly, and R. Kohavi. Mineset: An integrated system for data mining. In *SIGKDD Int. Conf. On Knowledge Discovery & Data Mining (KDD 1997), Newport Beach, CA*, pages 135–138, 1997.
98. J. Buckley and Y. Hayashi. Fuzzy-input output controller are universal approximators. *Fuzzy Sets and Systems*, 58:273–278, 1993.
99. A. Buja, D. F. Swayne, and D. Cook. Interactive high-dimensional data visualization. *Journal of Computational and Graphical Statistics*, 5(1):78–99, 1996.
100. W. Buntine. Operations for learning with graphical models. *Journal of Artificial Intelligence Research*, 2:159–225, 1994.
101. W. Buntine. Graphical models for discovering knowledge. In *Advances in Knowledge Discovery and Data Mining*, pages 59–81. MIT Press, Cambridge, MA, 1996.
102. E. Cantu-Paz et al, editor. *GECCO 2002: Proceedings of the Genetic and Evolutionary Computation Conference.* Morgan Kaufmann Publishers, San Francisco, CA, 2002.
103. S. Card, J. Mackinlay, and B. Shneiderman. *Readings in Information Visualization.* Morgan Kaufmann, 1999.
104. M. S. T. Carpendale, D. J. Cowperthwaite, and F. D. Fracchia. IEEE computer graphics and applications, special issue on information visualization. *IEEE Journal Press*, 17(4):42–51, July 1997.
105. D. B. Carr, E. J. Wegman, and Q. Luo. Explorn: Design considerations past and present. In *Technical Report, No. 129, Center for Computational Statistics, George Mason University*, 1996.
106. M. Casdagli and S. Eubank, editors. *Nonlinear Modeling and Forecasting.* Addison Wesley, 1992.
107. A. Casotto, F. Romeo, and A. Sangiovanni-Vincentelli. A parallel simulated annealing algorithm for the placement of macro-cells. *IEEE Transactions on Computer Aided Design*, 6:838–847, 1987.

108. E. Castillo, J. Gutierrez, and A. Hadi. *Expert Systems and Probabilistic Network Models*. Springer Verlag, New York, 1997.
109. B. Cestnik. Estimating probabilities: A crucial task in machine learning. In *Proceedings of the Ninth European Conference on Artificial Intelligence*, pages 147–149, London, 1990. Pitman.
110. J. Charles. AI and law enforcement. *IEEE Intelligent Systems*, 13:71–80, 1998.
111. C. Chatfield. Model uncertainty, data mining and statistical inference (with discussion). *Journal of the Royal Statistical Society, Series A*, 158:419–466, 1995.
112. C. Chatfield and A. J. Collins. *Introduction to Multivariate Analysis*. Chapman and Hall, London, 1980.
113. C. Chen. *Information Visualisation and Virtual Environments*. Springer-Verlag, London, 1999.
114. V. Cherkassky and F. Mulier. *Learning from data*. Wiley, 1998.
115. H. Chernoff. The use of faces to represent points in k-dimensional space graphically. *Journal of the American Statistical Association*, 68:361–368, 1973.
116. Z. Chi and H. Yan. ID3-derived fuzzy rules and optimized defuzzification for handwritten numeral recognition. *IEEE Transactions on Fuzzy Systems*, 4(1):24–31, 1996.
117. P. Clark and R. Boswell. Rule induction with CN2: Some recent improvements. In *Proceedings of the Fifth European Working Session on Learning*, pages 151–163, Berlin, 1991. Springer.
118. P. Clark and T. Niblett. The CN2 induction algorithm. *Machine Learning*, 3(4):261–283, 1989.
119. W. S. Cleveland. *Visualizing Data*. AT&T Bell Laboratories, Murray Hill, NJ, Hobart Press, Summit NJ, 1993.
120. E. F. Codd. *Providing OLAP (On-Line Analytic Processing) to User-Analysts: An IT Mandate*. E. F. Codd and Associates, 1994.
121. W. Cohen. Fast effective rule induction. In *Proceedings of the 12th International Conference on Machine Learning*, pages 115–123, 1995.
122. W. W. Cohen. Fast effective rule induction. *Proc. of the Twelfth International Conference on Machine Learning*, pages 115–123, 1995.
123. D. Collett. *Modelling Binary Data*. Chapman and Hall, London, 1991.
124. D. Collett and T. Lewis. The subjective nature of outlier rejection procedures. *Applied Statistics*, 25:228–37, 1976.
125. N. Collins, R. Eglese, and B. Golden. Simulated annealing – an annotated bibliography. *American Journal of Mathematical and Management Sciences*, 8:209–307, 1988.
126. G. Cooper and E. Herskovitz. A Bayesian method for the induction of probabilistic networks from data. *Machine Learning*, 9:309–347, 1992.
127. J. Cowie and W. Lehnert. Information extraction. *Communications of the ACM*, 39:80–91, 1996.
128. D. R. Cox. Regression models and life tables (with discussion). *Journal of the Royal Statistical Society* Series B, 74:187–220, 1972.
129. D. R. Cox. Role of models in statistical analysis. *Statistical Science*, 5:169–174, 1990.
130. D. R. Cox and E. J. Snell. *Applied statistics: principles and examples*. Chapman and Hall, 1981.
131. E. Cox. A fuzzy system for detecting anomalous behaviors in healthcare provider claims. In S. Goonatilake and P. Treleaven, editors, *Intelligent Systems for Finance and Business*, pages 111–134. Wiley, 1995.

132. M. L. Cramer. A representation for the adaptive generation of simple sequential programs. In *[228]*, pages 183–187, 1985.
133. M. W. Craven and J. W. Shavlick. Extracting tree-structured representation of trained neural networks. *Advances in Neural Information Processing Systems*, 8:24–30, 1996.
134. N. Cristianini and J. Shawe-Taylor. Support vector book website. http://www.support-vector.net.
135. N. Cristianini and J. Shawe-Taylor. *An introduction to Support Vector Machines*. Cambridge University Press, Cambridge, UK, 2000.
136. C. Cruz-Neira, D. Sandin, and T. DeFanti. Surround-screen projection-based virtual reality: the design and implementation of cave. *Proc. of Siggraph'93 Computer Graphics Conference*, pages 135–142, 1993.
137. G. Cybenko. Approximation by superpositions of a sigmoidal function. *Mathematics of Control, Signals, and Systems*, 2:303 – 314, 1989.
138. D. Dagupta and Z. Michalewicz, editors. *Evolutionary Algorithms in Engineering Applications*. Springer, Berlin, 1997.
139. C. Darwin. *On the Origin of Species*. Murray, London, 1859.
140. R. N. Dave and R. Krishnapuram. Robust clustering methods: a unified view. *IEEE Transactions on Fuzzy Systems*, 5(2):270–293, 1997.
141. Y. Davidor, H.-P. Schwefel, and R. Männer, editors. *Parallel Problem Solving from Nature (PPSN III)*, Lecture Notes in Computer Science 866. Springer, Berlin, 1994.
142. L. Davis, editor. *Genetic Algorithms and Simulated Annealing*. Research Notes in Artificial Intelligence. Morgan Kaufmann, Los Altos, CA, 1987.
143. L. Davis, editor. *Handbook of Genetic Algorithms*. Van Nostrand Reinhold, New York, 1991.
144. L. Davis and M. Steenstrup. Genetic algorithms and simulated annealing: An overview. In *[142]*, pages 1–11. Morgan Kaufmann, Los Altos, CA, 1987.
145. A. Davison and D. Hinkley. *Bootstrap methods and their application*. Cambridge University Press, Cambridge, 1997.
146. P. de Graaf, G. van den Eijkel, H. Vullings, and B. de Mol. A decision-driven design of a decision support system in anesthesia. *Artificial Intelligence in Medicine*, 11:141–153, 1997.
147. L. De Raedt, editor. *Advances in Inductive Logic Programming*. IOS Press, Amsterdam, 1996.
148. L. Dehaspe and H. Toivonen. Discovery of relational association rules. In Džeroski and Lavrač [163], pages 189–212.
149. A. P. Dempster, N. M. Laird, and D. B. Rubin. Maximum likelihood from incomplete data via the EM algorithm. *Journal of the Royal Statistical Society*, 39:1–38, 1977.
150. V. Dhar and R. Stein. *Seven methods for transforming corporate data into business intelligence*. Prentice Hall, 1997.
151. D. D'Humieres, M. R. Beasley, B. Huberman, and A. Libchaber. Chaotic states and routes to chaos in the forced pendulum. *Physical Review A*, 26:3483–3496, 1982.
152. K. I. Diamantaras and S. Y. Kung. *Principal component neural networks: theory and applications*. Wiley Interscience, 1996.
153. R. Didner. Intelligent systems at american express. In S. Goonatilake and P. Treleaven, editors, *Intelligent Systems for Finance and Business*, pages 31–38. Wiley, 1995.

154. R. Diekmann, R. Lüling, and J. Simon. Problem independent distributed simulated annealing and its applications. Technical report, University of Paderborn, Germany, 1992.
155. P. J. Diggle, K.-Y. Liang, and S. L. Zeger. *Analysis of Longitudinal Data*. Number 13 in Oxford Statistical Science Series. Oxford University Press, Oxford, 1994.
156. M. Dodge and R. Kitchin. *Atlas of Cyberspace*. Addison Wesley, Aug 2001.
157. N. R. Draper and H. Smith. *Applied Regression Analysis*. John Wiley & Sons, New York, 2nd edition, 1981.
158. R. Duda and P. Hart. *Pattern classification and scene analysis*. Wiley, New York, 1973.
159. G. Dueck. New optimization heuristics. the great deluge algorithm and the record-to-record travel. *Journal of Computational Physics*, 104:86–92, 1993.
160. G. Dueck and T. Scheuer. Threshold accepting: A general purpose optimization algorithm appearing superior to simulated annealing. *Journal of Computational Physics*, 90:161–175, 1990.
161. G. Dunn, B. S. Everitt, and A. Pickles. *Modelling Covariances and Latent Variables Using EQS*. Chapman and Hall, London, 1993.
162. S. Džeroski, B. Cestnik, and I. Petrovski. Using the m-estimate in rule induction. *Journal of Computing and Information Technology*, 1:37–46, 1993.
163. S. Džeroski and N. Lavrač, editors. *Relational Data Mining*. Springer-Verlag, Berlin, 2001.
164. A. Edwards. *Likelihood*. The John Hopkins University Press, Baltimore, 1992.
165. D. Edwards. *Introduction to Graphical Modelling*. Springer-Verlag, New York, 1995.
166. B. Efron and R. Tibshirani. *An Introduction to the Bootstrap*. Chapman & Hall, New York, 1993.
167. S. G. Eick. Data visualization sliders. In *Proc. ACM UIST*, pages 119–120, 1994.
168. M. Eigen. *Steps Towards Life*. Oxford University Press, Oxford, UK, 1992.
169. B. Epstein. *Partial Differential Equations: An Introduction*. McGraw-Hill, 1962.
170. Y. Ermoliev. Random optimization and stochastic programming. In *[383]*, pages 104–115. Springer, Berlin, 1970.
171. L. J. Eshelman, R. A. Caruna, and J. D. Schaffer. Biases in the crossover landscape. In *[462]*, pages 10–19, 1989.
172. M. Evans and T. Swartz. Methods for approximating integrals in statistics with special emphasis on Bayesian integration problems. *Statistical Science*, 10:254–272, 1995.
173. J. Farmer and J. Sidorowich. Exploiting chaos to predict the future and reduce noise. In *Evolution, Learning and Cognition*. World Scientific, 1988.
174. P. M. Fayers and D. J. Hand. Factor analysis, causal indicators, and quality of life. *Quality of Life Research*, 6:139–150, 1997.
175. U. Fayyad, S. Djorgovski, and N. Weir. From digitized images to online catalogs. *AI Magazine*, 17:51–66, 1996.
176. U. Fayyad and K. Irani. On the handling of continuous-valued attributes in decision tree generation. *Machine Learning*, 8:87–102, 1992.
177. U. Fayyad, G. Piatetsky-Shapiro, P. Smyth, and R. Uthurusamy, editors. *Advances in Knowledge Discovery and Data Mining*. MIT Press, 1996.
178. S. Feiner and C. Beshers. Visualizing n-dimensional virtual worlds with n-vision. *Computer Graphics*, 24(2):37–38, 1990.
179. R. Ferguson. Chemical processes optimization utilizing neural network systems. *Proc. of SICHEM'92*, 1992.

180. R. A. Fisher. The use of multiple measurements in taxonomic problems. *Annals of Eugenics*, 7:179–188, 1936.
181. K. Fishkin and M. C. Stone. Enhanced dynamic queries via movable filters. In *Proc. Human Factors in Computing Systems CHI '95 Conf., Denver, CO*, pages 415–420, 1995.
182. M. Fitzsimons, T. Khabaza, and C. Shearer. The application of rule induction and neural networks for television audience prediction. *Proc. of the ESOMAR/EMAC/AFM Symposium on Information Based Decision Making in Marketing*, pages 69–82, 1993.
183. P. Flach. Predicate invention in inductive data engineering. In P. Brazdil, editor, *Proceedings of the 6th European Conference on Machine Learning*, volume 667 of *Lecture Notes in Artificial Intelligence*, pages 83–94. Springer-Verlag, 1993.
184. P. Flach. *Simply Logical – intelligent reasoning by example.* John Wiley, 1994.
185. P. Flach, C. Giraud-Carrier, and J. Lloyd. Strongly typed inductive concept learning. In D. Page, editor, *Proceedings of the 8th International Conference on Inductive Logic Programming*, volume 1446 of *Lecture Notes in Artificial Intelligence*, pages 185–194. Springer-Verlag, 1998.
186. P. Flach and N. Lachiche. 1BC: A first-order Bayesian classifier. In S. Džeroski and P. Flach, editors, *Proceedings of the 9th International Workshop on Inductive Logic Programming*, volume 1634 of *Lecture Notes in Artificial Intelligence*, pages 92–103. Springer-Verlag, 1999.
187. P. Flach and N. Lachiche. Confirmation-guided discovery of first-order rules with Tertius. *Machine Learning*, 42(1/2):61–95, 2001.
188. D. B. Fogel. *Evolving artificial intelligence.* PhD thesis, University of California, San Diego, Nov. 1992.
189. D. B. Fogel, editor. *Evolutionary Computation: The Fossil Record.* IEEE Press, New York, 1998.
190. D. B. Fogel and W. Atmar, editors. *Proceedings of the First Annual Conference on Evolutionary Programming*, La Jolla, CA, 1992.
191. L. J. Fogel. Evolutionary programming in perspective: The top-down view. In *[561]*, pages 135–146. IEEE Press, New York, 1994.
192. L. J. Fogel, A. J. Owens, and M. J. Walsh. *Artificial Intelligence through Simulated Evolution.* John Wiley and Sons, New York, 1966.
193. P. Frasconi, M. Gori, M. Maggini, and G. Soda. Unified integration of explicit knowledge and learning by example in recurrent networks. *IEEE Trans. on Knowledge and Data Engineering*, 7(2):340 – 346, 1995.
194. A. M. Fraser and H. L. Swinney. Independent coordinates for strange attractors from mutual information. *Physical Review A*, 33(2):1134–1140, 1986.
195. R. M. Friedberg. A learning machine: Part I. *IBM Journal of Research and Development*, 2(1):2–13, January 1958.
196. R. M. Friedberg, B. Dunham, and J. H. North. A learning machine: Part II. *IBM Journal of Research and Development*, 3(3):282–287, July 1959.
197. J. Friedman and N. Fisher. Bump hunting in high-dimensional data. *Statistics and Computing*, 9:123–143, 1999.
198. J. H. Friedman. Multivariate adaptive regression splines. *Annals of Statistics*, 19:1–141, 1991.
199. J. H. Friedman. On bias, variance, 0/1-loss, and the curse-of-dimensionality. *Data Mining and Knowledge Discovery*, 1(1):55–77, 1997.
200. J. H. Friedman, J. L. Bentley, and R. A. Finkel. An algorithm for finding best matches in logarithmic expected time. *ACM Transactions on Mathematical Software*, 3:209–226, 1977.

201. N. Friedman, D. Geiger, and M. Goldszmidt. Bayesian network classifiers. *Machine Learning*, 29:131–163, 1997.
202. L. M. Fu. Rule generation from neural networks. *IEEE Transactions on Systems, Man, and Cybernetics*, 24:1114 – 1124, 1994.
203. C. Fujiki and J. Dickinson. Using the genetic algorithm to generate lisp source code to solve the prisoner's dilemma. In *[229]*, pages 236–240, 1991.
204. G. Furnas. Generalized fisheye views. In *Proc. Human Factors in Computing Systems CHI '86 Conf., Boston, MA*, pages 18–23, 1986.
205. G. W. Furnas and A. Buja. Prosections views: Dimensional inference through sections and projections. *Journal of Computational and Graphical Statistics*, 3(4):323–353, 1994.
206. J. Gao and H. Cai. On the structures and quantification of recurrence plots. *Physical Letters A*, 270:75–87, 2000.
207. D. Geiger and D. Heckerman. A characterization of Dirichlet distributions through local and global independence. *Annals of Statistics*, 25:1344–1368, 1997.
208. A. Gelman, J. Carlin, H. Stern, and D. B. Rubin. *Bayesian Data Analysis*. Chapman & Hall, London, 1995.
209. C. F. Gerald and P. O. Wheatley. *Applied Numerical Analysis*. Addison-Wesley, 1994.
210. R. Giachetti and R. Young. A parametric representation of fuzzy numbers and their arithmetic operators. *Fuzzy Sets and Systems*, 92(2):185–202, Oct. 1997.
211. A. Gifi. *Nonlinear Multivariate Analysis*. Wiley, Chichester, 1990.
212. C. Giles and M. Gori, editors. *Adaptive processing of sequences and data structures*. Springer-Verlag, 1998.
213. W. R. Gilks, S. Richardson, and D. J. Spiegelhalter. Introducing Markov Chain Monte Carlo. In *[214]*, pages 1–19. Chapman and Hall, 1996.
214. W. R. Gilks, S. Richardson, and D. J. Spiegelhalter, editors. *Markov chain monte carlo in practice*. Chapman and Hall, London, 1996.
215. W. R. Gilks and G. Roberts. Strategies for improving MCMC. In *[214]*, pages 89–114. Chapman and Hall, 1996.
216. B. Glasgow, A. Mandell, D. Binney, l Ghemri, and D. Fisher. Mita: an information-extraction approach to the analysis of free-form text in life insurance applications. *AI Magazine*, 19:59–72, 1998.
217. F. Glover. Tabu search – Part I. *ORSA-Journal on Computing*, 1:190–206, 1989.
218. C. Glymour, D. Madigan, D. Pregibon, and P. Smyth. Statistical themes and lessons for data mining. *Data Mining and Knowledge Discovery*, 1:11–28, 1997.
219. D. Goldberg. *Genetic Algorithms in Search, Optimization, and Machine Learning*. Addison-Wesley, Reading, MA, 1989.
220. J. Goldstein and S. F. Roth. Using aggregation and dynamic queries for exploring large data sets. In *Proc. Human Factors in Computing Systems CHI '94 Conf., Boston, MA*, pages 23–29, 1994.
221. I. Good. *The Estimation of Probability: An Essay on Modern Bayesian Methods*. MIT Press, Cambridge, MA, 1968.
222. J. C. Gower and D. J. Hand. *Biplots*. Number 54 in Monographs on Statistics and Applied Probability. Chapman and Hall, London, 1996.
223. C. Grebogi, E. Ott, and J. A. Yorke. Chaos, strange attractors and fractal basin boundaries in nonlinear dynamics. *Science*, 238:632–638, 1987.
224. J. R. Green. Population-based incremental learning as a simple versatile tool for engineering optimisation. In *Proceedings of the First International Conference on Evolutionary Computation and Its Applications (EvCA ´96)*, pages 258–269, 1996.

225. M. J. Greenacre. *Theory and Applications of Correspondence Analysis.* Academic Press, London, 1984.
226. D. R. Greenberg. Parallel simulated annealing techniques. *Physica D*, 42:293–306, 1990.
227. W. Greene. *Econometric Analysis.* Macmillan, New York, 2nd edition, 1993.
228. J. J. Grefenstette, editor. *Proceedings of the 1st International Conference on Genetic Algorithms and their Applications.* Lawrence Erlbaum Associates, Hillsdale, NJ, 1985.
229. J. J. Grefenstette, editor. *Proceedings of the 2nd International Conference on Genetic Algorithms.* Lawrence Erlbaum Associates, Hillsdale, NJ, 1987.
230. B. Grosz. Collaborative systems. *AI Magazine*, 17(2):67–86, 1996.
231. F. Gruau. On using syntactic constraints with genetic programming. In *[24]*, pages 377–394. MIT Press, 1996.
232. J. Guckenheimer. Noise in chaotic systems. *Nature*, 298:358–361, 1982.
233. I. Haimowitz and Kohane. Automated trend detection with alternate temporal hypotheses. *Proc. of the 13th International Joint Conference on Artificial Intelligence*, pages 146–151, 1993.
234. J. Han, Y. Fu, W. Wang, J. Chiang, W. Gong, K. Koperski, D. Li, and Y. Lu. DBMiner: a system for mining knowledge in large relational databases. In E. Simoudis, J. Han, and U. Fayyad, editors, *Proc. of the 2nd Int. Conf. on Knowledge Discovery and Data Mining*, pages 250–255. AAAI Press, 1996.
235. D. J. Hand. *Discrimination and Classification.* John Wiley & Sons, Chicester, 1981.
236. D. J. Hand. Patterns in statistical strategy. *Artificial Intelligence and Statistics*, pages 355–387, 1986.
237. D. J. Hand. Emergent themes in statistical expert systems research. In M. Schader and W. Gaul, editors, *Knowledge, Data and Computer-Assisted Decisions*, pages 279–288. Springer, 1990.
238. D. J. Hand. Deconstructing statistical questions (with discussion). *Journal of the Royal Statistical Society, Series A*, 157:317–356, 1994.
239. D. J. Hand. Discussion contribution to [111]. *Journal of the Royal Statistical Society*, 158:448, 1995.
240. D. J. Hand. Classification and computers: shifting the focus. In A. Prat, editor, *COMPSTAT - Proceedings in Computational Statistics*, pages 77–88. Physica-Verlag, 1996.
241. D. J. Hand. *Construction and assessment of classification rules.* John Wiley, Chichester, 1997.
242. D. J. Hand. Intelligent data analysis: Issues and opportunities. In *[355]*, pages 1–14, 1997.
243. D. J. Hand. Breaking misconceptions - statistics and its relationship to mathematics. *Journal of the Royal Statistical Society*, 47:245–250, 1998.
244. D. J. Hand, H. Mannila, and P. Smyth. *Principles of Data Mining.* MIT Press, 1999.
245. G. Handley. *Nonlinear and Dynamic Programming.* Addison-Wesley, Reading, MA, 1964.
246. B.-L. Hao. Symbolic dynamics and characterization of complexity. *Physica D*, 51:161–176, 1991.
247. M. Hao, M. Hsu, U. Dayal, S. F. Wei, T. Sprenger, and T. Holenstein. Market basket analysis visualization on a spherical surface. In *Visual Data Exploration and Analysis Conference, San Jose, CA*, 2001.

248. J. A. Hartigan and M. A. Wong. Algorithm AS 136. A k-means clustering algorithm. *Applied Statistics*, 28:100–108, 1979.
249. T. J. Hastie and R. J. Tibshirani. *Generalized Additive Models.* Number 43 in Monographs on Statistics and Applied Probability. Chapman and Hall, London, 1990.
250. R. L. Haupt and S. E. Haupt. *Practical Genetic Algorithms.* John Wiley & Sons, New York, NY, 1998.
251. S. Havre, B. Hetzler, L. Nowell, and P. Whitney. Themeriver: Visualizing thematic changes in large document collections. *Transactions on Visualization and Computer Graphics*, 8(1), 2002.
252. B. Hayes-Roth. An architecture for adaptive intelligent systems. *Artificial Intelligence*, 72:329–365, 1995.
253. S. Haykin. *Neural networks, a comprehensive foundation.* IEEE Press, 1994.
254. T. D. Haynes, D. A. Schoenefeld, and R. L. Wainwright. Type inheritance in strongly typed genetic programming. In *[24]*, pages 359–375. MIT Press, 1996.
255. M. Hearst. TileBars: Visualization of term distribution information in full text information access. In *Proc. of ACM Human Factors in Computing Systems CHI '95 Conf., Denver, CO*, pages 59–66, 1995.
256. D. O. Hebb. *The Organization of Behaviour: a neuropsychological theory.* Wiley, New York, 1949.
257. R. Hecht-Neilsen. *Neurocomputing.* Addison Wesley, 1990.
258. D. Heckerman. Bayesian networks for knowledge discovery. In *Advances in Knowledge Discovery and Data Mining*, pages 153–180. MIT Press, Cambridge, MA, 1996.
259. D. Heckerman, D. Geiger, and D. Chickering. Learning Bayesian networks: The combinations of knowledge and statistical data. *Machine Learning*, 20:197–243, 1995.
260. J. Hernández-Orallo and M. Ramírez-Quintana. A strong complete schema for inductive functional logic programming. In S. Džeroski and P. Flach, editors, *Proceedings of the 9th International Workshop on Inductive Logic Programming*, volume 1634 of *Lecture Notes in Artificial Intelligence*, pages 116–127. Springer-Verlag, 1999.
261. J. Hertz, A. Krogh, and R. G. Palmer. *Introduction to the theory of neural computation.* Addison-Wesley, 1991.
262. C. Higgins and R. Goodman. Learning fuzzy rule-based neural networks for control. In *Advances in Neural Information Processing Systems*, 5, pages 350–357, California, 1993. Morgan Kaufmann.
263. A. Hinneburg, D. Keim, and M. Wawryniuk. HD-Eye: Visual Mining of High-Dimensional Data. *IEEE Computer Graphics and Applications*, 19(5):22–31, 1999.
264. M. W. Hirsch and S. Smale. *Differential Equations, Dynamical Systems, and Linear Algebra.* Academic Press, San Diego CA, 1974.
265. F. Hoffmeiser and T. Bäck. Genetic algorithms and evolution strategies: Similarities and differences. Technical Report SYS-1/92, University of Dortmund, Germany, 1992.
266. H. Hofmann, A. Siebes, and A. Wilhelm. Visualizing association rules with interactive mosaic plots. In *Proc. SIGKDD Int. Conf. On Knowledge Discovery & Data Mining (KDD 2000), Boston, MA*, 2000.
267. J. H. Holland. *Adaptation in Natural and Artificial Systems.* Univ. of Michigan Press, Ann Arbor, Michigan, 1975.
268. J. H. Holland. *Adaptation in Natural and Artificial Systems.* MIT Press, Cambridge, MA, 1992. Reprint of [267].

269. R. Holte. Very simple classification rules perform well on most commonly used data sets. *Machine Learning*, 11:63–90, 1993.
270. T. Honkela, V. Pulkki, and T. Kohonen. Contextual relations of words in Grimm tales analyzed by self-organizing map. In *Proc. ICANN95*, pages 3–7, 1995.
271. W. Horn, S. Miksch, G. Egghart, C. Popow, and F. Paky. Effective data validation of high frequency data: Time-point, time-interval and time-based methods. *Computers in Biology and Medicine*, 27:389–409, 1997.
272. C. S. Hsu. *Cell-to-Cell Mapping.* Springer-Verlag, New York, 1987.
273. P. J. Huber. *Robust Statistics.* Wiley, 1981.
274. P. J. Huber. The annals of statistics. *Projection Pursuit*, 13(2):435–474, 1985.
275. P. J. Huber. From large to huge: a statistician's reactions to KDD and DM. In D. Heckerman, H. Mannila, D. Pregibon, and R. Uthurusamy, editors, *Proc. of the Third International Conference on Knowledge Discovery and Data Mining*, pages 304–308. AAAI Press, 1997.
276. J. N. Hwang, S. R. Lay, M. Maechler, R. D. Martin, and J. Schimert. Regression modeling in back-propagation and projection pursuit learning. *IEEE Trans. on Neural Networks*, 5(3):342 – 353, 1994.
277. P. Idestam-Almquist. *Generalization of clauses.* PhD thesis, Stockholm University, Department of Computer and Systems Sciences, 1993.
278. P. Idestam-Almquist. Generalization of clauses under implication. *Journal of Artificial Intelligence Research*, 3:467–489, 1995.
279. D. T. Inc. Dbminer. *http://www.dbminer.com*, 2001.
280. A. Inselberg and B. Dimsdale. Parallel coordinates: A tool for visualizing multidimensional geometry. In *Proc. Visualization 90, San Francisco, CA*, pages 361–370, 1990.
281. International Human Genome Sequencing Consortium. Initial sequencing and analysis of the human genome. *Nature*, pages 860–921, 2001.
282. H. Ishibuchi, K. Nozaki, N. Yamamoto, and H. Tanaka. Selecting fuzzy if-then rules for classification problems using genetic algorithms. *IEEE Transactions on Fuzzy Systems*, 3(3):260–270, 1995.
283. J. Iwanski and E. Bradley. Recurrence plots of experimental data: To embed or not to embed? *Chaos*, 8(4):861–871, 1998.
284. C. Jacob. Evolution programs evolved. In *[529]*, pages 42–51, 1996.
285. C. Jacob. Evolving evolution programs: Genetic programming and L-systems. In *[320]*, pages 107–115, 1996.
286. C. Jacob. *Principia Evolvica — Simulierte Evolution mit Mathematica.* dpunkt, Heidelberg, 1997.
287. C. Jacob. *Illustrating Evolutionary Computation with Mathematica.* Morgan Kaufmann, San Francisco, CA, 2001.
288. C. Janikow. Fuzzy decision trees: Issues and methods. *IEEE Transactions on Systems, Man, and Cybernetics - Part B: Cybernetics*, 28(1):1–14, 1998.
289. E. Jaynes. Information theory and statistical mechanics. *Phys. Rev.*, 106:620–630, 1957.
290. E. Jaynes. Prior probabilities. *IEEE Transactions on Systems, Science and Cybernetics*, SSC-4:227–241, 1968.
291. H. Jeffreys. *Theory of Probability.* Oxford University Press, Oxford, 3rd edition, 1961.
292. T. Joachims, D. Freitag, and T. Mitchell. Webwatcher: A tour guide for the world wide web. *Proc. of the 15th International Joint Conference on Artificial Intelligence*, pages 770–775, 1997.

293. T. Jochem and D. Pomerleau. Life in the fast lane: The evolution of an adaptive vehicle control system. *AI Magazine*, 17:11–50, 1996.
294. B. Johnson and B. Shneiderman. Treemaps: A space-filling approach to the visualization of hierarchical information. In *Proc. Visualization '91 Conf., San Diego, CA*, pages 284–291, 1991.
295. K. D. Jong. *An analysis of the behavior of a class of genetic adaptive systems.* PhD thesis, Department of Computer and Communication Sciences, University of Michigan, Ann Arbor, Nov. 1975.
296. K. D. Jong. On using genetic algorithms to search program spaces. In *[229]*, pages 210–216, 1987.
297. J.-N. Juang. *Applied System Identification.* Prentice Hall, Englewood Cliffs, N.J., 1994.
298. J. A. Kaandorp. *Fractal Modelling Growth and Form in Biology.* Springer, Berlin, 1994.
299. D. A. Keim. Designing pixel-oriented visualization techniques: Theory and applications. *Transactions on Visualization and Computer Graphics*, 6(1):59–78, Jan–Mar 2000.
300. D. A. Keim. Visual exploration of large databases. *Communications of the ACM*, 44(8):38–44, 2001.
301. D. A. Keim. Information visualization and visual data mining. *Transactions on Visualization and Computer Graphics*, 8(1):1–8, 2002.
302. D. A. Keim and H.-P. Kriegel. VisDB: Database exploration using multidimensional visualization. *Computer Graphics & Applications*, 6:40–49, Sept. 1994.
303. D. A. Keim, H.-P. Kriegel, and M. Ankerst. Recursive Pattern: A technique for visualizing very large amounts of data. In *Proc. Visualization 95, Atlanta, GA*, pages 279–286, 1995.
304. K. Kennedy, C. Bender, J. Connolly, J. Hennessy, M. Vernon, and L. Smarr. A nationwide parallel computing environment. *Communications of the ACM*, 40:63–72, 1997.
305. M. B. Kennel, R. Brown, and H. D. I. Abarbanel. Determining minimum embedding dimension using a geometrical construction. *Physical Review A*, 45:3403–3411, 1992.
306. K. Kinnear, editor. *Advances in Genetic Programming.* MIT Press, Cambridge, MA, 1994.
307. S. Kirkpatrick, C. G. Jr., and M. Vecchi. Optimization by simulated annealing. *Science*, 220:671–680, 1983.
308. M. Kirsten, S. Wrobel, and T. Horvath. Distance based approaches to relational learning and clustering. In Džeroski and Lavrač [163], pages 213–232.
309. G. Klir and B. Yuan, editors. *Fuzzy Sets, Fuzzy Logic, and Fuzzy Systems : selected papers by Lotfi A. Zadeh.* World Scientific, 1996.
310. E. Knorr and R. Ng. A unified notion of outliers: Properties and computation. In D. Heckerman, H. Mannila, D. Pregibon, and R. Uthurusamy, editors, *Proc of the 3rd Int. Conf. on Knowledge Discovery and Data Mining*, pages 219–22. AAAI Press, 1997.
311. T. Kohonen. *Self-Organizing Maps.* Springer-Verlag, Berlin, Heidelberg, 2001.
312. I. Kononenko. Combining decisions of multiple rules. In B. du Boulay and V. Sgurev, editors, *Artificial Intelligence V; Methodology, Systems, Applications.* Elsevier, Amsterdam, 1992.
313. J. R. Koza. Hierarchical genetic algorithms operating on populations of computer programs. In N. S. Sridharan, editor, *Proceedings of the 11th International Joint*

Conference on Artificial Intelligence, pages 768–774. Morgan Kaufmann Publishers, San Mateo, CA, 1989.

314. J. R. Koza. Genetic programming: A paradigm for genetically breeding populations of computer programs to solve problems. Technical Report STAN-CS-90-1314, Department of Computer Science, Stanford University, Stanford, CA, 1990.
315. J. R. Koza. *Genetic Programming - On the Programming of Computers by Means of Natural Selection.* MIT Press, Cambridge, MA, 1992.
316. J. R. Koza. *Genetic Programming II — Automatic Discovery of Reusable Programs.* MIT Press, Cambridge, MA, 1994.
317. J. R. Koza, D. Andre, F. H. Bennett, and M. A. Keane. *Genetic Programming III — Automatic Programming and Automatic Circuit Synthesis.* Morgan Kaufmann, San Francisco, CA, 1998.
318. J. R. Koza, W. Banzhaf, K. Chellapilla, K. Deb, M. Dorigo, D. B. Fogel, M. H. Garzon, D. E. Goldberg, H. Iba, and R. Riolo, editors. *Genetic Programming 1998: Proceedings of the Third Annual Conference.* Morgan Kaufmann, San Francisco, CA, 1998.
319. J. R. Koza, K. Deb, M. Dorigo, D. B. Fogel, M. Garzon, H. Iba, and R. Riolo, editors. *Genetic Programming 1997: Proceedings of the Second Annual Conference.* Morgan Kaufmann, San Francisco, CA, July 13-16 1997.
320. J. R. Koza, D. E. Goldberg, D. B. Fogel, and R. Riolo, editors. *Genetic Programming 1996: Proceedings of the First Annual Conference.* MIT Press, Cambridge, MA, July 28-31 1996.
321. R. Krishnapuram and J. Keller. A possibilistic approach to clustering. *IEEE Transactions on Fuzzy Systems*, 1(2):98–110, May 1993.
322. A. Krone and H. Kiendl. An evolutionary concept for generating relevant fuzzy rules from data. *International Journal of Knowledge-Based Intelligent Engineering Systems*, 1(4):207–213, Oct. 1997.
323. J. Kruskal and M. Wish. *Multidimensional Scaling.* Beverly Hills, CA, Sage Publications, 1978.
324. J. B. Kruskal. Multidimensional scaling by optimising goodness-of-fit to a nonmetric hypothesis. *Psychometrika*, 29:1–27, 1964.
325. J. B. Kruskal. Nonmetric multidimensional scaling: A numerical method. *Psychometrika*, 29:115–129, 1964.
326. W. J. Krzanowski. *Principles of Multivariate Analysis: A User's Perspective.* Number 3 in Oxford Statistical Science Series. Oxford University Press, Oxford, 1988.
327. S. Kundu and J. Chen. Fuzzy logic or Lukasiewicz logic: A clarification. *Fuzzy Sets and Systems*, 95:369–379, 1998.
328. K. Lagus, I. Karanta, and J. Ylä-Jääski. Paginating the generalized newspapers - a comparison of simulated annealing and a heuristic method. In *[529]*, pages 594–603, 1996.
329. P. Lai and C. Fyfe. Kernel and nonlinear canonical correlation analysis. *International Journal of Neural Systems*, 10(5):365–377, 2001.
330. J. Lamping, R. Rao, and P. Pirolli. A focus + context technique based on hyperbolic geometry for visualizing large hierarchies. In *Proc. Human Factors in Computing Systems CHI '95 Conf., Denver, CO*, pages 401–408, 1995.
331. W. Langdon, L. Spector, U.-M. O'Reilly, and P. Angeline, editors. *Advances in Genetic Programming 3.* MIT Press, Cambridge, MA, 1998.

332. W. B. Langdon. *Genetic Programming and Data Structures: Genetic Programming + Data Structures = Automatic Programming!* Kluwer Academic Publishers, Boston, 1998.
333. W. B. Langdon and R. Poli. *Foundation of Genetic Programming.* Springer Verlag, Berlin, 2002.
334. C. Larizza, A. Moglia, and A. Riva. M-HTP: A system for monitoring heart transplant patients. *Artificial Intelligence in Medicine*, 4:111–126, 1992.
335. J. Larsson and B. Hayes-Roth. Guardian: An intelligent autonomous agent for medical monitoring and diagnosis. *IEEE Intelligent Systems*, 13:58–64, 1998.
336. S. L. Lauritzen. Propagation of probabilities, means and variances in mixed graphical association models. *Journal of the American Statistical Association*, 87(420):1098–108, 1992.
337. S. L. Lauritzen. *Graphical Models.* Clarendon Press, Oxford, 1996.
338. P. S. Laursen. Problem-independent parallel simulated annealing using selection and migration. In *[141]*, pages 408–417, 1994.
339. N. Lavrač and S. Džeroski. *Inductive Logic Programming: Techniques and Applications.* Ellis Horwood, Chichester, 1994.
340. N. Lavrač, S. Džeroski, V. Pirnat, and V. Križman. The utility of background knowledge in learning medical diagnostic rules. *Applied Artificial Intelligence*, 7:273–293, 1993.
341. N. Lavrač, S. Džeroski, and M. Grobelnik. Learning nonrecursive definitions of relations with LINUS. In Y. Kodratoff, editor, *Proceedings of the Fifth European Working Session on Learning*, volume 482 of *Lecture Notes in Artificial Intelligence*, pages 265–281. Springer-Verlag, 1991.
342. N. Lavrač and P. Flach. An extended transformation approach to inductive logic programming. *ACM Transactions on Computational Logic*, 2(4):458–494, 2001.
343. N. Lavrač, P. Flach, B. Kavšek, and L. Todorovski. Adapting classification rule learning to subgroup discovery. In *Proceedings of the IEEE International Conference on Data Mining.* IEEE Press, 2002.
344. D. Lawrence and L. Pao. (http://ece-www.colorado.edu/~pao/research.html).
345. E. E. Leamer. *Specification searches: ad hoc inference with nonexperimental data.* Wiley, 1978.
346. J. LeBlanc, M. O. Ward, and N. Wittels. Exploring n-dimensional databases. In *Proc. Visualization '90, San Francisco, CA*, pages 230–239, 1990.
347. Y. Lecun, L. Bottou, Y. Bengio, and P. Haffner. Gradient-based learning applied to document recognition. *Proc. of the IEEE*, 86(11):2278–2324, 1998.
348. Y. Leung and M. Apperley. A review and taxonomy of distortion-oriented presentation techniques. In *Proc. Human Factors in Computing Systems CHI '94 Conf., Boston, MA*, pages 126–160, 1994.
349. H. Levkowitz. Color icons: Merging color and texture perception for integrated visualization of multiple parameters. In *Proc. Visualization 91, San Diego, CA*, pages 22–25, 1991.
350. T.-Y. Li and J. A. Yorke. Period three implies chaos. *American Mathematical Monthly*, 82:985–992, 1975.
351. D. Lind and B. Marcus. *An Introduction to Symbolic Dynamics and Coding.* Cambridge University Press, 1995.
352. X. Liu. Intelligent data analysis: issues and challenges. *The Knowledge Engineering Review*, 11:365–371, 1996.
353. X. Liu, G. Cheng, and J. Wu. Noise and uncertainty management in intelligent data modeling. *Proc. of the 12th National Conference on Artificial Intelligence (AAAI-94)*, pages 263–268, 1994.

354. X. Liu, G. Cheng, and J. Wu. AI for public health: Self-screening for eye diseases. *IEEE Intelligent Systems*, 13(5):28–35, 1998.
355. X. Liu, P. Cohen, and M. Berthold, editors. *Advances in Intelligent Data Analysis: Reasoning about Data*, volume 1280 of *Lecture Notes in Computer Science*, Berlin Heidelberg, 1997. Springer-Verlag.
356. L. Ljung. *System Identification: Theory for the user.* Prentice Hall, Englewood Cliffs, NJ, 1987.
357. E. N. Lorenz. Deterministic nonperiodic flow. *Journal of the Atmospheric Sciences*, 20:130–141, 1963.
358. J. Lukasiewicz. *Selected Works - Studies in Logic and the Foundations of Mathematics.* North-Holland, Amsterdam, 1970.
359. P. Lyman and H. Varian. How much information. *http://www.sims.berkeley.edu/how-much-info*, 2000.
360. R. S. MacKay and J. D. Meiss, editors. *Hamiltonian Dynamical Systems.* Adam Hilger, 1987.
361. J. D. Mackinlay, G. G. Robertson, and S. K. Card. The perspective wall: Detail and context smoothly integrated. In *Proc. Human Factors in Computing Systems CHI '91 Conf., New Orleans, LA*, pages 173–179, 1991.
362. J. MacQueen. Some methods for classification and analysis of multivariate data. In *5th Berkeley Symposium*, volume 1, pages 281–297, 1967.
363. E. Mamdani and S. Assilian. An experiment in linguistic synthesis with a fuzzy logic controller. *International Journal of Man-Machine Studies*, 7(1):1–13, 1975.
364. K. F. Man, K. S. Tang, and S. Kwong. *Genetic Algorithms.* Springer, London, 1999.
365. M. Mareš. *Computation over Fuzzy Quantities.* CRC Press, Boca Raton, 1994.
366. T. Masters. *Practical Neural networks Recipes in C++.* Academic Press, 1993.
367. N. Matic, I. Guyon, L. Bottou, J. Denker, and V. Vapnik. Computer aided cleaning of large databases for character recognition. *Proc. of the 11th Int. Conf. on Pattern Recognition*, pages 330–333, 1992.
368. P. McCullagh and J. A. Nelder. *Generalized Linear Models.* Number 37 in Monographs on Statistics and Applied Probability. Chapman and Hall, London, 2nd edition, 1989.
369. W. S. McCulloch and W. Pitts. A logical calculus of the ideas immanent in nervous activity. *Bulletin of Mathematical Biophysics*, 5:115–133, 1943.
370. G. McLachlan. *Discriminant Analysis and Statistical Pattern Recognition.* John Wiley, New York, 1992.
371. P. Mehra, L. Rendell, and B. Wah. Principled constructive induction. In *Proceedings of the 11th International Joint Conference on Artificial Intelligence*, pages 651–656. Morgan Kaufmann, 1989.
372. M. Mehta, R. Agrawal, and J. Rissanen. SLIQ: A fast scalable classifier for data mining. In *Conf. on Extending Database Technology (EDBT), Avignon, France*, 1996.
373. W. Meier, R. Weber, and H.-J. Zimmermann. Fuzzy data analysis - methods and industrial applications. *Fuzzy Sets and Systems*, 61:19–28, 1994.
374. N. Metropolis, A. Rosenbluth, M. Rosenbluth, A. Teller, and E. Teller. Equation of state calculations by fast computing machines. *Journal of Chemical Physics*, 21:1087–1092, 1953.
375. Z. Michalewicz. *Genetic Algorithms + Data Structures = Evolution Programs.* Springer, New York, 1992.
376. R. Michalski, J. Carbonell, and T. Mitchell, editors. *Machine Learning: An Artificial Intelligence Approach, Volume I.* Tioga, Palo Alto, CA, 1983.

377. D. Michie, S. Muggleton, D. Page, and A. Srinivasan. To the international computing community: A new East-West challenge. Technical report, Oxford University Computing laboratory, Oxford,UK, 1994.
378. D. Michie, D. Spiegelhalter, and C. Taylor, editors. *Machine Learning, Neural and Statistical Classification.* Ellis Horwood, Chichester, 1994.
379. M. Minsky and S. Papert. *Perceptrons.* MIT Press, Cambridge, MA, 1969.
380. M. Mitchell. *An Introduction to Genetic Algorithms.* MIT Press, Cambridge, MA, 1996.
381. T. Mitchell. Does machine learning really work? *AI Magazine*, 18(3):11–20, 1997.
382. T. Mitchell. *Machine Learning.* McGraw-Hill, Singapore, 1997.
383. N. N. Moiseev, editor. *Colloquium on methods of optimization.* Springer, Berlin, 1970.
384. D. J. Montana. Strongly typed genetic programming. *Evolutionary Computation*, 3(2):199–230, 1995.
385. J. Moody and C. J. Darken. Fast learning in networks of locally tuned processing units. *Neural computation*, 1(2):281–294, 1989.
386. D. Moore. Bayes for beginners: some reasons to hesitate. *The American Statistician*, 51(3):254–261, 1997.
387. D. Moore. *The Basic Practice of Statistics.* W. H. Freeman, 2000.
388. R. Moore. *Methods and applications of interval analysis.* Siam, Philadelphia, 1979.
389. N. Morgan and H. Bourlard. Neural networks for statistical recognition of continuous speech. *Proc. of IEEE*, 83(5):742–770, 1995.
390. N. Mori, H. Kita, and Y. Nishikawa. Adaptation to a changing environment by means of the thermodynamical genetic algorithm. In *[529]*, pages 513–522, 1996.
391. A. S. Morrison, M. M. Black, C. R. Lowe, B. MacMahon, and S. Yuasa. Some international differences in histology and survival in breast cancer. *International Journal of Cancer*, 11:261–267, 1973.
392. S. Mott. Inside dealing detection at the toronto stock exchange. In S. Goonatilake and P. Treleaven, editors, *Intelligent Systems for Finance and Business*, pages 135–144. Wiley, 1995.
393. S. Muggleton, editor. *Inductive Logic Programming.* Academic Press, London, 1992.
394. S. Muggleton and C. Feng. Efficient induction of logic programs. In S. Muggleton, editor, *Inductive Logic Programming*, pages 281–298. Academic Press, 1992.
395. H. Mühlenbein and D. Schlierkamp-Voosen. The science of breeding and its application to the breeder genetic algorithm (BGA). *Evolutionary Computation*, 1(4):335–360, 1993.
396. H. Mühlenbein and D. Schlierkamp-Voosen. Theory and application of the breeder genetic algorithm. In *[561]*, pages 182–193. IEEE Press, New York, 1994.
397. T. Munzner and P. Burchard. Visualizing the structure of the world wide web in 3D hyperbolic space. In *Proc. VRML '95 Symp, San Diego, CA*, pages 33–38, 1995.
398. G. Nakhaeizadeh. What Daimler-Benz has learned as an industrial partner from the machine learning project StatLog? In *Proceedings of Workshop on Applying Machine Learning in Practice*, pages 22–26, 1995.
399. G. Nakhaeizadeh and C. Taylor. *Machine Learning and Statistics; the interface.* John Wiley & Sons Inc., 1997.
400. K. Narendra and S. Mukhopadhyay. Adaptive control using neural networks and approximate models. *IEEE Trans. on Neural Networks*, 8(3):475–485, 1997.

401. D. Nauck, F. Klawonn, and R. Kruse. *Foundations of Neuro-Fuzzy Systems.* John Wiley, New York, 1997.
402. N. J. Nilsson. *Learning Machines.* McGraw-Hill, 1965.
403. V. Nissen. *Einführung in Evolutionäre Algorithmen.* Vieweg, Braunschweig, 1997.
404. A. D. Nola and A. Ventre, editors. *The Mathematics of Fuzzy Systems.* TÜV Rheinland, Köln, 1986.
405. V. Novák. Fuzzy control from the logical point of view. *Fuzzy Sets and Systems*, 66:159–173, 1994.
406. K. Nozaki, H. Ishibuchi, and H. Tanaka. Adaptive fuzzy rule-based classification systems. *IEEE Transactions on Fuzzy Systems*, 4(3):238–250, 1996.
407. A. O'Hagan. *Bayesian Inference.* Kendall's Advanced Theory of Statistics. Arnold, London, 1994.
408. R. W. Oldford and S. C. Peters. Implementation and study of statistical strategy. *Artificial Intelligence and Statistics*, pages 335–353, 1986.
409. R. O'Neill. Algorithm AS 47 – function minimization using a simplex procedure. *Applied Statistics*, 20:338–345, 1971.
410. A. V. Oppenheim and R. W. Schafer. *Discrete-Time Signal Processing.* Prentice Hall, 1989.
411. U.-M. O'Reilly and F. Oppacher. Programm search with a hierarchical variable length representation: Genetic programming, simulated annealing and hill climbing. In *[141]*, pages 397–406, 1991.
412. J. Ostriker and M. Norman. Cosmology of the early universe viewed through the new infrastructure. *Communications of the ACM*, 40:84–94, 1997.
413. N. Packard, J. Crutchfield, J. Farmer, and R. Shaw. Geometry from a time series. *Physical Review Letters*, 45:712, 1980.
414. T. S. Parker and L. O. Chua. *Practical Numerical Algorithms for Chaotic Systems.* Springer-Verlag, New York, 1989.
415. J. Pearl. *Probabilistic Reasoning in Intelligent Systems: Networks of plausible inference.* Morgan Kaufmann, San Mateo, CA, 1988.
416. B. A. Pearlmutter. Learning state space trajectories in recurrent neural networks. *Neural Computation*, 1:263–269, 1989.
417. M. Perkowitz and O. Etzioni. Adaptive web sites: an ai challenge. *Proc. of the 15th International Joint Conference on Artificial Intelligence*, pages 16–21, 1997.
418. K. Perlin and D. Fox. PAD: An alternative approach to the computer interface. In *Proc. SIGGRAPH, Anaheim, CA*, pages 57–64, 1993.
419. C. M. Perou et al. Molecular portraits of human breast tumours. *Nature*, 406:747–752, 2000.
420. G. Piatetsky-Shapiro and W. J. Frawley. *Knowledge Discovery in Databases.* AAAI Press / The MIT Press, 1991.
421. R. M. Pickett and G. G. Grinstein. Iconographic displays for visualizing multidimensional data. In *Proc. IEEE Conf. on Systems, Man and Cybernetics, IEEE Press, Piscataway, NJ*, pages 514–519, 1988.
422. F. J. Pineda. Recurrent back propagation and the dynamical approach to adaptive neural computation. *Neural Computation*, 1:161–172, 1989.
423. F. J. Pineda and J. C. Sommerer. Estimating generalized dimensions and choosing time delays: A fast algorithm. In *Time Series Prediction: Forecasting the Future and Understanding the Past.* Santa Fe Institute Studies in the Sciences of Complexity, Santa Fe, NM, 1993.
424. J. B. Pollack. Connectionism: past, present, and future. *Artificial Intelligence Review*, 3:3–20, 1989.

425. V. W. Porto, N. Saravanan, D. Waagen, and A. E. Eiben, editors. *Proceedings of the 7th Conference on Evolutionary Programming.* Springer, Berlin, 1998.
426. M. J. D. Powell. *Radial Basis Functions for multivariable interpretation: a review,* pages 143–167. Oxford Clarendon Press, 1987.
427. D. Pregibon. A DIY guide to statistical strategy. *Artificial Intelligence and Statistics,* pages 389–399, 1986.
428. B. Price, R. Baecker, and I. Small. A principled taxonomy of sofware visualization. *Journal of Visual Languages and Computing,* 4(3):211–266, 1993.
429. F. Provost and T. Fawcett. Robust classification for imprecise environments. *Machine Learning,* 42(3):203–231, 2001.
430. F. Provost and V. Kolluri. Scaling up inductive algorithms: An overview. In D. Heckerman, H. Mannila, D. Pregibon, and R. Uthurusamy, editors, *Proc. of the 3rd Int. Conf. on Knowledge Discovery and Data Mining,* pages 239–242. AAAI Press, 1997.
431. S. P. Pullen and B. W. Parkinson. System design under uncertainty: Evolutionary optimization of the gravity probe-b spacecraft. In *[141],* pages 598–607, 1994.
432. J. Quinlan. Discovering rules by induction from large collections of examples. In D. Michie, editor, *Expert Systems in the micro-electronic age.* Edinburgh University Press, 1979.
433. J. Quinlan. Induction of decision trees. *Machine Learning,* 1:81–106, 1986.
434. J. Quinlan. Learning logical definitions from relations. *Machine Learning,* 5:239–266, 1990.
435. J. R. Quinlan. *C4.5: Programs For Machine Learning.* Morgan Kaufmann, Los Altos, CA, 1993.
436. S. Raman and L. Patnaik. Optimization via evolutionary processes. In M. V. Zelkowitz, editor, *Advances in Computers,* volume 45, pages 156–196. Academic Press, San Diego, 1997.
437. M. Ramoni and P. Sebastiani. Learning Bayesian networks from incomplete databases. In *Proceedings of the Thirteen Conference on Uncertainty in Artificial Intelligence,* pages 401–408, San Mateo, CA, 1997. Morgan Kaufman.
438. M. Ramoni and P. Sebastiani. The use of exogenous knowledge to learn bayesian networks from incomplete databases. In *[355],* pages 539–548, 1997.
439. M. Ramoni and P. Sebastiani. Parameter estimation in Bayesian networks from incomplete databases. *Intelligent Data Analysis,* 2(1), 1998. (http://www.elsevier.nl/locate/ida).
440. R. Rao and S. K. Card. The table lens: Merging graphical and symbolic representation in an interactive focus+context visualization for tabular information. In *Proc. Human Factors in Computing Systems CHI '94 Conf., Boston, MA,* pages 318–322, 1994.
441. G. Rawlins, editor. *Foundations of Genetic Algorithms.* Morgan Kaufmann, San Mateo, CA, 1991.
442. I. Rechenberg. Cybernetic solution path of an experimental problem. Technical report, Royal Aircraft Establ., libr. transl. 1122, Hants, Farnborough, U.K., 1965.
443. I. Rechenberg. *Evolutionsstrategie: Optimierung technischer Systeme nach Prinzipien biologischer Evolution.* Frommann-Holzboog, Stuttgart, 1973.
444. I. Rechenberg. Evolution strategy: Nature's way of optimization. In *Optimization: Methods and Applications, Possibilities and Limitations,* volume 47 of *Lecture Notes in Engineering.* Springer, Berlin, 1989.
445. I. Rechenberg. *Evolutionsstrategie'94.* Frommann-Holzboog, Stuttgart, 1994.
446. R. Reed. Pruning algorithms: a survey. *IEEE Trans. on Neural Networks,* 4:740–746, 1993.

447. P. Riddle, R. Segal, and O. Etzioni. Representation design and brute-force induction in a Boeing manufacturing domain. *Applied Artificial Intelligence*, 8:125–147, 1994.
448. B. D. Ripley. *Pattern Recognition and Neural Networks.* Cambridge University Press, Cambridge, 1996.
449. J. K. Roberge. *Operational Amplifiers: Theory and Practice.* Wiley, New York, 1975.
450. G. G. Robertson, J. D. Mackinlay, and S. K. Card. Cone trees: Animated 3D visualizations of hierarchical information. In *Proc. Human Factors in Computing Systems CHI '91 Conf., New Orleans, LA*, pages 189–194, 1991.
451. F. Rosenblatt. The perceptron: a probabilistic model for information storage and organization in the brain. *Psychological Review*, 65:386–408, 1959.
452. F. Rosenblatt. *Principles of neurodynamic.* Spartan, 1962.
453. C. Rouveirol. Flattening and saturation: Two representation changes for generalization. *Machine Learning*, 14(2):219–232, 1994.
454. R. Royall. *Statistical evidence: a likelihood paradigm.* Chapman & Hall, London, 1997.
455. D. E. Rumelhart and J. L. McClelland. *Parallel Distributed Processing: Exploration in the Microstructure of Cognition.* MIT Press, Cambridge, MA, 1987.
456. A. Samuel. Some studies in machine learning using the game of checkers. In E. Feigenbaum and J. Feldman, editors, *Computers and Thought.* McGraw-Hill, New York, 1963.
457. T. D. Sanger. Optimal unsupervised learning in a single-layer linear feedforward neural network. *Neural Networks*, 2:459–473, 1989.
458. M. Sarkar and M. Brown. Graphical fisheye views. *Communications of the ACM*, 37(12):73–84, 1994.
459. T. Sauer, J. A. Yorke, and M. Casdagli. Embedology. *Journal of Statistical Physics*, 65:579–616, 1991.
460. G. Saunders, A. Gammerman, and V. Vovk. Ridge regression learning algorithm in dual variables. In *Proc. 15th International Conf. on Machine Learning*, pages 515–521. Morgan Kaufmann, San Francisco, CA, 1998.
461. Schaffer, Doug, Zuo, Zhengping, Bartram, Lyn, Dill, John, Dubs, Shelli, Greenberg, Saul, and Roseman. Comparing fisheye and full-zoom techniques for navigation of hierarchically clustered networks. In *Proc. Graphics Interface (GI '93), Toronto, Ontario, 1993, in: Canadian Information Processing Soc., Toronto, Ontario, Graphics Press, Cheshire, CT*, pages 87–96, 1993.
462. J. D. Schaffer, editor. *Proceedings of the 3rd International Conference on Genetic Algorithms and Their Applications.* Morgan Kaufmann, San Mateo, CA, 1989.
463. J. D. Schaffer and A. Morishima. An adaptive crossover distribution mechanism for genetic algorithms. In *[229]*, pages 36–40, 1987.
464. M. Schena, D. Shalon, R. W. Davis, and P. O. Brown. Quantitative monitoring of gene expression patterns with a complementary dna microarray. *Science*, 270:467–470, 1995.
465. B. Schölkopf, A. Smola, and K.-R. Müller. Nonlinear component analysis as a kernel eigenvalue problem. *Neural Computation*, 10:1299–1319, 1998. Technical Report No. 44, 1996, Max Planck Institut für biologische Kybernetik, Tübingen.
466. H. Schumann and W. Müller. *Visualisierung: Grundlagen und allgemeine Methoden.* Springer, 2000.
467. H.-P. Schwefel. *Numerical optimization of computer models.* Wiley, Chichester, 1981.

468. H.-P. Schwefel. *Evolution and Optimum Seeking.* John Wiley and Sons, New York, 1995.
469. D. W. Scott. *Multivariate Density Estimation.* Wiley and Sons, 1992.
470. P. Sebastiani and M. Ramoni. Bayesian selection of decomposable models with incomplete data. *Jornal of the Amererican Statistical Association*, 96(456):1375–1386, 2001.
471. T. Senator, H. Goldberg, J. Wooton, M. Cottini, A. Khan, C. Klinger, W. Llamas, M. Marrone, and R. Wong. The financial crimes enforcement network AI system. *AI Magazine*, 16(4):21–40, 1995.
472. J. Shafer, R. Agrawal, and M. Mehta. SPRINT: A scalable parallel classifier for data mining. *Conf. on Very Large Databases, Mumbay, India*, 1996.
473. Y. Shahar. A framework for knowledge-based temporal abstraction. *Artificial Intelligence*, 90:79–133, 1997.
474. E. Shapiro. Inductive inference of theories from facts. Technical report, Computer Science Department, Yale University, 1981.
475. E. Shapiro. *Algorithmic Program Debugging.* MIT Press, 1983.
476. J. Shawe-Taylor, P. Bartlett, R. C. Williamson, and M. Anthony. Structural risk minimization over data-dependent hierarchies. *IEEE Transactions on Information Theory*, 44(5):1926–1940, 1998.
477. J. Shawe-Taylor and N. Cristianini. On the generalisation of soft margin algorithms. *IEEE Transactions on Information Theory*, 48(10):2721–2735, 2002.
478. B. Shneiderman. Tree visualization with treemaps: A 2D space-filling approach. *ACM Transactions on Graphics*, 11(1):92–99, 1992.
479. B. Shneiderman. The eye have it: A task by data type taxonomy for information visualizations. In *Proc. Visual Languages*, 1996.
480. W. M. Siebert. *Circuits, Signals, and Systems.* MIT Press, 1986.
481. A. Silberschatz and A. Tuzhilin. User-assisted knowledge discovery: How much should the user be involved? In *Proc. of the ACM-SIGMOD'96 Workshop on Research Issues on Data Mining and Knowledge Discovery.* ACM, 1996.
482. R. Silipo, M. Gori, A. Taddei, M. Varanini, and C. Marchesi. Classification of arrhythmic events in ambulatory ECG, using artificial neural networks. *Computers and Biomedical Research*, 28:305–318, 1995.
483. R. Silipo and C. Marchesi. Artificial neural networks for automatic ECG analysis. *IEEE Trans. on Signal Processing*, 46:1417–1425, 1998.
484. E. Simoudis, R. Kerber, B. Livezey, and P. Miller. Developing customer vulnerability models using data mining techniques. *Proc. of Intelligent Data Analysis 95*, pages 181–185, 1995.
485. P. Simpson. Fuzzy min-max neural networks – Part 1: Classification. *IEEE Transactions on Neural Networks*, 3(5):776–786, Sept. 1992.
486. P. Simpson. Fuzzy min-max neural networks – Part 2: Clustering. *IEEE Transactions on Fuzzy Systems*, 1(1):32–45, Jan. 1993.
487. D. Smith and M. Broadwell. The pilot's associate - an overview. *Proc. of the SAE Aerotech Conference*, 1988.
488. A. Smola and B. Schölkopf. Kernel machines website. http://www.kernel-machines.org.
489. P. Smyth and R. Goodman. Rule induction using information theory. In G. Piatetsky and W. Frawley, editors, *Knowledge Discovery in Databases.* MIT Press, Cambridge, MA, 1990.
490. W. M. Spears and K. A. D. Jong. On the virtues of parametrized uniform crossover. In *[54]*, 1991.

491. D. F. Specht. Probabilistic neural networks. *Neural Networks*, 3:109–118, 1990.
492. L. Spector and E. D. Goodman, editors. *GECCO 2001: Proceedings of the Genetic and Evolutionary Computation Conference.* Morgan Kaufmann, San Francisco, CA, 2001.
493. B. Spence. *Information Visualization.* Pearson Education Higher Education publishers, UK, 2000.
494. R. Spence and M. Apperley. Data base navigation: An office environment for the professional. *Behaviour and Information Technology*, 1(1):43–54, 1982.
495. R. Spence, L. Tweedie, H. Dawkes, and H. Su. Visualization for functional design. In *Proc. Int. Symp. on Information Visualization (InfoVis '95)*, pages 4–10, 1995.
496. D. J. Spiegelhalter and S. L. Lauritzen. Sequential updating of conditional probabilities on directed graphical structures. *Networks*, 20:157–224, 1990.
497. A. Spoerri. InfoCrystal: A visual tool for information retrieval. In *Proc. Visualization '93, San Jose, CA*, pages 150–157, 1993.
498. J. Stasko, J. Domingue, M. Brown, and B. Price. *Software Visualization.* MIT Press, Cambridge, MA, 1998.
499. R. Stevens, P. Woodward, T. DeFanti, and C. Catlett. From the I-way to the national technology grid. *Communications of the ACM*, 40:51–60, 1997.
500. I. Stewart. *Does God Play Dice?: The Mathematics of Chaos.* Blackwell, Cambridge MA, 1989.
501. P. Stolorz and C. Dean. Quakefinder: a scalable data mining system for detecting earthquakes from space. In J. H. E Simoudis and U. Fayyad, editors, *Proc. of the 2nd Int. Conf. on Knowledge Discovery and Data Mining*, pages 208–213. AAAI Press, 1996.
502. C. Stolte, D. Tang, and P. Hanrahan. Polaris: A system for query, analysis and visualization of multi-dimensional relational databases. *Transactions on Visualization and Computer Graphics*, 8(1), 2002.
503. R. L. Streit and T. E. Luginbuhl. Maximum likelihood training of probabilistic neural networks. *IEEE Trans. on Neural Networks*, 5(5):764–783, 1994.
504. S. H. Strogatz. *Nonlinear Dynamics and Chaos.* Addison-Wesley, Reading, MA, 1994.
505. D. F. Swayne, D. Cook, and A. Buja. *User's Manual for XGobi: A Dynamic Graphics Program for Data Analysis.* Bellcore Technical Memorandum, 1992.
506. G. Syswerda. A study of reproduction in generational and steady-state genetic algorithms. In *[441].* Morgan Kaufmann, San Mateo, CA, 1991.
507. T. Takagi and M. Sugeno. Fuzzy identification of systems and its application to modeling and control. *IEEE Transactions on Systems, Man, and Cybernetics*, SMC-15(1):116–132, 1985.
508. F. Takens. Detecting strange attractors in fluid turbulence. In D. Rand and L.-S. Young, editors, *Dynamical Systems and Turbulence*, pages 366–381. Springer, Berlin, 1981.
509. G. Tayi and T. Ballou. Examining data quality. *Communications of the ACM*, 41:2, 1998.
510. P. C. Taylor and D. J. Hand. Finding 'superclassifications' with an acceptable misclassification rate. *Journal of Applied Statistics*, 26(5):579–590, 1999.
511. R. A. Thisted. *Elements of Statistical Computing: Numerical Computation.* Chapman and Hall, New York, 1988.
512. A. Thomas, D. J. Spiegelhalter, and W. R. Gilks. Bugs: A program to perform Bayesian inference using Gibbs Sampling. In *Bayesian Statistics 4*, pages 837–42. Clarendon Press, Oxford, 1992.

513. S. Thrun. Extracting rules from artificial neural networks with distributed representations. *Advances in Neural Information Processing Systems*, 7:505–512, 1995.
514. L. Tierney. *LispStat: An Object-Oriented Environment for Statistical Computing and Dynamic Graphics*. Wiley, New York, NY, 1991.
515. H. Tong. *Nonlinear Time Series Analysis: A Dynamical Systems Approach*. Oxford University Press, 1990.
516. L. Trulla, A. Giuliani, J. Zbilut, and C. Webber. Recurrence quantification analysis of the logistic equation with transients. *Physics Letters A*, 223:255–260, 1996.
517. E. R. Tufte. *The Visual Display of Quantitative Information*. Graphics Press, Cheshire, CT, 1983.
518. J. W. Tukey. *Exploratory Data Analysis*. Addison Wesley, 1977.
519. J. W. Tukey. An alphabet for statisticians' expert systems. In W. A. Gale, editor, *Artificial Intelligence and Statistics*, pages 401–409. Addison-Wesley, 1986.
520. P. E. Utgoff. Incremental induction of decision trees. *Machine Learning*, 4:161–186, 1989.
521. P. E. Utgoff, N. C. Berkman, and J. A. Clouse. Decision tree induction based on efficient tree restructuring. *Machine Learning*, 29:5–44, 1997.
522. J. van der Lubbe. *Information Theory*. Cambridge University Press, Cambridge, 1997.
523. C. van Rijsbergen. *Information Retrieval*. Dept. of Computer Science, University of Glasgow, 2nd edition, 1979. Available on-line at http://www.dcs.gla.ac.uk/Keith/Preface.html.
524. J. J. van Wijk and R. D. van Liere. Hyperslice. In *Proc. Visualization '93, San Jose, CA*, pages 119–125, 1993.
525. V. Vapnik. *Statistical Learning Theory*. Wiley, New York, 1998.
526. P. F. Velleman. *Data Desk 4.2: Data Description*. Data Desk, Ithaca, NY, 1992.
527. J. C. Venter et al. The sequence of the human genome. *Science*, 291:1304–1351, 2001.
528. R. V. Vidal. *Applied Simulated Annealing*. Springer, Berlin, 1993.
529. H.-M. Voigt, W. Ebeling, I. Rechenberg, and H.-P. Schwefel, editors. *Parallel Problem Solving from Nature (PPSN IV)*, Lecture Notes in Computer Science 1141. Springer, Berlin, 1996.
530. P. Walley. *Statistical Reasoning with Imprecise Probabilities*. Chapman and Hall, London, 1991.
531. L. Wang and J. Mendel. Generating fuzzy rules by learning from examples. *IEEE Transactions on Systems, Man, and Cybernetics*, 22(6):1414–1427, 1992.
532. L.-X. Wang. Fuzzy systems are universal approximators. In *International Conference on Fuzzy Systems*, pages 1163–1170. IEEE, 1992.
533. M. O. Ward. XmdvTool: Integrating multiple methods for visualizing multivariate data. In *Proc. Visualization 94, Washington, DC*, pages 326–336, 1994.
534. C. Ware. *Information Visualization: Perception for Design*. Morgan Kaufmann, 2000.
535. S. Weber. A general concept of fuzzy connectives, negations and implications based on t-norms and t-conorms. *Fuzzy Sets and Systems*, 11(2):113–134, Oct. 1983.
536. A. S. Weigend and N. S. Gershenfeld, editors. *Time Series Prediction: Forecasting the Future and Understanding the Past*. Santa Fe Institute Studies in the Sciences of Complexity, Santa Fe, NM, 1993.
537. S. M. Weiss and C. A. Kulikowski. *Computer Systems that Learn*. Morgan Kaufmann, 1991.

538. D. S. Weld and J. de Kleer, editors. *Readings in Qualitative Reasoning About Physical Systems.* Morgan Kaufmann, San Mateo CA, 1990.
539. D. Whitley, D. Goldberg, and E. Cantu-Paz, editors. *GECCO 2000: Proceedings of the Genetic and Evolutionary Computation Conference.* Morgan Kaufmann, San Francisco, CA, 2000.
540. J. C. Whittaker. *Graphical Models in Applied Multivariate Statistics.* Wiley, Chichester, 1990.
541. B. Widrow and M. Lehr. 30 years of adaptive neural networks: Perceptron, Madeline, and Back-Propagation. *Proceedings of the IEEE*, 78(9):1415–1442, 1990.
542. A. Wilhelm, A. Unwin, and M. Theus. Software for interactive statistical graphics - a review. In *Proc. Softstat'95 Conf., Heidelberg, Germany*, 1995.
543. S. Wilks. *Mathematical Statistics.* Wiley, New York, 1963.
544. R. J. Williams and D. Zipser. A learning algorithm for continually running fully recurrent neural networks. *Neural Computation*, 1:270–280, 1989.
545. M. Wineberg and F. Oppacher. A representation scheme to perform program induction in a canonical genetic algorithm. In *[141]*, pages 292–301, 1994.
546. J. A. Wise, J. J. Thomas, K. Pennock, D. Lantrip, M. Pottier, A. Schur, and V. Crow. Visualizing the non-visual: Spatial analysis and interaction with information from text documents. In *Proc. Symp. on Information Visualization, Atlanta, GA*, pages 51–58, 1995.
547. I. Witten and E. Frank. *Data Mining: Practical Machine Learning Tools and Techniques with Java Implementations.* Morgan Kaufman, 2000.
548. S. Wright. Correlation and causation. *Journal of Agricultural Research*, 20:557–585, 1921.
549. S. Wright. The method of path coefficients. *Annals of Mathematical Statistics*, 5:161–215, 1934.
550. T. Wright. *Statistical Methods and the Improvement of Data Quality.* Academic Press, 1983.
551. S. Wrobel. Inductive logic programming for knowledge discovery in databases. In Džeroski and Lavrač [163], pages 74–101.
552. J. Wu, G. Cheng, and X. Liu. Reasoning about outliers by modelling noisy data. In *[355]*, pages 549–558, 1997.
553. R. Yager, S. Ovchinnikov, R. Tong, and H. Ngugen, editors. *Fuzzy Sets and Applications.* Wiley, New York, 1987.
554. T. Yamane. *Statistics, An Introductory Analysis.* Harper & Row, 1967.
555. L. Yang. Interactive exploration of very large relational data sets through 3d dynamic projections. In *Proc. SIGKDD Int. Conf. On Knowledge Discovery & Data Mining (KDD 2000), Boston, MA*, 2000.
556. L. Zadeh. Fuzzy sets. *Information and Control*, 8:338–353, 1965.
557. L. Zadeh. Fuzzy logic and approximate reasoning. *Synthese*, 30:407–428, 1975.
558. L. Zadeh. Probability theory and fuzzy logic are complementary rather than competitive. *Technometrics*, 37(3):271–276, 1995.
559. L. Zadeh. Fuzzy logic and the calculi of fuzzy rules and fuzzy graphs: A precis. *Multi. Val. Logic*, 1:1–38, 1996.
560. J. M. Zurada. *Introduction to Artificial Neural Networks systems.* West Pub. Co., 1992.
561. J. M. Zurada, R. J. Marks, and C. J. Robinson, editors. *Computational Intelligence Imitating Life.* IEEE Press, New York, 1994.
562. IEEE Transactions on Fuzzy Systems. Journal of the IEEE Neural Networks Council, IEEE Press.

563. Sci.nonlinear Frequently Asked Questions (FAQ) (http://amath-www.colorado.edu/appm/faculty/jdm/faq.html).
564. Fuzzy Sets and Systems. Journal of the International Fuzzy Systems Association, Elsevier Science.

Index

LIVERPOOL
JOHN MOORES UNIVERSITY
AVRIL ROBARTS LRC
TITHEBARN STREET
LIVERPOOL L2 2ER
TEL. 0151 231 4022

Author Addresses

Luis Aburto
Data Analysis for Decision Support
José Domingo Cañas 2511
Santiago
Chile
luis.aburto@dads.cl

Michael R. Berthold
Data Analysis Research Lab
Tripos, Inc.
601 Gateway Blvd, Suite 720
South San Francisco, CA 94080
USA
berthold@tripos.com

Elizabeth Bradley
Department of Computer Science
University of Colorado
Boulder, CO 80309
USA
lizb@cs.colorado.edu

Nello Cristianini
Department of Statistics
University of California
Davis, CA 95616
USA
nello@support-vector.net

Gerard van den Eijkel
Telematica Instituut
Drienerlolaan 5
7522 NB Enschede
The Netherlands
gerard.vandeneijkel@telin.nl

Ad J. Feelders
Institute of Information and
Computing Sciences
Utrecht University
PO Box 80.089
3508TB Utrecht
The Netherlands
ad@cs.uu.nl

Peter A. Flach
Department of Computer Science
University of Bristol
Bristol, BS8 1UB
United Kingdom
Peter.Flach@bristol.ac.uk

David J. Hand
Department of Mathematics
Imperial College
London, SW7 2BZ
United Kingdom
d.j.hand@ic.ac.uk

Christian Jacob
Department of Computer Science
University of Calgary
Calgary, Alberta, T2N 1N4
Canada
jacob@cpsc.ucalgary.ca

Daniel A. Keim
Computer Science Institute
University of Konstanz
Universitätsstr. 10
78457 Konstanz
Germany
keim@informatik.uni-konstanz.de

Nada Lavrač
Department of Intelligent Systems
Jožef Stefan Institute
1000 Ljubljana
Slovenia
Nada.Lavrac@ijs.si

Xiaohui Liu
Department of Information Systems and Computing
Brunel University
Uxbridge, Middlesex, UB8 3PH
United Kingdom
Xiaohui.Liu@brunel.ac.uk

Marco Ramoni
Harvard Medical School
300 Longwood Avenue
Boston, MA 02115
USA
marco_ramoni@harvard.edu

Paola Sebastiani
Department of Mathematics and Statistics
University of Massachusetts
Amherst, MA 01003
USA
sebas@math.umass.edu

Mariano Silva
Egoland Consultores Asociados Ltda.
Domeyko 2469
Santiago Centro
Chile
mariano@egoland.cl

John Shawe-Taylor
Department of Computer Science
Royal Holloway, University of London
Egham, Surrey, TW20 0EX
United Kingdom
jst@cs.rhul.ac.uk

Rosaria Silipo
International Computer Science Institute (ICSI)
1947 Center Street, Suite 600
Berkeley, CA 94704
USA
rosaria@icsi.berkeley.edu

Paul C. Taylor
Business School
University of Hertfordshire
College Lane, Hatfield
Herts, AL10 9AB
United Kingdom
p.c.taylor@herts.ac.uk

Matthew Ward
Department of Computer Science
Worcester Polytechnic Institute
100 Institute Road
Worcester, MA 01609
USA
matt@cs.wpi.edu

Richard Weber
Department of Industrial Engineering
University of Chile
Chile
rweber@dii.uchile.cl